Lecture Notes in Mathematics 2227

More information about this series at http://www.springer.com/series/304

Pontus Giselsson • Anders Rantzer
Editors

Large-Scale and Distributed Optimization

 Springer

Editors
Pontus Giselsson
Department of Automatic Control
Lund University
Lund, Sweden

Anders Rantzer
Department of Automatic Control
Lund University
Lund, Sweden

MATLAB® is a registered trademark of The MathWorks, Inc., 1 Apple Hill Drive, Natick, MA 01760-2098, USA, http://www.mathworks.com.

ISSN 0075-8434   ISSN 1617-9692   (electronic)
Lecture Notes in Mathematics
ISBN 978-3-319-97477-4   ISBN 978-3-319-97478-1   (eBook)
https://doi.org/10.1007/978-3-319-97478-1

Library of Congress Control Number: 2018959407

This Springer imprint is published by the registered company Springer Nature Switzerland AG
The registered company address is: Gewerbestrasse 11, 6330 Cham, Switzerland

Preface

Mathematical optimization has over the last several decades emerged as a core component in a wide variety of technical fields. The optimization needs in these fields often drive the focus of the optimization society. Over the last decade, technical fields with big data problems have greatly influenced the focus of the optimization research community. These fields require optimization algorithms that scale well with problem dimension, that can exploit problem structure, and that can be efficiently implemented in modern computation environments.

In an effort to bring together leading mathematical optimization experts who have focused on these topics, the Linnaeus Center LCCC at Lund University organized in June 2017 a workshop on Large-Scale and Distributed Optimization. It was a 3 day single-track workshop with talks from about twenty distinguished speakers. In addition to the workshop, about twenty students and junior faculty members were invited to a 5 week focus period on the same topic that was held in conjunction with the workshop. This book is an offspring of the workshop and focus period as all participants were invited to contribute a chapter. We are happy to have received contributions from twelve participants.

We want to thank all workshop and focus period participants for making the events fruitful and interesting. Special thanks go to the scientific committee consisting of Heinz Bauschke, Amir Beck, Stephen Boyd, Mikael Johansson, Angelia Nedić, Asuman Ozdaglar, Panagiotis Patrinos, and Wotao Yin. We also want to thank Eva Westin for organizing all practical aspects of the meeting and Leif Andersson for helping to compile this book.

Lund, Sweden
May 2018

Pontus Giselsson
Anders Rantzer

Contents

Contributors

Robert Baier University of Bayeuth, Chair of Applied Mathematics, Bayreuth, Germany

University of Newcastle, School of Electrical Engineering and Computing, Callaghan, NSW, Australia

Anastasia Bayandina Moscow Institute of Physics and Technology, Dolgoprudny, Moscow Region, Russia

Skolkovo Institute of Science and Technology, Skolkovo Innovation Center, Moscow, Russia

Amir Beck School of Mathematical Sciences, Tel Aviv University, Tel Aviv, Israel

Philipp Braun University of Bayeuth, Chair of Applied Mathematics, Bayreuth, Germany

University of Newcastle, School of Electrical Engineering and Computing, Callaghan, NSW, Australia

Volkan Cevher École Polytechnique Fédérale de Lausanne (EPFL), Lausanne, Switzerland

Lijun Ding Operations Research and Information Engineering, Cornell University, Ithaca, NY, USA

Pavel Dvurechensky Weierstrass Institute for Applied Analysis and Stochastics, Berlin, Germany

Institute for Information Transmission Problems RAS, Moscow, Russia

Giovanni Fantuzzi Department of Aeronautics, Imperial College London, London, UK

Alexander Gasnikov Moscow Institute of Physics and Technology, Dolgoprudny, Moscow Region, Russia

Institute for Information Transmission Problems RAS, Moscow, Russia

Pontus Giselsson Department of Automatic Control, Lund University, Lund, Sweden

Lars Grüne University of Bayeuth, Chair of Applied Mathematics, Bayreuth, Germany

Anders Hansson Division of Automatic Control, Linkoping University, Linkoping, Sweden

Christopher M. Kellett University of Newcastle, School of Electrical Engineering and Computing, Callaghan, NSW, Australia

Puya Latafat KU Leuven, Department of Electrical Engineering (ESAT-STADIUS), Leuven, Belgium
IMT School for Advanced Studies Lucca, Lucca, Italy

D. Russell Luke Institute for Numerical and Applied Mathematics, University of Göttingen, Göttingen, Germany

Yura Malitsky Institute for Numerical and Applied Mathematics, University of Göttingen, Göttingen, Germany

Angelia Nedić School of Electrical, Computer, and Energy Engineering, Arizona State University, Tempe, AZ, USA

Alexander Olshevsky Department of Electrical Engineering, Boston University, Boston, MA, USA

Sina Khoshfetrat Pakazad C3 IoT, Redwood City, CA, USA

Dror Pan Faculty of Industrial Engineering and Management, Technion - Israel Institute of Technology, Haifa, Israel

Antonis Papachristodoulou Department of Engineering Science, University of Oxford, Oxford, UK

Panagiotis Patrinos KU Leuven, Department of Electrical Engineering (ESAT-STADIUS), Leuven, Belgium

Anders Rantzer Department of Automatic Control, Lund University, Lund, Sweden

Wei Shi School of Electrical, Computer, and Energy Engineering, Arizona State University, Tempe, AZ, USA

Fedor Stonyakin V.I. Vernadsky Crimean Federal University, Simferopol, Russia
Moscow Institute of Physics and Technolog, Dolgoprudny, Moscow Region, Russia

Alexander Titov Moscow Institute of Physics and Technology, Dolgoprudny, Moscow Region, Russia

Quoc Tran-Dinh Department of Statistics and Operations Research, The University of North Carolina at Chapel Hill (UNC), UNC-Chapel Hill, NC, USA

Madeleine Udell Operations Research and Information Engineering, Cornell University, Ithaca, NY, USA

Bằng Công Vũ École Polytechnique Fédérale de Lausanne (EPFL), Lausanne, Switzerland

Lin Xiao Microsoft Research, Redmond, WA, USA

Alp Yurtsever École Polytechnique Fédérale de Lausanne (EPFL), Lausanne, Switzerland

Yuchen Zhang Stanford University, Stanford, CA, USA

Yang Zheng Department of Engineering Science, University of Oxford, Oxford, UK

Chapter 1
Large-Scale and Distributed Optimization: An Introduction

Pontus Giselsson and Anders Rantzer

Abstract The recent explosion in size and complexity of datasets and the increased availability of computational resources has led us to what is sometimes called the big data era. In many big data fields, mathematical optimization has over the last decade emerged as a vital tool in extracting information from the data sets and creating predictors for unseen data. The large dimension of these data sets and the often parallel, distributed, or decentralized computational structures used for storing and handling the data, set new requirements on the optimization algorithms that solve these problems. This has led to a dramatic shift in focus in the optimization community over this period. Much effort has gone into developing algorithms that scale favorably with problem dimension and that can exploit structure in the problem as well as the computational environment. This is also the main focus of this book, which is comprised of individual chapters that further contribute to this development in different ways. In this introductory chapter, we describe the individual contributions, relate them to each other, and put them into a wider context.

Keywords Convex optimization · Monotone inclusions · Big data problems · Scalable methods · Operator splitting methods · Stochastic methods · Nonconvex methods

1.1 Context

1.1.1 The Traditional Standard

The traditional standard in convex optimization was to translate the optimization problem into a conic program and solve it using a primal-dual interior point method. The monograph [22] was instrumental in setting this standard. Appealing features

P. Giselsson (✉) · A. Rantzer
Department of Automatic Control, Lund University, Lund, Sweden
e-mail: pontusg@control.lth.se; rantzer@control.lth.se

© Springer Nature Switzerland AG 2018
P. Giselsson, A. Rantzer (eds.), *Large-Scale and Distributed Optimization*, Lecture Notes in Mathematics 2227,
https://doi.org/10.1007/978-3-319-97478-1_1

of this standard are that most convex problems can be translated to equivalent cone programs and that primal-dual interior point methods can solve cone problems efficiently and reliably to high accuracy. The result is a robust technology for solving convex optimization problems.

Shortcomings of this standard are twofold: First, the main computationally intense operation is that of solving a linear system of equations in every iteration. This operation grows unfavorably with problem dimension and is difficult to parallelize. Second, the reformulation to conic form may destroy exploitable structure in the original problem formulation. The first shortcoming is critical in the big data setting, where problems with millions or billions of decision variables are solved. Big data problems also often have simple structures that may be hard to exploit in conic form. These characteristics render the traditional standard inapplicable in the large-scale regime.

Much effort has over the last decade been devoted to developing algorithms that overcome the shortcomings of the traditional standard and that are suitable for solving large-scale problems. This book collects contributions that further contribute to this development. In this introductory chapter, we describe the contributions of the book and outline the underlying ideas. We also relate them to the literature and briefly discuss how they address the above shortcomings.

1.1.1.1 Chordal Structure

The second of the two shortcomings outlined above can be avoided by retaining the original problem formulation. For instance, for problems arising in model predictive control the resulting quadratic program can directly be solved using interior point methods. The problem has a very specific structure and a well known method to exploit this structure is to perform efficient matrix factorization via Riccati recursions when solving the linear system to find the search direction. In the contribution *Exploiting Chordality in Optimization Algorithms for Model Predictive Control*, Hansson and Pakazad show that the exploitable structure is a special case of a *chordal* structure, see [27] for more on chordal structures. Based on this observation, they propose other chordal structures that can be exploited to speed up computation by means of a message passing algorithm. They also combine different chordal structures in one problem to devise parallel algorithms for solving the linear system. This addresses the first shortcoming of the traditional standard.

1.1.2 The Composite Form

Much of the recent research efforts to overcome the shortcomings of the previous standard have been focused on algorithms that deal with problems on *composite form*:

$$\text{minimize} \quad f(x) + g(x), \tag{1.1}$$

with $x \in \mathbb{R}^n$ and $f, g : \mathbb{R}^n \to \mathbb{R} \cup \{\infty\}$ being proper, closed, and convex functions. The reason for splitting the problem into two functions is purely for algorithmic purposes and will be discussed later. The composite form is very general and captures, e.g., constrained optimization via indicator functions of sets. Provided some constraint qualification holds for (1.1), this problem is a special case of the inclusion problem

$$0 \in A(x) + B(x), \tag{1.2}$$

with A, B being maximal monotone operators. Letting $A = \partial f$ and $B = \partial g$, with $\partial \cdot$ denoting the subdifferential operator, gives (1.1). If the problem to be minimized is instead $f(x) + g(Lx)$ with L a linear operator, it can be advantageous to create another monotone inclusion problem by combining the primal and dual optimality conditions to get:

$$0 \in \begin{pmatrix} \partial f(x) + L^T y \\ \partial g^*(y) - Lx \end{pmatrix}, \tag{1.3}$$

where $g^* : \mathbb{R}^n \to \mathbb{R} \cup \{\infty\}$ is the convex conjugate function of g. This is a monotone inclusion of the form (1.2) with $A(x, y) = (\partial f(x), \partial g^*(y))$ and $B(x, y) = (L^T y, -Lx)$ and is the basis of many first-order primal-dual methods. Due to the generality of the formulation (1.2), considerable research effort has gone into developing efficient and scalable algorithms for solving such problems. See, e.g., [1, 25] for more on the connection between optimization problems and monotone inclusion problems and on algorithms to efficiently solve them.

The whole point of writing problems on the form (2) is that it could be much easier to perform efficient operations on A and B individually, than to handle A+B directly. This observation is the basis for so called *operator splitting methods*. (For instance, for feasibility problems involving two sets, it may be computationally efficient to project onto the individual sets, while projecting onto the intersection may be costly.) These operator splitting methods define an operator with two properties:

(1) The fixed-point set of the operator has a simple relationship with the solution set of (1.2).
(2) Sequences generated by repeated application of the operator, starting from any initial point, will converge to a fixed-point.

There are three main operator splitting operators for (1.2) that satisfy these properties; the Douglas-Rachford operator [10, 18], the forward-backward operator, and the forward-backward-forward operator [26]. The former one is applicable to general problems of the form (1.2), while the latter two require different smoothness assumptions (cocoercivity and Lipschitz continuity respectively) on one of the operators. See [1, 25] for definitions and more details on and these operators.

The key property for Douglas-Rachford splitting and forward-backward splitting is nonexpansiveness of the operator. For forward-backward-forward splitting, the key property is quasi-nonexpansiveness, which is nonexpansiveness towards each fixed-point. The fixed-point sets of these operators encode the solution set of (1.2), i.e., the operators satisfy property (1). Iteration of a (quasi-)nonexpansive operator may, however, not converge to a fixed-point—it can, e.g., be a rotation. This is dealt with by forming an averaged map of the (quasi-)nonexpansive operator with the identity operator. This operation retains the original fixed-point set and the averaged map has the additional property that it converges towards a fixed-point when iterated. Hence, the averaged map satisfies both property (1) and (2).

The (quasi-)nonexpansiveness of the operators come from similar properties of their main building blocks—resolvents (also called backward operators) and forward operators. In the convex optimization context, these reduce to proximal operators, that are evaluated by solving an optimization problem, and gradient step operators. See [1, 8, 23, 25] for more details.

Key to devising efficient algorithms for the big data setting is to combine an appropriate split of the problem (1.2) with one of the three operator splitting methods such that the computational cost of evaluating the algorithm building blocks grows favorably with problem dimension. The degrees of freedom in this choice are plentiful. Many algorithms exist that specify some degrees of freedom in that they (often in hindsight) can be viewed as intelligent applications of the above operator splitting methods to a specific problem formulation. For instance; the alternating direction method of multipliers (ADMM) [6, 13, 14] is Douglas-Rachford splitting applied to a specific problem formulation [11], the projected and proximal gradient methods are the direct convex optimization counterparts of forward-backward splitting, and many first-order primal-dual methods (e.g., [7]) are special cases of the either of these splittings applied to (1.3). These algorithms are sometimes analyzed directly, sometimes via the underlying operator splitting method.

In *Decomposition Methods for Large-Scale Semidefinite Programs with Chordal Aggregate Sparsity and Partial Orthogonality*, the authors Zheng, Fantuzzi, and Papachristodoulou improve the scalability of an ADMM method tailored to the homogeneous self-dual embedding [28] formulation of large-scale semidefinite programs. The performance improvement relies on the presence of sparsity in the data matrices. More specifically that the union of the sparsity patterns of a collection of data matrices is chordal or has a sparse chordal extension [27]. This enables for decomposing the large positive semidefinite constraint to smaller ones coupled with additional affine constraints. This can greatly improve performance since projection onto large semidefinite cones is costly. This approach bears similarities with the contribution by Hansson and Pakazad in the use of chordal structures to improve scalability.

In *Smooth Alternating Direction Methods for Fully Nonsmooth Constrained Convex Optimization*, Tran-Dinh and Cevher show how to apply the forward-backward splitting algorithm to nonsmooth convex optimization problems. They apply the algorithm to a smoothed version of the dual, where the dual is a composite problem consisting of two nonsmooth conjugate functions composed with linear mappings besides a linear term. The problem smoothing is achieved by replacing one of the conjugate functions by a smoothed version. The smoothed version is obtained by adding a strongly convex term to the function before taking the conjugate. Convergence in objective value is guaranteed for problems with bounded primal domain. To improve convergence, they utilize Nesterov acceleration [3, 19, 20], which is a well known and widely used method to improve the convergence of forward-backward splitting when applied to convex optimization problems.

For monotone inclusion problems with more than two operators, a product space trick [24] can be used to reduce the problem to involve two operators only. Each operator in the original formulation is assigned its own variable and together they define the first operator on a product space. The second operator is a consensus operator that ensures equality between all introduced variables at solutions. Even though evaluating the forward and backward steps on the product space boils down to evaluating the individual steps on their respective spaces, it may be desirable to solve the problem without lifting it to a higher dimensional space. Much effort has therefore recently been devoted to develop splitting operators for problems with more than two operators. This has been successful in the three-operator case, i.e., for problems of the form

$$0 \in A(x) + B(x) + C(x),$$

with additional assumptions on some of the maximal monotone operators. The forward Douglas-Rachford algorithm [9] solves such problems with A, B maximally monotone and C cocoercive, while Latafat and Patrinos present in *Primal-Dual Proximal Algorithms for Structured Convex Optimization: A Unifying Framework* another splitting operator, based on their previous work [17], that in addition requires B to be linear. Both these splitting operators satisfy the same properties as the two-operator splittings from before, namely: (1) that the fixed-points correspond to solutions of the underlying inclusion problem and (2) that when iterated, the resulting sequence converges to a fixed-point. Latafat and Patrinos apply in their contribution their splitting to a three-operator generalization of the primal-dual inclusion (1.3) with the underlying primal problem being a sum of three functions. They show that most available primal-dual first-order methods in the literature are special cases of this. In addition, they present novel primal-dual methods based on the same splitting and problem combination.

1.1.3   Randomized Methods

A fixed-point to the previously described operator splitting operators can be found by iterating the operator itself. For very large-scale problems, evaluating the operator might be computationally too expensive. One approach that has become increasingly popular for solving larger problems is to instead iterate a computationally less expensive stochastic approximation of the operator. These algorithms obey different dynamics than the nominal, but almost sure convergence towards a fixed-point of the underlying operator is typically guaranteed. The stochastic approximation may be to randomly select blocks of coordinates that are updated in every iteration, an approach that was popularized by Nesterov [21]. For separable problem structures, such updates are often computationally much less demanding than full operator updates and therefore applicable to larger problems. In *Block-Coordinate Primal-Dual Method for Nonsmooth Minimization Over Linear Constraints*, Luke and Malitsky propose and analyze a stochastic block-coordinate version of the primal-dual method [7]. They analyze problems with separable cost subject to the decision variable solving a least squares problem. This is a generalization of a linearly constrained separable problem to handle inconsistent linear constraints. They show that the problem template covers linear programming, composite minimization, distributed optimization using the product space formulation, and inverse problems and provide almost sure convergence of the method. In *Stochastic Forward Douglas-Rachford Splitting Method for Monotone Inclusions*, Cevher, Vũ, and Yurtsever propose and analyze a stochastic version of the forward Douglas-Rachford algorithm [9]. The stochasticity lies in the evaluation of the cocoercive term. Almost sure convergence is shown based on the assumptions that the stochastic approximation is unbiased and its variance sequence is summable. They also propose a novel stochastic primal-dual method using the same problem template as Latafat and Patrinos, but where neither formulation is a special case of the other.

In *Mirror Descent and Convex Optimization Problems With Non-Smooth Inequality Constraints*, Bayandina, Dvurechensky, Gasnikov, Stonyakin, and Titov analyze deterministic and stochastic variants of the mirror descent method. The mirror descent method is a nonlinear projected subgradient method [2]. It is not a special case of the before mentioned operator splitting methods and therefore cannot rely on their convergence analyses. In the stochastic setting, the authors of this contribution provide results on the number of iterations needed to, with a prespecified probability, reach a certain accuracy of the objective value and constraint violation. The assumption on the stochasticity is that the stochastic gradients are unbiased and bounded almost surely.

1.1.4 Consensus Optimization

Stochastic approximations can naturally be computed in optimization problems of the form (1.1) if $f(x) = \frac{1}{n} \sum_{i=1}^{n} f_i(x)$ and n is large. Such problems are refereed to as empirical risk minimization problems or consensus problems and are very common, e.g., in machine learning. A stochastic approximation of the gradient of f can be obtained by randomly selecting a set of indices and compute the sum of gradients of the corresponding functions. This results in an unbiased gradient approximation that is computationally much less expensive than a full gradient evaluation and serves as the basis in incremental aggregated gradient methods, see, e.g., [4] for a survey. In *Frank-Wolfe Style Algorithms for Large Scale Optimization*, Ding and Udell show that the Frank-Wolfe method [12, 15], which minimizes convex differentiable functions over compact sets, converges also when the true gradient is replaced by a stochastic approximation of this form. They also present, and show convergence for, a Frank-Wolfe variation for problems where it is too costly to store the full decision variable. This may happen for large-scale optimization problems with matrix variables. They use a storage saving matrix sketching procedure to approximate the true decision variable. The underlying assumption for the sketching procedure to be beneficial is the existence of some structure of the matrix at the solution, in this case low rank.

1.1.4.1 Distributed Algorithms

Another problem for large-scale problems is that the problem data might be difficult to store on one machine. For consensus optimization problems and empirical risk minimization problems, the amount of data may be huge and spread over different machines in computer clusters or distributed over a network of computing agents. Therefore, it is desirable to employ parallel or distributed algorithms that can fully exploit the available computational power. In this setting, the above stochastic algorithms will for consensus optimization wake up computing units based on if that unit is responsible for the gradients associated with some of the randomly chosen function indices. This approach will potentially leave units idle from time to time. The idle time can be reduced if instead employing a deterministic synchronized method where the local computations are allocated evenly. In *Decentralized Consensus Optimization and Resource Allocation*, Nedić, Olshevsky, and Shi provide an overview of distributed deterministic synchronous first-order methods for solving the consensus optimization problem over a distributed network of computing agents. The algorithms are based on the communication assumption that only neighboring agents can communicate. The also show a mirror, or dual, relationship between the consensus optimization problem and the resource allocation problem. Based in this relationship, distributed algorithms for the resource allocation problem are also presented. In *Communication-Efficient Distributed Optimization of Self-Concordant Empirical Loss*, Zhang and Xiao take a different approach to solving specific consensus optimization problems under the empirical risk minimization framework. They assume that all computing units can communicate with a central entity and take off from the observation that many distributed algorithms based on first-order methods require many iterations to converge. This requires extensive communication between computational unit, which may be a severe bottleneck. To reduce the communication burden, they propose to employ a Newton method and to solve the linear system for the search direction using a preconditioned conjugate gradient method. They show that under a local function similarity assumption, the convergence in terms of number of iterations is faster than for typical first-order methods.

1.1.5 Nonconvex Optimization

The discussion so far in this introduction has been on methods to solve convex optimization problems. Recently, also nonconvex formulations have gained considerable attention. This allows for constructing more complex models that can capture more complicated relationships. The use of nonconvex regularizers and cost functions in machine learning and statistical estimation has increased. Another notable example in which nonconvex optimization is used is in training of deep neural networks. Deep learning has seen a tremendous success over the last couple of years with applications, e.g., in speech recognition and object classification in images and videos.

In *Numerical Construction of Nonsmooth Control Lyapunov Functions*, another application of nonconvex optimization is presented where Baier, Braun, Grüne and Kellett show how to construct nonsmooth control Lyapunov functions by solving a mixed-integer linear program. The existence of such a Lyapunov function guarantees asymptotic stabilizability of an equilibrium of the underlying dynamical system.

One difference between nonconvex and convex problems is that local minima need not be global minima in nonconvex optimization. Therefore, there is a distinction in nonconvex optimization between global and local methods. Global methods solve the problem to global optimality, while local methods typically find a critical point. In the big data setting, global methods are often intractable which is why local nonconvex methods have gained in popularity. In *Convergence of an Inexact Majorization-Minimization Method for Solving a Class of Composite Optimization Problems*, Beck and Pan analyze an inexact majorization minimization algorithm for nonconvex optimization. Majorization-minimization is a general framework that is the basis for, e.g., the proximal-gradient method. In every iteration, it constructs a so-called *consistent majorizer* of the function to be minimized. The majorizer is then (approximately) minimized to find the next iterate. The authors provide many examples of consistent majorizers in important nonconvex settings and show, e.g., that all accumulation points of the generated algorithm sequence are (strongly) stationary points of the optimization problem.

1.1.6 Omissions

There are, of course, many research directions in large-scale and distributed optimization that have not or have only briefly been touched upon in this book. Below, we describe two such directions that are receiving considerable attention at the moment. The first is asynchronous methods. To solve very large-scale problems over a distributed computer network or in a computer cluster, the ideal situation would be to let the individual computing units immediately push the results of their computations to their neighbors or the master, without waiting for synchronization. Then, on completion, immediately start on a new task. This would yield a higher utilization of available computing resources. The analysis of such algorithms is, however, more complicated although some models have been around for some time [5]. Another topic of recent interest is that of methods for training deep neural networks. The training problem is nonconvex and the local optimization methods are typically based on variations of the stochastic subgradient method with adaptive scaling. The popularity of the application gives a very wide user-base for well performing methods. One notable example is the Adam algorithm [16] that has received considerable attention lately.

1.2   Conclusion and Outlook

The rise of the big data era with its enormous amount of applications that need optimization has had a tremendous impact on the mathematical optimization society. It has steered much of the focus from traditional methods with linear system solves as main computational step to more scalable and parallelizable methods. Much attention has been directed towards convex methods, while recently nonconvex methods has gained in popularity since they have been shown to push state-of-the-art in many application domains further than what is possible with convex methods. Although there is still room for much algorithmic improvement in the large-scale convex setting, a big challenge for the optimization community is to drive the nonconvex setting towards a technology. This development needs to be performed close to the application domains, since nonconvex optimization problem modeling and solution algorithm development are not as separable as in the convex setting.

References

1. H.H. Bauschke, P.L. Combettes, *Convex Analysis and Monotone Operator Theory in Hilbert Spaces*, 2nd edn. (Springer, New York, 2017)
2. A. Beck, M. Teboulle, Mirror descent and nonlinear projected subgradient methods for convex optimization. Oper. Res. Lett. **31**(3), 167–175 (2003). ISSN: 0167-6377
3. A. Beck, M. Teboulle, A fast iterative shrinkage-thresholding algorithm for linear inverse problems. SIAM J. Imag. Sci. **2**(1), 183–202 (2009)
4. D.P. Bertsekas, Incremental Aggregated Proximal and Augmented Lagrangian Algorithms, Sept. 2015, arXiv:1509.09257
5. D.P. Bertsekas, J.N. Tsitsiklis, *Parallel and Distributed Computation: Numerical Methods* (Prentice-Hall, Upper Saddle River, NJ, 1989)
6. S. Boyd, N. Parikh, E. Chu, B. Peleato, J. Eckstein, Distributed optimization and statistical learning via the alternating direction method of multipliers. Found. Trends Mach. Learn. **3**(1), 1–122 (2011)

7. A. Chambolle, T. Pock, A first-order primal-dual algorithm for convex problems with applications to imaging. J. Math. Imaging Vision **40**(1), 120–145 (2011)
8. P.L. Combettes, J.-C. Pesquet, Proximal splitting methods in signal processing, in *Fixed-Point Algorithms for Inverse Problems in Science and Engineering*, ed. by H.H. Bauschke, R. S. Burachik, P.L. Combettes, V. Elser, D.R. Luke, H. Wolkowicz (Springer, New York, 2011), pp. 185–212
9. D. Davis, W. Yin, A three-operator splitting scheme and its optimization applications. Set-Valued Var. Anal. **25**(4), 829–858 (2017)
10. J. Douglas, H.H. Rachford, On the numerical solution of heat conduction problems in two and three space variables. Trans. Am. Math. Soc. **82**, 421–439 (1956)
11. J. Eckstein, Splitting methods for monotone operators with applications to parallel optimization. PhD thesis, MIT, 1989
12. M. Frank, P. Wolfe, An algorithm for quadratic programming. Naval Res. Log. Q. **3**, 95–110 (1956)
13. D. Gabay, B. Mercier, A dual algorithm for the solution of nonlinear variational problems via finite element approximation. Comput. Math. Appl. **2**(1), 17–40 (1976)
14. R. Glowinski, A. Marroco, Sur l'approximation, par éléments finis d'ordre un, et la résolution, par pénalisation-dualité d'une classe de problémes de dirichlet non linéaires. ESAIM: Mathematical Modelling and Numerical Analysis - Modélisation Mathématique et Analyse Numérique **9**, 41–76 (1975)
15. M. Jaggi, Revisiting Frank-Wolfe: projection-free sparse convex optimization, in *Proceedings of the 30th International Conference on Machine Learning*, Atlanta, ed. by S. Dasgupta, D. McAllester. Proceedings of Machine Learning Research, vol. 28 of number 1, pp. 427–435 (2013)
16. D.P. Kingma, J. Ba, Adam: a method for stochastic optimization, Dec. 2014. arXiv:1412.6980
17. P. Latafat, P. Patrinos, Asymmetric forward–backward–adjoint splitting for solving monotone inclusions involving three operators. Comput. Optim. Appl. **68**(1), 57–93 (2017)
18. P.L. Lions, B. Mercier, Splitting algorithms for the sum of two nonlinear operators. SIAM J. Numer. Anal. **16**(6), 964–979 (1979)
19. Y. Nesterov, A method of solving a convex programming problem with convergence rate O $(1/k^2)$. Sov. Math. Dokl. **27**(2), 372–376 (1983)
20. Y. Nesterov, *Introductory Lectures on Convex Optimization: A Basic Course*, 1st edn. (Springer, Boston, 2003)
21. Y. Nesterov, Efficiency of coordinate descent methods on huge-scale optimization problems. SIAM J. Optim. **22**(2), 341–362 (2012)
22. Y. Nesterov, A. Nemirovskii, *Interior-Point Polynomial Algorithms in Convex Programming* (Society for Industrial and Applied Mathematics, Philadelphia, PA, 1994)
23. N. Parikh, S. Boyd, Proximal algorithms. Found. Trends Optim. **1**(3), 123–231 (2014)
24. G. Pierra, Decomposition through formalization in a product space. Math. Program. **28**(1), 96–115 (1984)
25. E.K. Ryu, S. Boyd, A primer on monotone operator methods. Appl. Comput. Math. **15**(1), 3–43 (2016)
26. P. Tseng, A modified forward-backward splitting method for maximal monotone mappings. SIAM J. Control Optim. **38**(2), 431–446 (2000)
27. L. Vandenberghe, M.S. Andersen, Chordal graphs and semidefinite optimization. Found. Trends Optim. **1**(4), 241–433 (2015)
28. Y. Ye, M.J. Todd, S. Mizuno, An o($\sqrt{n}L$)-iteration homogeneous and self-dual linear programming algorithm. Math. Oper. Res. **19**(1), 53–67 (1994)

Chapter 2
Exploiting Chordality in Optimization Algorithms for Model Predictive Control

Anders Hansson and Sina Khoshfetrat Pakazad

Abstract In this chapter we show that chordal structure can be used to devise efficient optimization methods for many common model predictive control problems. The chordal structure is used both for computing search directions efficiently as well as for distributing all the other computations in an interior-point method for solving the problem. The chordal structure can stem both from the sequential nature of the problem as well as from distributed formulations of the problem related to scenario trees or other formulations. The framework enables efficient parallel computations.

Keywords Model predictive control · Quadratic programming · Chordal graphs · Message passing · Dynamic programming · Parallel computations

AMS Subject Classifications 90C25, 90C35, 90C39, 90C51

2.1 Introduction

Model Predictive Control (MPC) is an important class of controllers that are being employed more and more in industry, [25]. It has its root going back to [6]. The success is mainly because it can handle constraints on control signals and/or states in a systematic way. In the early years its applicability was limited to slow processes, since an optimization problem has to be solved at each sampling instant. Tremendous amount of research has been spent on overcoming this limitation.

A. Hansson (✉)
Division of Automatic Control, Linköping University, Linköping, Sweden
e-mail: anders.g.hansson@liu.se

S. K. Pakazad
C3 IoT, Redwood city, CA, USA
e-mail: sina.pakazad@c3iot.com

© Springer Nature Switzerland AG 2018
P. Giselsson, A. Rantzer (eds.), *Large-Scale and Distributed Optimization*, Lecture Notes in Mathematics 2227,
https://doi.org/10.1007/978-3-319-97478-1_2

11

One avenue has been what is called explicit MPC, [2], where the optimization problem is solved parametrically off-line. Another avenue has been to exploit the inherent structure of the optimization problems stemming from MPC, [1, 3, 4, 7–9, 11, 13, 14, 16, 17, 19, 23, 26–31]. Typically this has been to use Riccati recursions to efficiently compute search directions for Interior Point (IP) methods or active set methods to solve the optimization problem. In this paper we will argue that the important structures that have been exploited can all be summarized as *chordal structure*. Because of this the same structure exploiting software can be used to speed up all computations for MPC. This is irrespective of what MPC formulation is considered and irrespective of what type of optimization algorithm is used. We assume that the reader is familiar with the receding horizon strategy of MPC and we will only discuss the associated constrained finite-time optimal control problems. We will from now on refer to the associated problem as the MPC problem. We will mostly assume quadratic cost and linear dynamics and inequality constraints. Even if not all problems fall into this category, problems with quadratic objective and linear constraints are often solved as subproblems in solvers. A preliminary version of this chapter has been published in [15].

The remaining part of the paper is organized as follows. We will in Sect. 2.2 present and discuss some results related to IP methods and to parametric quadratic optimization problems. In Sect. 2.3 we will then discuss the classical formulation of MPC, and how the problem can be solved using an IP method. Specifically we will discuss how the equations for the search directions can be distributed over the clique tree. The well-known backward dynamic programming solution will be derived as a special case. We will see that we can also do forward dynamic programming, combinations of forward and backward dynamic programming, and even dynamic programming in parallel. In Sect. 2.4 we will discuss regularized MPC. In Sect. 2.5 we will discuss stochastic MPC. In Sect. 2.6 we will discuss distributed MPC, and finally in Sect. 2.7 we will give some conclusions, discuss generalizations of our results and directions for future research.

Notation

We denote with $\mathbf{R}$ the set of real numbers, with $\mathbf{R}^n$ the set of n-dimensional real-valued vectors and with $\mathbf{R}^{m \times n}$ the set of real-valued matrices with m rows and n columns. We denote by $\mathbf{N}$ the set of natural numbers and by $\mathbf{N}_n$ the subset $\{1, 2, \ldots, n\}$ of $\mathbf{N}$. For a vector $x \in \mathbf{R}^n$ the matrix $X = \mathbf{diag}(x)$ is a diagonal matrix with the components of x on the diagonal. For two matrices A and B the matrix $A \oplus B$ is a block-diagonal matrix with A as the 1,1-block and B as the 2,2-block. For a symmetric matrix A the notation $A(\succeq) \succ 0$ is equivalent to A being positive (semi)-definite.

2.2   Convex Optimization

Consider the following convex optimization problem

$$\min_{x} \quad F_1(x) + \cdots + F_N(x), \tag{2.1}$$

where $F_i : \mathbf{R}^n \to \mathbf{R}$ for all $i = 1, \ldots, N$. We assume that each function F_i is only dependent on a small subset of elements of x. Let us denote the ordered set of these indexes by $J_i \subseteq \mathbf{N}_n$. We can then rewrite the problem in (2.1), as

$$\min_{x} \quad \bar{F}_1(E_{J_1}x) + \cdots + \bar{F}_N(E_{J_N}x), \tag{2.2}$$

where E_{J_i} is a 0–1 matrix that is obtained from an identity matrix of order n by deleting the rows indexed by $\mathbf{N}_n \setminus J_i$. The functions $\bar{F}_i : \mathbf{R}^{|J_i|} \to \mathbf{R}$ are lower dimensional descriptions of F_is such that $F_i(x) = \bar{F}_i(E_{J_i}x)$ for all $x \in \mathbf{R}^n$ and $i \in \mathbf{N}_N$. For details on how this structure can be exploited using message passing the reader is referred to [18].

A brief summary is that we may define a so-called sparsity graph for the above optimization problem with n nodes and edges between two nodes j and k if x_j and x_k appear in the same term $\bar{F}_i$. We assume that this graph is chordal, i.e., every cycle of length four our more has a chord.[1] The maximal complete subgraphs of a graph are called its cliques. If the original graph is chordal then there exists a tree of the cliques called the clique tree which is such that it enjoys the clique intersection property. This property is that all elements in the intersection of two cliques C_i and C_j should be elements of the cliques on the path between the cliques C_i and C_j. It is then possible to use the clique tree as a computational tree where we non-uniquely assign terms of the objective function to each clique in such a way that all the variables of the term in the function are elements of the clique. After this we may solve the optimization problem distributedly over the clique tree by starting with leafs and for each leaf solve a parametric optimization problem, where we optimize with the respect to the variables of the leaf problem which are not variables of the parent of the leaf in the clique tree. The optimization should be done parametrically with respect to all the variables that are shared with the parent. After this the optimal objective function value of the leaf can be expressed as a function of the variables that are shared with the parent. This function is sent to the parent and added to its objective function term. The leaf has been pruned away, and then the optimization can continue with the parent assuming all its children has also carried out their local optimizations. Eventually we reach the root of the tree, where the remaining variables are optimized. Then we can finally go down the tree

---

[1] In case the graph is not chordal we make a chordal embedding, i.e., we add edges to the graph until it becomes chordal. This corresponds to saying that some of the $\bar{F}_i$ depend on variables that they do not depend on.

and recover all optimal variables. This is based on the fact that we have stored the parametric optimal solutions in the nodes of the clique tree.

2.2.1 Interior-Point Methods

All the MPC problems that we encounter in this chapter are special cases of Quadratic Programs (QPs). We will now discuss how such problems can be solved using IP methods, [32]. Consider the QP

$$\min_{z} \frac{1}{2} z^T Q z + q^T z \tag{2.3}$$

$$\text{s.t. } \mathcal{A}z = b \tag{2.4}$$

$$\mathcal{D}z \le e \tag{2.5}$$

where $Q \succeq 0$, i.e., positive semidefinite, where $\mathcal{A}$ has full row rank, and where the matrices and vectors are of compatible dimensions. The inequality in (2.5) is component-wise inequality. The Karush-Kuhn-Tucker (KKT) optimality conditions for this problem is

$$\begin{bmatrix} Q & \mathcal{A}^T & \mathcal{D}^T & \\ \mathcal{A} & & & \\ \mathcal{D} & & & I \\ & & M & \end{bmatrix} \begin{bmatrix} z \\ \lambda \\ \mu \\ s \end{bmatrix} = \begin{bmatrix} -q \\ b \\ e \\ 0 \end{bmatrix} \tag{2.6}$$

and $(\mu, s) \ge 0$, where $M = \mathbf{diag}(\mu)$. Blank entries in a matrix are the same as zero entries. Above λ and μ are the Lagrange multipliers for the equality and inequality constraints, respectively. The vector s is the slack variable for the inequality constraints. In IP methods one linearizes the above equations to obtain equations for search directions:

$$\begin{bmatrix} Q & \mathcal{A}^T & \mathcal{D}^T & \\ \mathcal{A} & & & \\ \mathcal{D} & & & I \\ & & S & M \end{bmatrix} \begin{bmatrix} \Delta z \\ \Delta \lambda \\ \Delta \mu \\ \Delta s \end{bmatrix} = \begin{bmatrix} r_z \\ r_\lambda \\ r_\mu \\ r_s \end{bmatrix} \tag{2.7}$$

where $S = \mathbf{diag}(s)$, and where $r = (r_z, r_\lambda, r_\mu, r_s)$ is some residual vector that depends on what IP method is used. The quantities r, S and M depend on the value of the current iterate in the IP method. From the last two rows above we have $\Delta s = r_\mu - \mathcal{D}\Delta z$ and $\Delta \mu = S^{-1}(r_s - M\Delta s)$. After substitution of these expressions into the first two rows we obtain

$$\begin{bmatrix} Q + \mathcal{D}^T S^{-1} M \mathcal{D} & \mathcal{A}^T \\ \mathcal{A} & \end{bmatrix} \begin{bmatrix} \Delta z \\ \Delta \lambda \end{bmatrix} = \begin{bmatrix} r_z - \mathcal{D}^T S^{-1} (r_s - M r_\mu) \\ r_\lambda \end{bmatrix}. \tag{2.8}$$

We notice that the search directions are obtained by solving an indefinite symmetric linear system of equations. The indefinite matrix is referred to as the KKT matrix, and it is invertible if and only if

$$Q_s = Q + \mathcal{D}^T S^{-1} M \mathcal{D} \tag{2.9}$$

is positive definite on the null-space of $\mathcal{A}$. Notice that the KKT matrix for the search directions can be interpreted as the optimality conditions of a QP with only equality constraints, where the quadratic weight is modified such that it is larger the closer the iterates are to the boundary of the constraints. In case this QP is loosely coupled with chordal structure message passing over a clique tree can be used to compute the search directions in a distributed way as described in [18]. The key to this will be to solve parametric QPs, which is the next topic. Before finishing this section we remark that for active set methods similar QPs also have to be solved. There will however be additional equality constraints depending on what constraints are active at the current iterate.

2.2.2 Parametric QPs

Consider the quadratic optimization problem

$$\min_z \frac{1}{2} z^T M z + m^T z \tag{2.10}$$

$$\text{s.t. } Cz = d \tag{2.11}$$

with C full row rank and $M \succeq 0$. The KKT conditions for the optimal solution are:

$$\begin{bmatrix} M & C^T \\ C & \end{bmatrix} \begin{bmatrix} z \\ \lambda \end{bmatrix} = \begin{bmatrix} -m \\ d \end{bmatrix}.$$

These equations have a unique solution if and only if $M + C^T C \succ 0$. Now consider the partitioning of the above problem defined by

$$M = \begin{bmatrix} Q & S \\ S^T & R \end{bmatrix}; \quad C = \begin{bmatrix} A & B \\ & D \end{bmatrix}; \quad d = \begin{bmatrix} e \\ f \end{bmatrix}; \quad m = \begin{bmatrix} q \\ r \end{bmatrix}; \quad z = \begin{bmatrix} x \\ y \end{bmatrix}$$

with A full row rank. We then assume that we want to solve the problem

$$\min_{x} \frac{1}{2} \begin{bmatrix} x \\ y \end{bmatrix}^T \begin{bmatrix} Q & S \\ S^T & \end{bmatrix} \begin{bmatrix} x \\ y \end{bmatrix} + q^T x \tag{2.12}$$

$$\text{s.t. } Ax + By = e \tag{2.13}$$

parametrically with respect to all y. We call this a leaf problem. The KKT conditions for this problem are

$$\begin{bmatrix} Q & A^T \\ A & \end{bmatrix} \begin{bmatrix} x \\ \mu \end{bmatrix} = \begin{bmatrix} -q - Sy \\ e - By \end{bmatrix}.$$

Notice that the solution x will be affine in y, and hence when it is substituted back into the objective function we obtain a quadratic message in y, see [18] for why this is relevant in message passing algorithms. The 1,1-block of $M + C^T C$ is $Q + A^T A$, which by the Schur complement formula is positive definite. Hence the leaf problem has a unique solution. If we then substitute the solution of the leaf into the overall problem, we will have a unique solution also for this problem, since the overall problem has a unique solution. Because of this, every leaf in the message passing algorithm will have a problem with unique solutions assuming that the overall problem has a unique solution. This also goes for all nodes, since they will become leaves as other leaves are pruned away.

Notice that it is always possible to make sure that the matrix A has full rank for a leaf by pre-processing of the inequality constraints. In case A does not have full row rank, perform a rank-revealing factorization such that the constraints can be written

$$\begin{bmatrix} \bar{A}_1 \\ 0 \end{bmatrix} x + \begin{bmatrix} \bar{B}_1 \\ \bar{B}_2 \end{bmatrix} y = \begin{bmatrix} \bar{e}_1 \\ \bar{e}_2 \end{bmatrix}$$

and append the constraint $\bar{B}_2 y = \bar{e}_2$ to belong to

$$Dy = f.$$

This can be done recursively over the so-called clique tree so that the parametric QPs for each node satisfy the rank condition, [18]. This is handled differently in [24] for a similar problem.

We will now see how chordal sparsity and distributed computations can be used to solve optimization problems arising in MPC efficiently.

2.3 Classical MPC

A classical MPC problem can be cast in the form

$$\min_{u} \frac{1}{2} \sum_{k=0}^{N-1} \begin{bmatrix} x_k \\ u_k \end{bmatrix}^T Q \begin{bmatrix} x_k \\ u_k \end{bmatrix} + \frac{1}{2} x_N^T S x_N \tag{2.14}$$

$$\text{s.t. } x_{k+1} = A x_k + B u_k + v_k, \quad x_0 = \bar{x} \tag{2.15}$$

$$C x_k + D u_k \leq e_k \tag{2.16}$$

with A, B, C, D, Q, S, $\bar{x}$, e_k, and v_k given, and where $u = (u_0, u_1, \ldots, u_{N-1})$ are the optimization variables. The dimensions of the control signal u_k and the state vector x_k are m and n, respectively. The number of inequality constraints are q for each time index k. The dimensions of all other quantities are defined to be consistent with this. We assume that $Q \succeq 0$ and that $S \succeq 0$. This is a convex quadratic optimization problem. When the inequality constraints are not present it is a classical Linear Quadratic (LQ) control problem. It is of course possible to extend the problem formulation to time-varying dynamics, inequality constraints and weights. Also the extension to a linear term in the objective function is straight forward.

2.3.1 Quadratic Program

The classical formulation in (2.14)–(2.16) is equivalent to (2.3)–(2.5) with $q = 0$,

$$z = (x_0, u_0, x_1, u_1 \ldots, x_{N-1}, u_{N-1}, x_N)$$

$$\lambda = (\lambda_0, \lambda_1, \ldots, \lambda_N)$$

$$b = (\bar{x}, v_0, v_1, \ldots, v_{N-1})$$

$$e = (e_0, e_1, \ldots, e_{N-1})$$

and

$$\mathcal{A} = \begin{bmatrix} I & & & & & \\ -A & -B & I & & & \\ & & -A & -B & I & \\ & & & & \ddots & \\ & & & & & -A & -B & I \end{bmatrix}$$

$$\mathcal{D} = \begin{bmatrix} C & D \end{bmatrix} \oplus \begin{bmatrix} C & D \end{bmatrix} \oplus \ldots \oplus \begin{bmatrix} C & D \end{bmatrix}$$

$$\mathcal{Q} = Q \oplus Q \oplus \cdots \oplus Q \oplus S.$$

We see that the data matrices are banded. Hence, sparse linear system solvers could be used when solving the KKT equations for search directions in an IP method, but we will see that the structure within the bands can be further utilized. Also we notice that the matrix Q_s in (2.9) has the same structure as Q. Therefore the KKT matrix for the search directions can be interpreted as the optimality conditions of an unconstrained LQ control problem for the search directions, where the weights are modified such that they are larger the closer the iterates are to the boundary of the constraints. Since inequality constraints do not affect the structure of the KKT matrix we will from now on not consider them when we discuss the different MPC problems.

The classical formulation without any constraints is usually solved using a backward Riccati recursion. We will see how this can be obtained from the general techniques presented above. This is the same derivation that is usually done using backward dynamic programming. We will also investigate forward dynamic programming, and finally we will see how the computations can be parallelized.

2.3.2 Backward Dynamic Programming

When looking at (2.14)–(2.15) it can be put in the almost separable formulation in (2.2) by defining

$$\bar{F}_1(x_0, u_0, x_1) = \mathcal{I}_\mathcal{D}(x_0) + \frac{1}{2}\begin{bmatrix} x_0 \\ u_0 \end{bmatrix}^T Q \begin{bmatrix} x_0 \\ u_0 \end{bmatrix} + \mathcal{I}_{\mathcal{C}_0}(x_0, u_0, x_1)$$

$$\bar{F}_{k+1}(x_k, u_k, x_{k+1}) = \frac{1}{2}\begin{bmatrix} x_k \\ u_k \end{bmatrix}^T Q \begin{bmatrix} x_k \\ u_k \end{bmatrix} + \mathcal{I}_{\mathcal{C}_k}(x_k, u_k, x_{k+1}), k = 1, \ldots, N-2$$

$$\bar{F}_N(x_{N-1}, u_{N-1}, x_N) = \frac{1}{2}\begin{bmatrix} x_{N-1} \\ u_{N-1} \end{bmatrix}^T Q \begin{bmatrix} x_{N-1} \\ u_{N-1} \end{bmatrix}$$

$$+ \mathcal{I}_{\mathcal{C}_{N-1}}(x_{N-1}, u_{N-1}, x_N) + \frac{1}{2}x_N^T S x_N$$

where $\mathcal{I}_{\mathcal{C}_k}(x_k, u_k, x_{k+1})$ is the indicator function for the set

$$\mathcal{C}_k = \{(x_k, u_k, x_{k+1}) \mid x_{k+1} = Ax_k + Bu_k\}$$

and where $\mathcal{I}_\mathcal{D}(x_0)$ is the indicator function for the set

$$\mathcal{D} = \{x_0 \mid x_0 = \bar{x}\}.$$

We have assumed that $v_k = 0$. It should be stressed that the derivations done below easily can be extended to the general case. The sparsity graph for this problem is

Fig. 2.1 Sparsity graph for the problem in (2.14)–(2.15) to the left and its corresponding clique tree to the right

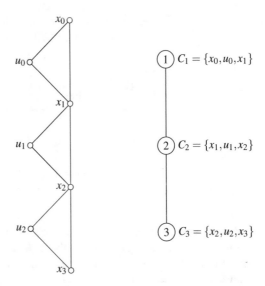

depicted in Fig. 2.1 for the case of $N = 3$, see [18] for the definition of a sparsity graph and its cliques. For ease of notation we label the nodes with the states and control signals.[2] The cliques for this graph are

$$C_{k+1} = \{x_k, u_k, x_{k+1}\}, \quad k = 0, \ldots, N-1.$$

To obtain a backward dynamic problem formulation we define a clique tree by taking C_1 as root as seen in Fig. 2.1. Clique trees and how they can be used as computational trees are described in more detail in [18]. We then assign $\bar{F}_k$ to C_k. This is the only information that has to be provided to a general purpose software for solving loosely coupled quadratic programs.

One may of course derive the well-known Riccati-recursion based solution from what has been defined above. The k:th problem to solve is

$$\min_{u_k} \frac{1}{2} \begin{bmatrix} x_k \\ u_k \end{bmatrix}^T Q \begin{bmatrix} x_k \\ u_k \end{bmatrix} + m_{k+1,k}(x_{k+1})$$

$$\text{s.t. } x_{k+1} = Ax_k + Bu_k$$

---

[2]Here we use a super-node for all components of a state and a control signal, respectively. In case there is further structure in the dynamic equations such that not all components of the control signal and the states are coupled, then more detailed modeling could potentially be beneficial.

for given x_k starting with $k = N - 1$ going down to $k = 0$, where $m_{N,N-1}(x_N) = \frac{1}{2}x_N^T S x_N$. The optimality conditions are for $k = N - 1$

$$\begin{bmatrix} S & I \\ Q_2 & -B^T \\ I & -B \end{bmatrix} \begin{bmatrix} x_N \\ u_{N-1} \\ \lambda_N \end{bmatrix} = \begin{bmatrix} 0 \\ -Q_{12}^T x_{N-1} \\ A x_{N-1} \end{bmatrix}.$$

with

$$Q = \begin{bmatrix} Q_1 & Q_{12} \\ Q_{12}^T & Q_2 \end{bmatrix}$$

We notice that no pre-processing of constraints is needed since $\begin{bmatrix} I & -B \end{bmatrix}$ has full row rank. Therefore by the results in Section 2.2.2, if the overall problem has a unique solution, so does the above problem. It is however not easy to give conditions when the overall problem has a unique solution. The optimality conditions above are equivalent to $\lambda_N = -S x_N$, $x_N = A x_{N-1} + B u_{N-1}$ and the equation

$$G_{N-1} u_{N-1} = -H_{N-1}^T x_{N-1}$$

where $G_{N-1} = Q_2 + B^T S B$ and $H_{N-1} = Q_{12} + A^T S B$. Let $F_{N-1} = Q_1 + A^T S A$. Then, if $\begin{bmatrix} A & B \end{bmatrix}$ has full row rank, Q is positive definite on the null-space of $\begin{bmatrix} A & B \end{bmatrix}$, and if $S \succ 0$, it holds that[3]

$$\begin{bmatrix} F_{N-1} & H_{N-1} \\ H_{N-1}^T & G_{N-1} \end{bmatrix} = Q + \begin{bmatrix} A & B \end{bmatrix}^T S \begin{bmatrix} A & B \end{bmatrix} \succ 0.$$

Therefore G_{N-1} is positive definite by the Schur complement formula, and hence there is a unique solution

$$u_{N-1} = -G_{N-1}^{-1} H_{N-1}^T x_{N-1}.$$

Back-substitution of this into the objective function shows that

$$m_{N-1,N-2}(x_{N-1}) = \frac{1}{2}x_{N-1}^T S_{N-1} x_{N-1}$$

---

[3]Notice that the assumptions are not necessary for the block matrix to be positive definite. Moreover, for the case when it is only positive semidefinite, we still have a solution u_{N-1}, but it is not unique. One may use pseudo inverse to obtain one solution. This follows from the generalized Schur complement formula. The full row rank assumption is equivalent to (A, B) not having any uncontrollable modes corresponding to zero eigenvalues. The positive definiteness of Q on the null-space of $\begin{bmatrix} A & B \end{bmatrix}$ is equivalent to $C(zI - A)^{-1}B + D$ not having any zeros at the origin where $Q = \begin{bmatrix} C & D \end{bmatrix}^T \begin{bmatrix} C & D \end{bmatrix}$ is a full rank factorization.

where $S_{N-1} = F_{N-1} - H_{N-1}G_{N-1}^{-1}H_{N-1}$ which is the first step of the well-known Riccati recursion. Notice that S_{N-1} is positive definite by the Schur complement formula. Repeating the above steps for the remaining problems shows that the overall solution can be obtained using the Riccati recursion. Notice that a general purpose solver instead factorizes the local optimality conditions at each step. It is well-known that the Riccati recursion provides a factorization of the overall KKT matrix, [28]. It can be shown that the message passing algorithm does the same, [18]. The main point of this chapter is, however, that there is no need to derive Riccati recursions or have any interpretations as factorizations. This becomes even more evident when we look at not so well-studied MPC formulations, where the corresponding structure makes it possible to see how Riccati recursions can be used are only revealed after cumbersome manipulations of the KKT equations. In some cases it is not even possible to derive Riccati recursions, which is the point of the next subsection.

2.3.3 Forward Dynamic Programming

Instead of taking C_1 as root as we did in the previous subsection we can also choose C_N as root. We then obtain a forward dynamic programming formulation. We assign the functions to the cliques in the same way. The initial problem to solve is

$$\min_{u_0} \frac{1}{2}\begin{bmatrix} x_0 \\ u_0 \end{bmatrix}^T Q \begin{bmatrix} x_0 \\ u_0 \end{bmatrix}$$
$$\text{s.t. } x_1 = Ax_0 + Bu_0, \quad x_0 = \bar{x}$$

parametrically for all possible values of x_1. Here we realize that the constraints for (x_0, u_0) do not satisfy the full row rank assumption, i.e.,

$$\begin{bmatrix} I \\ A \ B \end{bmatrix}$$

does not have full row rank. Therefore pre-processing is required, which can be done using e.g., a QR-factorization on B. This will result in constraints on x_1 that should be passed to the next problem, and then this procedure should be repeated. Because of this, there is no such clean solution procedure for the forward approach as for the backward approach, and particularly no Riccati recursion based approach. However, the general message passing approach indeed works.

2.3.4 Parallel Computation

In the previous cases we had a tree that was a chain. It was then possible to let either of the end cliques be the root of the tree. However, nothing stops us from picking up any one of the middle cliques as the root. This would result in two branches, and it would then be possible to solve the problems in the two branches in parallel, one branch using the backward approach, and one using the forward approach. This does however not generalize to more than two parallel branches. If we want to have three or more we need to proceed differently.

To this end, let us consider a simple example where $N = 6$. Let us also assume that we want to solve this problem using two computational agents such that each would perform independently, and hence in parallel. For this, we define dummy variables $\bar{u}_0$ and $\bar{u}_1$ and constrain them as

$$\bar{u}_0 = x_3, \quad \bar{u}_1 = x_6.$$

This is similar to what is done in [23] to obtain parallel computations. We also define the following sets

$$
\begin{aligned}
\mathcal{C}_{-1} &= \{x_0 : x_0 = \bar{x}\} \\
\mathcal{C}_k &= \{(x_k, u_k, x_{k+1}) : x_{k+1} = Ax_k + Bu_k\}; \ k = 0, 1 \\
\mathcal{C}_2 &= \{(x_2, u_2, \bar{u}_0) : \bar{u}_0 = Ax_2 + Bu_2\} \\
\mathcal{C}_k &= \{(x_k, u_k, x_{k+1}) : x_{k+1} = Ax_k + Bu_k\}; \ k = 3, 4 \quad\quad (2.17)\\
\mathcal{C}_5 &= \{(x_5, u_5, \bar{u}_1) : \bar{u}_1 = Ax_5 + Bu_5\} \\
\mathcal{D}_0 &= \{(x_3, \bar{u}_0) : \bar{u}_0 = x_3\} \\
\mathcal{D}_1 &= \{(x_6, \bar{u}_1) : \bar{u}_1 = x_6\}.
\end{aligned}
$$

Then the problem in (2.14)–(2.15) can be equivalently written as

$$
\min_u \frac{1}{2} \sum_{k=0}^{1} \begin{bmatrix} x_k \\ u_k \end{bmatrix}^T Q \begin{bmatrix} x_k \\ u_k \end{bmatrix} + \mathcal{I}_{\mathcal{C}_k}\{x_k, u_k, x_{k+1}\} \quad\quad (2.18)
$$

$$
+ \frac{1}{2} \begin{bmatrix} x_2 \\ u_2 \end{bmatrix}^T Q \begin{bmatrix} x_2 \\ u_2 \end{bmatrix} + \mathcal{I}_{\mathcal{C}_2}\{x_2, u_2, \bar{u}_0\}
$$

$$
+ \frac{1}{2} \sum_{k=3}^{4} \begin{bmatrix} x_k \\ u_k \end{bmatrix}^T Q \begin{bmatrix} x_k \\ u_k \end{bmatrix} + \mathcal{I}_{\mathcal{C}_k}\{x_k, u_k, x_{k+1}\}
$$

$$
+ \frac{1}{2} \begin{bmatrix} x_5 \\ u_5 \end{bmatrix}^T Q \begin{bmatrix} x_5 \\ u_5 \end{bmatrix} + \mathcal{I}_{\mathcal{C}_5}\{x_5, u_5, \bar{u}_1\} + \frac{1}{2}\bar{u}_1^T S \bar{u}_1
$$

$$
+ \mathcal{I}_{\mathcal{C}_{-1}}\{x_0\} + \mathcal{I}_{\mathcal{D}_0}\{x_3, \bar{u}_0\} + \mathcal{I}_{\mathcal{D}_1}\{x_6, \bar{u}_1\}
$$

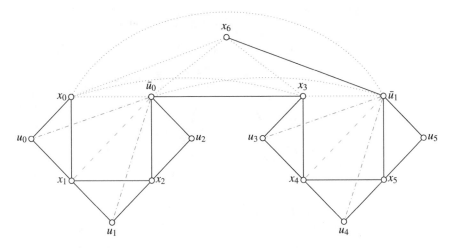

Fig. 2.2 A modified sparsity graph for the problem in (2.18). The initial sparsity graph, without any modification, is marked with solid lines

where $\mathcal{I}_{\mathcal{X}}(x)$ is the indicator function for the set $\mathcal{X}$. Notice that it is important to define $\mathcal{C}_2$ in terms of $\bar{u}_0$ and not in terms of x_3, and similarly for $\mathcal{C}_5$. This trick will allow us to have two independent computational agents. The reason for this will be clear later on.

Let us consider the sparsity graph for the problem in (2.18), which is depicted in Fig. 2.2, marked with solid lines. In order to obtain a clique tree that facilitates parallel computations, we first add edges, marked with dotted lines, between x_0, $\bar{u}_0$, x_3, $\bar{u}_1$ and x_6 such that they form a *maximal* complete subgraph in the graph. The original graph was chordal, but adding the dotted edges destroyed this. Therefore we make a chordal embedding by adding the dashed edges.[4] We actually add even more edges, which correspond to merging cliques. These are the dash-dotted edges. The reason we do this is that we do not need computational agents for more cliques than the ones we get after the merging. A clique tree which corresponds to the modified sparsity graph in Fig. 2.2, is illustrated in Fig. 2.3. This clique tree obviously enables parallel computations. The different terms in (2.18) are assigned such that rows one and two are assigned to the left branch, rows three and four to the right branch and the last row to the root.

Notice that in this particular example we obtained a clique tree with two parallel branches. However, we can generalize to several parallel branches by introducing more dummy variables and constraints. Also it is worth pointing out that the subproblem which is assigned to the root of clique tree can be seen as an LQ problem and hence we can use the procedure discussed above recursively. This is similar to

---

[4]The added edges corresponds to saying that terms in the objective function are functions of variables which they are actually not.

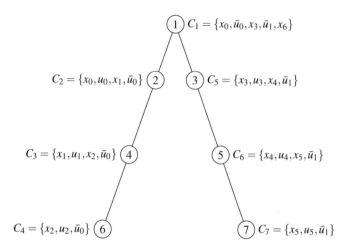

Fig. 2.3 Corresponding clique tree for the modified sparsity graph shown in Fig. 2.2

what is presented in [23]. This is obtained in the example above by not connecting all of x_0, $\bar{u}_0$, x_3, $\bar{u}_1$ and x_6. Instead one should connect all of x_0, $\bar{u}_0$ and x_3 and then all of x_3, $\bar{u}_1$ and x_6 separately. This will then split the clique C_1 into two cliques in the clique tree. The dynamics for this LQ problem is defined by the sets $\mathcal{D}_k$, $k = 0, 1$. The incremental costs for this problem will be the messages sent by the children of these cliques. Notice that we do not really have to know that the resulting problem will be an LQ problem—that is just an interpretation. We only need to know how to split the root clique into one for each parallel branch. If we want four parallel branches the clique tree will be like in Fig. 2.4.

2.3.5 Merging of Cliques

It is not always the case that one has one agent or processor available for each and every clique in the clique tree. What then can be done is to merge cliques until there are as many cliques as there are processors. Let us consider the clique tree in Fig. 2.4. We can merge the cliques in each and every parallel branch into one clique. The resulting clique tree will then be a chain as depicted in Fig. 2.5. This could have been done even before the clique tree was constructed. However, it is beneficial for each of the four agents in the example to utilize the additional structure within their cliques, i.e., that they have an internal chain structure. This information would have been lost in case the cliques were merged before the clique tree was formed.

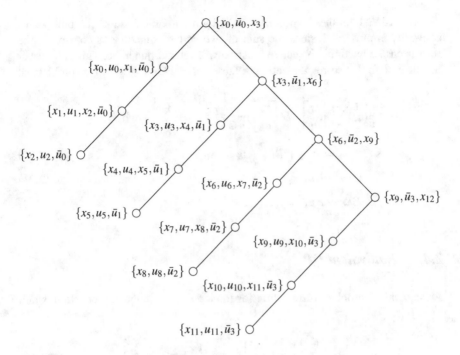

Fig. 2.4 Clique tree with four parallel branches

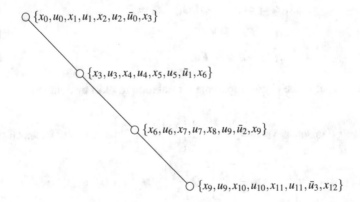

Fig. 2.5 Merged clique tree

2.4 Regularized MPC

A regularized MPC problem is obtained from the above MPC problem by adding a regularization term to the objective function. Typically this is term is either proportional to the squared l^2 (Euclidean) norm, so-called Tikhonov-regularization, or proportional to the l^1 norm, so-called Lasso-regularization. In both cases convexity is preserved, since the sum of two convex functions is convex. In the former case a quadratic objective function will remain quadratic. This is however not the case for Lasso-regularization. A fairly general Lasso-regularized problem is:

$$\min_u \frac{1}{2} \sum_{k=0}^{N-1} \begin{bmatrix} x_k \\ u_k \end{bmatrix}^T Q \begin{bmatrix} x_k \\ u_k \end{bmatrix} + \frac{1}{2} x_N^T S x_N + \sum_{k=0}^{N-1} \| y_k \|_1 \tag{2.19}$$

$$\text{s.t. } x_{k+1} = A x_k + B u_k + v_k, \quad x_0 = \bar{x} \tag{2.20}$$

$$C x_k + D u_k \leq e_k \tag{2.21}$$

$$E x_k + F u_k = y_k. \tag{2.22}$$

2.4.1   Equivalent QP

An equivalent problem formulation for the case of no inequality constraints and $v_k = 0$ is:

$$\min_u \frac{1}{2} \sum_{k=0}^{N-1} \begin{bmatrix} x_k \\ u_k \end{bmatrix}^T Q \begin{bmatrix} x_k \\ u_k \end{bmatrix} + \frac{1}{2} x_N^T S x_N + \sum_{k=0}^{N-1} t_k \tag{2.23}$$

$$\text{s.t. } x_{k+1} = A x_k + B u_k, \quad x_0 = \bar{x} \tag{2.24}$$

$$t_k \geq E x_k + F u_k \tag{2.25}$$

$$t_k \geq -E x_k - F u_k. \tag{2.26}$$

This can be put in the almost separable formulation in (2.2) by defining

$$\bar{F}_1(x_0, u_0, t_0, x_1) = \mathcal{I}_\mathcal{D}(x_0) + \frac{1}{2} \begin{bmatrix} x_0 \\ u_0 \end{bmatrix}^T Q \begin{bmatrix} x_0 \\ u_0 \end{bmatrix} + t_0 + \mathcal{I}_{\mathcal{C}_0}(x_0, u_0, t_0, x_1)$$

$$\bar{F}_{k+1}(x_k, u_k, t_k, x_{k+1}) = \frac{1}{2} \begin{bmatrix} x_k \\ u_k \end{bmatrix}^T Q \begin{bmatrix} x_k \\ u_k \end{bmatrix} + t_k + \mathcal{I}_{\mathcal{C}_k}(x_k, u_k, t_k, x_{k+1}), k \in \mathbf{N}_{N-2}$$

$$\bar{F}_N(x_{N-1}, u_{N-1}, t_{N-1} x_N) = \frac{1}{2} \begin{bmatrix} x_{N-1} \\ u_{N-1} \end{bmatrix}^T Q \begin{bmatrix} x_{N-1} \\ u_{N-1} \end{bmatrix} + t_{N-1}$$

$$+ \mathcal{I}_{\mathcal{C}_{N-1}}(x_{N-1}, u_{N-1}, t_{N-1}, x_N) + \frac{1}{2} x_N^T S x_N$$

where $\mathcal{I}_{\mathcal{C}_k}(x_k, u_k, t_k, x_{k+1})$ is the indicator function for the set

$$\mathcal{C}_k = \{(x_k, u_k, x_{k+1}) \mid x_{k+1} = A x_k + B u_k; \ t_k \geq E x_k + F u_k; \ t_k \geq -E x_k - F u_k\}$$

and where $\mathcal{I}_D(x_0) = \{x_0 \mid x_0 = \bar{x}\}$. It should be stressed that the derivations done below easily can be extended to the general case. The sparsity graph for this problem is very similar to the one for the classical formulation. The cliques for this graph are

$$C_{k+1} = \{x_k, u_k, t_k, x_{k+1}\}, \quad k = 0, \ldots, N - 1.$$

To obtain a backward dynamic problem formulation we define a clique tree by taking C_1 as root similarly as for the classical formulation. We then assign $\bar{F}_k$ to C_k. This is the only information that has to be provided to a general purpose software for solving loosely coupled quadratic programs. We can do the forward dynamic programming formulation as well as a parallel formulation.

2.5 Stochastic MPC

We will in this section consider a stochastic MPC problem based on a scenario tree description. Several other authors have investigated how the structure stemming from scenario trees can be exploited, e.g., [10, 12, 21, 22]. The total number of scenarios is $M = d^r$, where d is the number of stochastic events that can take place at each time stage k, and where r is the number of time stages for which we consider stochastic events to take place. The outcome of the stochastic events are the different values of A_k^j, B_k^j and v_k^j. Notice that for values of $k < r$ several of these quantities are the same. The optimization problem is

$$\min_u \sum_{j=1}^M \omega_j \left(\frac{1}{2} \sum_{k=0}^{N-1} \begin{bmatrix} x_k^j \\ u_k^j \end{bmatrix}^T Q \begin{bmatrix} x_k^j \\ u_k^j \end{bmatrix} + \frac{1}{2}(x_N^j)^T S x_N^j \right) \tag{2.27}$$

$$\text{s.t. } x_{k+1}^j = A_k^j x_k^j + B_k^j u_k^j + v_k^j, \quad x_0^j = \bar{x} \tag{2.28}$$

$$\bar{C}u = 0 \tag{2.29}$$

where the index j refers to the j:th scenario. Here we define $u = (u^1, u^2, \ldots, u^M)$ with $u^j = (u_0^j, u_1^j, \ldots, u_{N-1}^j)$, and

$$\bar{C} = \begin{bmatrix} C_{1,2} & -C_{1,2} & & & \\ & C_{2,3} & -C_{2,3} & & \\ & & \ddots & & \ddots & \\ & & & C_{M-1,M} & -C_{M-1,M} \end{bmatrix}$$

with

$$C_{j,j+1} = \begin{bmatrix} I & 0 \end{bmatrix}$$

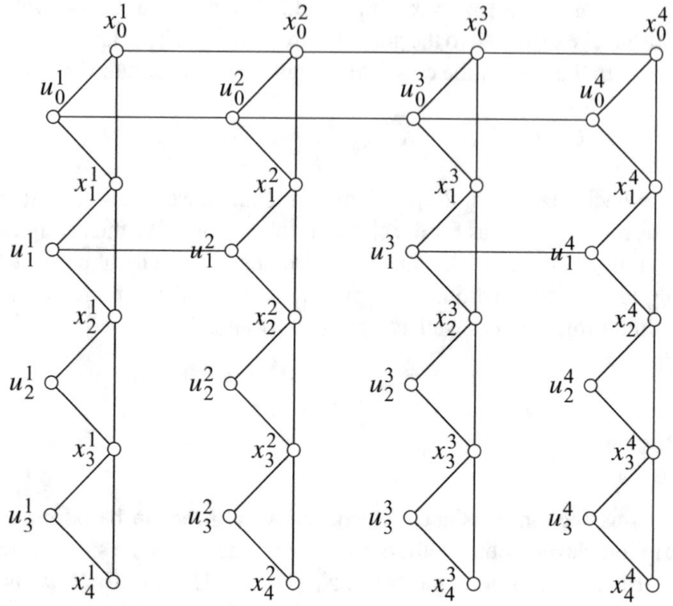

Fig. 2.6 Sparsity graph for the problem in (2.27)–(2.29)

where I is an identity matrix of dimension m times the number of nodes that scenarios j and $j+1$ have in common. The value of ω_j is the probability of scenario j. The constraint $\bar{C}u = 0$ is the so-called non-ancipativity constraint. Instead of saying that each initial state x_0^j is equal to $\bar{x}$ we instead consider the equivalent formulation $x_0^1 = \bar{x}$ and $x_0^j = x_0^{j+1}$, for $j \in \mathbf{N}_{M-1}$.

We show in Fig. 2.6 the sparsity graph for the case of $d = r = 2$ and $N = 4$. Then M is equal to 4. We realize that this graph is not chordal. A chordal embedding is obtained by adding edges such that $C_0 = \{x_0^1, x_0^2, x_0^3, x_0^4\}$ is a complete graph. Also edges should be added such that $C_1^1 = \{x_0^1, u_0^1, x_1^1, x_0^2, u_0^2, x_1^2\}$ and $C_1^3 = \{x_0^3, u_0^3, x_1^3, x_0^4, u_0^4, x_1^4\}$ are complete graphs. A clique tree for this chordal embedding is shown in Fig. 2.7, where $C_{k+1}^j = \{x_k^j, u_k^j, x_{k+1}^j\}$ with $k \in \mathbf{N}_{N-1}$ The assignments of functions are for $C_0 = \{x_0^1, x_0^2, x_0^3, x_0^4\}$

$$\bar{F}_0(x_0^1, x_0^2, x_0^3, x_0^4) = \mathcal{I}_{\mathcal{D}}(x_0^1) + \sum_{j=1}^{M-1} \mathcal{I}_{\mathcal{E}}(x_0^j, x_0^{j+1})$$

where $\mathcal{D} = \{x \mid x = \bar{x}\}$ and $\mathcal{E} = \{(x, y) \mid x = y\}$. For C_1^1 we assign

$$\bar{F}_1^1(x_0^1, u_0^1, x_1^1, x_0^2, u_0^2, x_1^2) = \sum_{j=1}^{2} \omega_j \frac{1}{2} \begin{bmatrix} x_0^j \\ u_0^j \end{bmatrix}^T Q \begin{bmatrix} x_0^j \\ u_0^j \end{bmatrix} + \mathcal{I}_{C_0^j}(x_0^j, u_0^j, x_1^j) + \mathcal{I}_{\mathcal{E}}(u_0^1, u_0^2),$$

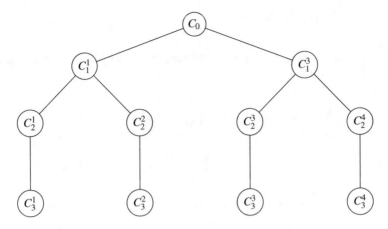

Fig. 2.7 Clique tree for the problem in (2.27)–(2.29)

for C_1^3 we assign

$$\bar{F}_1^3(x_0^3, u_0^3, x_1^3, x_0^4, u_0^4, x_1^4) = \sum_{j=3}^{4} \omega_j \frac{1}{2} \begin{bmatrix} x_0^j \\ u_0^j \end{bmatrix}^T Q \begin{bmatrix} x_0^j \\ u_0^j \end{bmatrix} + \mathcal{I}_{C_0^j}(x_0^j, u_0^j, x_1^j) + \mathcal{I}_{\mathcal{E}}(u_0^3, u_0^4),$$

for C_{k+1}^j, where $k \in \mathbf{N}_{N-1}$ and $j \in \mathbf{N}_M$, we assign

$$\bar{F}_{k+1}^j(x_k^j, u_k^j, x_{k+1}^j) = \omega_j \frac{1}{2} \begin{bmatrix} x_k^j \\ u_k^j \end{bmatrix}^T Q \begin{bmatrix} x_k^j \\ u_k^j \end{bmatrix} + \mathcal{I}_{C_k^j}(x_k^j, u_k^j, x_{k+1}^j)$$

and for C_N^j, where $j \in \mathbf{N}_M$, we assign

$$\bar{F}_N^j(x_{N-1}^j, u_{N-1}^j, x_N^j)$$

$$= \omega_j \frac{1}{2} \begin{bmatrix} x_{N-1}^j \\ u_{N-1}^j \end{bmatrix}^T Q \begin{bmatrix} x_{N-1}^j \\ u_{N-1}^j \end{bmatrix} + \omega_j \frac{1}{2}(x_N^j)^T S x_N^j + \mathcal{I}_{C_{N-1}^j}(x_{N-1}^j, u_{N-1}^j, x_N^j)$$

where $\mathcal{I}_{C_k^j}(x_k^j, u_k^j, x_{k+1}^j)$ is the indicator function for the set

$$C_k^j = \left\{ (x_k^j, u_k^j, x_{k+1}^j) \mid x_{k+1}^j = A_k^j x_k^j + B_k^j u_k^j \right\}.$$

It is possible to introduce even further parallelism by combining the above formulation with a parallel formulation in time as described in Sect. 2.3.4.

2.6   Distributed MPC

There are many ways to define distributed MPC problems. We like to think of them in the following format:

$$\min_u \frac{1}{2} \sum_{i=1}^{M} \sum_{k=0}^{N-1} \begin{bmatrix} x_k^i \\ u_k^i \end{bmatrix}^T Q^i \begin{bmatrix} x_k^i \\ u_k^i \end{bmatrix} + \frac{1}{2} \left(x_N^i \right)^T S^i x_N^i$$

$$\text{s.t. } x_{k+1}^i = A^{(i,i)} x_k^i + B^{(i,i)} u_k^i + \sum_{j \in \mathcal{N}_i} A^{(i,j)} x_k^j + B^{(i,j)} u_k^j + v_k^i, \quad x_0^i = \bar{x}^i$$

$$C^i x_k^i + D^i u_k^i \le e_k^i$$

for $i \in \mathbf{N}_M$, where $\mathcal{N}_i \subset \mathbf{N}_M \setminus \{i\}$. We see that the only coupling in the problem is in the dynamic constraints through the summation over $\mathcal{N}_i$, which typically contains few elements. If one consider a sparsity graph for the above problem one will realize that it is not necessarily chordal. Some heuristic method, such as presented in [5, 20], can most likely be applied successfully in many cases to obtain a sparse chordal embedding of the sparsity graph. From this a clique tree can be computed using other algorithms presented in [5, 20]. See also [18] for a more detailed discussion on how to compute clique trees.

We may consider distributed problems that are stochastic as well. Also extensions to parallelism in time is possible.

2.7   Conclusions

We have in this chapter shown how it is possible to make use of the inherent chordal structure of many MPC formulations in order to exploit IP methods that make use of any chordal structure to distribute its computations over several computational agents that can work in parallel. We have seen how the classical backward Riccati recursion can be seen as a special case of this, albeit not a parallel recursion, but serial. We have also discussed distributed MPC and stochastic MPC over scenario trees. The latter formulation can probably be extended also to robust MPC over scenario trees. Then the subproblems will be quadratic feasibility problems and not quadratic programs. Also it should be possible to consider sum-of-norms regularized MPC. We also believe that it is possible to exploit structure in MPC coming from spatial discretization of PDEs using chordal sparsity. How to carry out these extensions is left for future work.

Acknowledgements   The authors want to thank Daniel Axehill and Isak Nielsen for interesting discussions regarding parallel computations for Riccati recursions. Shervin Parvini Ahmadi has contributed with a figure. This research has been supported by WASP, which is gratefully acknowledged.

References

1. M. Åkerblad, A. Hansson, Efficient solution of second order cone program for model predictive control. Int. J. Control. **77**(1), 55–77 (2004)
2. A. Alessio, A. Bemporad, A survey on explicit model predictive control, in *Nonlinear Model Predictive Control: Towards New Challenging Applications*, ed. by L. Magni, D.M. Raimondo, F. Allgoüwer (Springer, Berlin, 2009), pp. 345–369. ISBN: 978-3-642-01094-1
3. E. Arnold, H. Puta, An SQP-type solution method for constrained discrete-time optimal control problems, in *Computational Optimal Control*, ed. by R. Bulirsch, D. Kraft. International Series of Numerical Mathematics, vol. 115 (Birkhaüuser, Basel, 1994), pp. 127–136
4. D. Axehill, L. Vandenberghe, A. Hansson, Convex relaxations for mixed integer predictive control. Automatica **46**, 1540–1545 (2010)
5. T. Cormen, C. Leiserson, R. Rivest, C. Stein, *Introduction to Algorithms* (MIT Press, Cambridge, 2001). ISBN: 9780262032933
6. C.R. Cutler, B.L. Ramaker, Dynamic matrix control—a computer control algorithm, in *Proceedings of the AIChE National Meeting*, Huston (1979)
7. M. Diehl, H.J. Ferreau, N. Haverbeke, Efficient numerical methods for nonlinear MPC and moving horizon estimation, in *Nonlinear Model Predictive Control: Towards New Challenging Applications*, ed. by L. Magni, D.M. Raimondo, F. Allgoüwer (Springer, Berlin, 2009), pp. 391–417. ISBN: 978-3-642-01094-1
8. A. Domahidi, A.U. Zgraggen, M.N. Zeilinger, M. Morari, C.N. Jones, Efficient interior point methods for multistage problems arising in receding horizon control, in *51st IEEE Conference on Decision and Control*, Maui (2012), pp. 668–674
9. G. Frison, Algorithms and methods for fast model predictive control. Ph.D. thesis, Technical University of Denmark, 2015
10. G. Frison, D. Kouzoupis, M. Diehl, J.B. Jorgensen, A high-performance Riccati based solver for tree-structured quadratic programs, in *Proceedings of the 20th IFAC World Congress* (2017), pp. 14964–14970
11. T. Glad, H. Jonson, A method for state and control constrained linear quadratic control problems, in *Proceedings of the 9th IFAC World Congress*, Budapest (1984)
12. J. Gondzio, A. Grothey, Parallel interior-point solver for structured quadratic programs: applications to financial planning problems. Ann. Oper Res. **152**, 319–339 (2007)
13. V. Gopal, L.T. Biegler, Large scale inequality constrained optimization and control. IEEE Control Syst. Mag. **18**(6), 59–68 (1998)
14. A. Hansson, A primal-dual interior-point method for robust optimal control of linear discrete-time systems. IEEE Trans. Autom. Control **45**(9), 1639–1655 (2000)
15. A. Hansson, S.K. Pakazad, Exploiting chordality in optimization algorithms for model predictive control (2017). arXiv:1711.10254
16. J.L. Jerez, E.C. Kerrigan, G.A. Constantinides, A sparse condensed QP formulation for control of LTH systems. Automatica **48**(5), 999–1002 (2012)
17. J.B. Jorgensen, Moving horizon estimation and control. PhD thesis, Technical University of Denmark, 2004
18. S. Khoshfetrat Pakazad, A. Hansson, M.S. Andersen, I. Nielsen, Distributed primal–dual interior-point methods for solving tree-structured coupled convex problems using message-passing. Optim. Methods Softw. **32**, 1–35 (2016)
19. E. Klintberg, Structure exploiting optimization methods for model predictive control. Ph.D. thesis, Chalmers University of Technology, 2017
20. D. Koller, N. Friedman, *Probabilistic Graphical Models: Principles and Techniques* (MIT Press, Cambridge, 2009)
21. C. Leidereiter, A. Potschka, H.G. Bock, Dual decomposition of QPs in scenario tree NMPC, in *Proceedings of the 2015 European Control Conference* (2015), pp. 1608–1613

22. R. Marti, S. Lucia, D. Sarabia, R. Paulen, S. Engell, C. de Prada, An efficient distributed algorithm for multi-stage robust nonlinear predictive control, in *Proceedings of the 2015 European Control Conference* (2015), pp. 2664–2669
23. I. Nielsen, Structure-exploiting numerical algorithms for optimal control. PhD thesis, Linköping University, 2017
24. I. Nielsen, A. Axehill. An O(log N) parallel algorithm for Newton step computation in model predictive control, in *IFAC World Congress* pp. 10505–10511, Cape Town (2014)
25. S.J. Qin, T.A. Badgwell, A survey of industrial model predictive control technology. Control Eng. Pract. **11**, 722–764 (2003)
26. C.V. Rao, S.J. Wright, J.B. Rawlings, Application of interior-point methods to model predictive control. Preprint ANL/MCS-P664-0597, Mathematics and Computer Science Division, Argonne National Laboratory, May 1997
27. M.C. Steinbach, A structured interior point SQP method for nonlinear optimal control problems, in *Computational Optimal Control*, ed. by R. Bulirsch, D. Kraft. International Series of Numerical Mathematics, vol. 115 (Birkhaüuser, Basel, 1994), pp. 213–222
28. L. Vandenberghe, S. Boyd, M. Nouralishahi, Robust linear programming and optimal control. Internal Report, Department of Electrical Engineering, University of California, Los Angeles (2001)
29. Y. Wang, S. Boyd, Fast model predictive control using online optimization. IEEE Trans. Control Syst. Technol. **18**(2), 267–278 (2010)
30. S.J. Wright, Interior-point methods for optimal control of discrete-time systems. J. Optim. Theory Appl. **77**, 161–187 (1993)
31. S.J. Wright, Applying new optimization algorithms to model predictive control. Chemical Process Control-V (1996)
32. S.J. Wright, *Primal-Dual InteriorPoint Methods* (SIAM, Philadelphia, 1997)

Chapter 3
Decomposition Methods for Large-Scale Semidefinite Programs with Chordal Aggregate Sparsity and Partial Orthogonality

Yang Zheng, Giovanni Fantuzzi, and Antonis Papachristodoulou

Abstract Many semidefinite programs (SDPs) arising in practical applications have useful structural properties that can be exploited at the algorithmic level. In this chapter, we review two decomposition frameworks for large-scale SDPs characterized by either chordal aggregate sparsity or partial orthogonality. Chordal aggregate sparsity allows one to decompose the positive semidefinite matrix variable in the SDP, while partial orthogonality enables the decomposition of the affine constraints. The decomposition frameworks are particularly suitable for the application of first-order algorithms. We describe how the decomposition strategies enable one to speed up the iterations of a first-order algorithm, based on the alternating direction method of multipliers, for the solution of the homogeneous self-dual embedding of a primal-dual pair of SDPs. Precisely, we give an overview of two structure-exploiting algorithms for semidefinite programming, which have been implemented in the open-source MATLAB solver CDCS. Numerical experiments on a range of large-scale SDPs demonstrate that the decomposition methods described in this chapter promise significant computational gains.

Keywords Large-scale semidefinite programs · Chordal decomposition · Partial orthogonality · Operator-splitting algorithms · Decomposition methods

AMS Subject Classifications 90C06, 90C25, 49M27

Y. Zheng (✉) · A. Papachristodoulou
Department of Engineering Science, University of Oxford, Oxford, UK
e-mail: yang.zheng@eng.ox.ac.uk; antonis@eng.ox.ac.uk

G. Fantuzzi
Department of Aeronautics, Imperial College London, London, UK
e-mail: giovanni.fantuzzi10@imperial.ac.uk

© Springer Nature Switzerland AG 2018
P. Giselsson, A. Rantzer (eds.), *Large-Scale and Distributed Optimization*, Lecture Notes in Mathematics 2227,
https://doi.org/10.1007/978-3-319-97478-1_3

3.1 Introduction

Semidefinite programs (SDPs) are a type of convex optimization problems that
arise in many fields, for example control theory, combinatorics, machine learning,
operations research, and fluid dynamics [9, 12, 15, 40]. SDPs generalize other
common types of optimization problems such as linear and second-order cone
programs [8], and have attracted considerable attention because many nonlinear
constraints admit numerically-tractable SDP reformulations or relaxations [37]. The
standard primal and dual forms of an SDP are, respectively,

$$
\begin{aligned}
\min_{X} \quad & \langle C, X \rangle \\
\text{subject to} \quad & \langle A_i, X \rangle = b_i, \quad i = 1, \ldots, m, \\
& X \in \mathbb{S}_+^n,
\end{aligned}
\tag{3.1}
$$

and

$$
\begin{aligned}
\max_{y, Z} \quad & \langle b, y \rangle \\
\text{subject to} \quad & Z + \sum_{y=1}^{m} y_i A_i = C, \\
& Z \in \mathbb{S}_+^n,
\end{aligned}
\tag{3.2}
$$

where X is the primal variable, y and Z are the dual variables, and the vector $b \in \mathbb{R}^m$
and the matrices $C, A_1, \ldots, A_m \in \mathbb{S}^n$ are given problem data. In (3.1), (3.2),
and throughout this chapter, $\mathbb{R}^m$ denotes the usual m-dimensional Euclidean space,
$\mathbb{S}^n$ is the space of $n \times n$ symmetric matrices, and $\mathbb{S}_+^n$ represents the cone of
positive semidefinite (PSD) matrices. The notation $\langle \cdot, \cdot \rangle$ denotes the appropriate
inner product: $\langle x, y \rangle = x^\mathsf{T} y$ for $x, y \in \mathbb{R}^m$ and $\langle X, Y \rangle = \mathrm{trace}(XY)$ for $X, Y \in \mathbb{S}^n$.

It is well-known that in theory SDPs can be solved in polynomial time using
interior-point methods (IPMs). At the time of writing, however, these are only
practical for small- to medium-sized problem instances [3, 20]: Memory or CPU
time constraints prevent the solution of (3.1)–(3.2) when n is larger than a few
hundred and m is larger than a few thousand using a regular PC. Improving the
scalability of current SDP solvers therefore remains an active area of research [4],
with particular emphasis being put on taking advantage of structural properties
pertaining to specific problem classes. This chapter gives an overview of promising
recent developments that exploit two kinds of structural properties, namely chordal
aggregate sparsity and partial orthogonality, in semidefinite programming.

Since the data $b, C, A_1, \ldots, A_m$ are sparse in many large-scale SDPs encountered
in applications [9], perhaps the most obvious approach to improve computational
efficiency is to try and exploit this sparsity [14]. The main challenge in this respect is
that the optimal solution X^* to (3.1) or the inverse of the optimal solution Z^* to (3.2)

(required to compute the gradient and Hessian of the dual barrier function [5]) can be dense even when the problem data are extremely sparse. Nonetheless, one can take advantage of sparsity if the aggregate sparsity pattern of an SDP—that is, the union of the sparsity patterns of the matrices C, A_1, ..., A_m—is *chordal*, or has a sparse *chordal extension* (precise definitions of these properties will be given in Sect. 3.2). In these cases, Grone's theorem [19] and Agler's theorem [1] enable the decomposition of the large PSD matrix variables in (3.1)–(3.2) into smaller PSD matrices, coupled by an additional set of affine constraints. Such a reformulation is attractive because common implementations of IPMs can handle multiple small PSD matrices very efficiently. This observation leads to the development of so-called *domain-* and *range-space* decomposition techniques [17, 22], which are implemented in the MATLAB package SparseCoLO [16].

One drawback of these sparsity-based decomposition methods is that the added equality constraints introduced by the application of Grone's and Agler's theorems can offset the benefit of working with smaller PSD cones. To overcome this problem, it has been suggested that chordal sparsity can be exploited directly at the algorithmic level to develop specialized interior-point solvers: Fukuda et al. developed a primal-dual path-following method for sparse SDPs [17]; Burer proposed a nonsymmetric primal-dual method using Cholesky factors of the dual variable Z and maximum determinant completion of the primal variable X [11]; Andersen et al. developed fast recursive algorithms for SDPs with chordal sparsity [5]. Another promising solution is to abandon IPMs altogether, and solve the decomposed SDP utilizing first-order algorithms instead. These only aim to achieve a solution of moderate accuracy, but scale more favourably than IPMs and can be implemented on highly-parallel computer architectures with relative ease. For these reasons, Sun et al. proposed a first-order operator splitting method for conic optimization with partially separable structure [34], and Madani et al. developed a highly-parallelizable first-order algorithm for sparse SDPs with inequality constraints [26]. Both approaches offer fast first-order algorithms built on Grone's and Agler's theorems [1, 19].

The second limitation of the decomposition methods of Refs. [17, 22] is that, of course, they can only be applied to SDPs whose aggregate sparsity patterns are chordal or admit a sparse chordal extension. One notable class of problems for which the data matrices are individually but not aggregately sparse is that of SDPs arising from sum-of-squares (SOS) programming. SOS programming is a powerful relaxation technique to handle NP-hard polynomial optimization problems [23, 31], with far-reaching applications in systems analysis and control theory [29]. In this case, the aggregate sparsity pattern is fully dense (see, e.g., [42, Section III-B] and [43, Section 3]). While it is possible to tackle large-scale SOS programs using further relaxation techniques based on diagonally dominant matrices (see the DSOS and SDSOS techniques [2]), algorithms that enable their solution without introducing additional conservativeness are highly desired.

In this chapter, we describe two recent methods [43, 44] that address the two aforementioned disadvantages of the domain- and range-space decomposition techniques of Refs. [17, 22]. The first method [44] applies to SDPs with chordal aggregate sparsity patterns and is based on a decomposition framework for the PSD cone, which resembles that of Refs. [17, 22] but is more suitable for the application of fast first-order methods. The second algorithm [43] specializes in SDPs that enjoy a structural property that we call partial orthogonality: Loosely speaking, the data matrices $A_1, \ldots, A_m$ in (3.1)–(3.2) are mutually orthogonal up to a small submatrix (a precise definition will be given in Sect. 3.5). This property is inherent (but of course not exclusive) to SOS programs [43] and enables a decomposition of the affine constraints of the SDP, rather than of the PSD variable. More precisely, it leads to a diagonal plus low-rank representation of a certain matrix associated with the affine constraints. The computational backbone of both methods we describe is a variant of a classical first-order operator splitting method known as the alternating direction method of multipliers (ADMM), tailored to solve the homogeneous self-dual embedding of a primal-dual pair of conic programs [28].

The rest of this chapter is organized as follows. Section 3.2 briefly introduces the PSD matrix decomposition techniques based on chordal sparsity. The homogeneous self-dual embedding of the primal-dual pair of SDPs (3.1)–(3.2) and the ADMM algorithm of Ref. [28] are reviewed in Sect. 3.3. We present a fast ADMM algorithm for SDPs with chordal sparsity in Sect. 3.4, and another fast ADMM algorithm for SDPs with partial orthogonality in Sect. 3.5. Section 3.6 describes the MATLAB solver CDCS (Cone Decomposition Conic Solver) [41], which includes an implementation of both algorithms, and reports the results of numerical experiments on a set of benchmark problems. Concluding remarks and suggestions for future research are given in Sect. 3.7.

3.2   Matrix Decomposition Using Chordal Sparsity

As mentioned in the introduction, one of the decomposition methods for SDPs (3.1)–(3.2) described in this chapter exploits the sparsity of the problem data. The key to this approach is the description of sparse matrices using graphs, so here we review some essential concepts from graph theory and their relation to sparse matrices.

3.2.1   Essential Notions from Graph Theory

An undirected graph $\mathcal{G}(\mathcal{V}, \mathcal{E})$ is defined as a set of nodes $\mathcal{V} = \{1, 2, \ldots, n\}$ plus a set of edges (connections between nodes) $\mathcal{E} \subseteq \mathcal{V} \times \mathcal{V}$. For example, Fig. 3.1a illustrates a graph with nodes $\mathcal{V} = \{1, 2, 3, 4, 5, 6\}$ and edges $\mathcal{E} = \{(1, 2), (1, 3), (1, 6), (2, 4), (2, 5), (3, 4), (3, 5)\}$. A subset of nodes $\mathcal{C} \subseteq \mathcal{V}$ is called a *clique* if any pair of distinct nodes in $\mathcal{C}$ is connected by an edge, i.e., $(i, j) \in \mathcal{E}, \forall i, j \in \mathcal{C}$. If a clique $\mathcal{C}$ is not a subset of any other clique, we refer to it as a *maximal clique*. In Fig. 3.1b, there are four maximal cliques: $\mathcal{C}_1 = \{1, 2, 3\}, \mathcal{C}_2 = \{2, 3, 5\}, \mathcal{C}_3 = \{2, 3, 4\}$ and

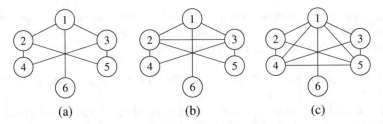

Fig. 3.1 (**a**) A simple nonchordal graph: The cycles $\{1, 2, 4, 3\}$, $\{1, 2, 5, 3\}$, and $\{2, 4, 3, 5\}$ have length four but no chords. (**b**) A chordal extension of the nonchordal graph in (**a**), obtained by adding edge $(2, 3)$. (**c**) A different chordal extension of the nonchordal graph in (**a**), obtained by adding edges $(1, 4)$, $(1, 5)$, and $(4, 5)$.

$\mathcal{C}_4 = \{1, 6\}$. We denote the number of nodes in a clique $\mathcal{C}$ by $|\mathcal{C}|$, while $\mathcal{C}(i)$ indicates its i-th node when sorted in the natural ordering.

A cycle of length k is a subset of nodes $\{v_1, v_2, \ldots, v_k\} \subseteq \mathcal{V}$ such that $(v_k, v_1) \in \mathcal{E}$ and $(v_i, v_{i+1}) \in \mathcal{E}$ for $i = 1, \ldots, k - 1$. Any additional edge joining two nonconsecutive nodes in a cycle is called a *chord*, and an undirected graph $\mathcal{G}$ is called *chordal* if every cycle of length greater than three has one chord. Of course, given a nonchordal graph $\mathcal{G}(\mathcal{V}, \mathcal{E})$, one can always construct a chordal graph $\mathcal{G}'(\mathcal{V}, \mathcal{E}')$ by adding suitable edges to the original edge set $\mathcal{E}$. This process, known as *chordal extension*, is not unique as illustrated by Fig. 3.1. The graph in Fig. 3.1a is not chordal: Adding the edge $(2, 3)$ results in the chordal graph depicted in Fig. 3.1b, while adding edges $(1, 4)$, $(1, 5)$, and $(4, 5)$ yields the chordal graph in Fig. 3.1c. Note that computing the chordal extension with the minimum number of additional edges is NP-hard in general [38], but sufficiently good extensions can often be found efficiently using heuristic approaches [36]. More examples can be found in [36, 44].

3.2.2 Graphs, Sparse Matrices, and Chordal Decomposition

Given an undirected graph $\mathcal{G}(\mathcal{V}, \mathcal{E})$, let $\mathcal{E}^* = \mathcal{E} \cup \{(i, i), \forall i \in \mathcal{V}\}$ be the extended set of edges to which all self loops have been added. This can be used to define the space of sparse matrices

$$\mathbb{S}^n(\mathcal{E}, 0) := \{X \in \mathbb{S}^n : (i, j) \notin \mathcal{E}^* \Rightarrow X_{ij} = 0\}. \tag{3.3}$$

Note that this definition does not preclude $X_{ij} = 0$ if $(i, j) \in \mathcal{E}^*$, so a matrix $X \in \mathbb{S}^n(\mathcal{E}, 0)$ can be sparser than allowed by the underlying graph representation.

Two subsets of this sparse matrix space will be useful in the following. The first is the subspace of positive semidefinite sparse matrices,

$$\mathbb{S}^n_+(\mathcal{E}, 0) := \{X \in \mathbb{S}^n(\mathcal{E}, 0) : X \in \mathbb{S}^n_+\} \equiv \mathbb{S}^n(\mathcal{E}, 0) \cap \mathbb{S}^n_+. \tag{3.4}$$

The second is the subspace of sparse matrices that can be completed into a positive semidefinite matrix by filling in the zero entries (sometimes called partial semidefinite matrices), defined according to

$$\mathbb{S}^n_+(\mathcal{E}, ?) := \{X \in \mathbb{S}^n(\mathcal{E}, 0) : \exists M \in \mathbb{S}^n_+ \mid (i, j) \in \mathcal{E}^* \Rightarrow M_{ij} = X_{ij}\}. \tag{3.5}$$

Note that $\mathbb{S}^n_+(\mathcal{E}, ?)$ can be viewed as the projection of the space of positive semidefinite matrices $\mathbb{S}^n_+$ onto $\mathbb{S}^n(\mathcal{E}, 0)$ with respect to the Frobenius norm (the natural norm induced by the trace inner product). Moreover, it is not difficult to check that for any graph $\mathcal{G}(\mathcal{V}, \mathcal{E})$ the sets $\mathbb{S}^n_+(\mathcal{E}, ?)$ and $\mathbb{S}^n_+(\mathcal{E}, 0)$ are a pair of dual cones with respect to the ambient space $\mathbb{S}^n(\mathcal{E}, 0)$ [36].

When the graph $\mathcal{G}(\mathcal{V}, \mathcal{E})$ that defines the space $\mathbb{S}^n(\mathcal{E}, 0)$ is chordal, we say that $X \in \mathbb{S}^n(\mathcal{E}, 0)$ has a chordal sparsity pattern. In this case, the cones $\mathbb{S}^n_+(\mathcal{E}, ?)$ and $\mathbb{S}^n_+(\mathcal{E}, 0)$ enjoy a particularly useful property: Membership to the cone can be expressed in terms of PSD constraints on the submatrices corresponding to the maximal cliques of the underlying graph. Replacing the condition of cone membership with this equivalent set of PSD constraints is known as chordal decomposition. To make this concept precise, we will need some further notation: Given any maximal clique $\mathcal{C}$ of $\mathcal{G}(\mathcal{V}, \mathcal{E})$, consider the matrix $E_\mathcal{C} \in \mathbb{R}^{|\mathcal{C}| \times n}$ with entries defined by

$$(E_\mathcal{C})_{ij} = \begin{cases} 1, & \text{if } \mathcal{C}(i) = j, \\ 0, & \text{otherwise.} \end{cases}$$

For any matrix $X \in \mathbb{S}^n(\mathcal{E}, 0)$, the submatrix corresponding to the maximal clique $\mathcal{C}$ can therefore be represented as $E_\mathcal{C} X E_\mathcal{C}^\mathsf{T} \in \mathbb{S}^{|\mathcal{C}|}$. Moreover, note that the operation $E_\mathcal{C}^\mathsf{T} Y E_\mathcal{C}$ "inflates" a $|\mathcal{C}| \times |\mathcal{C}|$ matrix Y into a sparse $n \times n$ matrix with nonzero entries in the rows/columns specified by the indices of the clique $\mathcal{C}$.

We can now state the conditions for membership to the cones $\mathbb{S}^n_+(\mathcal{E}, ?)$ and $\mathbb{S}^n_+(\mathcal{E}, 0)$ when the sparsity pattern is chordal more precisely using Grone's [19] and Agler's [1] theorems. These have historically been proven individually, but can be derived from each other using duality [36].

Theorem 1 (Grone's Theorem [19]) *Let $\mathcal{G}(\mathcal{V}, \mathcal{E})$ be a chordal graph with maximal cliques $\{\mathcal{C}_1, \mathcal{C}_2, \ldots, \mathcal{C}_p\}$. Then, $X \in \mathbb{S}^n_+(\mathcal{E}, ?)$ if and only if $X_k := E_{\mathcal{C}_k} X E_{\mathcal{C}_k}^\mathsf{T} \in \mathbb{S}^{|\mathcal{C}_k|}_+$ for all $k = 1, \ldots, p$.*

Theorem 2 (Agler's Theorem [1]) *Let $\mathcal{G}(\mathcal{V}, \mathcal{E})$ be a chordal graph with maximal cliques $\{\mathcal{C}_1, \mathcal{C}_2, \ldots, \mathcal{C}_p\}$. Then, $Z \in \mathbb{S}^n_+(\mathcal{E}, 0)$ if and only if there exist matrices $Z_k \in \mathbb{S}^{|\mathcal{C}_k|}_+$ for $k = 1, \ldots, p$ such that $Z = \sum_{k=1}^p E_{\mathcal{C}_k}^\mathsf{T} Z_k E_{\mathcal{C}_k}$.*

3.3 Homogeneous Self-dual Embedding and ADMM

As anticipated in the introduction, the computational engine of the scalable SDP solver described in this chapter is a specialized ADMM algorithm for the solution of the homogeneous self-dual embedding of conic programs. To make this chapter self-contained, in this section we review the main ideas behind the homogeneous self-dual embedding of SDPs and summarize the ADMM algorithm for its solution developed in [28].

3.3.1 Homogeneous Self-dual Embedding

An elegant method to solve the primal-dual pair of SDPs (3.1)–(3.2) that is also able to detect primal or dual infeasibility is to embed it into a homogeneous self-dual feasibility problem [39]. Solving the latter amounts to finding a nonzero point in the intersection of a convex cone and an affine space. Once such a nonzero point is found, one can either recover an optimal solution for the primal-dual pair (3.1)–(3.2), or construct a certificate of primal or dual infeasibility (more on this is outlined below).

To formulate the homogeneous self-dual embedding of (3.1)–(3.2), it is convenient to consider their vectorized form. Specifically, let $\text{vec} : \mathbb{S}^n \to \mathbb{R}^{n^2}$ be the operator that maps a matrix to the stack of its columns,[1] and define the vectorized data

$$c := \text{vec}(C), \quad A := \left[\text{vec}(A_1) \ \ldots \ \text{vec}(A_m)\right]^\mathsf{T}. \tag{3.6}$$

It is assumed that $A_1, \ldots, A_m$ are linearly independent matrices, and consequently that A has full row rank. Upon defining the set of vectorized PSD matrices as

$$S := \{x \in \mathbb{R}^{n^2} : \text{vec}^{-1}(x) \in \mathbb{S}^n_+\}, \tag{3.7}$$

the primal SDP (3.1) can be rewritten as

$$\min_{x} \quad c^\mathsf{T} x$$

$$\text{subject to} \quad Ax = b, \tag{3.8}$$

$$x \in S,$$

---

[1] We use the vec operator for both conceptual and notational simplicity. Practical implementation of the methods described in this chapter should take advantage of the symmetry of all matrices. This can be done using symmetric vectorization through the operator svec, which maps a matrix $X \in \mathbb{S}^n$ to the vector $\text{svec}(X) := (X_{11}, \sqrt{2}X_{21}, \ldots, \sqrt{2}X_{n1}, X_{22}, \sqrt{2}X_{32}, \ldots, X_{nn}) \in \mathbb{R}^{n(n+1)/2}$. All statements made in this chapter hold when vec is replaced by svec.

while the dual SDP (3.2) becomes

$$
\max_{y,z} \quad b^\mathsf{T} y
$$

$$
\text{subject to} \quad z + A^\mathsf{T} y = c, \tag{3.9}
$$

$$
z \in \mathcal{S}.
$$

When strong duality holds, an optimal solution of (3.8)–(3.9) or a certificate of infeasibility can be recovered from any nonzero solution of the homogeneous linear system

$$
\begin{bmatrix} z \\ s \\ \kappa \end{bmatrix} = \begin{bmatrix} 0 & -A^\mathsf{T} & c \\ A & 0 & -b \\ -c^\mathsf{T} & b^\mathsf{T} & 0 \end{bmatrix} \begin{bmatrix} x \\ y \\ \tau \end{bmatrix}, \tag{3.10}
$$

provided that it also satisfies

$$
(x, y, \tau) \in \mathcal{S} \times \mathbb{R}^m \times \mathbb{R}_+, \qquad (z, s, \kappa) \in \mathcal{S} \times \{0\}^m \times \mathbb{R}_+. \tag{3.11}
$$

Note that the linear system (3.10) and the conic constraint (3.11) embeds the optimality conditions of (3.8)–(3.9): When $\tau = 1$ and $\kappa = 0$ they reduce to the KKT conditions

$$
Ax = b, \quad x \in \mathcal{S}, \quad z + A^\mathsf{T} y = c, \quad z \in \mathcal{S}, \quad c^\mathsf{T} x = b^\mathsf{T} y. \tag{3.12}
$$

(These are necessary and sufficient optimality conditions under the assumption of strong duality.)

In addition, τ and κ are required to be nonnegative and complementary variables, i.e., at most one of them is nonzero [39]. If $\tau > 0$ and $\kappa = 0$ the point

$$
x^* = \frac{x}{\tau}, \quad y^* = \frac{y}{\tau}, \quad z^* = \frac{z}{\tau},
$$

satisfies the KKT conditions (3.12) and is a primal-dual optimal solution of (3.8)–(3.9). If $\tau = 0$ and $\kappa > 0$, instead, a certificate of primal or dual infeasibility can be constructed depending on the values of $c^\mathsf{T} x$ and $b^\mathsf{T} y$. We refer the interested reader to [28] and references therein for more details.

For notational convenience, we finally rewrite the linear system (3.10) and the conditions in (3.11) in a compact form by defining the vectors and matrix

$$
u := \begin{bmatrix} x \\ y \\ \tau \end{bmatrix}, \quad v := \begin{bmatrix} z \\ s \\ \kappa \end{bmatrix}, \quad Q := \begin{bmatrix} 0 & -A^\mathsf{T} & c \\ A & 0 & -b \\ -c^\mathsf{T} & b^\mathsf{T} & 0 \end{bmatrix}, \tag{3.13}
$$

and the cones

$$K := S \times \mathbb{R}^m \times \mathbb{R}_+, \qquad K^* := S \times \{0\}^m \times \mathbb{R}_+. \tag{3.14}$$

Then, an optimal point for (3.8)–(3.9) or a certificate of infeasibility can be recovered from a nonzero solution of the homogeneous self-dual feasibility problem

$$
\begin{aligned}
\text{find} \quad & (u, v) \\
\text{subject to} \quad & v = Qu, \\
& (u, v) \in K \times K^*.
\end{aligned}
\tag{3.15}
$$

3.3.2 A Tailored ADMM Algorithm

It has been shown in Ref. [28] that problem (3.15) can be solved with a simplified version of the classical ADMM algorithm (see e.g., [10]) by virtue of its self-dual character. The k-th iteration of this tailored ADMM algorithm consists of the following three steps:

$$w^{(k)} = (I + Q)^{-1} \left(u^{(k-1)} + v^{(k-1)} \right), \tag{3.16a}$$

$$u^{(k)} = \mathbb{P}_K \left(w^{(k)} - v^{(k-1)} \right), \tag{3.16b}$$

$$v^{(k)} = v^{(k-1)} - w^{(k)} + u^{(k)}. \tag{3.16c}$$

Here and in the following $\mathbb{P}_K$ denotes the projection onto the cone K and the superscript (k) indicates that a variable has been fixed to its value after the k-th iteration.

Since the last step is computationally trivial, practical implementations of the algorithm require an efficient computation of (3.16a) and (3.16b). An efficient C implementation that handles generic conic programs with linear, second-order, semidefinite and exponential cones is available in the solver SCS [27].

In this chapter, we show that when one is interested in solving SDPs with chordal aggregate sparsity or partial orthogonality (which are very common in certain practical applications), the computational efficiency of (3.16a) and (3.16b) can be improved further. In particular, chordal sparsity allows one to speed up the conic projection (3.16b) because one replaces a large PSD cone with smaller ones. Following Refs. [44–46], we refer to this procedure as cone decomposition. Partial orthogonality, instead, can be exploited to reduce the size of matrix to be inverted (or factorized) when solving the linear system (3.16a). Since (3.16a) can be interpreted as a projection onto the affine constraints of (3.8)–(3.9), in this chapter we will slightly abuse terminology and say that partial orthogonality allows an affine decomposition.

Remark 1 Recall from (3.13) that $u = \begin{pmatrix} x \\ y \\ \tau \end{pmatrix}$ and $v = \begin{pmatrix} z \\ s \\ \kappa \end{pmatrix}$ in (3.15), and that x and z are vectorized symmetric matrices. It is not difficult to check that if the corresponding entries in the initial guesses $u^{(0)}$ and $v^{(0)}$ for algorithm (3.16a)–(3.16c) are vectorized symmetric matrices, then the same is true for all subsequent iterates $w^{(k)}$, $u^{(k)}$, and $v^{(k)}$, $k = 1, 2, \ldots$. Indeed, steps (3.16b) and (3.16c) preserve the vectorized matrix structure provided that the vector $w^{(k)}$ obtained with (3.16a) possesses it. Since step (3.16a) is equivalent to the solution of a linear system of the form

$$\begin{bmatrix} I & -A^\mathsf{T} & c \\ A & I & -b \\ -c^\mathsf{T} & b^\mathsf{T} & I \end{bmatrix} \begin{bmatrix} w_x \\ w_y \\ w_\tau \end{bmatrix} = \begin{bmatrix} \xi_x \\ \xi_y \\ \xi_\tau \end{bmatrix},$$

where ξ_x is a vectorized symmetric matrix, one needs to verify that w_x is so too. Now, the first block of equations implies that

$$w_x = A^\mathsf{T} w_y - c w_\tau + \xi_x,$$

which—recalling the definitions of c and A from (3.6)—is satisfied if and only if

$$\mathrm{vec}^{-1}(w_x) = \mathrm{vec}^{-1}(\xi_x) - \mathrm{vec}^{-1}(c w_\tau) + \mathrm{vec}^{-1}(A^\mathsf{T} w_y)$$

$$= \mathrm{vec}^{-1}(\xi_x) - w_\tau C + \sum_{i=1}^{m} (w_y)_i A_i.$$

In the last expression, C, $A_1, \ldots, A_m$ are the symmetric data matrices of the original SDP and $(w_y)_i$ denotes the i-th entry of the vector w_y. Thus, the vector w_x must be a vectorized symmetric matrix, as required.

Remark 2 Algorithm (3.16a)–(3.16c) is not the only one available to solve the convex feasibility problem (3.15), and other methods could be employed. In particular, one could utilize other operator-splitting first-order methods, including Douglas-Rachford iterations [24], Spingarn's method [32], and Dykstra's method [6]. We have not tried these, but in light of the equivalence between ADMM and many operator-splitting methods [10], it is not unreasonable to expect that the ADMM-based algorithm (3.16a)–(3.16c) will perform at least as efficiently as many other alternatives.

3.4 Cone Decomposition in Sparse SDPs

Let us consider the case in which SDPs (3.1) and (3.2) have an aggregate sparsity pattern defined by the graph $\mathcal{G}(\mathcal{V}, \mathcal{E})$, meaning that

$$C \in \mathbb{S}^n(\mathcal{E}, 0) \quad \text{and} \quad A_i \in \mathbb{S}^n(\mathcal{E}, 0), \quad i = 1, \ldots, m.$$

Without loss of generality, we assume that $\mathcal{G}(\mathcal{V}, \mathcal{E})$ is chordal with a set of maximal cliques $\mathcal{C}_1, \mathcal{C}_2, \ldots, \mathcal{C}_p$ (for sparse nonchordal SDPs, we assume that a suitable sparse chordal extension has been found).

Aggregate sparsity implies that the dual variable Z in (3.2) satisfies $Z \in \mathbb{S}^n(\mathcal{E}, 0)$. As for the primal SDP (3.1), although the variable X can be dense, only the entries X_{ij} defined by the extended edge set $\mathcal{E}^*$ appear in the equality constraints and the cost function, while the remaining entries only ensure that X is PSD. Consequently, it suffices to consider $X \in \mathbb{S}^n_+(\mathcal{E}, ?)$. We can then apply Theorems 1 and 2 to rewrite (3.1) and (3.2), respectively, as

$$
\begin{aligned}
\min_{X, X_1, \ldots, X_p} \quad & \langle C, X \rangle \\
\text{subject to} \quad & \langle A_i, X \rangle = b_i, \qquad i = 1, \ldots, m, \\
& X_k = E_{\mathcal{C}_k} X E_{\mathcal{C}_k}^{\mathsf{T}}, \quad k = 1, \ldots, p, \\
& X_k \in \mathbb{S}^{|\mathcal{C}_k|}_+, \qquad k = 1, \ldots, p,
\end{aligned}
\tag{3.17}
$$

and

$$
\begin{aligned}
\max_{y, Z_1, \ldots, Z_p, V_1, \ldots, V_p} \quad & \langle b, y \rangle \\
\text{subject to} \quad & \sum_{k=1}^{p} E_{\mathcal{C}_k}^{\mathsf{T}} V_k E_{\mathcal{C}_k} + \sum_{i=1}^{m} y_i A_i = C, \\
& Z_k = V_k, \qquad k = 1, \ldots, p, \\
& Z_k \in \mathbb{S}^{|\mathcal{C}_k|}_+, \qquad k = 1, \ldots, p.
\end{aligned}
\tag{3.18}
$$

We refer to (3.17) and (3.18) as the *cone decomposition* of a primal-dual pair of SDPs with chordal sparsity. Note that this cone decomposition is similar to the domain- and range-space decompositions developed in [17] but for one key feature: We do not eliminate the variables X in (3.17) and we introduce slack variables $V_k, k = 1, 2, \ldots, p$ in (3.18). This is essential if the conic and the affine constraints are to be separated effectively when using operator-splitting algorithms; see [44–46] for more detailed discussions. Also, a standard argument based on the Lagrange function and Lagrange multipliers [8, Chapter 5] reveals that the decomposed problems (3.17) and (3.18) are the dual of each other. This fact, which may seem surprising at first, is a consequence of the duality between the original problems (3.1) and (3.2), and the more subtle dual relationship between Grone's and Agler's theorems (cf. Ref. [36]).

As in Sect. 3.3, to formulate and solve the HSDE of problems (3.17) and (3.18) we use vectorized variables. Precisely, we let

$$x := \mathrm{vec}(X), \; x_k := \mathrm{vec}(X_k), \; z_k := \mathrm{vec}(Z_k), \; v_k := \mathrm{vec}(V_k), \; k = 1, \ldots, p,$$

and define matrices

$$H_k := E_{\mathcal{C}_k} \otimes E_{\mathcal{C}_k}, \quad k = 1, \ldots, p,$$

such that

$$x_k = \text{vec}(X_k) = \text{vec}(E_{\mathcal{C}_k} X E_{\mathcal{C}_k}^\mathsf{T}) = H_k x.$$

In other words, the matrices $H_1, \ldots, H_p$ project x onto the subvectors $x_1, \ldots, x_p$, respectively. Moreover, we denote the constraints $X_k \in \mathbb{S}_+^{|\mathcal{C}_k|}$ by $x_k \in \mathcal{S}_k$ (the formal definition of $\mathcal{S}_k$ is analogous to that of the set $\mathcal{S}$; see (3.7) in Sect. 3.3.1). We then group the vectorized variables according to

$$\hat{x} = \begin{bmatrix} x \\ x_1 \\ \vdots \\ x_p \end{bmatrix}, \quad \hat{y} = \begin{bmatrix} y \\ v_1 \\ \vdots \\ v_p \end{bmatrix}, \quad \hat{z} = \begin{bmatrix} 0 \\ z_1 \\ \vdots \\ z_p \end{bmatrix},$$

define $H^\mathsf{T} := \begin{bmatrix} H_1^\mathsf{T} & \ldots & H_p^\mathsf{T} \end{bmatrix}$ and $\hat{\mathcal{S}} := \mathcal{S}_1 \times \mathcal{S}_2 \times \cdots \times \mathcal{S}_p$, and augment the problem data matrices according to

$$\hat{c} = \begin{bmatrix} c \\ 0 \end{bmatrix}, \quad \hat{b} = \begin{bmatrix} b \\ 0 \end{bmatrix}, \quad \hat{A} = \begin{bmatrix} A & 0 \\ H & -I \end{bmatrix}.$$

With these definitions, we can rewrite (3.17) and (3.18) in the compact vectorized forms

$$\begin{array}{ll} \min_{\hat{x}} \quad \hat{c}^\mathsf{T} \hat{x} & \qquad\qquad \max_{\hat{y}, \hat{z}} \quad \hat{b}^\mathsf{T} \hat{y} \\[2ex] \text{subject to} \quad \hat{A}\hat{x} = \hat{b}, & \qquad\qquad \text{subject to} \quad \hat{z} + \hat{A}^\mathsf{T} \hat{y} = \hat{c}, \qquad (3.19\text{a,b}) \\[2ex] \hspace{3.5em} \hat{x} \in \mathbb{R}^{n^2} \times \hat{\mathcal{S}}, & \qquad\qquad \hspace{3.5em} \hat{z} \in \{0\}^{n^2} \times \hat{\mathcal{S}}. \end{array}$$

These are a standard pair of primal-dual conic programs, and can be solved using the ADMM algorithm (3.16). Of course, the same could be done for the original pair of SDPs, but at a higher computational cost as we shall now demonstrate.

First, consider step (3.16a). An apparent difficulty is that the size of matrix Q is much larger for the decomposed problem (3.19) than that for the original problem, due to the introduction of the variables X_k and V_k in (3.17) and (3.18). However, Q is also highly structured and sparse. Using block elimination and the matrix inversion lemma [8, Appendix C.4.3], it is shown in Ref. [45] that step (3.16a) for the decomposed problem requires a set of relatively inexpensive matrix-vector multiplications and the solution of a linear system of equations with coefficient matrix

$$I + A \left(I + \frac{1}{2} D \right)^{-1} A^\mathsf{T} \in \mathbb{S}^m, \tag{3.20}$$

where $D = \sum_{k=1}^{p} H_k^\mathsf{T} H_k$ is diagonal. Note that the matrix in (3.20) only depends on the problem data and, consequently, its preferred factorization can be computed and cached before iterating the ADMM algorithm. Applying the ADMM algorithm of Ref. [28] directly to (3.1)–(3.2) also requires the solution of a linear system of the same size. It is reasonable to assume that solving the linear system (including the factorization step) bears the most computational burden, while further matrix-vector products are comparatively inexpensive. Step (3.16a) therefore has the same leading-order cost irrespective of whether algorithm (3.16a)–(3.16c) is applied to the original pair of SDPs (3.1)–(3.2) or their decomposed counterparts (3.17)–(3.18).

Consider now the conic projection (3.16b). When the ADMM algorithm described in Sect. (3.3.2) is applied to the original pair of SDPs (3.1)–(3.2), the operator $\mathbb{P}_{\mathcal{K}}$ requires the projection of a (large) $n \times n$ matrix on the PSD cone. This can be achieved via an eigenvalue decomposition, which to leading order requires $O(n^3)$ flops. When applied to the decomposed problem (3.19), instead, one needs to project a vector onto the cone

$$\hat{\mathcal{K}} = \mathbb{R}^{n^2} \times \mathcal{S}_1 \times \mathcal{S}_2 \times \cdots \times \mathcal{S}_p \times \mathbb{R}^{m+n_d} \times \mathbb{R}_+,$$

where $n_d = \sum_{k=1}^{p} |\mathcal{C}_k|^2$ is the length of the vector v. Projecting onto the sub-cones $\mathbb{R}^{n^2}, \mathbb{R}^{m+n_d}, \mathbb{R}_+$ is trivial, and the computational burden rests mostly on the projections onto the vectorized PSD cones $\mathcal{S}_1, \mathcal{S}_2, \ldots, \mathcal{S}_p$. The size of each of these cones only depends on the size of the corresponding maximal clique of the chordal sparsity pattern of the problem data. Consequently, if the largest maximal clique is small compared to the original problem size n, each of the PSD projections is much less expensive computationally than projecting the original $n \times n$ matrix. Moreover, each projection is independent of the others, so the computation can be parallelized. Therefore, step (3.16b) can be carried out more efficiently when the ADMM is applied to the decomposed problem (3.19) instead of the original pair of SDPs (3.1)–(3.2).

In conclusion, SDPs with chordal aggregate sparsity can be solved very efficiently using the ADMM algorithm (3.16a)–(3.16c) thanks to the computational saving in the conic projection step (3.16b). Numerical results that demonstrate this in practice will be presented in Sect. 3.6, and we refer the interested reader to Refs. [44–46] for more details.

3.5   Affine Decomposition in SDPs with Partial Orthogonality

As anticipated in the introduction to this chapter, there are some large-scale SDPs whose aggregate sparsity patterns (after chordal extension if necessary) are almost full, so the chordal decomposition method presented in Sect. 3.4 brings little to no advantage. Notable examples are the SDPs arising from general SOS programming: In this case, the individual data matrices $C, A_1, \ldots, A_m$ are extremely sparse, but their aggregate sparsity pattern is full. However, it is easily shown [43] that when SOS programs are formulated in the usual monomial basis the data matrices are partially orthogonal (this property is defined precisely below).

Motivated by applications in SOS programming, we therefore consider SDPs characterized by partial orthogonality and discuss how to improve the computational efficiency of the ADMM steps (3.16a)–(3.16c) for this class of SDPs. We will show that partial orthogonality allows computational improvements in step (3.16a). For reasons that will become apparent below, and with a slight abuse of terminology, we say that partial orthogonality enables an affine decomposition of (3.16a), which in some respects parallels the cone decomposition of (3.16b) allowed by the chordal sparsity.

Let us make these ideas more precise. Consider the coefficient matrix A in the vectorized problems (3.8)–(3.9). We say that an SDP satisfies partial orthogonality if there exists a column permutation matrix P such that $AP = \begin{bmatrix} A_1 & A_2 \end{bmatrix}$ with $A_1 \in \mathbb{R}^{m \times t_1}$, $A_2 \in \mathbb{R}^{m \times t_2}$ and $A_2 A_2^\mathsf{T} = D$ a diagonal matrix. In this case,

$$AA^\mathsf{T} = APP^\mathsf{T}A^\mathsf{T} = \begin{bmatrix} A_1 & A_2 \end{bmatrix} \begin{bmatrix} A_1^\mathsf{T} \\ A_2^\mathsf{T} \end{bmatrix} = A_1 A_1^\mathsf{T} + D. \tag{3.21}$$

For the SDPs resulting from SOS representations, one usually has $t_1 \ll \min\{t_2, m\}$, so that the product AA^T is of the "diagonal plus low-rank" form. In this chapter, we refer to the partition (3.21) as *affine decomposition*.

Partial orthogonality can be exploited to gain significant computational savings in the ADMM step (3.16a). Recalling the definition of Q, this step requires the solution of the linear system of equations

$$\begin{bmatrix} I & -A^\mathsf{T} & c \\ A & I & -b \\ -c^\mathsf{T} & b^\mathsf{T} & 1 \end{bmatrix} \begin{bmatrix} w_1 \\ w_2 \\ w_3 \end{bmatrix} = \begin{bmatrix} \theta_1 \\ \theta_2 \\ \theta_3 \end{bmatrix}. \tag{3.22}$$

We denote

$$M := \begin{bmatrix} I & -A^\mathsf{T} \\ A & I \end{bmatrix}, \quad \zeta := \begin{bmatrix} c \\ -b \end{bmatrix}.$$

Then, by eliminating w_3 from the first and second block-equations in (3.22), we have

$$(M + \zeta\zeta^T)\begin{bmatrix} w_1 \\ w_2 \end{bmatrix} = \begin{bmatrix} \theta_1 \\ \theta_2 \end{bmatrix} - \theta_3\zeta, \tag{3.23a}$$

$$w_3 = \theta_3 + c^T w_1 - b^T w_2. \tag{3.23b}$$

We then apply the matrix inversion lemma [8, Appendix C.4.3] to (3.23a), leading to

$$\begin{bmatrix} w_1 \\ w_2 \end{bmatrix} = \left[I - \frac{(M^{-1}\zeta)\zeta^T}{1 + \zeta^T(M^{-1}\zeta)} \right] M^{-1} \begin{bmatrix} \theta_1 - c\theta_3 \\ \theta_2 + b\theta_3 \end{bmatrix}. \tag{3.24}$$

The vector $(M^{-1}\zeta)/(1 + \zeta^T M^{-1}\zeta)$ depends only on the problem data and can be computed before starting the ADMM algorithm, so multiplication by the first matrix on the right-hand side of (3.24) at each iteration can be implemented only using vector-vector operations. The core of the computation is therefore to solve a linear system of the form

$$\begin{bmatrix} I & -A^T \\ A & I \end{bmatrix} \begin{bmatrix} \sigma_1 \\ \sigma_2 \end{bmatrix} = \begin{bmatrix} \hat{\theta}_1 \\ \hat{\theta}_2 \end{bmatrix}. \tag{3.25}$$

Again, by eliminating σ_1 from the second block-equation in (3.25), we obtain

$$\sigma_1 = \hat{\theta}_1 + A^T\sigma_2, \tag{3.26a}$$

$$(I + AA^T)\sigma_2 = -A\hat{\theta}_1 + \hat{\theta}_2. \tag{3.26b}$$

At this stage, we can use the property of partial orthogonality (3.21): There exists a diagonal matrix $J := I + D$ such that $I + AA^T = J + A_1 A_1^T$. In the context of typical SOS programs, we know that $A_1 \in \mathbb{R}^{m \times t_1}$ with $t_1 \ll m$. Therefore, it is convenient to apply the matrix inversion lemma again to (3.26b), resulting in

$$(I + AA^T)^{-1} = (J + A_1 A_1^T)^{-1}$$
$$= J^{-1} - J^{-1}A_1(I + A_1^T J^{-1} A_1)^{-1} A_1^T J^{-1}.$$

Since J is diagonal, its inverse is trivial to compute. Then, σ_1 and σ_2 in (3.26) are available by solving a $t_1 \times t_1$ linear system with coefficient matrix

$$I + A_1^T J^{-1} A_1 \in \mathbb{S}^{t_1}, \tag{3.27}$$

plus relatively inexpensive matrix-vector, vector-vector, and scalar-vector operations. Furthermore, the matrix $I + A_1^T J^{-1} A_1$ is the same at all iterations and its preferred factorization can be computed and cached before iterating steps (3.16a)–(3.16c). Once σ_1 and σ_2 have been computed, the solution of (3.22) can be recovered using vector-vector and scalar-vector operations.

In summary, for SDPs with partial orthogonality, we only need to invert or factorize the $t_1 \times t_1$ matrix shown in (3.27), in contrast to the usual $m \times m$ matrix (e.g., $I + AA^\mathsf{T} \in \mathbb{R}^{m \times m}$, or see (3.20)). If $t_1 \ll m$, which is true for typical SOS programs, then the affine decomposition can yield significant computational saving in the ADMM step (3.16a). More details can be found in Ref. [43].

3.6 Numerical Simulations

In this section we present numerical results for CDCS (cone decomposition conic solver), a MATLAB package that provides an efficient implementation of the cone decomposition and affine decomposition strategies described above. For the cone decomposition strategy, we show results on selected benchmark problems from SDPLIB [7] and some large and sparse SDPs with nonchordal sparsity patterns from [5]. For the affine decomposition strategy, we report results on the SDPs arising from SOS relaxations of constrained polynomial optimizations. We compared the results to the interior-point solver SeDuMi [33], and to the first-order solver SCS (the direct implementation was called) [27]. We called CDCS and SCS with termination tolerance 10^{-3} and limited the maximum number of iterations to 2000. Default values were used for all other parameters. SeDuMi was called with its default parameters. All experiments were carried out on a PC with a 2.8 GHz Intel Core i7 CPU and 8 GB of RAM.

3.6.1 CDCS

CDCS is the first open-source first-order conic solver that exploits chordal decomposition for the PSD cones and affine decomposition for the equality constraints. Infeasible problems can be detected thanks to the usage of homogeneous self-dual embedding. CDCS supports Cartesian products of the following standard cones: $\mathbb{R}^n$, non-negative orthant, second-order cones, and PSD cones. In order to save memory and increase computational efficiency, CDCS uses the symmetric vectorization for all variables (cf. footnote 1 on page 7). The current implementation is written in MATLAB and can be downloaded from

https://github.com/oxfordcontrol/CDCS.

Different solver options are available. The default solver, `hsde`, implements the sparsity-exploiting algorithm described in Sect. 3.4. Changing the solver option to `sos` exploits partial orthogonality. We will refer to these two solver options as CDCS-hsde and CDCS-sos, respectively. CDCS also includes `primal` and `dual` options, which implement the primal-only and dual-only ADMM algorithms described in Ref. [44]. CDCS can be called directly from MATLAB's command window or through the optimization modeling toolboxes YALMIP [25] and SOSTOOLS [30].

3.6.2 Cone Decomposition: The hsde Option

To illustrate the benefits brought by the cone decomposition strategy, we consider three benchmark SDPs from SDPLIB [7] (maxG11, maxG32 and qpG11) and three large-scale sparse SDPs from [5] (rs35, rs200, rs228). The SDPLIB problems are from practical applications (max cut problems and relaxations of box-constrained quadratic programs), while those of [5] are random SDPs with aggregate sparsity pattern coming from the University of Florida Sparse Matrix Collection [13].

Table 3.1 reports the dimensions of these problems and chordal decomposition details, while the aggregate sparsity patterns of these problems are illustrated in Fig. 3.2. Note that although the size of the PSD cone for problems maxG32, qpG11, rs35, rs200 and rs228 is over 1000, the chordal extensions of the underlying sparsity patterns are very sparse and the maximum clique size is much smaller than the original cone size. We expect CDCS-hsde to perform well on these problems since working with smaller PSD cones makes the conic projection step more efficient.

The numerical results for these sparse SDPs are summarized in Tables 3.2 and 3.3. As shown in Table 3.2, CDCS-hsde was faster than either SeDuMi or SCS for all the problems we considered. In particular, CDCS-hsde is able to return an approximate solution of maxG11, maxG32, qpG11, rs200 or rs228 in less than 100s, providing a speed up of approximately $10\times, 67\times, 79\times, 126\times$ and $64\times$ over SCS for each problem respectively. Table 3.3 lists the average CPU time per iteration for CDCS-hsde and SCS, giving a fairer comparison of the performance of CDCS-hsde and SCS because any dependence on the exact stopping conditions used by each solver is removed. CDCS-hsde was faster than SCS for all problems, and—perhaps not unexpectedly—the computational gains become more significant when the size of the largest maximal clique is much smaller than the original

Table 3.1 Summary of chordal decomposition for the chordal extensions of the large-scale sparse SDPs tested in this chapter

	maxG11	maxG32	qpG11	rs35	rs200	rs228
Original cone size, n	800	2000	1600	2003	3025	1919
Affine constraints, m	800	2000	800	200	200	200
Number of cliques, p	598	1499	1405	588	1635	783
Maximum clique size	24	60	24	418	102	92
Minimum clique size	5	5	1	5	4	3

Problems are taken from Refs. [5, 7]

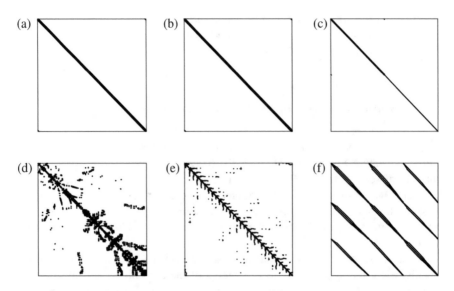

Fig. 3.2 Aggregate sparsity patterns of the large-scale sparse SDPs considered in this chapter; see Table 3.1 for the matrix dimensions. (**a**) maxG11. (**b**) maxG32. (**c**) qpG11. (**d**) rs35. (**e**) rs200. (**f**) rs228

Table 3.2 Numerical results for the SDPs described in Table 3.1

	maxG11	maxG32	qpG11	rs35	rs200	rs228
SeDuMi						
Total CPU time (s)	49.7	954.3	458.2	1120.9	4003.7	1291.2
Iterations	13	14	14	17	17	21
Objective	629.2	1567.6	2448.7	25.3	99.7	64.7
SCS						
Total CPU time (s)	91.3	2414.6	1070.4	2425.5	9259.1	2187.9
Pre-time (s)	0.1	0.5	0.3	0.5	1.0	0.4
Iterations	1080	2000[a]	2000[a]	2000[a]	2000[a]	2000[a]
Objective	629.2	1567.6	2449.2	25.1	81.9	62.1
CDCS-hsde						
Total CPU time (s)	9.5	36.3	13.5	258.4	73.4	34.4
Pre-time (s)	2.2	12.8	2.2	5.2	8.2	3.4
Iterations	109	134	172	265	256	130
Objective	629.8	1570.1	2453.1	25.9	98.6	64.7

[a]Maximum number of iterations reached

PSD cone size. For problems maxG32, qpG11, rs200, and rs228 the average CPU time of CDCS-hsde improves on SCS by approximately 6.7×, 8.2×, 18.5×, and 4.5×, respectively. Finally, although first-order algorithms are only meant to provide solutions of moderate accuracy, the objective value returned by CDCS-hsde

Table 3.3 Average CPU time per ADMM iteration (in seconds) for the SDPs described in Table 3.1

	maxG11	maxG32	qpG11	rs35	rs200	rs228
SCS	0.085	1.21	0.54	1.21	4.63	1.09
CDCS-hsde	0.067	0.18	0.066	0.96	0.25	0.24

was always within 2.5% of the optimal value computed using SeDuMi. This gap may be considered acceptable in practical applications. Of course, as with all first-order algorithms accuracy could be improved further by setting tighter convergence tolerances at the expense of longer computation time.

3.6.3 Affine Decomposition: The *sos* Option

To show the benefits brought by the exploitation of partial orthogonality, we next consider a series of SOS relaxations of constrained polynomial optimization problems (POPs). Note that partial orthogonality is an inherent structural property of SOS programs. In particular, we consider the constrained POP

$$\min_{x} \quad \sum_{1 \leq i < j \leq n} (x_i x_j + x_i^2 x_j - x_j^3 - x_i^2 x_j^2)$$

$$\text{subject to} \quad \sum_{i=1}^{n} x_i^2 \leq 1. \tag{3.28}$$

We recast (3.28) into an SDP using the second Lasserre relaxation and GloptiPoly [21].

The numerical results are summarized in Tables 3.4 and 3.5. Table 3.4 reports the CPU time (in seconds) required by the solvers considered in this chapter to solve the SDP relaxations as the number of variables n in (3.28) was increased. As we can see, CDCS-sos is the fastest method for all the instances we tested thanks to partial orthogonality. For large-scale POPs ($n \geq 25$), the number of constraints in the resulting SDP is over 20,000, and the interior-point solver SeDuMi terminated prematurely due to RAM limitations. For all values of n the number t of non-orthogonal constraints is much smaller than the number m of constraints (see Table 3.4), so we expect CDCS-sos to perform well thanks to the affine decomposition described in Sect. 3.5. Indeed, for $n \geq 25$ CDCS-sos was approximately twice as fast as SCS. Finally, Table 3.5 shows that although first-order methods only aim to provide solutions of moderate accuracy, the optimal objective value returned by CDCS-sos and SCS was always within 0.5% of the high-accuracy optimal value computed using SeDuMi (when available). As remarked at

Table 3.4 CPU time (in seconds) to solve the SDP relaxations of (3.28)

n	Dimensions			CPU time (s)		
	N	m	t	SeDuMi	SCS	CDCS-sos
10	66	1000	66	2.7	0.4	0.4
15	136	3875	136	162.8	2.4	2.1
20	231	10,625	120	2730.8	10.4	8.2
25	351	23,750	351	a	45.8	28.8
30	496	46,375	496	a	156.1	85.3
35	666	82,250	666	a	455.4	222.2
40	861	135,750	861	a	1182.8	526.6

N is the size of the largest PSD cone, m is the number of constraints, t is the size of the matrix factorized by CDCS-sos
[a] The problem could not be solved due to memory limitations

Table 3.5 Terminal objective value by SeDuMi, SCS and CDCS-sos for the SDP relaxation of (3.28)

n	SeDuMi	SCS		CDCS-sos	
	Objective	Objective	Accuracy	Objective	Accuracy
10	−9.114	−9.124	0.10%	−9.090	0.27%
15	−14.118	−14.090	0.20%	−14.138	0.14%
20	−19.120	−19.165	0.24%	−19.078	0.22%
25	a	−24.174	b	−24.164	b
30	a	−29.087	b	−29.145	b
35	a	−34.053	b	−34.075	b
40	a	−39.146	b	−39.114	b

[a] The problem could not be solved due to memory limitations
[b] No comparison is possible due to the lack of a reference accurate optimal value

the end of Sect. 3.6.2 such a small difference can be considered negligible in many practical applications.

3.7   Conclusion

In this chapter, we have presented an overview of two recent approaches that exploit two structural properties of large-scale sparse semidefinite programs, namely chordal aggregate sparsity and partial orthogonality. Chordal sparsity is common to many SDPs encountered in practice, while partial orthogonality is an inherent characteristic of SOS programs. The first approach, developed in Refs. [44–46] and called cone decomposition in this chapter, allows one to decompose the original, large PSD cone into smaller ones. The second strategy, proposed in [43] and referred to as affine decomposition in this chapter, relies on a "diagonal plus low-rank" representation of a matrix related to the affine constraints of the SDP. By utilizing these two decompositions, the computational efficiency of the tailored ADMM algorithm for conic programs developed in Ref. [28] can be improved significantly. The proposed methods have been implemented in the MATLAB solver CDCS, and we have illustrated their efficiency on a set of benchmark test problems.

Looking ahead, further software development seems essential. Compared to the current version of CDCS, improvements at the implementation level are possible: Many steps of the algorithms presented in this chapter can be carried out in parallel, so one can take full advantage of distributed and/or parallel computing architectures. Algorithmic refinements to include acceleration techniques such as those proposed in Refs. [18, 35] also promise to improve the known poor convergence performance (in terms of numbers of iterations) of ADMM-based algorithms.

Finally, the computational gains obtained through the decomposition methods described in this chapter motivate the search for other structural properties that can be exploited. The identification of properties that characterize generic SDPs seems difficult, but one possible way forward is to focus on the special classes of problems arising in particular fields (e.g., control theory, machine learning, combinatorics, fluid dynamics). In fact, it is reasonable to expect that discipline-specific structural properties exist, which can be exploited to develop advanced and efficient discipline-specific algorithms. This and the software developments mentioned above will be essential to provide efficient and reliable computational tools to solve large-scale SDPs across all fields in which semidefinite programming has found applications.

Acknowledgements The authors would like to thank Prof. Paul Goulart at the University of Oxford and Dr. Andrew Wynn at Imperial College London for their insightful comments. Y. Zheng is supported by the Clarendon Scholarship and the Jason Hu Scholarship. G. Fantuzzi was supported by an EPSRC studentship, Award Reference 1864077, and A. Papachristodoulou is supported by EPSRC grant EP/M002454/1.

References

1. J. Agler, W. Helton, S. McCullough, L. Rodman, Positive semidefinite matrices with a given sparsity pattern. Linear Algebra Appl. **107**, 101–149 (1988)
2. A.A. Ahmadi, A. Majumdar, DSOS and SDSOS optimization: more tractable alternatives to sum of squares and semidefinite optimization. arXiv:1706.02586 (2017, Preprint)
3. F. Alizadeh, J.-P.A. Haeberly, M.L. Overton, Primal-dual interior-point methods for semidefinite programming: convergence rates, stability and numerical results. SIAM J. Optim. **8**(3), 746–768 (1998)
4. M.S. Andersen, J. Dahl, L. Vandenberghe, Implementation of nonsymmetric interior point methods for linear optimization over sparse matrix cones. Math. Program. Comput. **2**(3–4), 167–201 (2010)
5. M. Andersen, J. Dahl, Z. Liu, L. Vandenberghe, Interior-point methods for large-scale cone programming, in *Optimization for Machine Learning* (MIT Press, Cambridge, 2011), pp. 55–83
6. H.H. Bauschke, J.M. Borwein, Dykstra's alternating projection algorithm for two sets. J. Approx. Theory **79**(3), 418–443 (1994)

7. B. Borchers, SDPLIB 1.2, a library of semidefinite programming test problems. Optim. Methods Softw. **11**(1–4), 683–690 (1999)
8. S. Boyd, L. Vandenberghe, *Convex Optimization* (Cambridge University Press, Cambridge, 2004)
9. S. Boyd, L. El Ghaoui, E. Feron, V. Balakrishnan, *Linear Matrix Inequalities in System and Control Theory* (SIAM, Philadelphia, 1994)
10. S. Boyd, N. Parikh, E. Chu, B. Peleato, J. Eckstein, Distributed optimization and statistical learning via the alternating direction method of multipliers. Found. Trends Mach. Learn. **3**(1), 1–122 (2011)
11. S. Burer, Semidefinite programming in the space of partial positive semidefinite matrices. SIAM J. Optim. **14**(1), 139–172 (2003)
12. S.I. Chernyshenko, P.J. Goulart, D. Huang, A. Papachristodoulou, Polynomial sum of squares in fluid dynamics: a review with a look ahead. Philos. Trans. R. Soc. A **372**(2020), 20130350 (2014)
13. T.A. Davis, Y. Hu, The University of Florida sparse matrix collection. ACM Trans. Math. Softw. **38**(1), 1 (2011)
14. E. De Klerk, Exploiting special structure in semidefinite programming: a survey of theory and applications. Eur. J. Oper. Res. **201**(1), 1–10 (2010)
15. G. Fantuzzi, A. Wynn, Optimal bounds with semidefinite programming: an application to stress driven shear flows. Phys. Rev. E **93**(4), 043308 (2016)
16. K. Fujisawa, S. Kim, M. Kojima, Y. Okamoto, M. Yamashita, User's manual for Sparse- CoLO: conversion methods for sparse conic-form linear optimization problems. Technical report, Research Report B-453, Tokyo Institute of Technology, Tokyo (2009)
17. M. Fukuda, M. Kojima, K. Murota, K. Nakata, Exploiting sparsity in semidefinite programming via matrix completion I: general framework. SIAM J. Optim. **11**(3), 647–674 (2001)
18. P. Giselsson, M. Faült, S. Boyd, Line search for averaged operator iteration, in *2016 IEEE 55th Conference on Decision and Control (CDC)* (IEEE, Piscataway, 2016), pp. 1015–1022
19. R. Grone, C.R. Johnson, E.M. Saó, H. Wolkowicz, Positive definite completions of partial Hermitian matrices. Linear Algebra Appl. **58**, 109–124 (1984)
20. C. Helmberg, F Rendl, R.J. Vanderbei, H. Wolkowicz, An interior-point method for semidefinite programming. SIAM J. Optim. **6**(2), 342–361 (1996)
21. D. Henrion, J.-B. Lasserre, GloptiPoly: global optimization over polynomials with MATLAB and SeDuMi. ACM Tans. Math. Softw. **29**(2), 165–194 (2003)
22. S. Kim, M. Kojima, M. Mevissen, M. Yamashita, Exploiting sparsity in linear and nonlinear matrix inequalities via positive semidefinite matrix completion. Math. Program. **129**(1), 33–68 (2011)
23. J.B. Lasserre, Global optimization with polynomials and the problem of moments. SIAM J. Optim. **11**(3), 796–817 (2001)
24. P.-L. Lions, B. Mercier, Splitting algorithms for the sum of two nonlinear operators. SIAM J. Numer. Anal. **16**(6), 964–979 (1979)
25. J. Lofberg, YALMIP: a toolbox for modeling and optimization in MATLAB, in *2004 IEEE International Symposium on Computer Aided Control Systems Design* (IEEE, Piscataway, 2005), pp. 284–289
26. R. Madani, A. Kalbat, J. Lavaei, ADMM for sparse semidefinite programming with applications to optimal power flow problem, in *Proceedings of 54th IEEE Conference Decision Control* (2015), pp. 5932–5939
27. B. O'Donoghue, E. Chu, N. Parikh, S. Boyd, SCS: splitting conic solver version 1.2.6. (2016), https://github.com/cvxgrp/scs
28. B. O'Donoghue, E. Chu, N. Parikh, S. Boyd, Conic optimization via operator splitting and homogeneous self-dual embedding. J. Optim. Theory Appl. **169**(3), 1042–1068 (2016)
29. A. Papachristodoulou, S. Prajna, A tutorial on sum of squares techniques for systems analysis, in *American Control Conference (ACC)* (IEEE, Piscataway, 2005), pp. 2686–2700
30. A. Papachristodoulou, J. Anderson, G. Valmorbida, S. Prajna, P. Seiler, P. Parrilo, SOS-TOOLS version 3.00 sum of squares optimization toolbox for MATLAB. arXiv:1310.4716 (2013, Preprint)

31. P.A. Parrilo, Semidefinite programming relaxations for semialgebraic problems. Math. Program. Ser. B **96**(2), 293–320 (2003)
32. J.E. Spingarn, Applications of the method of partial inverses to convex programming: decomposition. Math. Program. **32**(2), 199–223 (1985)
33. J.F. Sturm, Using SeDuMi 1.02, a MATLAB toolbox for optimization over symmetric cones. Optim. Methods Softw. **11**(1–4), 625–653 (1999)
34. Y. Sun, M.S. Andersen, L. Vandenberghe, Decomposition in conic optimization with partially separable structure. SIAM J. Optim. **24**(2), 873–897 (2014)
35. A. Themelis, P. Patrinos, SuperMann: a superlinearly convergent algorithm for finding fixed points of nonexpansive operators. arXiv:1609.06955 (2016, Preprint)
36. L. Vandenberghe, M.S. Andersen, Chordal graphs and semidefinite optimization. Found. Trends Optim. **1**(4), 241–433 (2014)
37. L. Vandenberghe, S. Boyd, Semidefinite programming. SIAM Rev. **38**(1), 49–95 (1996)
38. M. Yannakakis, Computing the minimum fill-in is NP-complete. SIAM J. Algebraic Discrete Methods **2**, 77–79 (1981)
39. Y. Ye, M.J. Todd, S. Mizuno, An $O\sqrt{n}L$-iteration homogeneous and self-dual linear programming algorithm. Math. Oper. Res. **19**(1), 53–67 (1994)
40. R.Y. Zhang, C. Josz, S. Sojoudi, Conic optimization theory: convexification techniques and numerical algorithms. arXiv:1709.08841 (2017, Preprint)
41. Y. Zheng, G. Fantuzzi, A. Papachristodoulou, P. Goulart, A. Wynn, CDCS: cone decomposition conic solver version 1.1 (2016). https://github.com/OxfordControl/CDCS
42. Y. Zheng, G. Fantuzzi, A. Papachristodoulou, Exploiting sparsity in the coefficient matching conditions in sum-of-squares programming using ADMM. IEEE Control Syst. Lett. **1**(1), 80–85 (2017). ISSN: 2475-1456
43. Y. Zheng, G. Fantuzzi, A. Papachristodoulou, Fast ADMM for sum-of-squares programs using partial orthogonality. arXiv:1708.04174 (2017, Preprint)
44. Y. Zheng, G. Fantuzzi, A. Papachristodoulou, P. Goulart, A. Wynn, Chordal decomposition in operator-splitting methods for sparse semidefinite programs. arXiv:1707.05058 (2017, Preprint)
45. Y. Zheng, G. Fantuzzi, A. Papachristodoulou, P. Goulart, A. Wynn, Fast ADMM for homogeneous self-dual embedding of sparse SDPs, in *Proceedings of the 20th IFAC World Congress* (2017), pp. 8741–8746
46. Y. Zheng, G. Fantuzzi, A. Papachristodoulou, P. Goulart, A. Wynn, Fast ADMM for semidefinite programs with chordal sparsity, in *American Control Conference (ACC)* (IEEE, Piscataway, 2017), pp. 3335–3340

Chapter 4
Smoothing Alternating Direction Methods for Fully Nonsmooth Constrained Convex Optimization

Quoc Tran-Dinh and Volkan Cevher

Abstract We propose two new alternating direction methods to solve "fully" nonsmooth constrained convex problems. Our algorithms have the best known worst-case iteration-complexity guarantee under mild assumptions for both the objective residual and feasibility gap. Through theoretical analysis, we show how to update all the algorithmic parameters automatically with clear impact on the convergence performance. We also provide a representative numerical example showing the advantages of our methods over the classical alternating direction methods using a well-known feasibility problem.

Keywords Gap reduction technique · Smooth alternating minimization · Smooth alternating direction method of multipliers · Homotopy smoothing technique · Nonsmooth constrained convex optimization

AMS Subject Classifications 90C25, 90C06, 90-08

Q. Tran-Dinh (✉)
Department of Statistics and Operations Research, The University of North Carolina at Chapel Hill (UNC), Chapel Hill, NC, USA
e-mail: quoctd@email.unc.edu

V. Cevher
Laboratory for Information and Inference Systems (LIONS), École Polytechnique Fédérale de Lausanne (EPFL), Lausanne, Switzerland
e-mail: volkan.cevher@epfl.ch

© Springer Nature Switzerland AG 2018
P. Giselsson, A. Rantzer (eds.), *Large-Scale and Distributed Optimization*, Lecture Notes in Mathematics 2227,
https://doi.org/10.1007/978-3-319-97478-1_4

4.1   Introduction

In this paper, we aim at developing new optimization algorithms to solve nonsmooth constrained convex optimization problems of the form:

$$f^\star := \begin{cases} \min\limits_{x:=(u,v)\in\mathbb{R}^p} & \{f(x) := g(u) + h(v)\}, \\ \text{s.t.} & Au + Bv = c, \end{cases} \qquad (4.1)$$

where $g : \mathbb{R}^{p_1} \to \mathbb{R} \cup \{+\infty\}$ and $h : \mathbb{R}^{p_2} \to \mathbb{R} \cup \{+\infty\}$ are two proper, closed, and convex functions, $A \in \mathbb{R}^{n\times p_1}$, $B \in \mathbb{R}^{n\times p_2}$, $c \in \mathbb{R}^n$ are given, and $p := p_1 + p_2$. Although our proposed methods can solve (4.1) with both smooth and nonsmooth objective functions, we are more interested in the case where both f and g are nonsmooth. In this case, we refer to (4.1) as a "fully" nonsmooth problem since, except for convexity, we do not require any structure assumptions on g and h such as Lipschitz continuous gradient or strong convexity. Problem (4.1) covers many prominent applications such as convex feasibility problems [2], support vector machine [5], matrix completion [8], basis pursuit [33], among many others.

Associated with the primal problem (4.1), we also look at the dual problem:

$$d^\star := \min_{\lambda\in\mathbb{R}^n} \{d(\lambda) := g^*(A^\top\lambda) + h^*(B^\top\lambda) - \langle c, \lambda\rangle\}, \qquad (4.2)$$

where g^* and h^* are the Fenchel conjugates [30] of g and h, respectively; d is the dual function; λ is the dual variable; and $d^\star$ denotes the dual optimal value. The convex template (4.1) also manifests itself when we apply convex splitting techniques to decompose the composite objective f into two terms g and h that are coupled via linear constraints. It can also include convex constraints on u and v via indicator functions.

This paper develops a new primal-dual algorithmic framework to solve (4.1) which processes g and h in an alternating fashion to obtain approximately numerical solutions. The alternating optimization approach has regained popularity due to its ability to decentralize data, decompose problem components, and distribute computation in large-scale problems. The underlying theory for the classical alternating optimization methods, such as the alternating direction method of multipliers (ADMM) or the alternating minimization algorithm (AMA), is mature as they have their roots from the splitting methods in monotone inclusions and other classical approaches, such as forward-backward splitting, Douglas-Rachford splitting, Dykstra projections, and Hauzageau's methods [1, 2].

Alternating optimization strategies often provide computational advantages as compared to processing both terms jointly. This approach leads to several methods and variants for solving (4.1) as can be found in the literature, see, e.g., [4, 7, 10–13, 15, 17, 19–22, 24, 28, 31, 32, 34, 37–39]. Among those, ADMM and AMA are the most popular ones. Unlike the standard AMA and ADMM methods and their variants mentioned here, we focus on the case that the objective functions g and

h are nonsmooth and the sum f does not have a "tractable" proximal operator. As an example, in convex feasibility problems, we aim at finding a common point in the intersection of many convex sets. This problem can be formulated into a nonsmooth constrained convex problem (4.1) as we are targeting here. The "full" nonsmoothness of (4.1) creates some fundamental drawbacks for numerical algorithms. First, algorithms that require gradients of the objective function are not applicable. Second, evaluating a proximal operator of the full objective function f becomes impractical. Third, methods using penalty or augmented Lagrangian functions are often inefficient due to complicated subproblems and tuning parameters. A more thorough discussion on our approach and existing methods is postponed to Sect. 4.7. In this paper, we overcome these drawbacks by proposing a combination of different techniques in optimization for solving (4.1).

Our Contributions Our main contribution can be summarized as follows:

(a) (*Theory*) We introduce *a split-gap reduction technique* as a new framework for deriving new alternating direction methods. Our framework unifies the model-based gap reduction technique of [35], smoothing techniques, and the powerful forward-backward and Douglas-Rachford splitting techniques. We establish explicit relations between primal weighting strategy, the parameter choices, and the global convergence rate of the algorithms in our framework.
(b) (*Algorithms and convergence guarantees*) We propose two new smoothing alternating direction optimization algorithms: smoothing alternating minimization algorithm (SAMA), and smoothing alternating direction method of multipliers (SADMM). We derive update rules for all algorithmic parameters including penalty parameters in a heuristic-free fashion. We rigorously characterize the convergence rate of our algorithms for both the objective residual $f(\bar{x}^k) - f^\star$ and the feasibility gap $\|A\bar{u}^k + B\bar{v}^k - c\|$. To the best of our knowledge, this is the best known global convergence rate that can be achieved under mildest assumptions in the literature.
(c) (*Special cases*) We also illustrate that our technique can exploit additional assumptions on A or B, g and h, whenever they are available.

Let us emphasize the following important points of our contribution.

1. (*Mild assumptions*) We only assume that g and h are proper, closed, and convex, the solution set of (4.1) is nonempty, and Slater's condition holds. We also require a technical assumption on the boundedness of the domain of g and h. However, this assumption can be removed by using Lemma 1. Therefore, our methods can solve a broad class of convex optimization problems covered by (4.1).
2. (*Computational complexity*) Our smoothing AMA algorithm essentially has the same per-iteration complexity as the standard AMA [37]. Similarly, our smoothing ADMM has essentially the same per-iteration complexity as the standard ADMM [5]. Although we require additional computation for accelerated steps and averaging, this computation only requires vector-vector additions and scalar-vector multiplications, whose cost is negligible.

3. (*Parameter update*) Our algorithms are heuristic-free in the sense that we update all the parameters automatically at each iteration including the so-called penalty parameter in alternating direction methods [3, 23, 27]. This solves the major drawback in augmented Lagrangian-based methods. We argue that this key feature is important in parallel and distributed implementation, when tuning parameters is impossible to carry out. Intriguingly, our algorithms update their penalty parameters in a decreasing fashion in stark contrast to the classical algorithms.
4. (*Convergence guarantees*) The proposed methods achieve the best known global convergence rate on the primal problem (4.1) as well as on the dual one (4.2) under required assumptions. Moreover, we can explicitly show how the choice of algorithmic parameters can trade-off the convergence guarantee of the objective residual $f(\bar{x}^k) - f^\star$ and the primal feasibility gap $\|A\bar{u}^k + B\bar{v}^k - c\|$ in the worst case.

Paper Organization Section 4.2 briefly presents a primal-dual formulation of problem (4.1) under basic assumptions, and characterizes its optimality condition. Section 4.3 deals with a smoothing technique for the primal-dual gap function. Section 4.4 presents a smoothing AMA algorithm and analyzes its convergence. The strongly convex case is also studied in this section. Section 4.5 is devoted to developing a smoothing ADMM algorithm and analyzes its convergence. Section 4.6 presents numerical experiments to verify the performance of our algorithms. We conclude with a discussion of our results in the context of existing work. For clarity of exposition, several technical and new proofs are moved to the Appendix.

Notation In the sequel, we refer to (4.1) as the primal problem. We work on the real and finite dimensional spaces $\mathbb{R}^p$ and $\mathbb{R}^n$, endowed with the inner product $\langle x, \lambda \rangle$ and the standard Euclidean norm $\| \cdot \|$. We use the superscript $\top$ for both the transpose and adjoint operators. For a convex function f, we use ∂f for its subdifferential, and f^* for its Fenchel conjugate. For a convex set $\mathcal{X}$, we use $\delta_{\mathcal{X}}$ for its indicator function, and $\mathrm{ri}(\mathcal{X})$ for its relative interior. We also use $\mathbb{R}_{++}$ for the set of positive real numbers.

For any proper, closed, and convex function $\varphi : \mathbb{R}^p \to \mathbb{R} \cup \{+\infty\}$, the proximal operator is defined as follows:

$$\mathrm{prox}_\varphi(x) := \underset{z}{\mathrm{argmin}} \left\{ \varphi(z) + (1/2)\|z - x\|^2 \right\}. \tag{4.3}$$

Generally, computing prox_φ is intractable. However, if prox_φ can be efficiently computed in a closed form or in polynomial time, then we say that φ has a *tractable* proximity operator. Several examples can be found, e.g., in [2, 29].

4.2 Preliminaries: Lagrangian Primal-Dual Formulation

This section briefly describes the primal-dual formulation of (4.1) and our fundamental assumptions.

4.2.1 The Dual Problem

Let $x := (u, v) \equiv (u^{\top}, v^{\top})^{\top} \in \mathbb{R}^p$ be the primal variable, dom $(f) := \text{dom}(g) \times \text{dom}(h)$, and $\mathcal{D} := \{(u, v) \in \text{dom}(f) \mid Au + Bv = c\}$ be the feasible set of (4.1). We define the Lagrange function of (4.1) associated with $Au + Bv = c$ as $\mathcal{L}(x, \lambda) := g(u) + h(v) - \langle \lambda, Au + Bv - c \rangle$, where $\lambda \in \mathbb{R}^n$ is the Lagrange multiplier. We recall the dual problem (4.2) of (4.1) here:

$$d^{\star} := \min_{\lambda \in \mathbb{R}^n} \left\{ d(\lambda) := \max_u \left\{ \langle A^{\top}\lambda, u \rangle - g(u) \right\} + \max_v \left\{ \langle B^{\top}\lambda, v \rangle - h(v) \right\} - c^{\top}\lambda \right\}, \quad (4.4)$$

where d is the dual function, and two terms can be individually computed as

$$\begin{cases} \varphi(\lambda) := \max_{u \in \text{dom}(g)} \left\{ \langle A^{\top}\lambda, u \rangle - g(u) \right\} = g^*(A^{\top}\lambda), \\ \psi(\lambda) := \max_{v \in \text{dom}(h)} \left\{ \langle B^{\top}\lambda, v \rangle - h(v) \right\} - c^{\top}\lambda = h^*(B^{\top}\lambda) - c^{\top}\lambda. \end{cases} \quad (4.5)$$

Let us denote by $u^*(\lambda)$ and $v^*(\lambda)$ one solution of these subproblems, respectively, if they exist. In this case, using the optimality condition, we have $A^{\top}\lambda \in \partial g(u^*(\lambda))$, which is equivalent to $u^*(\lambda) \in \partial g^*(A^{\top}\lambda)$. Similarly, $B^{\top}\lambda \in \partial h(v^*(\lambda))$, which is equivalent to $v^*(\lambda) \in \partial h^*(B^{\top}\lambda)$. These dual components are convex, but generally nonsmooth. Subgradient or bundle-type methods for directly solving (4.4) are generally inefficient [25, 26].

4.2.2 Basic Assumptions

Let us denote by $\mathcal{X}^{\star}$ the solution set of (4.1). We say that the *Slater condition* holds for (4.1) if we have

$$\text{ri}(\text{dom}(f)) \cap \{(u, v) \in \mathbb{R}^p \mid Au + Bv = c\} \neq \emptyset, \quad (4.6)$$

where $\text{ri}(\mathcal{X})$ is the relative interior of $\mathcal{X}$ (see [30]).

For the primal-dual pair (4.1) and (4.4), we require the following assumption:

Assumption 1 *The functions g and h are proper, closed, and convex. The solution set $\mathcal{X}^{\star}$ of (4.1) is nonempty. Either dom(f) is polyhedral or the Slater condition (4.6) holds.*

Compared to existing methods for solving (4.1) in the literature [4, 7, 10–13, 15, 17, 19–22, 24, 28, 31, 32, 34, 37–39], this assumption is perhaps the mildest one so far. We do not require any strong convexity, error bound, regularity, or Lipschitz gradient assumptions on g and h.

4.2.3   Zero Duality Gap

Under Assumption 1, the solution set $\Lambda^\star$ of the dual problem (4.4) is nonempty and bounded. Moreover, *strong duality* holds, i.e., $f^\star + d^\star = 0$. From the classical duality theory, we have $f(x) + d(\lambda) \geq 0$ for any feasible primal-dual point (x, λ). Hence, the duality gap function G is defined by

$$G(w) := f(x) + d(\lambda) \geq 0, \quad \forall x \in \mathcal{D}, \ \forall \lambda \in \mathbb{R}^n, \tag{4.7}$$

where $w := (x, \lambda)$. Clearly, $G(w^\star) = 0$ (zero duality gap) for any primal-dual solution $w^\star := (x^\star, \lambda^\star) \in \mathcal{X}^\star \times \Lambda^\star$. In addition, $w^\star$ is a saddle point of the Lagrange function; that is $\mathcal{L}(x^\star, \lambda) \leq \mathcal{L}(x^\star, \lambda^\star) = f^\star = -d^\star \leq \mathcal{L}(x, \lambda^\star)$ for all $x \in \text{dom}(f)$ and $\lambda \in \mathbb{R}^n$. The optimality condition of (4.1) can be written as

$$Au^\star + Bv^\star = c, \quad A^\top \lambda^\star \in \partial g(u^\star), \quad \text{and} \quad B^\top \lambda^\star \in \partial h(v^\star). \tag{4.8}$$

4.2.4   Technical Assumption

Apart from Assumption 1, the methods we will develop in the following sections require the following boundedness assumption:

Assumption 2 *Both* dom (g) *and* dom (h) *are bounded.*

According to [2, Corollary 17.19], the boundedness of dom (g) and dom (h) is equivalent to the Lipschitz continuity of the conjugates g^* and h^*, respectively. Assumption 2 also theoretically restricts the class of problems in (4.1) that we can solve. However, if Assumption 2 does not hold, then we can always add an artificial constraint $\|x\| \leq R$ to (4.1) (or $\|u\| \leq R$ and $\|v\| \leq R$) so that Assumption 2 is satisfied for this modified problem, where $R \in (0, +\infty)$. Under a proper choice of R, this problem is equivalent to (4.1) as showed in the following lemma.

Lemma 1 *Consider two constrained convex optimization problems:*

$$(\text{P}_\infty) \ \ f^\star := \min_{x \in \mathcal{D}} f(x) \quad and \quad (\text{P}_R) \ \ \bar{f}^\star := \min_{x \in \mathcal{D}} \{f(x) \mid \|x\| \leq R\},$$

where f *is defined in* (4.1), $\mathcal{D} := \{x = (u, v) \mid Au + Bv = c, \ u \in \text{dom}(g),$
$v \in \text{dom}(h)\}$ *is the feasible set of* (4.1), *and* $R \in (0, +\infty)$.

If $x^\star$ *is a solution of* (P_∞), *and* $\|x^\star\| \leq R$, *then it is a solution of* (P_R). *Conversely, if* $\bar{x}^\star$ *is a solution of* (P_R) *and* $\|\bar{x}^\star\| < R$, *then it is a solution of* (P_∞).

Proof It is obvious that if $x^\star$ is a solution of (P_∞), and $\|x^\star\| \leq R$, then it is a solution of (P_R). Conversely, if $\bar{x}^\star$ is a solution of (P_R), then we have $f(\bar{x}^\star) \leq f(x)$ for all $x \in \mathcal{D}$ and $\|x\| \leq R$. Take any $x \in \mathcal{D} \backslash \mathbb{B}_R$, where $\mathbb{B}_R := \{x \in \mathbb{R}^p \mid \|x\| \leq R\}$ is a ball centered at the origin with radius R. Since $\bar{x}^\star \in \text{int}(\mathbb{B}_R)$, the interior of $\mathbb{B}_R$, there exists $\hat{x}$ on the open segment $(\bar{x}^\star, x)$ such that $\hat{x} = (1 - \tau)\bar{x}^\star + \tau x$

and $\hat{x} \in \mathcal{D} \cap \mathbb{B}_R$, where $\tau \in (0, 1)$. In this case, by convexity of f, we have $f(\bar{x}^\star) \leq f(\hat{x}) = f((1-\tau)\bar{x}^\star + \tau x) \leq (1-\tau)f(x^\star) + \tau f(x)$. Since $\tau \in (0, 1)$, this inequality implies $f(\bar{x}^\star) \leq f(x)$. Therefore, $\bar{x}^\star$ is a solution of (P_∞). □

As suggested by Lemma 1, if we add artificial bounds $\|u\| \leq R$ and $\|v\| \leq R$ to (4.1), then the resulting problem is equivalent to

$$\min_{u,v} \left\{ \hat{g}(u) + \hat{h}(v) \mid Au + Bv = c \right\},$$

where $\hat{g} := g + \delta_{\mathbb{B}_R}$, $\hat{h} := h + \delta_{\mathbb{B}_R}$, and $\delta_{\mathbb{B}_R}$ is the indicator function of the closed ball $\mathbb{B}_R := \{z \mid \|z\| \leq R\}$. This problem has the same form as (4.1). Under Assumption 2, the following quantity:

$$D_f := \sup_{u \in \mathrm{dom}(g),\ \hat{v}, v \in \mathrm{dom}(h)} \left\{ \max \left\{ \|Au + Bv - c\|, \|Au + B(2\hat{v} - v) - c\| \right\} \right\} \quad (4.9)$$

is bounded, i.e., $0 \leq D_f < +\infty$.

Note that, in our algorithms below, since we do not require D_f as an input of the algorithms, this quantity can be heuristically estimated after we terminate the algorithms, and estimate the corresponding artificial radius R based on iteration sequences obtained from the algorithms (see Remark 1).

4.3 Smoothing the Primal-Dual Gap Function

The dual function d defined by (4.4) is convex, but it is generally nonsmooth. Our key idea is to replace the component g^* in (4.5) with a new smoothed approximation g_γ^* to derive new algorithms.

Let us consider the domain $\mathcal{U} := \mathrm{dom}(g)$ of g. Associated with $\mathcal{U}$, we choose a proximity function ω, i.e., ω is continuous and strongly convex with the convexity parameter $\mu_\omega = 1 > 0$, and $\mathcal{U} \subseteq \mathrm{dom}(\omega)$. In addition, we assume that ω is smooth, and its gradient is Lipschitz continuous with the Lipschitz constant $L_\omega \in [0, +\infty)$.

Given ω, we define the associated Bregman distance

$$b_\mathcal{U}(u, \hat{u}) := \omega(u) - \omega(\hat{u}) - \langle \nabla\omega(\hat{u}), u - \hat{u} \rangle. \quad (4.10)$$

Let $\bar{u}_c := \mathrm{argmin}_u\, \omega(u)$ be the prox-center of ω, which exists and is unique. We consider the function $b_\mathcal{U}(\cdot, \bar{u}^c)$. Clearly, $b_\mathcal{U}(\cdot, \bar{u}^c)$ is smooth and strongly convex with the convexity parameter $\mu_b = \mu_\omega = 1$. Its gradient $\nabla_1 b_\mathcal{U}(u, \bar{u}^c) = \nabla\omega(u) - \nabla\omega(\bar{u}^c)$ is Lipschitz continuous with the Lipschitz constant $L_b = L_\omega \geq \mu_\omega = 1$. In addition, $b_\mathcal{U}(\bar{u}^c, \bar{u}^c) = 0$ and $\nabla_1 b_\mathcal{U}(\bar{u}^c, \bar{u}^c) = 0$.

Given $b_{\mathcal{U}}(\cdot, \bar{u}^c)$, and the conjugate g^* of g, we define

$$g_\gamma^*(z) := \max_{u \in \mathbb{R}^{p_1}} \left\{ \langle z, u \rangle - g(u) - \gamma b_{\mathcal{U}}(u, \bar{u}^c) \right\}, \tag{4.11}$$

where $\gamma > 0$ is a smoothness parameter. We denote by $u_\gamma^*(z)$ the solution of the maximization problem in (4.11), i.e.:

$$u_\gamma^*(z) := \arg \max_{u \in \mathbb{R}^{p_1}} \left\{ \langle z, u \rangle - g(u) - \gamma b_{\mathcal{U}}(u, \bar{u}^c) \right\}, \tag{4.12}$$

which is well-defined and unique. Clearly, $\nabla g_\gamma^*(z) = u_\gamma^*(z)$ is the gradient of g_γ^*, which has $(1/\gamma)$-Lipschitz gradient. Hence, g_γ^* is $(1/\gamma)$-smooth [2].

Let g_γ^* and ψ be defined by (4.11) and (4.5), respectively, and $\beta > 0$. We consider

$$\begin{cases} d_\gamma(\lambda) & := g_\gamma^*(A^\top \lambda) + \left(h^*(B^\top \lambda) - \langle c, \lambda \rangle \right) = \varphi_\gamma(\lambda) + \psi(\lambda), \\ f_\beta(x) & := g(u) + h(v) + \frac{1}{2\beta} \|Au + Bv - c\|^2, \\ G_{\gamma\beta}(w) & := f_\beta(x) + d_\gamma(\lambda). \end{cases} \tag{4.13}$$

If $\gamma \downarrow 0^+$, then we have $d_\gamma(\lambda) \to d(\lambda)$. Hence, d_γ is a smoothed approximation of d, but it is not fully smooth due to possible nonsmoothness of ψ. For any feasible point $x = (u, v) \in \mathcal{D}$, we have $f_\beta(x) = f(x)$. Here, f_β can be considered as an approximation to f near the feasible set $\mathcal{D}$. Hence, the smoothed gap function $G_{\gamma\beta}$ is an approximation of the duality gap function G in (4.7). Moreover, the smoothed gap function $G_{\gamma\beta}$ is convex. The following lemma shows us how to use $G_{\gamma\beta}$ to characterize the primal-dual solutions for (4.1)–(4.2), whose proof is in section "Proof of Lemma 2: The Primal-Dual Bounds" in Appendix.

Lemma 2 *For any $\bar{x}^k := (\bar{u}^k, \bar{v}^k) \in \mathrm{dom}\,(f)$ and $\bar{\lambda}^k \in \mathbb{R}^n$, it holds that*

$$- \|\lambda^\star\| \|A\bar{u}^k + B\bar{v}^k - c\| \le f(\bar{x}^k) - f^\star \le f(\bar{x}^k) + d(\bar{\lambda}^k). \tag{4.14}$$

Let $\{\bar{w}^k\}$ be an arbitrary sequence in $\mathrm{dom}\,(f) \times \mathbb{R}^n$ and $\{(\gamma_k, \beta_k)\}$ be a sequence in $\mathbb{R}_{++}^2$. Then, the following estimates hold:

$$\begin{cases} f(\bar{x}^k) - f^\star & \le S_k(\bar{w}^k), \\ \|A\bar{u}^k + B\bar{v}^k - c\| & \le 2\beta_k \|\lambda^\star\| + \sqrt{2\beta_k S_k(\bar{w}^k)}, \\ d(\bar{\lambda}^k) - d^\star & \le 2\beta_k \|\lambda^\star\|^2 + \|\lambda^\star\| \sqrt{2\beta_k S_k(\bar{w}^k)} + S_k(\bar{w}^k), \end{cases} \tag{4.15}$$

where $S_k(\bar{w}^k) := G_{\gamma_k \beta_k}(\bar{w}^k) + \gamma_k b_{\mathcal{U}}(u^\star, \bar{u}^c)$, which requires the values of $G_{\gamma\beta}$.

Computing exactly a primal-dual solution $(x^\star, \lambda^\star)$ is impractical. Hence, our objective is to find an approximation $(\bar{x}^k, \bar{\lambda}^k)$ to $(x^\star, \lambda^\star)$ in the following sense:

Definition 1 Given an accuracy $\varepsilon > 0$, a primal-dual point $(\bar{x}^k, \bar{\lambda}^k) \in \mathrm{dom}(f) \times \mathbb{R}^n$ is said to be an ε-solution of (4.1)–(4.2) if

$$f(\bar{x}^k) - f^\star \le \varepsilon, \quad \|A\bar{u}^k + B\bar{v}^k - c\| \le \varepsilon, \quad \text{and} \quad d(\bar{\lambda}^k) - d^\star \le \varepsilon.$$

We use the same accuracy parameter ε for each of these terms for simplicity.

We note that by combining $\|A\bar{u}^k + B\bar{v}^k - c\| \le \varepsilon$ and (4.14), we can guarantee a lower abound $f(\bar{x}^k) - f^\star \ge -\|\lambda^\star\|\varepsilon$. In addition, the domain $\mathrm{dom}(f)$ is usually simple (e.g., box, ball, cone, or simplex) so that the constraint $\bar{x}^k \in \mathrm{dom}(f)$ can be guaranteed via a closed form projection onto $\mathrm{dom}(f)$.

The goal is to generate a primal-dual sequence $\{\bar{w}^k\}$ and a parameter sequence $\{(\gamma_k, \beta_k)\}$ in Lemma 2 such that $\{G_{\gamma_k \beta_k}(\bar{w}^k)\}$ converges to 0 and $\{(\gamma_k, \beta_k)\}$ also converges to zero. Moreover, the convergence rate of $f(\bar{x}^k) - f^\star$ and $\|A\bar{u}^k + B\bar{v}^k - c\|$ depends on the convergence rate of $\{G_{\gamma_k \beta_k}(\bar{w}^k)\}$ and $\{(\gamma_k, \beta_k)\}$.

4.4 Smoothing Alternating Minimization Algorithm (SAMA)

We propose a new alternating direction method via the application of the accelerated forward-backward splitting to the smoothed gap function. We describe SAMA in three subsections: main steps, initialization, and parameter updates.

4.4.1 Main Steps

At the iteration $k \ge 0$, given $\hat{\lambda}^k \in \mathbb{R}^n$ and the parameters $\gamma_{k+1} > 0$ and $\eta_k > 0$, the main steps of our SAMA consists of two primal alternating direction steps and one dual ascend step as follows:

$$\begin{cases} \hat{u}^{k+1} := \underset{u \in \mathrm{dom}(g)}{\mathrm{argmin}} \left\{ g(u) - \langle A^\top \hat{\lambda}^k, u \rangle + \gamma_{k+1} b_{\mathcal{U}}(u, \bar{u}^c) \right\}, \\ \hat{v}^{k+1} := \underset{v \in \mathrm{dom}(h)}{\mathrm{argmin}} \left\{ h(v) - \langle B^\top \hat{\lambda}^k, v \rangle + \dfrac{\eta_k}{2} \|A\hat{u}^{k+1} + Bv - c\|^2 \right\}, \quad \text{(SAMA)} \\ \bar{\lambda}^{k+1} := \hat{\lambda}^k - \eta_k (A\hat{u}^{k+1} + B\hat{v}^{k+1} - c), \end{cases}$$

where γ_{k+1} and η_k are referred to as the smoothness and the penalty parameter, respectively, and $\bar{u}_c$ is the prox-center of ω in (4.10).

The subproblems in SAMA can often be computed in a closed form. Let us describe two cases. First, if $b_{\mathcal{U}}(\cdot, \bar{u}^c) := (1/2)\| \cdot -\bar{u}^c\|^2$, the standard Euclidean distance, then computing $\hat{u}^{k+1}$ reduces to computing the proximal operator of g, i.e.,

$$\hat{u}^{k+1} = \mathrm{prox}_{\gamma_{k+1}^{-1} g}\left(\bar{u}_c + \gamma_{k+1}^{-1} A^\top \hat{\lambda}^k\right).$$

Second, if we have $B = \mathbb{I}$ or B is orthonormal, then computing $\hat{v}^{k+1}$ reduces to computing the proximal operator of h, i.e.,

$$\hat{v}^{k+1} = \text{prox}_{\eta_k^{-1}h}\big(B^\top(c - A\hat{u}^{k+1}) + \eta_k^{-1}B^\top\hat{\lambda}^k\big).$$

By inspection, it is easy to see that SAMA is an analog of the classical AMA (cf., (4.46)). The first subproblem, due to (4.11), corresponds to the forward step while the last two lines correspond to the backward step. Moreover, if we set $\gamma_{k+1} = 0$ and $\hat{\lambda}^{k+1} = \bar{\lambda}^{k+1}$, SAMA becomes AMA. However, in contrast to the AMA, the SAMA also features a dual acceleration and a primal weighted averaging step:

$$\begin{cases} \hat{\lambda}_k & := (1 - \tau_k)\bar{\lambda}^k + \tau_k\lambda_k^*, \qquad\qquad \text{(dual acceleration)} \\ (\bar{u}^{k+1}, \bar{v}^{k+1}) & := (1 - \tau_k)(\bar{u}^k, \bar{v}^k) + \tau_k(\hat{u}^{k+1}, \hat{v}^{k+1}), \text{ (weighted averaging)} \end{cases} \qquad (4.16)$$

where $\lambda_k^* := \beta_k^{-1}(c - A\bar{u}^k - B\bar{v}^k)$, and $\tau_k \in (0, 1)$ is a given step size. As we will prove in Theorem 1 below, these dual acceleration and primal weighted averaging steps allow us to achieve a better convergence rate on both the primal and the dual spaces compared to standard AMA methods [17].

The following lemma provides conditions showing that the sequence $\{(\bar{x}^k, \bar{\lambda}^k)\}$ generated by (SAMA)–(4.16) maintains the non-monotone gap reduction condition introduced in [36]. The proof of this lemma can be found in section "Proof of Lemma 3: Gap Reduction Condition" in Appendix.

Lemma 3 *Let $\{\bar{w}^k\}$ with $\bar{w}^k := (\bar{u}^k, \bar{v}^k, \bar{\lambda}^k)$ be the sequence generated by (SAMA)–(4.16). If $\tau_k \in (0, 1]$ and $\gamma_k, \beta_k, \eta_k \in \mathbb{R}_{++}$ satisfy the following conditions:*

$$\begin{aligned} (1 + L_b^{-1}\tau_k)\gamma_{k+1} &\geq \gamma_k, \qquad\qquad \beta_{k+1} \geq (1 - \tau_k)\beta_k, \\ (1 - \tau_k^2)\gamma_{k+1}\beta_k &\geq 2\|A\|^2\tau_k^2, \quad and \quad 2\|A\|^2\eta_k = \gamma_{k+1}, \end{aligned} \qquad (4.17)$$

then the following non-monotone gap reduction condition holds:

$$G_{\gamma_{k+1}\beta_{k+1}}(\bar{w}^{k+1}) \leq (1 - \tau_k)G_{\gamma_k\beta_k}(\bar{w}^k) + \frac{\eta_k\tau_k^2}{4}D_f^2, \qquad (4.18)$$

where $G_{\gamma_k\beta_k}$ is defined by (4.13) and D_f is defined by (4.9).

4.4.2 Initialization

We note that we can initialize the algorithm at any starting point $\bar{w}^1 := (\bar{u}^1, \bar{v}^1, \bar{\lambda}^1)$. However, the convergence bounds will depend on $G_{\gamma_1\beta_1}(\bar{w}^1)$. In order to provide transparent convergence results, we propose to use the following initialization in

Lemma 4, whose proof is given in section "Proof of Lemma 4: Bound on $G_{\gamma\beta}$ for the First Iteration" in Appendix.

Lemma 4 *Given* $\hat{\lambda}^0 \in \mathbb{R}^m$, $\gamma_1 > 0$, *and* $\eta_0 > 0$, *let* $(\bar{u}^1, \bar{v}^1, \bar{\lambda}^1)$ *be computed by*

$$
\begin{cases}
\bar{u}^1 := \underset{u \in \text{dom}(g)}{\text{argmin}} \left\{ g(u) - \langle A^\top \hat{\lambda}^0, u \rangle + \gamma_1 b_{\mathcal{U}}(u, \bar{u}^c) \right\}, \\
\bar{v}^1 := \underset{v \in \text{dom}(h)}{\text{argmin}} \left\{ h(v) - \langle B^\top \hat{\lambda}^0, v \rangle + \dfrac{\eta_0}{2} \| A\bar{u}^1 + Bv - c \|^2 \right\}, \\
\bar{\lambda}^1 := \hat{\lambda}^0 - \eta_0 (A\bar{u}^1 + B\bar{v}^1 - c).
\end{cases} \tag{4.19}
$$

Then, for any $\beta_1 > 0$, $\bar{w}^1 := (\bar{u}^1, \bar{v}^1, \bar{\lambda}^1)$, *and* $G_{\gamma\beta}$ *defined by (4.13) satisfy*

$$
G_{\gamma_1 \beta_1}(\bar{w}^1) \leq \frac{\eta_0}{4} D_f^2 + \frac{1}{2\eta_0^2} \left[\frac{1}{\beta_1} - \frac{(5\gamma_1 - 2\eta_0 \|A\|^2)\eta_0}{2\gamma_1} \right] \| \bar{\lambda}^1 - \hat{\lambda}^0 \|^2 \\
+ \eta_0^{-1} \langle \hat{\lambda}^0, \bar{\lambda}^1 - \hat{\lambda}^0 \rangle. \tag{4.20}
$$

Consequently, if we choose γ_1, β_1, *and* η_0 *such that* $5\gamma_1 > 2\eta_0 \|A\|^2$ *and* $\beta_1 \geq \frac{2\gamma_1}{(5\gamma_1 - 2\eta_0 \|A\|^2)\eta_0}$, *then* $G_{\gamma_1 \beta_1}(\bar{w}^1) \leq \frac{\eta_0}{4} D_f^2 + \eta_0^{-1} \langle \hat{\lambda}^0, \bar{\lambda}^1 - \hat{\lambda}^0 \rangle$.

4.4.3 Updating the Parameters

For simplicity of presentation, we choose ω as $\omega(u) := \frac{1}{2} \| u - \bar{u}^c \|^2$ for a fixed $\bar{u}_c \in \text{dom}(g)$. In this case, $b_{\mathcal{U}}(\cdot, \bar{u}_c)$ defined by (4.10) becomes $b_{\mathcal{U}}(\cdot, \bar{u}_c) = \frac{1}{2} \| \cdot - \bar{u}_c \|^2$. Hence, we can update $\tau_k, \gamma_k, \beta_k$ and η_k such that the equality in the conditions (4.17) holds. The following lemma provides **one possibility** to update these parameters whose proof is given in section "Proof of Lemma 5: Parameter Updates" in Appendix.

Lemma 5 *Let* $b_{\mathcal{U}}$ *be chosen such that* $b_{\mathcal{U}}(\cdot, \bar{u}_c) := \frac{1}{2} \| \cdot - \bar{u}_c \|^2$ *for a fixed* $\bar{u}_c \in \text{dom}(g)$, *and* $\gamma_1 > 0$. *Then, for* $k \geq 1$, *if* $\tau_k, \gamma_k, \beta_k$, *and* η_k *are updated by*

$$
\tau_k := \frac{3}{k+4}, \quad \gamma_k := \frac{5\gamma_1}{k+4}, \quad \beta_k := \frac{18\|A\|^2(k+5)}{5\gamma_1(k+1)(k+7)}, \quad \text{and} \quad \eta_k := \frac{5\gamma_1}{2\|A\|^2(k+5)}, \tag{4.21}
$$

then they satisfy conditions (4.17). Moreover, the convergence rate of $\{\tau_k\}$ *is optimal, and* $\beta_k \leq \frac{18\|A\|^2}{5\gamma_1(k+1)}$.

Let us comment here on our weighting strategy and its relation to [12], which places emphasis on the later iterates in averaging by using $\omega_i = i + 1$ as described by (4.45) in Sect. 4.7. In our updates, we consider another weighting scheme (4.45) that places even more emphasis. For this purpose, we use $\omega_i = (i + 1)(i + 2)$ and rewrite (4.45) in a way to mimic the averaging step in (4.16): $\bar{x}^{k+1} = \frac{1}{k+4}\bar{x}^k + \frac{3}{k+4}x^{k+1}$. Hence, our particular primal weighting scheme (SAMA) uses $\tau_k = \frac{3}{k+4}$.

4.4.4   The New Smoothing AMA Algorithm

Since λ_k^* in the first line of (4.16) requires one matrix-vector multiplication (Au, Bv), we can combine the third line of SAMA and the second line of (4.16) to compute λ_k^* recursively as

$$\lambda_{k+1}^* := \beta_{k+1}^{-1}\big[(1 - \tau_k)\beta_k\lambda_k^* + \tau_k\eta_k^{-1}(\bar{\lambda}^{k+1} - \hat{\lambda}^k)\big]. \tag{4.22}$$

Consequently, each iteration of Algorithm 1 below requires one matrix-vector multiplication (Au, Bv) and one corresponding adjoint operation $(A^\top\lambda, B^\top\lambda)$. Hence, the per-iteration complexity of (SAMA) and the standard AMA (4.46) are essentially the same. Finally, we can combine the main steps (SAMA), (4.16), (4.22), and the update rule (4.21) to complete the smoothing alternating minimization algorithm (SAMA) in Algorithm 1.

Algorithm 1 Smoothing alternating minimization algorithm (SAMA)

Initialization:
1: Fix $\bar{u}_c \in \mathrm{dom}\,(g)$. Choose $\hat{\lambda}^0 \in \mathbb{R}^n$ and $\gamma_1 > 0$.
2: Set $\eta_0 := \frac{\gamma_1}{2\|A\|^2}$ and $\beta_1 := \frac{27\|A\|^2}{20\gamma_1}$.
3: Compute $\bar{u}^1 := \mathrm{prox}_{\gamma_1^{-1}g}\big(\bar{u}_c + \gamma_1^{-1}A^\top\hat{\lambda}^0\big)$.
4: Solve $\bar{v}^1 := \arg\min_v \big\{h(v) - \langle\hat{\lambda}^0, Bv\rangle + \frac{\eta_0}{2}\|A\bar{u}^1 + Bv - c\|^2\big\}$.
5: Update $\bar{\lambda}^1 := \hat{\lambda}^0 - \eta_0(A\bar{u}^1 + B\bar{v}^1 - c)$ and $\lambda_1^* := \beta_1^{-1}(c - A\bar{u}^1 - B\bar{v}^1)$.

Iteration: For $k = 1$ to $k_{\max}$, perform:
6: Compute $\tau_k := \frac{3}{k+4}$, $\gamma_{k+1} := \frac{5\gamma_1}{k+5}$, $\beta_k := \frac{18\|A\|^2(k+5)}{5\gamma_1(k+1)(k+7)}$ and $\eta_k := \frac{5\gamma_1}{2\|A\|^2(k+5)}$.
7: Set $\hat{\lambda}^k := (1 - \tau_k)\bar{\lambda}^k + \tau_k\lambda_k^*$.
8: Compute $\hat{u}^{k+1} := \mathrm{prox}_{\gamma_{k+1}^{-1}g}\big(\bar{u}_c + \gamma_{k+1}^{-1}A^\top\hat{\lambda}^k\big)$.
9: Solve $\hat{v}^{k+1} := \arg\min_v \big\{h(v) - \langle\hat{\lambda}^k, Bv\rangle + \frac{\eta_k}{2}\|A\hat{u}^{k+1} + Bv - c\|^2\big\}$.
10: Update $\bar{\lambda}^{k+1} := \hat{\lambda}^k - \eta_k(A\hat{u}^{k+1} + B\hat{v}^{k+1} - c)$.
11: Compute $\lambda_{k+1}^* := \beta_{k+1}^{-1}\big[(1 - \tau_k)\beta_k\lambda_k^* + \tau_k\eta_k^{-1}(\bar{\lambda}^{k+1} - \hat{\lambda}^k)\big]$.
12: Update $\bar{u}^{k+1} := (1 - \tau_k)\bar{u}^k + \tau_k\hat{u}^{k+1}$ and $\bar{v}^{k+1} := (1 - \tau_k)\bar{v}^k + \tau_k\hat{v}^{k+1}$.
End for

We can view Algorithm 1 as a primal-dual method, where we apply Nesterov's accelerated method to the smoothed dual problem while using a weighted averaging scheme $\bar{x}^k = \left(\sum_{i=0}^k \omega_i \right)^{-1} \sum_{i=0}^k \omega_i \hat{x}^i$ for the primal variables. However, Algorithm 1 aims at solving the nonsmooth problem (4.1) without any additional assumption on g and h except for the finiteness of D_f in (4.9).

4.4.5 Convergence Analysis

We prove in section "Proof of Theorem 1: Convergence of Algorithm 1" in Appendix the convergence and the worst-case iteration-complexity of Algorithm 1 in Theorem 1.

Theorem 1 *Assume that $b_\mathcal{U}$ is chosen as $b_\mathcal{U}(\cdot, \bar{u}_c) := \frac{1}{2} \| \cdot - \bar{u}_c \|^2$ for any fixed $\bar{u}_c \in \mathrm{dom}\,(g)$. Let $\{\bar{w}^k\}$ be the sequence generated by Algorithm 1. Then, for any $\gamma_1 > 0$, the following estimates hold*

$$
\begin{cases}
f(\bar{x}^k) - f^\star & \leq \frac{5\gamma_1}{(k+4)} \left(\frac{\|\bar{u}^c - u^\star\|^2}{2} + \frac{9D_f^2}{8\|A\|^2(k+3)} \right), \\[2mm]
\|A\bar{u}^k + B\bar{v}^k - c\| & \leq \frac{36\|A\|^2\|\lambda^\star\|}{5\gamma_1(k+1)} + \frac{6\|A\|}{(k+1)} \sqrt{\frac{\|\bar{u}^c - u^\star\|^2}{2} + \frac{9D_f^2}{8\|A\|^2(k+7)}}, \\[2mm]
d(\bar{\lambda}^k) - d^\star & \leq \frac{36\|A\|^2\|\lambda^\star\|^2}{5\gamma_1(k+1)} + \frac{6\|A\|\|\lambda^\star\|}{(k+1)} \sqrt{\frac{\|\bar{u}^c - u^\star\|^2}{2} + \frac{9D_f^2}{8\|A\|^2(k+7)}} \\[2mm]
& + \frac{5\gamma_1}{(k+4)} \left(\frac{\|\bar{u}^c - u^\star\|^2}{2} + \frac{9D_f^2}{8\|A\|^2(k+3)} \right),
\end{cases}
\tag{4.23}
$$

where D_f are defined by (4.9). As a consequence, if we choose $\gamma_1 := \|A\|$, then the worst-case iteration-complexity of Algorithm 1 to achieve an ε-primal-dual solution $(\bar{x}^k, \bar{\lambda}^k)$ of (4.1) and (4.2) in the sense of Definition 1 is $\mathcal{O}\left(\varepsilon^{-1}\right)$.

Theorem 1 shows that the convergence rate of Algorithm 1 consists of two parts. While the first part depends on $\|\bar{u}^c - u^\star\|^2$ which is only $\mathcal{O}(1/k)$, the second part depending on D_f is up to $\mathcal{O}(1/k^2)$. We can obtain the convergence rate of the feasibility gap $\|A\bar{u}^k + B\bar{v}^k - c\|$ from the dual convergence as done in [17]. However, this rate is only $\mathcal{O}(1/\sqrt{k})$ when the rate on the dual objective residual $d(\bar{\lambda}^k) - d^\star$ is $\mathcal{O}(1/k)$.

Remark 1 If Assumption 2 fails to hold, then artificial constraints $\|u\| \leq R$ and/or $\|v\| \leq R$ must be added to (4.1). Since Algorithm 1 does not require R as an input, we can estimate R after we terminate this algorithm. Theoretically, the sequence $\{(\bar{u}^k, \bar{v}^k)\}$ generated by Algorithm 1 converges to $x^\star = (u^\star, v^\star)$ a solution of (4.1). Hence, by Lemma 1, R can roughly be estimated as $R > \sup_k \{\|\bar{u}^k\|, \|\bar{v}^k\|\}$. Note that, in this case, the objective function of the subproblems in u and v from (SAMA) is also changed from g to $g + \delta_{\mathbb{B}_R}$, and from h to $h + \delta_{\mathbb{B}_R}$, respectively. Practically, by assuming that R is sufficiently large so that $\|u\| \leq R$ and $\|v\| \leq R$ are inactive, we can discard the term $\delta_{\mathbb{B}_R}(u)$, and $\delta_{\mathbb{B}_R}(v)$. Therefore, the computation of $\hat{u}^{k+1}$ and $\hat{v}^{k+1}$ at Step 8 and Step 9, respectively, of Algorithm 1 is unchanged.

4.4.6 Special Case: g is Strongly Convex

We now consider a special case of the constrained problem (4.1) when g is strongly convex. If g is strongly convex with the convexity parameter $\mu_g > 0$, then we can modify Algorithm 1 so that $d(\bar{\lambda}^k) - d^\star \leq \mathcal{O}(\frac{1}{k^2})$ in terms of the dual objective function as shown in [17]. However, the convergence rate in terms of the primal objective residual $f(\bar{x}^k) - f^\star$ and the primal feasibility gap $\|A\bar{u}^k + B\bar{v}^k - c\|$ we can prove is worse than $\mathcal{O}(\frac{1}{k^2})$.

Let us consider again the dual function φ defined by (4.5). Since g is strongly convex with the strong convexity parameter $\mu_g > 0$, $\nabla\varphi$ is Lipschitz continuous with the Lipschitz constant $L_\varphi := \frac{\|A\|^2}{\mu_g}$. We modify Algorithm 1 in order to obtain a new variant that captures the strong convexity of g and removes the smoothness parameter γ_k. By a similar analysis as in Lemma 3, we can show in section "Proof of Corollary 1: Strong Convexity of g" in Appendix that if the following conditions hold

$$\beta_{k+1} \geq (1 - \tau_k)\beta_k \quad \text{and} \quad \eta_k\left(\frac{3}{2} + \tau_k - \frac{\|A\|^2\eta_k}{\mu_g}\right) \geq \frac{\tau_k^2}{(1 - \tau_k)\beta_k}, \tag{4.24}$$

then

$$G_{\beta_{k+1}}(\bar{w}^{k+1}) \leq (1 - \tau_k)G_{\beta_k}(\bar{w}^k) + \frac{\tau_k^2\eta_k D_f^2}{4}, \tag{4.25}$$

where $G_{\beta_k}(\bar{w}^k) := f_{\beta_k}(\bar{x}^k) + d(\bar{\lambda}^k)$. The first iterate $\bar{u}^1$ in (4.19) can be computed as

$$\bar{u}^1 := \underset{u \in \text{dom}(f)}{\text{argmin}} \left\{g(u) + \langle\hat{\lambda}^0, Au\rangle\right\}. \tag{4.26}$$

Using (4.26) and new update rules for the parameters in Algorithm 1, we obtain a new variant of Algorithm 1. The following corollary shows the convergence of this variant, whose proof is also moved to section "Proof of Corollary 1: Strong Convexity of g" in Appendix.

Corollary 1 *Let $\{\bar{w}^k\}$ be the sequence generated by Algorithm 1 using (4.26) and the update rules*

$$\tau_k := \frac{3}{k+4}, \quad \eta_k := \frac{\mu_g}{2\|A\|^2}, \quad \text{and} \quad \beta_k := \frac{2\|A\|^2\tau_k^2}{\mu_g(1 - \tau_k^2)} = \frac{18\|A\|^2}{\mu_g(k+1)(k+7)}. \tag{4.27}$$

Then, the following estimates hold

$$
\begin{cases}
f(\bar{x}^k) - f^\star & \leq \dfrac{9\mu_g D_f^2}{16\|A\|^2(k+3)} = \mathcal{O}\left(\dfrac{1}{k}\right), \\[2ex]
\|A\bar{u}^k + B\bar{v}^k - c\| & \leq \dfrac{36\|A\|^2\|\lambda^\star\|}{\mu_g(k+1)(k+7)} + \dfrac{9D_f}{2\sqrt{(k+1)(k+3)(k+7)}} = \mathcal{O}\left(\dfrac{1}{k^{3/2}}\right).
\end{cases}
\tag{4.28}
$$

Alternatively, if we use the following update rules in Algorithm 1

$$
\tau_k := \frac{3}{k+4}, \quad \eta_k := \frac{\mu_g \tau_k}{\|A\|^2}, \quad \text{and} \quad \beta_k := \frac{2\|A\|^2 \tau_k}{3\mu_g(1-\tau_k)} = \frac{2\|A\|^2}{\mu_g(k+1)},
\tag{4.29}
$$

then

$$
\begin{cases}
f(\bar{x}^k) - f^\star & \leq \dfrac{27\mu_g D_f^2}{4\|A\|^2(k+3)^2} = \mathcal{O}\left(\dfrac{1}{k^2}\right), \\[2ex]
\|A\bar{u}^k + B\bar{v}^k - c\| & \leq \dfrac{4\|A\|^2\|\lambda^\star\|}{\mu_g(k+1)} + \dfrac{3\sqrt{3}}{(k+3)}\dfrac{D_f}{\sqrt{k+1}} = \mathcal{O}\left(\dfrac{1}{k}\right).
\end{cases}
\tag{4.30}
$$

Here, D_f is defined by (4.9). In both cases, the guarantee of the primal-dual gap function $G(\bar{w}^k) := f(\bar{x}^k) + d(\bar{y}^k)$ is

$$
G(\bar{w}^k) + \frac{1}{2\beta_k}\|A\bar{u}^k + B\bar{v}^k - c\|^2 \leq \frac{9\mu_g D_f^2}{4\|A\|^2(k+3)},
\tag{4.31}
$$

where β_k is given by either (4.27) or (4.29).

We note that, similar to [17], if we modify Step 11 of Algorithm 1 by $\lambda_{k+1}^\star := \lambda_k^\star + \frac{1}{\tau_k}(\bar{\lambda}^{k+1} - \hat{\lambda}^k)$, then we can prove the $\mathcal{O}(\frac{1}{k^2})$-convergence rate for the dual objective residual $d(\lambda^k) - d^\star$ in Algorithm 1 under the strong convexity of g.

4.4.7 Composite Convex Minimization with Linear Operators

A common composite convex minimization formulation in image processing and machine learning [2] is the following problem:

$$
\min_{u \in \mathbb{R}_1^p} \{f(u) := g(u) + h(Fu - y)\},
\tag{4.32}
$$

where g and h are two proper, closed and convex functions (possibly nonsmooth), F is a linear operator from $\mathbb{R}^{p_1}$ to $\mathbb{R}^n$, and $y \in \mathbb{R}^n$ is a given observation vector. We are more interested in the case that g and h are nonsmooth but are equipped with a tractable proximal operator. For example, g and h are both the ℓ_1-norm.

Classical AMA and ADMM methods can solve (4.32) but do not have an $\mathcal{O}(1/k)$ - theoretical convergence rate guarantee without additional smoothness-type, properly proximal terms, or strong convexity-type assumption on g and h. In addition, the ADMM still requires to solve the subproblem at the second line of (4.44) iteratively when F is not orthogonal.

If we introduce a new variable $v := Fu - y$, then we can reformulate (4.32) into (4.1) with $A = F$ and $B = -\mathbb{I}$. In this case, we can apply both Algorithms 1 and 2 (in Sect. 4.5) to solve the resulting problem without additional assumption on g and h except for the boundedness of D_f. However, we only focus on Algorithm 1, which only requires the proximal operator of g and h. The main step of this algorithmic variant can be written explicitly as

$$
\begin{cases}
\hat{u}^{k+1} := \text{prox}_{\gamma_{k+1}^{-1}g}\left(\bar{u}_c + \gamma_{k+1}^{-1}F^\top\hat{\lambda}^k\right), \\
\hat{v}^{k+1} := \text{prox}_{\eta_k^{-1}h}\left(F\hat{u}^{k+1} - y - \eta_k^{-1}\hat{\lambda}^k\right).
\end{cases}
$$

Substituting this step into Algorithm 1, we obtain a new variant for solving (4.32) using only the proximal operator of g and h, and matrix-vector multiplications.

4.5 The New Smoothing ADMM Method

For completeness, we present a new alternating direction method of multipliers (ADMM) algorithm for solving (4.1) by applying Douglas-Rachford splitting method to the smoothed dual problem. Our new algorithm, dubbed the smoothing ADMM (SADMM), features similar optimal convergence rate guarantees as SAMA. See Sect. 4.7 for further discussion.

4.5.1 The Main Steps of the Smoothing ADMM Method

The main step of our SADMM scheme is as follows. Given $\hat{\lambda}^k \in \mathbb{R}^n$, $\hat{v}^k \in \text{dom}(h)$ and the parameters $\gamma_{k+1} > 0$, $\rho_k > 0$ and $\eta_k > 0$, we compute $(\hat{u}^{k+1}, \hat{v}^{k+1}, \bar{\lambda}^{k+1})$ as follows:

$$
\begin{cases}
\hat{u}^{k+1} := \underset{u \in \text{dom}(g)}{\text{argmin}}\left\{g(u; \gamma_{k+1}) - \langle A^\top\hat{\lambda}^k, u\rangle + \dfrac{\rho_k}{2}\|Au + B\hat{v}^k - c\|^2\right\}, \\
\hat{v}^{k+1} := \underset{v \in \text{dom}(h)}{\text{arg min}}\left\{h(v) - \langle B^\top\hat{\lambda}^k, v\rangle + \dfrac{\eta_k}{2}\|A\hat{u}^{k+1} + Bv - c\|^2\right\}, \quad \text{(SADMM)} \\
\bar{\lambda}^{k+1} := \hat{\lambda}^k - \eta_k\left(A\hat{u}^{k+1} + B\hat{v}^{k+1} - c\right),
\end{cases}
$$

where $g(u; \gamma) := g(u) + \gamma b_{\mathcal{U}}(u, \bar{u}^c)$. This scheme is different from the standard ADMM scheme (4.44) at two points. First, $\hat{u}^{k+1}$ is computed from the regularized subproblem with $g(\cdot; \gamma)$ instead of g. Second, we use different penalty parameters ρ_k and η_k compared to the standard ADMM scheme (4.44) in Sect. 4.7. The complexity of computing $\hat{u}^{k+1}$ in (SADMM) is essentially the same as computing u^{k+1} in the standard ADMM scheme (4.44) below.

As a special case, if $A = \mathbb{I}$, the identity operator, or A is orthonormal, then we can choose $b_{\mathcal{U}}(\cdot, \bar{u}^c) = (1/2)\| \cdot - \bar{u}^c \|^2$ to obtain a closed form solution of $\hat{u}^{k+1}$ as

$$\hat{u}^{k+1} := \mathrm{prox}_{(\rho_k+\gamma_{k+1})^{-1}g}\left((\rho_k + \gamma_{k+1})^{-1}\left(\gamma_{k+1}\bar{u}^c + A^\top(\hat{\lambda}^k - \rho_k(B\bar{v}^k - c))\right)\right).$$

In addition to (SADMM), our algorithm also requires additional steps

$$\begin{cases} \hat{\lambda}_k & := (1 - \tau_k)\bar{\lambda}^k + \tau_k\lambda_k^*, & \text{(dual acceleration)} \\ (\bar{u}^{k+1}, \bar{v}^{k+1}) & := (1 - \tau_k)(\bar{u}^k, \bar{v}^k) + \tau_k(\hat{u}^{k+1}, \hat{v}^{k+1}), & \text{(weighted averaging)} \end{cases} \quad (4.33)$$

as in Algorithm 1, where $\lambda_k^* := \beta_k^{-1}(c - A\bar{u}^k - B\bar{v}^k)$, and $\tau_k \in (0, 1)$ is a step size.

We prove in section "Proof of Lemma 6: Gap Reduction Condition" in Appendix the following lemma, which provides conditions on the parameters to guarantee the gap reduction condition.

Lemma 6 Let $\{\bar{w}^k\}$ with $\bar{w}^k := (\bar{u}^k, \bar{v}^k, \bar{\lambda}^k)$ be the sequence generated by (SADMM)–(4.33). If $\tau_k \in (0, 1)$ and $\gamma_k, \beta_k, \rho_k, \eta_k \in \mathbb{R}_{++}$ satisfy

$$\begin{cases} (1 - \tau_k)(1 + 2\tau_k)\eta_k\beta_k \geq 2\tau_k^2, & \gamma_{k+1} \geq \left(\dfrac{3-2\tau_k}{3-(2-L_b^{-1})\tau_k}\right)\gamma_k, \\ \beta_{k+1} \geq (1 - \tau_k)\beta_k, \text{ and } \gamma_{k+1} \geq \|A\|^2\left(\eta_k + \dfrac{\rho_k}{\tau_k}\right), \end{cases} \quad (4.34)$$

then the following non-monotone gap reduction condition holds

$$G_{\gamma_{k+1}\beta_{k+1}}(\bar{w}^{k+1}) \leq (1 - \tau_k)G_{\gamma_k\beta_k}(\bar{w}^k) + \left(\frac{\tau_k^2\eta_k}{4} + \frac{\tau_k\rho_k}{2}\right)D_f^2, \quad (4.35)$$

where $G_{\gamma_k\beta_k}$ is defined by (4.13), and D_f is defined by (4.9).

4.5.2   Updating Parameters

The second step of our algorithmic design is to derive an update rule for the parameters to satisfy the conditions (4.34). Lemma 7 shows **one possibility** to update these parameters, whose proof is given in section "Proof of Lemma 7: Parameter Updates" in Appendix.

Lemma 7 *Let $b_{\mathcal{U}}$ be chosen such that $b_{\mathcal{U}}(\cdot, \bar{u}_c) := \frac{1}{2}\| \cdot -\bar{u}_c\|^2$ for a fixed $\bar{u}_c \in$ dom (g), and $\gamma_1 > 0$. Then, for $k \geq 1$, τ_k, γ_k, β_k, ρ_k, and η_k updated by*

$$
\begin{aligned}
\tau_k &:= \frac{3}{k+4}, & \gamma_k &:= \frac{3\gamma_1}{k+2}, & \beta_k &:= \frac{6\|A\|^2(k+3)}{\gamma_1(k+1)(k+10)}, \\
\rho_k &:= \frac{9\gamma_1}{2\|A\|^2(k+3)(k+4)}, & \eta_k &:= \frac{3\gamma_1}{2\|A\|^2(k+3)},
\end{aligned}
\tag{4.36}
$$

satisfy (4.34). Moreover, $\beta_k \leq \frac{9\|A\|^2}{5\gamma_1(k+1)}$, and the convergence rate of $\{\tau_k\}$ is optimal.

We note that we have freedom to choose γ_1 in order to trade-off the upper-bound of the primal objective residual $f(\bar{x}^k) - f^\star$ and the primal feasibility gap $\|A\bar{u}^k + B\bar{v}^k - c\|$ as in Algorithm 1.

4.5.3 The Smoothing ADMM Algorithm

Similar to Algorithm 1, we can combine the third line of (SADMM) and the second line of (4.33) to update $\lambda_k^\star$. In this case, the arithmetic cost-per-iteration of Algorithm 2 is essentially the same as in the standard ADMM scheme (4.44). We also use $\bar{w}^1 = (\bar{u}^1, \bar{v}^1, \bar{\lambda}^1)$ computed by (4.19) at the first iteration. By putting (4.19), (4.36), (SADMM), (4.33) and (4.22) together, we obtain a complete SADMM algorithm as presented in Algorithm 2.

4.5.4 Convergence Analysis

The following theorem with its proof being in section "Proof of Theorem 2: Convergence of Algorithm 2" in Appendix shows the worst-case iteration-complexity of Algorithm 2.

Theorem 2 *Assume that $b_{\mathcal{U}}$ is chosen as $b_{\mathcal{U}}(\cdot, \bar{u}_c) := \frac{1}{2}\| \cdot -\bar{u}_c\|^2$ for a fixed $\bar{u}_c \in$ dom (g). Let $\{(\bar{u}^k, \bar{v}^k, \bar{\lambda}^k)\}$ be the sequence generated by Algorithm 2. Then the following estimates hold*

$$
\begin{cases}
f(\bar{x}^k) - f^\star \leq \dfrac{3\gamma_1}{(k+2)}\left[\dfrac{\|\bar{u}^c - u^\star\|^2}{2} + \dfrac{27D_f^2}{8\|A\|^2(k+3)}\right], \\[4mm]
\|A\bar{u}^k + B\bar{v}^k - c\| \leq \dfrac{18\|A\|^2\|\lambda^\star\|}{5\gamma_1(k+1)} + \dfrac{6\|A\|}{(k+1)}\sqrt{\|\bar{u}^c - u^\star\|^2 + \dfrac{27D_f^2}{8\|A\|^2(k+10)}},
\end{cases}
\tag{4.37}
$$

where D_f is given by (4.9). If $\gamma_1 := \|A\|$, then the worst-case iteration-complexity of Algorithm 2 to achieve an ε—solution $\bar{x}^k$ of (4.1) is $\mathcal{O}\left(\varepsilon^{-1}\right)$.

As can be seen from Theorem 2, the term $\frac{6\|A\|}{(k+1)}\left(\frac{\|\bar{u}^c - u^\star\|^2}{2} + \frac{27D_f^2}{8\|A\|^2(k+10)}\right)^{1/2}$ in (4.37) does not depend on the choice of γ_1. If we decrease γ_1, then the upper

---

Algorithm 2 Smoothing alternating direction method of multipliers (SADMM)

Initialization:

1: Fix $\bar{u}_c \in \mathrm{dom}\,(g)$. Choose $\hat{\lambda}^0 \in \mathbb{R}^n$ and $\gamma_1 > 0$.

2: Set $\eta_0 := \frac{\gamma_1}{2\|A\|^2}$ and $\beta_1 := \frac{12\|A\|^2}{11\gamma_1}$.

3: Compute $\bar{u}^1 := \mathrm{prox}_{\gamma_1^{-1}g}\big(\bar{u}_c + \gamma_1^{-1}A^\top\hat{\lambda}^0\big)$.

4: Solve $\bar{v}^1 := \arg\min_v \big\{ h(v) - \langle \hat{\lambda}^0, Bv \rangle + \frac{\eta_0}{2}\|A\bar{u}^1 + Bv - c\|^2 \big\}$. Set $\hat{v}^1 := \bar{v}^1$.

5: Update $\bar{\lambda}^1 := \hat{\lambda}^0 - \eta_0(A\bar{u}^1 + B\bar{v}^1 - c)$ and $\lambda_1^* := \beta_1^{-1}(c - A\bar{u}^1 - B\bar{v}^1)$.

Iteration: For $k = 1$ to $k_{\max}$, perform:

6: Compute $\tau_k := \frac{3}{k+4}$, $\gamma_{k+1} := \frac{3\gamma_1}{k+3}$, $\beta_k := \frac{6\|A\|^2(k+3)}{\gamma_1(k+1)(k+10)}$. Then, set $\eta_k := \frac{3\gamma_1}{2\|A\|^2(k+3)}$ and $\rho_k := \frac{9\gamma_1}{2\|A\|^2(k+3)(k+4)}$.

7: Set $\hat{\lambda}^k := (1 - \tau_k)\bar{\lambda}^k + \tau_k\lambda_k^*$.

8: Solve $\hat{u}^{k+1} := \arg\min_u\big\{ g(u) - \langle \hat{\lambda}^k, Au \rangle + \frac{\rho_k}{2}\|Au + B\hat{v}^k - c\|^2 + \gamma_{k+1}b_{\mathcal{U}}(u, \bar{u}^c) \big\}$.

9: Solve $\hat{v}^{k+1} := \arg\min_v \big\{ h(v) - \langle \hat{\lambda}^k, Bv \rangle + \frac{\eta_k}{2}\|A\hat{u}^{k+1} + Bv - c\|^2 \big\}$.

10: Update $\bar{\lambda}^{k+1} := \hat{\lambda}^k - \eta_k(A\hat{u}^{k+1} + B\hat{v}^{k+1} - c)$.

11: Compute $\lambda_{k+1}^* := \beta_{k+1}^{-1}\big[(1 - \tau_k)\beta_k\lambda_k^* + \tau_k\eta_k^{-1}(\bar{\lambda}^{k+1} - \hat{\lambda}^k)\big]$.

12: Update $\bar{u}^{k+1} := (1 - \tau_k)\bar{u}^k + \tau_k\hat{u}^{k+1}$ and $\bar{v}^{k+1} := (1 - \tau_k)\bar{v}^k + \tau_k\hat{v}^{k+1}$.

End for

---

bound of $f(\bar{x}^k) - f^\star$ decreases, while the upper bound of $\|A\bar{u}^k + B\bar{v}^k - c\|$ increases, and vice versa. Hence, γ_1 trades off these worse-case bounds. The convergence rate guarantee on the dual objective residual can be easily obtained from the last bound of (4.15).

4.5.5 SAMA vs. SADMM

There are at least two cases, where SAMA theoretically gains advantages over SADMM. First, if A is non-orthogonal. The u-subproblem in (SAMA) can be computed by using prox_g, while in SADMM, the nonorthogonal operator A prevents us from using prox_g. Second, if g is block separable, i.e., $g(u) := \sum_{i=1}^s g_i(u_i)$, then we can choose $g(u; \gamma) := \sum_{i=1}^s \big[g_i(u_i) + \frac{\gamma}{2}\|u_i - \bar{u}_i^c\|^2 \big]$, which can be evaluated in parallel. This is not preserved in SADMM. Indeed, for SADMM, the subproblem in u still has the quadratic term $\frac{\rho_k}{2}\|Au + B\hat{v}^k - c\|^2$, which makes it nonseparable even if g is separable.

4.6 Numerical Evidence

We illustrate a "geometric invariant" property of Algorithms 1 and 2 for solving the distance minimization problem (4.39). This problem is classical but solving it efficiently remains an interesting research topic. Various algorithms have been proposed including Douglas-Rachford (DR) splitting, Dykstra's projection, and Hauzageau's method [1, 2]. In this section, we compare our algorithms with these methods.

We consider the following convex feasibility problem with two convex sets:

$$\text{Find } \lambda^\star \text{ such that: } \lambda^\star \in \mathcal{C}_1 \cap \mathcal{C}_2, \tag{4.38}$$

where $\mathcal{C}_1$ and $\mathcal{C}_2$ are two nonempty, closed, and convex sets in $\mathbb{R}^p$. Problem (4.38) may not have solution. Hence, instead of solving (4.38), we consider a problem of finding the best substitution for a point in the intersection $\mathcal{C}_1 \cap \mathcal{C}_2$ even if it is empty. Such a problem can be formulated as

$$d^\star := \min_{\lambda \in \mathbb{R}^n} \left\{ d(\lambda) := d_{\mathcal{C}_1}(\lambda) + d_{\mathcal{C}_2}(\lambda) \right\}, \tag{4.39}$$

where $d_\mathcal{C}$ is the Euclidean distance to the set $\mathcal{C}$. Unlike (4.38), the optimal value d^* of (4.39) is always finite as long as $\mathcal{C}_1$ and $\mathcal{C}_2$ are nonempty. Moreover, $d^\star = \text{dist}(\mathcal{C}_1, \mathcal{C}_2)$, the distance between $\mathcal{C}_1$ and $\mathcal{C}_2$. Hence, if $\mathcal{C}_1 \cap \mathcal{C}_2 \neq \emptyset$, then $d^\star = 0$, see, e.g., [6].

According to [6], our primal template (4.1) for (4.39) then takes the following form

$$\min_{u,v} \left\{ s_{\mathcal{C}_1}(u) + s_{\mathcal{C}_2}(v) \mid u + v = 0, u \in \mathbb{B}_1, v \in \mathbb{B}_1 \right\}, \tag{4.40}$$

where $s_{\mathcal{C}_i}$ is the support function of $\mathcal{C}_i$ for $i = 1, 2$, and $\mathbb{B}_r := \{ w \mid \|w\| \le r \}$ for $r > 0$.

Clearly, (4.40) is fully nonsmooth, since $s_{\mathcal{C}_i}$ is convex and nonsmooth for $i = 1, 2$. In addition, (4.40) satisfies Assumption 2. Here, we can even increase the constraint radius, currently 1, to a sufficiently large number such that the constraints $u, v \in \mathbb{B}_r$ of each subproblems in (4.44), (SAMA) and (SADMM) are inactive without changing the underlying problem. In this particular setting, we can choose the prox-center points for u and v as zero since they actually obtain the optimal solution.

If we apply ADMM to solve (4.40), then it can be written explicitly as

$$\begin{cases} u^{k+1} := \text{prox}_{\rho^{-1} s_{\mathcal{C}_1}}(\lambda^k - v^k) = \lambda^k - v^k - \rho^{-1} \pi_{\mathcal{C}_1}\left(\rho(\lambda^k - v^k) \right), \\ v^{k+1} := \text{prox}_{\rho^{-1} s_{\mathcal{C}_2}}(\lambda^k - u^{k+1}) = \lambda^k - u^{k+1} - \rho^{-1} \pi_{\mathcal{C}_2}\left(\rho(\lambda^k - u^{k+1}) \right), \\ \lambda^{k+1} := \lambda^k - (u^{k+1} + v^{k+1}), \end{cases}$$

where π_{C_i} is the projection onto C_i for $i = 1, 2$, and $\rho > 0$ is the penalty parameter. Clearly, multiplying this expression by ρ and using the same notation, we obtain

$$\begin{cases} u^{k+1} := \lambda^k - v^k - \pi_{C_1}\left(\lambda^k - v^k\right), \\ v^{k+1} := \lambda^k - u^{k+1} - \pi_{C_2}\left(\lambda^k - u^{k+1}\right), \\ \lambda^{k+1} := \lambda^k - (u^{k+1} + v^{k+1}), \end{cases} \tag{4.41}$$

which shows that this scheme is independent of any parameter ρ. With an elementary transformation, we can write (4.41) as a Douglas-Rachford (DR) splitting scheme

$$\begin{cases} z^k := z^{k-1} + \pi_{C_1}\left(2\lambda^k - z^{k-1}\right) - \lambda^k, \\ \lambda^{k+1} := \pi_{C_2}(z^k). \end{cases} \tag{4.42}$$

To recover u^k and v^k from z^k and λ^k, we can use $u^k := \lambda^{k-1} - z^k$ and $v^k := z^{k-1} - \lambda^k$.

Now, if we apply our SAMA to solve (4.40) using $b_{\mathcal{U}}(u, \bar{u}^c) := (1/2)\|u - \bar{u}^c\|^2$, the two main steps of SAMA becomes

$$\begin{cases} \hat{u}^{k+1} := \text{prox}_{\gamma_{k+1}^{-1} s_{C_1}} (\bar{u}^c + \gamma_{k+1}^{-1}\hat{\lambda}^k) = \gamma_{k+1}^{-1}\hat{\lambda}^k + \bar{u}^c - \gamma_{k+1}^{-1}\pi_{C_1}\left(\hat{\lambda}^k + \gamma_{k+1}\bar{u}^c\right), \\ \hat{v}^{k+1} := \text{prox}_{\eta_k^{-1} s_{C_2}} (\eta_k^{-1}\hat{\lambda}^k - \hat{u}^{k+1}) = \eta_k^{-1}\hat{\lambda}^k - \hat{u}^{k+1} - \eta_k^{-1}\pi_{C_2}\left(\hat{\lambda}^k - \eta_k\hat{u}^{k+1}\right). \end{cases} \tag{4.43}$$

Clearly, the standard AMA is not applicable to solve (4.40) due to the lack of strong convexity. The standard ADMM applying to (4.40) becomes the alternative projection scheme (4.42) for solving (4.38). This scheme can be arbitrarily slow if the geometry between two sets C_1 and C_2 is ill-posed (see below).

To observe an interesting convergence behavior, we test Dykstra's projection, Hauzageau's method, and the ADMM (4.41) (or its DR form (4.42)), and compare them with our algorithms in the following configuration.

We first choose $C_i := \{u \in \mathbb{R}^n \mid \langle \mathbf{a}_i, u \rangle \leq b_i\}$ for $i = 1, 2$ as two half-planes in $\mathbb{R}^n$, where $b_1 = b_2 = 0$. Here, the normal vectors are $\mathbf{a}_1 := (\epsilon, \cdots, \epsilon, -1, \cdots, -1)^\top$, and $\mathbf{a}_2 := (0, \cdots, 0, 1, \cdots, 1)^\top$, where $\epsilon > 0$ is a positive angle. The tangent angle ϵ is repeated $\lfloor n/2 \rfloor$ times in $\mathbf{a}_1$, and the zero is repeated $\lfloor n/2 \rfloor$ times in $\mathbf{a}_2$, where $n = 1000$. The starting point is chosen as $u^0 := (1, \cdots, 1)^\top$. By varying ϵ, we can observe the convergence behavior of these five methods.

We note that Dykstra's and Hauzageau's algorithms directly solve the dual problem (4.39), while our methods and ADMM solve both the primal and dual problems (4.40) and (4.39). We compare these algorithms on the absolute dual objective residual $d(\lambda) - d^\star$ of (4.39).

Figure 4.1 shows the convergence of five algorithms with different choices of ϵ.

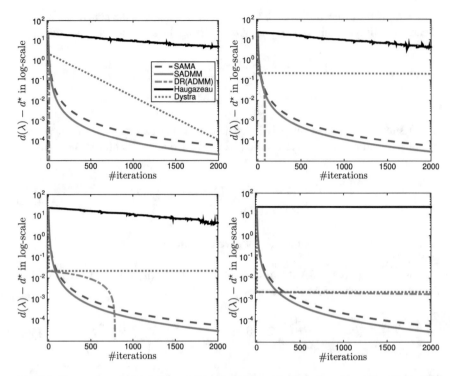

Fig. 4.1 The convergence behavior of five algorithms with different values of ϵ. These plots correspond to $\epsilon = 10^{-1}$ (left-top), 10^{-2} (right-top), 10^{-3} (left-bottom) or 10^{-4} (right-bottom)

We observe Hauzageau's and Dykstra's methods are slow, but Hauzageau's method is extremely slow. The speed of ADMM (or DR splitting) strongly depends on the geometry of the sets, in particular, the tangent angle between two sets. For large values of ϵ, these methods work well, but they become arbitrarily slow when ϵ is decreasing. The objective value of this method drops quickly to a certain level, then is saturated, and makes a very slow progress toward to the optimal value as seen in Fig. 4.1. Since the ADMM scheme (4.41) is independent of its penalty parameter, this is the best performance we can achieve for solving (4.39). Both SAMA and SADMM have almost identical convergence rate for different values of ϵ. These convergence rate reflects the theoretical guarantee, which is $\mathcal{O}(1/k)$ as predicted by our theoretical results.

4.7 Discussion

We have developed a rigorous alternating direction optimization framework for solving constrained convex optimization problems. Our approach is built upon the model-based gap reduction (MGR) technique in [35], and unifies five main ideas: smoothing, gap reduction, alternating direction, acceleration/averaging, and homotopy. By splitting the gap, we have developed two new smooth alternating optimization algorithms: SAMA and SADMM with rigorous convergence guarantees. One important feature of these methods is a heuristic-free parameter update, which has not been proved yet in the literature for AMA and ADMM as we discuss below:

(a) **Alternating direction method of multipliers** (ADMM). The ADMM algorithm can be viewed as the Douglas-Rachford splitting applied to the optimality condition of the dual problem (4.2). As a result, the standard ADMM algorithm generates a primal sequence $\{(u^k, v^k)\}$ together with a multiplier sequence $\{\lambda^k\}$ as

$$
\begin{cases}
u^{k+1} := \underset{u \in \text{dom}(g)}{\text{argmin}} \left\{ g(u) - \langle \lambda^k, Au \rangle + \frac{\eta_k}{2} \|Au + Bv^k - c\|_2^2 \right\} \\
v^{k+1} := \underset{v \in \text{dom}(h)}{\text{argmin}} \left\{ h(v) - \langle \lambda^k, Bv \rangle + \frac{\eta_k}{2} \|Au^{k+1} + Bv - c\|_2^2 \right\} \\
\lambda^{k+1} := \lambda^k - \eta_k (Au^{k+1} + Bv^{k+1} - c),
\end{cases}
\tag{4.44}
$$

where k denotes the iteration count and $\eta_k > 0$ is a penalty parameter. This basic method is closely related to or equivalent to many other algorithms, such as Spingarn's method of partial inverses, Dykstra's alternating projections, Bregman's iterative algorithms, and can also be motivated from the augmented Lagrangian perspective [5].

The ADMM algorithm serves as a good general-purpose tool for optimization problems arising in the analysis and processing of modern massive datasets. Indeed, its implementations have received a significant amount of engineering effort both in research and in industry. As a result, its global convergence rate characterizations for the template (4.1) is an active research topic, see, e.g., [10–13, 15, 17, 19, 22, 28, 31, 38], and the references quoted therein.

In the constrained setting of (4.1), a global convergence characterization specifically means the following: The algorithm provides us $\bar{x}^k = (\bar{u}^k, \bar{v}^k)$ and we determine the number of iterations k necessary to obtain $f(\bar{x}^k) - f^\star \leq \epsilon_f$ and $\|A\bar{u}^k + B\bar{v}^k - c\| \leq \epsilon_c$ for some fixed accuracy ϵ_f for the objective and for some—possibly another—fixed accuracy ϵ_c for the linear constraint. Separating constraint feasibility is crucial so that the primal convergence has any significance otherwise we can trivially have $f^\star - f(\bar{x}^k) \leq 0$ for some infeasible iterate $\bar{x}^k$.

A key theoretical strategy for obtaining global convergence rates for alternating direction methods is ergodic averaging [10–12, 19, 22, 24, 28, 31, 38]. For instance, as opposed to working with the primal-sequence $x^k := (u^k, v^k)$ from (4.44) directly, we instead choose a sequence of weights $\{\omega_k\} \subset (0, +\infty)$ and then average as follows

$$
\bar{x}^k := \left(\sum_{i=0}^{k} \omega_i \right)^{-1} \sum_{i=0}^{k} \omega_i x^i.
\tag{4.45}
$$

The averaged sequence $\bar{x}^k$ then makes it theoretically elementary to obtain the desired type of convergence rate characterizations for (4.1).

Indeed, existing literature critically relies on such weighting strategies in order to obtain global convergence guarantees. For instance, He and Yuan in [19] prove an $\mathcal{O}(1/k)$-convergence rate of their ADMM scheme (4.44) by using the form (4.45) with $\omega_i := 1$ but for both primal and dual variables x as well as λ simultaneously. They provided their guarantee in terms of a gap function for an associated variational inequality for (4.1) and assumed the boundedness on both primal and dual domains. This result is further extended by other authors to different variants of ADMM, including [18, 34, 39]. The same rate is obtained in [12] for a relaxed ADMM variant with similar assumptions along with a weighting strategy that emphasizes the latter iterations by using $\omega_i := k + 1$ in (4.45).

We should note that there are also weighted global convergence characterizations for ADMM, such as $f(\bar{x}^k) - f^\star + \rho\|A\bar{u}^k + B\bar{v}^k - c\|$ for some fixed $\rho > 0$ by Shefi and Teboulle [31]. The authors added proximal terms to the u- and v-subproblems and imposed conditions on three parameters to achieve the $\mathcal{O}(1/k)$-convergence rate jointly between the objective residual and feasibility gap. Intriguingly, this type of convergence rate guarantee does not necessarily imply the $\mathcal{O}(1/k)$-convergence separately on the primal objective residual and feasibility gap as indicated in [31, Theorem 5.2] without additional assumptions.

Interestingly, making additional assumptions on the template is quite common [12, 14, 16, 17]. For instance, the authors in [28] studied a linearized ADMM variant of (4.44) and proved the $\mathcal{O}(1/k)$-rate separately, but required the Lipschitz gradient assumption on either g or h in (4.1). In addition, the authors in [17] require strong convexity on both g and h. In contrast, the authors [14] require the strong convexity of either g or h but need A or B to be full rank as well. In [39] the authors proposed an asynchronous ADMM and showed the $\mathcal{O}(1/k)$ rate on the averaging sequence for a special case of (4.1) where $h = 0$, which trivially has Lipschitz gradient.

Unsurprisingly, these assumptions again limit the applicability of the algorithmic guarantees when, for instance, g and h are non-Lipschitz gradient loss functions or fully non-smooth regularizers, as in Poisson imaging, robust principal component analysis (RPCA), and graphical model learning [9]. Several recent results rely on other type of assumptions such as error bounds, metric regularity, or the well-known Kurdyka-Lojasiewicz condition [7, 20, 21]. Although these conditions cover a wide range of application models, it is unfortunately very hard to verify some quantities related to these assumptions in practice. Other times, the additional assumptions obviate the ADMM choice as they can allow application of a simpler algorithm:

(b) **Alternating minimization algorithm** (AMA). The AMA algorithm, given below, is guaranteed to converge when g is strongly convex or g^* has Lipschitz gradient [17]:

$$
\begin{cases}
\hat{u}^{k+1} := \underset{u \in \text{dom}(g)}{\text{argmin}} \ \{g(u) - \langle\hat{\lambda}^k, Au\rangle\}, \\[2mm]
\hat{v}^{k+1} := \underset{v \in \text{dom}(h)}{\text{argmin}} \ \{h(v) - \langle\hat{\lambda}^k, Bv\rangle + \dfrac{\eta_k}{2}\|A\hat{u}^{k+1} + Bv - c\|_2^2\}, \\[2mm]
\hat{\lambda}^{k+1} := \hat{\lambda}^k - \eta_k\big(A\hat{u}^{k+1} + B\hat{v}^{k+1} - c\big),
\end{cases} \tag{4.46}
$$

where $\eta_k > 0$ is a penalty parameter.

One can view AMA as the forward-backward splitting algorithm applied to the optimality condition of the dual problem (4.2) (cf., [17, 37]). Alternatively, we can motivate the algorithm by using one Lagrange dual step and one augmented Lagrangian dual step between two blocks of variables u and v [4, 32, 37]. Computationally, (4.46) is arguably easier than (4.44). However, it often requires stronger assumptions than ADMM to guarantee convergence [17, 37]. The most obvious assumption is the strong convexity of g.

Acknowledgements QTD's work was supported in part by the NSF-grant No. DMS-1619884, USA. VC's work was supported by European Research Council (ERC) under the European Union's Horizon 2020 research and innovation programme (grant agreement no 725594—time-data). The authors would like to acknowledge Dr. C.B., Vu, and Dr. V.Q. Nguyen with their help on verifying the technical proofs and the numerical experiment. The authors also thank Mr. Ahmet Alacaoglu, Mr. Nhan Pham, and Ms. Yuzixuan Zhu for their careful proofreading.

Appendix: Proofs of Technical Results

This appendix provides full proofs of technical results presented in the main text.

Proof of Lemma 2: The Primal-Dual Bounds

First, using the fact that $-d(\lambda) \leq -d^\star = f^\star \leq \mathcal{L}(x, \lambda^\star) = f(x) + \langle \lambda^\star, Au + Bv - c \rangle \leq f(x) + \|\lambda^\star\| \|Au + Bv - c\|$, we get

$$-\|\lambda^\star\| \|Au + Bv - c\| \leq f(x) - f^\star \leq f(x) + d(\lambda), \qquad (4.47)$$

which is exactly the lower bound (4.14).

Next, since $A^\top \lambda^\star \in \partial g(u^\star)$ due to (4.8), by Fenchel-Young's inequality, we have $g(u^\star) + g^*(A^\top \lambda^\star) = \langle A^\top \lambda^\star, u^\star \rangle$, which implies $g^*(A^\top \lambda^\star) = \langle A^\top \lambda^\star, u^\star \rangle - g(\hat{u}^\star)$. Using this relation and the definition of φ_γ, we have

$$\varphi_\gamma(\lambda) := \max\left\{ \langle A^\top \lambda, u \rangle - g(u) - \gamma b_\mathcal{U}(u, \bar{u}^c) \right\} \geq \langle A^\top \lambda, u^\star \rangle - g(u^\star) - \gamma b_\mathcal{U}(u^\star, \bar{u}^c)$$

$$= \langle A^\top \lambda^\star, u^\star \rangle - g(u^\star) + \langle A^\top(\lambda - \lambda^\star), u^\star \rangle - \gamma b_\mathcal{U}(u^\star, \bar{u}^c)$$

$$= g^*(A^\top \lambda^\star) + \langle A^\top(\lambda - \lambda^\star), u^\star \rangle - \gamma b_\mathcal{U}(u^\star, \bar{u}^c)$$

$$= \varphi(\lambda^\star) + \langle \lambda - \lambda^\star, Au^\star \rangle - \gamma b_\mathcal{U}(u^\star, \bar{u}^c).$$

Alternatively, we have $\psi(\lambda) \geq \psi(\lambda^\star) + \langle \nabla \psi(\lambda^\star), \lambda - \lambda^\star \rangle$, where $\nabla \psi(\lambda^\star) = B \nabla h^*(B^\top \lambda^\star) - c = Bv^\star - c$ due to the last relation in (4.8), where $\nabla h^*(B^\top \lambda^\star) \in$

$\partial h^*(B^\top \lambda^*)$ is one subgradient of ∂h^*. Hence, $\psi(\lambda) \geq \psi(\lambda^*) + \langle \lambda - \lambda^*, Bv^* - c\rangle$. Adding this inequality to the last estimation with the fact that $d_\gamma = \varphi_\gamma + \psi$ and $d = \varphi + \psi$, we obtain

$$d_\gamma(\lambda) \geq d(\lambda^*) + \langle \lambda - \lambda^*, Au^* + Bv^* - c\rangle - \gamma b_{\mathcal{U}}(u^*, \bar{u}^c) \overset{(4.8)}{=} d^* - \gamma b_{\mathcal{U}}(u^*, \bar{u}^c) \qquad (4.48)$$

Using this inequality with $d^* = -f^*$ and the definition (4.13) of f_β we have

$$f(x) - f^* \overset{(4.13)+(4.48)}{\leq} f_\beta(x) + d_\gamma(\lambda) + \gamma b_{\mathcal{U}}(u^*, \bar{u}^c) - \frac{1}{2\beta}\|Au + Bv - c\|^2 \qquad (4.49)$$

$$= G_{\gamma\beta}(w) + \gamma b_{\mathcal{U}}(u^*, \bar{u}^c) - \frac{1}{2\beta}\|Au + Bv - c\|^2.$$

Let $S := G_{\gamma\beta}(w) + \gamma b_{\mathcal{U}}(u^*, \bar{u}^c)$. Then, by dropping the last term $-\frac{1}{2\beta}\|Au + Bv - c\|^2$ in (4.49), we obtain the first inequality of (4.15).

Let $t := \|Au + Bv - c\|$. Using again (4.47) and (4.49), we can see that $\frac{1}{2\beta}t^2 - \|\lambda^*\|t - S \leq 0$. Solving this quadratic inequation w.r.t. t and noting that $t \geq 0$, we obtain the second bound of (4.15). The last estimate of (4.15) is a direct consequence of (4.49), the first one of (4.15). Finally, from (4.47), we have $f(x) \geq f^* - \|\lambda^*\|\|Au + Bv - c\|$. Substituting this into (4.49) we get $d(\lambda) - d^* - \|\lambda^*\|\|Au + Bv - c\| \leq S - \frac{1}{2\beta}\|Au + Bv - c\|^2$, which implies

$$d(\lambda) - d^* \leq S - (1/(2\beta))\|Au + Bv - c\|^2 + \|\lambda^*\|\|Au + Bv - c\|.$$

By discarding $-(1/(2\beta))\|Au + Bv - c\|^2$ and using the second estimate of (4.15) into the last estimate, we obtain the last inequality of (4.15). $\qquad \square$

Convergence Analysis of Algorithm 1

We provide a full proof of Lemmas and Theorems related to the convergence of Algorithm 1. First, we prove the following key lemma, which will be used to prove Lemma 3.

Lemma 8 *Let $\bar{\lambda}^{k+1}$ be generated by (SAMA). Then*

$$d_{\gamma_{k+1}}(\bar{\lambda}^{k+1}) \leq (1 - \tau_k)d_{\gamma_{k+1}}(\bar{\lambda}^k) + \tau_k \hat{\ell}_{\gamma_{k+1}}(\lambda) + \frac{1}{\eta_k}\langle \bar{\lambda}^{k+1} - \hat{\lambda}^k, (1 - \tau_k)\bar{\lambda}^k + \tau_k\lambda - \hat{\lambda}^k\rangle$$

$$- \left(\frac{1}{\eta_k} - \frac{\|A\|^2}{2\gamma_{k+1}}\right)\|\bar{\lambda}^{k+1} - \hat{\lambda}^k\|^2 - \frac{(1 - \tau_k)\gamma_{k+1}}{2}\|u^*_{\gamma_{k+1}}(A^\top\bar{\lambda}^k) - \hat{u}^{k+1}\|^2, \qquad (4.50)$$

where

$$\hat{\ell}_{\gamma_{k+1}}(\lambda) := \varphi_{\gamma_{k+1}}(\hat{\lambda}^k) + \langle \nabla \varphi_{\gamma_{k+1}}(\hat{\lambda}^k), \lambda - \hat{\lambda}^k \rangle + \psi(\lambda)$$
$$\leq d_{\gamma_{k+1}}(\lambda) - \frac{\gamma_{k+1}}{2} \| u^*_{\gamma_{k+1}}(A^\top \lambda) - \hat{u}^{k+1} \|^2. \tag{4.51}$$

In addition, for any z, γ_k, $\gamma_{k+1} > 0$, *the function* g^*_γ *defined by* (4.11) *satisfies*

$$g^*_{\gamma_{k+1}}(z) \leq g^*_{\gamma_k}(z) + (\gamma_k - \gamma_{k+1}) b_{\mathcal{U}}(u^*_{\gamma_{k+1}}(z), \bar{u}^c). \tag{4.52}$$

Proof First, it is well-known that SAMA is equivalent to the proximal-gradient step applying to the smoothed dual problem

$$\min_\lambda \left\{ \varphi_{\gamma_{k+1}}(\lambda) + \psi(\lambda) : \lambda \in \mathbb{R}^n \right\}.$$

This proximal-gradient step can be presented as

$$\bar{\lambda}^{k+1} := \operatorname{prox}_{\eta_k \psi} \left(\hat{\lambda}^k - \eta_k \nabla \varphi_{\gamma_{k+1}}(\hat{\lambda}^k) \right).$$

We write down the optimality condition of this corresponding minimization problem of this step as

$$0 \in \partial \psi(\bar{\lambda}^{k+1}) + \nabla \varphi_{\gamma_{k+1}}(\hat{\lambda}^k) + \eta_k^{-1}(\bar{\lambda}^{k+1} - \hat{\lambda}^k).$$

Using this condition and the convexity of ψ, for any $\nabla \psi(\bar{\lambda}^{k+1}) \in \partial \psi(\bar{\lambda}^{k+1})$, we have

$$\psi(\bar{\lambda}^{k+1}) \leq \psi(\lambda) + \langle \nabla \psi(\bar{\lambda}^{k+1}), \bar{\lambda}^{k+1} - \lambda \rangle$$
$$= \psi(\lambda) + \langle \nabla \varphi_{\gamma_{k+1}}(\hat{\lambda}^k), \lambda - \bar{\lambda}^{k+1} \rangle + \eta_k^{-1} \langle \bar{\lambda}^{k+1} - \hat{\lambda}^k, \lambda - \bar{\lambda}^{k+1} \rangle. \tag{4.53}$$

Next, by the definition $\varphi_\gamma(\lambda) := g^*_\gamma(A^\top \lambda)$, we can show from (4.11) that $\hat{u}^{k+1} = u^*_{\gamma_{k+1}}(A^\top \hat{\lambda}^k)$. Since g^*_γ is $(1/\gamma)$-Lipschitz gradient continuous, we have

$$\frac{\gamma}{2} \| \nabla g^*_\gamma(z) - \nabla g^*_\gamma(\hat{z}) \|^2 \leq g^*_\gamma(z) - g^*_\gamma(\hat{z}) - \langle \nabla g^*_\gamma(\hat{z}), z - \hat{z} \rangle \leq \frac{1}{2\gamma} \| z - \hat{z} \|^2.$$

Using this inequality with $\gamma := \gamma_{k+1}$, $\nabla g^*_{\gamma_{k+1}}(A^\top \lambda) = u^*_{\gamma_{k+1}}(A^\top \lambda)$, $\nabla g^*_{\gamma_{k+1}}(A^\top \hat{\lambda}^k) = u^*_{\gamma_{k+1}}(A^\top \hat{\lambda}^k) = \hat{u}^{k+1}$, and $\nabla \varphi_{\gamma_{k+1}}(\lambda) = A \nabla g^*_{\gamma_{k+1}}(A^\top \lambda)$, we have

$$\frac{\gamma_{k+1}}{2} \| u^*_{\gamma_{k+1}}(A^\top \lambda) - \hat{u}^{k+1} \|^2 \leq \varphi_{\gamma_{k+1}}(\lambda) - \varphi_{\gamma_{k+1}}(\hat{\lambda}^k) - \langle \nabla \varphi_{\gamma_{k+1}}(\hat{\lambda}^k), \lambda - \hat{\lambda}^k \rangle$$
$$\leq \frac{1}{2\gamma_{k+1}} \| A^\top (\lambda - \hat{\lambda}^k) \|^2 \leq \frac{\|A\|^2}{2\gamma_{k+1}} \| \lambda - \hat{\lambda}^k \|^2. \tag{4.54}$$

Using (4.54) with $\lambda = \bar{\lambda}^{k+1}$, we have

$$\varphi_{\gamma_{k+1}}(\bar{\lambda}^{k+1}) \leq \varphi_{\gamma_{k+1}}(\hat{\lambda}^k) + \langle \nabla \varphi_{\gamma_{k+1}}(\hat{\lambda}^k), \bar{\lambda}^{k+1} - \hat{\lambda}^k \rangle + \frac{\|A\|^2}{2\gamma_{k+1}} \|\bar{\lambda}^{k+1} - \hat{\lambda}^k\|^2.$$

Summing up this inequality and (4.53), then using the definition of $\hat{\ell}_{\gamma_{k+1}}(\lambda)$ in (4.51), we obtain

$$d_{\gamma_{k+1}}(\bar{\lambda}^{k+1}) \leq \hat{\ell}_{\gamma_{k+1}}(\lambda) + \frac{1}{\eta_k}\langle \bar{\lambda}^{k+1} - \hat{\lambda}^k, \lambda - \hat{\lambda}^k \rangle - \left(\frac{1}{\eta_k} - \frac{\|A\|^2}{2\gamma_{k+1}}\right)\|\bar{\lambda}^{k+1} - \hat{\lambda}^k\|^2. \qquad (4.55)$$

Here, the second inequality in (4.51) follows from the right-hand side of (4.54).

Now, using (4.55) with $\lambda := \bar{\lambda}^k$, then combining with (4.51), we get

$$d_{\gamma_{k+1}}(\bar{\lambda}^{k+1}) \leq d_{\gamma_{k+1}}(\bar{\lambda}^k) + \frac{1}{\eta_k}\langle \bar{\lambda}^{k+1} - \hat{\lambda}^k, \bar{\lambda}^k - \hat{\lambda}^k \rangle - \left(\frac{1}{\eta_k} - \frac{\|A\|^2}{2\gamma_{k+1}}\right)\|\bar{\lambda}^{k+1} - \hat{\lambda}^k\|^2$$
$$- \frac{\gamma_{k+1}}{2}\|u_{\gamma_{k+1}}^*(A^\top\bar{\lambda}^k) - \hat{u}^{k+1}\|^2.$$

Multiplying the last inequality by $1 - \tau_k \in [0, 1]$ and (4.55) by $\tau_k \in [0, 1]$, then summing up the results, we obtain (4.50).

Finally, from (4.11), since $g_\gamma^*(z) := \max_u\{P(u, \gamma; z) := \langle z, u \rangle - g(u) - \gamma b_{\mathcal{U}}(u; \bar{u}^c)\}$, is the maximization of P over u indexing in γ and z, which is concave in u and linear in γ, we have $g_\gamma^*(z)$ is convex w.r.t. $\gamma > 0$. Moreover, $\frac{dg_\gamma^*(z)}{d\gamma} = -b_{\mathcal{U}}(u_\gamma^*(z), \bar{u}^c)$. Hence, using the convexity of g_γ^* w.r.t. $\gamma > 0$, we have $g_{\gamma_k}^*(z) \geq g_{\gamma_{k+1}}^*(z) - (\gamma_k - \gamma_{k+1})b_{\mathcal{U}}(u_\gamma^*(z), \bar{u}^c)$, which is indeed (4.52). $\qquad\square$

Proof of Lemma 4: Bound on $G_{\gamma\beta}$ for the First Iteration

Since $\bar{w}^1 := (\bar{u}^1, \bar{v}^1, \bar{\lambda}^1)$ is updated by (4.19), similar to (SAMA), we can use (4.55) with $k = 0$, $\lambda := \hat{\lambda}^0$ and $\hat{\ell}_{\gamma_1}(\hat{\lambda}^0) \leq d_{\gamma_1}(\hat{\lambda}^0)$ to obtain

$$d_{\gamma_1}(\bar{\lambda}^1) \leq d_{\gamma_1}(\hat{\lambda}^0) - \left(\frac{1}{\eta_0} - \frac{\|A\|^2}{2\gamma_1}\right)\|\bar{\lambda}^1 - \hat{\lambda}^0\|^2. \qquad (4.56)$$

Since $\bar{v}^1$ solves the second problem in (4.19) and $v^*(\hat{\lambda}^0) \in \text{dom}(h)$, we have

$$h(v^*(\hat{\lambda}^0)) - \langle \hat{\lambda}^0, Bv^*(\hat{\lambda}^0) \rangle + \frac{\eta_0}{2}\|A\bar{u}^1 + Bv^*(\hat{\lambda}^0) - c\|^2 \geq h(\bar{v}^1)$$
$$- \langle \hat{\lambda}^0, B\bar{v}^1 \rangle + \frac{\eta_0}{2}\|A\bar{u}^1 + B\bar{v}^1 - c\|^2 + \frac{\eta_0}{2}\|B(v^*(\hat{\lambda}^0) - \bar{v}^1)\|^2.$$

Using D_f in (4.9), this inequality implies

$$h^*(B^\top \hat{\lambda}^0) \leq \langle \hat{\lambda}^0, B\bar{v}^1 \rangle - h(\bar{v}^1) - \frac{\eta_0}{2}\|A\bar{u}^1 + B\bar{v}^1 - c\|^2 + \frac{\eta_0}{2}\|A\bar{u}^1 + B\bar{v}^1 - c\|D_f. \quad (4.57)$$

Using the definition of d_γ, we further estimate (4.56) using (4.57) as follows:

$$
\begin{aligned}
d_{\gamma_1}(\bar{\lambda}^1) &\overset{(4.56)}{\leq} \varphi_{\gamma_1}(\hat{\lambda}^0) + \psi(\hat{\lambda}^0) - \left(\frac{1}{\eta_0} - \frac{\|A\|^2}{2\gamma_1}\right)\|\bar{\lambda}^1 - \hat{\lambda}^0\|^2 \\
&\overset{(4.11)}{=} \langle A\bar{u}^1, \hat{\lambda}^0 \rangle - g(\bar{u}^1) - \gamma_1 b_\mathcal{U}(\bar{u}^1, \bar{u}^c) + \psi(\hat{\lambda}^0) - \left(\frac{1}{\eta_0} - \frac{\|A\|^2}{2\gamma_1}\right)\|\bar{\lambda}^1 - \hat{\lambda}^0\|^2 \\
&\overset{(4.57)}{\leq} \langle \hat{\lambda}^0, A\bar{u}^1 + B\bar{v}^1 - c \rangle - g(\bar{u}^1) - h(\bar{v}^1) - \gamma_1 b_\mathcal{U}(\bar{u}^1, \bar{u}^c) \\
&\quad - \frac{\eta_0}{2}\|A\bar{u}^1 + B\bar{v}^1 - c\|^2 - \left(\frac{1}{\eta_0} - \frac{\|A\|^2}{2\gamma_1}\right)\|\bar{\lambda}^1 - \hat{\lambda}^0\|^2 + \frac{\eta_0}{2}\|A\bar{u}^1 + B\bar{v}^1 - c\|D_f \\
&\leq -f_{\beta_1}(\bar{x}^1) + \frac{1}{2\eta_0^2}\left[\frac{1}{\beta_1} - \frac{5\eta_0}{2} + \frac{\|A\|^2\eta_0^2}{\gamma_1}\right]\|\bar{\lambda}^1 - \hat{\lambda}^0\|^2 + \frac{1}{\eta_0}\langle \hat{\lambda}^0, \bar{\lambda}^1 - \hat{\lambda}^0 \rangle + \frac{\eta_0}{4}D_f^2.
\end{aligned}
$$

Since $G_{\gamma_1\beta_1}(\bar{w}^1) = f_{\beta_1}(\bar{x}^1) + d_{\gamma_1}(\bar{\lambda}^1)$, we obtain (4.20) from the last inequality. If $\beta_1 \geq \frac{2\gamma_1}{\eta_0(5\gamma_1 - 2\|A\|^2\eta_0)}$, then (4.20) leads to $G_{\gamma_1\beta_1}(\bar{w}^1) \leq \frac{\eta_0}{4}D_f^2 + \frac{1}{\eta_0}\langle \hat{\lambda}^0, \bar{\lambda}^1 - \hat{\lambda}^0 \rangle$. $\square$

Proof of Lemma 3: Gap Reduction Condition

For notational simplicity, we first define the following abbreviations

$$
\begin{cases}
\bar{z}^k &:= A\bar{u}^k + B\bar{v}^k - c \\
\hat{z}^{k+1} &:= A\hat{u}^{k+1} + B\hat{v}^{k+1} - c \\
\bar{u}^*_{k+1} &:= u^*_{\gamma_{k+1}}(A^\top\bar{\lambda}^k) \text{ the solution of (4.11) at } \bar{\lambda}^k, \\
\hat{v}^*_k &:= v^*(\hat{\lambda}^k) \in \partial h^*(A^\top\hat{\lambda}^k) \text{ a subgradient of } h^* \text{ defined by (4.5) at } A^\top\hat{\lambda}^k, \text{ and} \\
D_k &:= \|A\hat{u}^{k+1} + B(2\hat{v}^*_k - \hat{v}^{k+1}) - c\|.
\end{cases}
$$

From SAMA, we have $\bar{\lambda}^{k+1} - \hat{\lambda}^k = \eta_k(c - A\hat{u}^{k+1} - B\hat{v}^{k+1}) = -\eta_k\hat{z}^{k+1}$. In addition, by (4.16), we have $\hat{\lambda}^k = (1 - \tau_k)\bar{\lambda}^k + \tau_k\lambda^*_k$, which leads to $(1 - \tau_k)\bar{\lambda}^k + \tau_k\hat{\lambda}^k - \hat{\lambda}^k = \tau_k(\hat{\lambda}^k - \lambda^*_k)$. Using these expressions into (4.50) with $\lambda := \hat{\lambda}^k$, and then using (4.51) with $\hat{\ell}_{\gamma_{k+1}}(\hat{\lambda}^k) \leq d_{\gamma_{k+1}}(\hat{\lambda}^k)$, we obtain

$$
\begin{aligned}
d_{\gamma_{k+1}}(\bar{\lambda}^{k+1}) &\leq (1 - \tau_k)d_{\gamma_{k+1}}(\bar{\lambda}^k) + \tau_k d_{\gamma_{k+1}}(\hat{\lambda}^k) + \tau_k\langle \hat{z}^{k+1}, \lambda^*_k - \hat{\lambda}^k \rangle \\
&\quad - \eta_k\left(1 - \frac{\eta_k\|A\|^2}{2\gamma_{k+1}}\right)\|\hat{z}^{k+1}\|^2 - (1 - \tau_k)\frac{\gamma_{k+1}}{2}\|\bar{u}^*_{k+1} - \hat{u}^{k+1}\|^2.
\end{aligned} \quad (4.58)
$$

By (4.52) with the fact that $\varphi_\gamma(\lambda) := g^*_\gamma(A^\top\lambda)$, for any $\gamma_{k+1} > 0$ and $\gamma_k > 0$, we have

$$\varphi_{\gamma_{k+1}}(\bar{\lambda}^k) \leq \varphi_{\gamma_k}(\bar{\lambda}^k) + (\gamma_k - \gamma_{k+1})b_\mathcal{U}(\bar{u}^*_{k+1}, \bar{u}_c).$$

Using this inequality and the fact that $d_\gamma := \varphi_\gamma + \psi$, we have

$$d_{\gamma_{k+1}}(\bar{\lambda}^k) \leq d_{\gamma_k}(\bar{\lambda}^k) + (\gamma_k - \gamma_{k+1})b_{\mathcal{U}}(\bar{u}_{k+1}^*, \bar{u}_c). \tag{4.59}$$

Next, using $\hat{v}^{k+1}$ from SAMA and its optimality condition, we can show that

$$h^*(B^\top \hat{\lambda}^k) - \tfrac{\eta_k}{2}\|A\hat{u}^{k+1} + B\hat{v}_k^* - c\|^2 = \langle B^\top \hat{\lambda}^k, \hat{v}_k^* \rangle - h(\hat{v}_k^*) - \tfrac{\eta_k}{2}\|A\hat{u}^{k+1} + B\hat{v}_k^* - c\|^2$$
$$\leq \langle B^\top \hat{\lambda}^k, \hat{v}^{k+1} \rangle - h(\hat{v}^{k+1}) - \tfrac{\eta_k}{2}\|A\hat{u}^{k+1} + B\hat{v}^{k+1} - c\|^2 - \tfrac{\eta_k}{2}\|B(\hat{v}_k^* - \hat{v}^{k+1})\|^2.$$

Since $\psi(\lambda) := h^*(B^\top \lambda) - c^\top \lambda$, this inequality leads to

$$\psi(\hat{\lambda}^k) \leq \langle B^\top \hat{\lambda}^k, \hat{v}^{k+1} \rangle - \langle c, \hat{\lambda}^k \rangle - h(\hat{v}^{k+1}) - \tfrac{\eta_k}{2}\|\hat{z}^{k+1}\|^2$$
$$- \tfrac{\eta_k}{2}\langle \hat{z}^{k+1}, A\hat{u}^{k+1} + B(2\hat{v}_k^* - \hat{v}^{k+1}) - c \rangle$$
$$\leq \langle \hat{\lambda}^k, B\hat{v}^{k+1} - c \rangle - h(\hat{v}^{k+1}) - \tfrac{\eta_k}{2}\|\hat{z}^{k+1}\|^2 + \tfrac{\eta_k}{2}\|\hat{z}^{k+1}\|D_k.$$

Now, by this estimate, $d_{\gamma_{k+1}} = \varphi_{\gamma_{k+1}} + \psi$ and SAMA, we can derive

$$d_{\gamma_{k+1}}(\hat{\lambda}^k) \leq \varphi_{\gamma_{k+1}}(\hat{\lambda}^k) - h(\hat{v}^{k+1}) + \langle \hat{\lambda}^k, B\hat{v}^{k+1} - c \rangle - \tfrac{\eta_k}{2}\|\hat{z}^{k+1}\|^2 + \tfrac{\eta_k}{2}\|\hat{z}^{k+1}\|D_k$$
$$= -f(\hat{x}^{k+1}) + \langle \hat{\lambda}^k, \hat{z}^{k+1} \rangle - \tfrac{\eta_k}{2}\|\hat{z}^{k+1}\|^2 + \tfrac{\eta_k}{2}\|\hat{z}^{k+1}\|D_k - \gamma_{k+1}b_{\mathcal{U}}(\hat{u}^{k+1}, \bar{u}^c).$$

Combining this inequality, (4.58) and (4.59), we obtain

$$d_{\gamma_{k+1}}(\bar{\lambda}^{k+1}) \leq (1 - \tau_k)d_{\gamma_k}(\bar{\lambda}^k) - \tau_k f(\hat{x}^{k+1}) + \tau_k \langle \lambda_k^*, \hat{z}^{k+1} \rangle$$
$$- \eta_k\left(1 + \tfrac{\tau_k}{2} - \tfrac{\|A\|^2 \eta_k}{2\gamma_{k+1}}\right)\|\hat{z}^{k+1}\|^2$$
$$- \tau_k \gamma_{k+1}b_{\mathcal{U}}(\hat{u}^{k+1}, \bar{u}^c) + (1 - \tau_k)(\gamma_k - \gamma_{k+1})b_{\mathcal{U}}(\bar{u}_{k+1}^*, \bar{u}_c)$$
$$- (1 - \tau_k)\tfrac{\gamma_{k+1}}{2}\|\bar{u}_{k+1}^* - \hat{u}^{k+1}\|^2 + \tfrac{\tau_k \eta_k}{2}\|\hat{z}^{k+1}\|D_k. \tag{4.60}$$

Now, using the definition G_k, we have

$$G_k(\bar{w}^k) := f_{\beta_k}(\bar{x}^k) + d_{\gamma_k}(\bar{\lambda}^k) = f(\bar{x}^k) + d_{\gamma_k}(\bar{\lambda}^k) + \tfrac{1}{2\beta_k}\|A\bar{u}^k + B\bar{v}^k - c\|^2$$
$$= f(\bar{x}^k) + d_{\gamma_k}(\bar{\lambda}^k) + \tfrac{1}{2\beta_k}\|\bar{z}^k\|^2.$$

Let us define $\Delta G_k := (1 - \tau_k)G_k(\bar{w}^k) - G_{k+1}(\bar{w}^{k+1})$. Then, we can show that

$$\Delta G_k = (1 - \tau_k)f(\bar{x}^k) + (1 - \tau_k)d_{\gamma_k}(\bar{\lambda}^k) - f(\bar{x}^{k+1}) - d_{\gamma_{k+1}}(\bar{\lambda}^{k+1})$$
$$+ \tfrac{(1-\tau_k)}{2\beta_k}\|\bar{z}^k\|^2 - \tfrac{1}{2\beta_{k+1}}\|\bar{z}^{k+1}\|^2. \tag{4.61}$$

By (4.16), we have $\bar{z}^{k+1} = (1 - \tau_k)\bar{z}^k + \tau_k \hat{z}^{k+1}$. Using this expression and the condition $\beta_{k+1} \geq (1 - \tau_k)\beta_k$ in (4.17), we can easily show that

$$\tfrac{(1-\tau_k)}{2\beta_k}\|\bar{z}^k\|^2 - \tfrac{1}{2\beta_{k+1}}\|\bar{z}^{k+1}\|^2 \geq -\tfrac{\tau_k}{\beta_k}\langle \hat{z}^{k+1}, \bar{z}^k \rangle - \tfrac{\tau_k^2}{2\beta_k(1-\tau_k)}\|\hat{z}^{k+1}\|^2.$$

Substituting this inequality into (4.61), and using the convexity of f, we further get

$$\Delta G_k \geq (1-\tau_k)d_{\gamma_k}(\bar{\lambda}^k) - d_{\gamma_{k+1}}(\bar{\lambda}^{k+1}) - \tau_k f(\hat{x}^{k+1})$$
$$-\frac{\tau_k}{\beta_k}\langle \hat{z}^{k+1}, \bar{z}^k \rangle - \frac{\tau_k^2}{2(1-\tau_k)\beta_k}\|\hat{z}^{k+1}\|^2. \tag{4.62}$$

Substituting (4.60) into (4.62) and using $\lambda_k^* := \frac{1}{\beta_k}(c - A\bar{u}^k - B\bar{v}^k) = -\frac{1}{\beta_k}\bar{z}^k$, we obtain

$$\Delta G_k \geq \left[\eta_k\left(1 + \frac{\tau_k}{2} - \frac{\|A\|^2\eta_k}{2\gamma_{k+1}}\right) - \frac{\tau_k^2}{2(1-\tau_k)\beta_k}\right]\|\hat{z}^{k+1}\|^2 + R_k - \frac{\tau_k\eta_k}{2}\|\hat{z}^{k+1}\|D_k. \tag{4.63}$$

where

$$R_k := \frac{1-\tau_k}{2}\gamma_{k+1}\|\bar{u}_{k+1}^* - \hat{u}^{k+1}\|^2 + \tau_k\gamma_{k+1}b_{\mathcal{U}}(\hat{u}^{k+1}, \bar{u}^c) - (1-\tau_k)(\gamma_k - \gamma_{k+1})b_{\mathcal{U}}(\bar{u}_{k+1}^*, \bar{u}^c).$$

Furthermore, we have

$$\frac{\eta_k}{4}\|\hat{z}^{k+1}\|^2 - \frac{\tau_k\eta_k}{2}\|\hat{z}^{k+1}\|D_k = \frac{\eta_k}{4}\left[\|z^{k+1}\| - \tau_k D_k\right]^2 - \frac{\eta_k\tau_k^2 D_k^2}{4} \geq -\frac{\eta_k\tau_k^2 D_k^2}{4}.$$

Using this estimate into (4.63), we finally get

$$\Delta G_k \geq \left[\eta_k\left(\frac{3}{4} + \frac{\tau_k}{2} - \frac{\|A\|^2\eta_k}{2\gamma_{k+1}}\right) - \frac{\tau_k^2}{2(1-\tau_k)\beta_k}\right]\|\hat{z}^{k+1}\|^2 + R_k - \frac{\eta_k\tau_k^2 D_k^2}{4}. \tag{4.64}$$

Next step, we estimate R_k. Let $\bar{a}_k := \bar{u}_{k+1}^* - \bar{u}_c$, $\hat{a}_k := \hat{u}^{k+1} - \bar{u}_c$. Using the smoothness of $b_{\mathcal{U}}$, we can estimate R_k explicitly as

$$2\gamma_{k+1}^{-1}R_k \geq (1-\tau_k)\|\bar{a}_k - \hat{a}_k\|^2 - (1-\tau_k)(\gamma_{k+1}^{-1}\gamma_k - 1)L_b\|\bar{a}_k\|^2 + \tau_k\|\hat{a}_k\|^2$$
$$= \|\hat{a}^k - (1-\tau_k)\bar{a}_k\|^2 + (1-\tau_k)\left(\tau_k - (\gamma_{k+1}^{-1}\gamma_k - 1)L_b\right)\|\bar{a}_k\|^2. \tag{4.65}$$

By the condition $(1 + L_b^{-1}\tau_k)\gamma_{k+1} \geq \gamma_k$ in (4.17), we have $\tau_k - (\gamma_{k+1}^{-1}\gamma_k - 1)L_b \geq 0$. Using this condition in (4.65), we obtain $R_k \geq 0$. Finally, by (4.9) we can show that $D_k \leq D_f$. Using this inequality, $R_k \geq 0$, and the second condition of (4.17), we can show from (4.63) that $\Delta G_k \geq -\frac{\eta_k\tau_k^2}{4}D_f^2$, which implies (4.18). $\qquad\square$

Proof of Lemma 5: Parameter Updates

The tightest update for γ_k and β_k is $\gamma_{k+1} := \frac{\gamma_k}{\tau_k+1}$ and $\beta_{k+1} := (1-\tau_k)\beta_k$ due to (4.17). Using these updates in the third condition in (4.17) leads to $\frac{(1-\tau_{k+1})^2}{(1+\tau_{k+1})\tau_{k+1}^2} \geq$

$\frac{1-\tau_k}{\tau_k^2}$. By directly checking this condition, we can see that $\tau_k = \mathcal{O}(1/k)$ which is the optimal choice.

Clearly, if we choose $\tau_k := \frac{3}{k+4}$, then $0 < \tau_k < 1$ for $k \geq 0$ and $\tau_0 = 3/4$. Next, we choose $\gamma_{k+1} := \frac{\gamma_k}{1+\tau_k/3} \geq \frac{\gamma_k}{1+\tau_k}$. Substituting $\tau_k = \frac{3}{k+4}$ into this formula we have $\gamma_{k+1} = \left(\frac{k+4}{k+5}\right)\gamma_k$. By induction, we obtain $\gamma_{k+1} = \frac{5\gamma_1}{k+5}$. This implies $\eta_k = \frac{5\gamma_1}{2\|A\|^2(k+5)}$. With $\tau_k = \frac{3}{k+4}$ and $\gamma_{k+1} = \frac{5\gamma_1}{k+5}$, we choose β_k from the third condition of (4.17) as $\beta_k = \frac{2\|A\|^2\tau_k^2}{(1-\tau_k^2)\gamma_{k+1}} = \frac{18\|A\|^2(k+5)}{5\gamma_1(k+1)(k+7)}$ for $k \geq 1$. Using the value of τ_k and β_k, we need to check the second condition $\beta_{k+1} \geq (1-\tau_k)\beta_k$ of (4.17). Indeed, this condition is equivalent to $2k^2 + 28k + 88 \geq 0$, which is true for all $k \geq 0$. From the update rule of β_k, it is obvious that $\beta_k \leq \frac{18\|A\|^2}{5\gamma_1(k+1)}$. □

Proof of Theorem 1: Convergence of Algorithm 1

We estimate the term $\tau_k^2\eta_k$ in (4.18) as

$$\tau_k^2\eta_k = \frac{45\gamma_1}{2\|A\|^2(k+4)^2(k+5)} < \frac{45\gamma_1}{2\|A\|^2(k+4)(k+5)} - (1-\tau_k)\frac{45\gamma_1}{2\|A\|^2(k+3)(k+4)}.$$

Combing this estimate and (4.18), we get

$$G_{k+1}(\bar{w}^{k+1}) - \frac{45\gamma_1 D_f^2}{8\|A\|^2(k+4)(k+5)} \leq (1-\tau_k)\left[G_k(\bar{w}^k) - \frac{45\gamma_1 D_f^2}{8\|A\|^2(k+3)(k+4)}\right].$$

By induction, we have $G_k(\bar{w}^k) - \frac{45\gamma_1 D_f^2}{8\|A\|^2(k+3)(k+4)} \leq \omega_k[G_1(\bar{w}^1) - \frac{9\gamma_1}{32\|A\|^2}D_f^2] \leq 0$ whenever $G_1(\bar{w}^1) \leq \frac{3\gamma_1}{4\|A\|^2}D_f$, where $\omega_k := \prod_{i=1}^{k-1}(1-\tau_i)$. Hence, we finally get

$$G_k(\bar{w}^k) \leq \frac{45\gamma_1 D_f^2}{8\|A\|^2(k+3)(k+4)}. \tag{4.66}$$

Since $\eta_0 = \frac{\gamma_1}{2\|A\|^2}$, it satisfies the condition $5\gamma_1 > 2\eta_0\|A\|^2$ in Lemma 4. In addition, from Lemma 5, we have $\beta_1 = \frac{27\|A\|^2}{20\gamma_1} > \frac{\|A\|^2}{\gamma_1}$, which satisfies the second condition in Lemma 4. We also note that $\beta_k \leq \frac{18\|A\|^2}{5\gamma_1(k+1)}$. If we take $\hat{\lambda}^0 = \mathbf{0}^m$, then Lemma 4 shows that $G_{\gamma_1\beta_1}(\bar{w}^1) \leq \frac{\eta_0}{2}D_f^2 = \frac{\gamma_1}{4\|A\|^2}D_f^2 < \frac{9\gamma_1}{32\|A\|^2}D_f^2$. Using this estimate and (4.66) into Lemma 2, we obtain (4.23). Finally, if we choose $\gamma_1 := \|A\|$, then we obtain the worst-case iteration-complexity of Algorithm 1 is $\mathcal{O}(\varepsilon^{-1})$. □

Proof of Corollary 1: Strong Convexity of g

First, we show that if condition (4.24) hold, then (4.25) holds. Since $\nabla \varphi$ given by (4.5) is Lipschitz continuous with $L_{d_0^g} := \mu_g^{-1} \|A\|^2$, similar to the proof of Lemma 3, we have

$$\Delta G_{\beta_k} \geq \left[\eta_k \left(\frac{3}{4} + \frac{\tau_k}{2} - \frac{\eta_k \|A\|^2}{2\mu_g} \right) - \frac{\tau_k^2}{2(1-\tau_k)\beta_k} \right] \|\hat{z}^{k+1}\|^2 - \frac{\tau_k^2 \eta_k}{4} D_f^2, \qquad (4.67)$$

where $\Delta G_{\beta_k} := (1-\tau_k) G_{\beta_k}(\bar{w}^k) - G_{\beta_{k+1}}(\bar{w}^{k+1})$. Under the condition (4.24), (4.67) implies (4.25).

The update rule (4.27) is in fact derived from (4.24). We finally prove the bounds (4.28). First, we consider the product $\tau_k^2 \eta_k$. By (4.27) we have

$$\tau_k^2 \eta_k = \frac{9\mu_g}{2\|A\|^2(k+4)^2} < \frac{9\mu_g}{2\|A\|^2(k+3)(k+4)}$$

$$= \frac{9\mu_g}{4\|A\|^2(k+4)} - (1-\tau_k)\frac{9\mu_g}{4\|A\|^2(k+3)}$$

By induction, it follows from (4.25) and this last expression that:

$$G_{\beta_k}(\bar{w}^k) - \frac{9\mu_g D_f^2}{16\|A\|^2(k+3)} \leq \omega_k \left(G_{\beta_1}(\bar{w}^1) - \frac{9\mu_g D_f^2}{64\|A\|^2} \right) \leq 0, \qquad (4.68)$$

whenever $G_{\beta_1}(\bar{w}^1) \leq \frac{9\mu_g D_f^2}{64\|A\|^2}$. Since $\bar{u}^1$ is given by (4.26), with the same argument as the proof of Lemma 4, we can show that if $\frac{1}{\beta_1} \leq \frac{5\eta_0}{2} - \frac{\|A\|^2\eta_0^2}{\mu_g}$, then $G_{\beta_1}(\bar{w}^1) \leq \frac{\eta_0}{4} D_f^2$. However, from the update rule (4.27), we can see that $\eta_0 = \frac{\mu_g}{2\|A\|^2}$ and $\beta_1 = \frac{18\|A\|^2}{16\mu_g}$. Using these quantities, we can clearly show that $\frac{1}{\beta_1} \leq \frac{5\eta_0}{2} - \frac{\|A\|^2\eta_0^2}{\mu_g} = \frac{\mu_g}{\|A\|^2}$. Moreover, $G_{\beta_1}(\bar{w}^1) \leq \frac{\eta_0}{4} D_f^2 < \frac{9\mu_g}{64\|A\|^2} D_f^2$. Hence, (4.68) holds. Finally, it remains to use Lemma 2 to obtain (4.28). The second part in (4.30) is proved similarly. The estimate (4.31) is a direct consequence of (4.68). $\qquad \square$

Convergence Analysis of Algorithm 2

This appendix provides full proof of Lemmas and Theorems related to the convergence of Algorithm 2.

Proof of Lemma 6: Gap Reduction Condition

We first require the following key lemma to analyze the convergence of our SADMM scheme, whose proof is similar to (4.55) and we omit the details here.

Lemma 9 *Let $\bar{\lambda}^{k+1}$ be generated by SADMM. Then, for $\lambda \in \mathbb{R}^n$, one has*

$$d_{\gamma_{k+1}}(\bar{\lambda}^{k+1}) \leq \tilde{\ell}_{\gamma_{k+1}}(\lambda) + \frac{1}{\eta_k}\langle \bar{\lambda}^{k+1} - \hat{\lambda}^k, \lambda - \hat{\lambda}^k \rangle - \frac{1}{\eta_k}\|\hat{\lambda}^k - \bar{\lambda}^{k+1}\|^2 + \frac{\|A\|^2}{2\gamma_{k+1}}\|\tilde{\lambda}^k - \bar{\lambda}^{k+1}\|^2,$$

where $\tilde{\lambda}^k := \hat{\lambda}^k - \rho_k(A\hat{u}^{k+1} + B\hat{v}^k - c)$ and $\tilde{\ell}_\gamma(\lambda) := \varphi_\gamma(\tilde{\lambda}^k) + \langle \nabla\varphi_\gamma(\tilde{\lambda}^k), \lambda - \tilde{\lambda}^k \rangle + \psi(\lambda)$.

Now, we can prove Lemma 6. We still use the same notations as in the proof of Lemma 3. In addition, let us denote by $\hat{u}^*_{k+1} := u^*_{\gamma_{k+1}}(A^\top\hat{\lambda}^k)$ and $\bar{u}^*_{k+1} := u^*_{\gamma_{k+1}}(A^\top\bar{\lambda}^k)$ given in (4.12), $\check{z}^k := A\hat{u}^{k+1} + B\hat{v}^k - c$ and $\check{D}_k := \|A\hat{u}^*_{k+1} + B\hat{v}^k - c\|$.

First, since $\varphi_\gamma(\tilde{\lambda}^k) + \langle \nabla\varphi_\gamma(\tilde{\lambda}^k), \lambda - \tilde{\lambda}^k \rangle \leq \varphi_\gamma(\lambda)$, it follows from Lemma 9 that

$$\begin{aligned} d_{\gamma_{k+1}}(\bar{\lambda}^{k+1}) \leq{}& d_{\gamma_{k+1}}(\lambda) + \frac{1}{\eta_k}\langle \bar{\lambda}^{k+1} - \hat{\lambda}^k, \lambda - \hat{\lambda}^k \rangle - \frac{1}{\eta_k}\|\hat{\lambda}^k - \bar{\lambda}^{k+1}\|^2 \\ &+ \frac{\|A\|^2}{2\gamma_{k+1}}\|\tilde{\lambda}^k - \bar{\lambda}^{k+1}\|^2. \end{aligned} \tag{4.69}$$

Next, using [26, Theorem 2.1.5 (2.1.10)] with g^*_γ defined in (4.11) and $\lambda := (1 - \tau_k)\bar{\lambda}^k + \tau_k\hat{\lambda}^k$ for any $\tau_k \in [0, 1]$, we have

$$\varphi_{\gamma_{k+1}}(\lambda) \leq (1 - \tau_k)\varphi_{\gamma_{k+1}}(\bar{\lambda}^k) + \tau_k\varphi_{\gamma_{k+1}}(\hat{\lambda}^k) - \frac{\tau_k(1 - \tau_k)\gamma_{k+1}}{2}\|\hat{u}^*_{k+1} - \bar{u}^*_{k+1}\|^2. \tag{4.70}$$

Since ψ is convex, we also have $\psi(\lambda) \leq (1 - \tau_k)\psi(\bar{\lambda}^k) + \tau_k\psi(\hat{\lambda}^k)$ and $\lambda - \hat{\lambda}^k = (1 - \tau_k)\bar{\lambda}^k + \tau_k\hat{\lambda}^k - \hat{\lambda}^k = \tau_k(\hat{\lambda}^k - \lambda^*_k)$ due to (4.33). Combining these expressions, the definition $d_\gamma := \varphi_\gamma + \psi$, (4.69), and (4.70), we can derive

$$\begin{aligned} d_{\gamma_{k+1}}(\bar{\lambda}^{k+1}) \leq{}& (1 - \tau_k)d_{\gamma_{k+1}}(\bar{\lambda}^k) + \tau_k d_{\gamma_{k+1}}(\hat{\lambda}^k) + \frac{\tau_k}{\eta_k}\langle \bar{\lambda}^{k+1} - \hat{\lambda}^k, \hat{\lambda}^k - \lambda^*_k \rangle \\ &- \frac{1}{\eta_k}\|\bar{\lambda}^{k+1} - \hat{\lambda}^k\|^2 + \frac{\|A\|^2}{2\gamma_{k+1}}\|\bar{\lambda}^{k+1} - \tilde{\lambda}^k\|^2 \\ &- (1 - \tau_k)\tau_k\frac{\gamma_{k+1}}{2}\|\bar{u}^*_{k+1} - \hat{u}^*_{k+1}\|^2. \end{aligned} \tag{4.71}$$

On the one hand, since $\hat{u}^{k+1}$ is the solution of the first convex subproblem in SADMM, using its optimality condition, we can show that

$$\begin{aligned} \varphi_{\gamma_{k+1}}(\hat{\lambda}^k) - \frac{\rho_k}{2}\check{D}^2_k ={}& \langle \hat{\lambda}^k, A\hat{u}^*_{k+1} \rangle - g(\hat{u}^*_{k+1}) - \gamma_{k+1}b_{\mathcal{U}}(\hat{u}^*_{k+1}, \bar{u}^c) - \frac{\rho_k}{2}\check{D}^2_k \\ \leq{}& \langle \hat{\lambda}^k, A\hat{u}^{k+1} \rangle - g(\hat{u}^{k+1}) - \frac{\rho_k}{2}\|\check{z}^k\|^2 - \gamma_{k+1}b_{\mathcal{U}}(\hat{u}^{k+1}, \bar{u}_c) \\ &- \frac{\rho_k}{2}\|A(\hat{u}^*_{k+1} - \hat{u}^{k+1})\|^2 - \frac{\gamma_{k+1}}{2}\|\hat{u}^*_{k+1} - \hat{u}^{k+1}\|^2. \end{aligned} \tag{4.72}$$

On the other hand, similar to the proof of Lemma 3, we can show that

$$\psi(\hat{\lambda}^k) \leq \langle \hat{\lambda}^k, B\hat{v}^{k+1} - c \rangle - h(\hat{v}^{k+1}) - \tfrac{\eta_k}{2}\|\hat{z}^{k+1}\|^2 - \tfrac{\eta_k}{2}\langle \hat{z}^{k+1}, A\hat{u}^{k+1} + B(2\hat{v}_k^* - \hat{v}^{k+1}) - c \rangle$$
$$\leq \langle \hat{\lambda}^k, B\hat{v}^{k+1} - c \rangle - h(\hat{v}^{k+1}) - \tfrac{\eta_k}{2}\|\hat{z}^{k+1}\|^2 + \tfrac{\eta_k}{2}\|\hat{z}^{k+1}\|D_k. \tag{4.73}$$

Combining (4.72) and (4.73) and noting that $d_\gamma := \varphi_\gamma + \psi$, we have

$$d_{\gamma_{k+1}}(\hat{\lambda}^k) \leq \langle \hat{\lambda}^k, \hat{z}^{k+1} \rangle - f(\hat{x}^{k+1}) - \tfrac{\eta_k}{2}\|\hat{z}^{k+1}\|^2 - \tfrac{\rho_k}{2}\|\check{z}^k\|^2 - \gamma_{k+1}b_U(\hat{u}^{k+1}, \bar{u}_c)$$
$$- \tfrac{\rho_k}{2}\|A(\hat{u}_{k+1}^* - \hat{u}^{k+1})\|^2 - \tfrac{\gamma_{k+1}}{2}\|\hat{u}_{k+1}^* - \hat{u}^{k+1}\|^2 + \tfrac{\eta_k}{2}\|\hat{z}^{k+1}\|D_k + \tfrac{\rho_k}{2}\check{D}_k^2. \tag{4.74}$$

Next, using the strong convexity of b_U with $\mu_{b_U} = 1$, we can show that

$$\tfrac{\gamma_{k+1}}{2}\|\hat{u}_{k+1}^* - \hat{u}^{k+1}\|^2 + \gamma_{k+1}b_U(\hat{u}^{k+1}, \bar{u}_c) \geq \tfrac{\gamma_{k+1}}{4}\|\hat{u}_{k+1}^* - \bar{u}_c\|^2. \tag{4.75}$$

Combining (4.71), (4.59), (4.74) and (4.75), we can derive

$$d_{\gamma_{k+1}}(\bar{\lambda}^{k+1}) \leq (1 - \tau_k)d_{\gamma_k}(\bar{\lambda}^k) + \tfrac{\tau_k}{\eta_k}\langle \bar{\lambda}^{k+1} - \hat{\lambda}^k, \hat{\lambda}^k - \lambda_k^* \rangle$$
$$- \tfrac{1}{\eta_k}\|\bar{\lambda}^{k+1} - \hat{\lambda}^k\|^2 + \tfrac{\|A\|^2}{2\gamma_{k+1}}\|\bar{\lambda}^{k+1} - \tilde{\lambda}^k\|^2$$
$$- \tau_k f(\hat{x}^{k+1}) + \tau_k \langle \hat{\lambda}^k, \hat{z}^{k+1} \rangle - \tfrac{\tau_k \eta_k}{2}\|\hat{z}^{k+1}\|^2 - \tfrac{\tau_k \rho_k}{2}\|\check{z}^k\|^2 \tag{4.76}$$
$$- \tfrac{\tau_k \gamma_{k+1}}{4}\|\hat{u}_{k+1}^* - \bar{u}_c\|^2 - (1 - \tau_k)\tau_k \tfrac{\gamma_{k+1}}{2}\|\hat{u}_{k+1}^* - \bar{u}_{k+1}^*\|^2$$
$$+ (1 - \tau_k)(\gamma_k - \gamma_{k+1})b_U(\bar{u}_{k+1}^*, \bar{u}^c) + \tfrac{\tau_k \eta_k}{2}\|\hat{z}^{k+1}\|D_k + \tfrac{\tau_k \rho_k}{2}\check{D}_k^2.$$

$$\hat{R}_k := \tfrac{\gamma_{k+1}}{2}(1 - \tau_k)\tau_k \|\hat{u}_{k+1}^* - \bar{u}_{k+1}^*\|^2 + \tfrac{\gamma_{k+1}}{4}\tau_k \|\hat{u}_{k+1}^* - \bar{u}_c\|^2$$
$$- (1 - \tau_k)(\gamma_k - \gamma_{k+1})b_U(\bar{u}_{k+1}^*, \bar{u}^c). \tag{4.77}$$

From SADMM, we have $\bar{\lambda}^{k+1} - \hat{\lambda}^k = -\eta_k \hat{z}^{k+1}$ and $\tilde{\lambda}^k - \hat{\lambda}^k = -\rho_k \check{z}^k$. Plugging these expressions and (4.77) into (4.76) we can simplify this estimate as

$$d_{\gamma_{k+1}}(\bar{\lambda}^{k+1}) \leq (1 - \tau_k)d_{\gamma_k}(\bar{\lambda}^k) + \tau_k \langle \hat{z}^{k+1}, \lambda_k^* \rangle - \tau_k f(\hat{x}^{k+1}) - \tfrac{(1 + \tau_k)\eta_k}{2}\|\hat{z}^{k+1}\|^2$$
$$- \tfrac{1}{\eta_k}\|\bar{\lambda}^{k+1} - \hat{\lambda}^k\|^2 + \tfrac{\|A\|^2}{2\gamma_{k+1}}\|\bar{\lambda}^{k+1} - \tilde{\lambda}^k\|^2 - \tfrac{\tau_k}{2\rho_k}\|\tilde{\lambda}^k - \hat{\lambda}^k\|^2 - \hat{R}_k \tag{4.78}$$
$$+ \tfrac{\tau_k \eta_k}{2}\|\hat{z}^{k+1}\|D_k + \tfrac{\tau_k \rho_k}{2}\check{D}_k^2.$$

Using again the elementary inequality $\nu\|a\|^2 + \kappa\|b\|^2 \geq \tfrac{\nu\kappa}{\nu+\kappa}\|a - b\|^2$, under the condition $\gamma_{k+1} \geq \|A\|^2\left(\eta_k + \tfrac{\rho_k}{\tau_k}\right)$ in (4.34), we can show that

$$\tfrac{1}{2\eta_k}\|\bar{\lambda}^{k+1} - \hat{\lambda}^k\|^2 + \tfrac{\tau_k}{2\rho_k}\|\tilde{\lambda}^k - \hat{\lambda}^k\|^2 - \tfrac{\|A\|^2}{2\gamma_{k+1}}\|\bar{\lambda}^{k+1} - \tilde{\lambda}^k\|^2 \geq 0. \tag{4.79}$$

On the other hand, similar to the proof of Lemma 3, we can show that $\frac{\eta_k}{4}\|\hat{z}^{k+1}\|^2 -$
$\frac{\tau_k\eta_k}{2}\|\hat{z}^{k+1}\|D_k \geq -\frac{\eta_k\tau_k^2}{4}D_k^2$. Using this inequality, (4.79), and $\lambda_k^* = -\frac{1}{\beta_k}\bar{z}^k$, we can
simplify (4.78) as

$$
\begin{aligned}
d_{\gamma_{k+1}}(\bar{\lambda}^{k+1})\leq\ & (1-\tau_k)d_{\gamma_k}(\bar{\lambda}^k) - \tfrac{\tau_k}{\beta_k}\langle\hat{z}^{k+1},\bar{z}^k\rangle - \tau_k f(\hat{x}^{k+1}) - \eta_k\left(\tfrac{1}{4}+\tfrac{\tau_k}{2}\right)\|\hat{z}^{k+1}\|^2 \\
& - \hat{R}_k + \left(\tfrac{\eta_k\tau_k^2}{4}D_k^2 + \tfrac{\tau_k\rho_k}{2}\check{D}_k^2\right).
\end{aligned}
\tag{4.80}
$$

Since $\beta_{k+1} \geq (1-\tau_k)\beta_k$ due to (4.34), similar to the proof of (4.62) we have

$$
\begin{aligned}
\Delta G_k \geq\ & (1-\tau_k)d_{\gamma_k}(\bar{\lambda}^k) - d_{\gamma_{k+1}}(\bar{\lambda}^{k+1}) - \tau_k f(\hat{x}^{k+1}) \\
& - \tfrac{\tau_k}{\beta_k}\langle\hat{z}^{k+1},\bar{z}^k\rangle - \tfrac{\tau_k^2}{2(1-\tau_k)\beta_k}\|\hat{z}^{k+1}\|^2.
\end{aligned}
\tag{4.81}
$$

Combining (4.80) and (4.81), we get

$$
\Delta G_k \geq \frac{1}{2}\left[\left(\frac{1}{2}+\tau_k\right)\eta_k - \frac{\tau_k^2}{(1-\tau_k)\beta_k}\right]\|\hat{z}^{k+1}\|^2 + \hat{R}_k - \left(\frac{\eta_k\tau_k^2}{4}D_k^2 + \frac{\tau_k\rho_k}{2}\check{D}_k^2\right).
\tag{4.82}
$$

Next, we estimate $\hat{R}_k$ defined by (4.77) as follows. We define $\bar{a}_k := \bar{u}_{k+1}^* - \bar{u}_c$, $\hat{a}_k := \hat{u}_{k+1}^* - \bar{u}_c$. Using $b_{\mathcal{U}}(\bar{u}_{k+1}^*,\bar{u}^c) \leq \frac{L_b}{2}\|\bar{u}_{k+1}^* - \bar{u}^c\|^2$, we can write $\hat{R}_k$ explicitly as

$$
\begin{aligned}
\frac{2\hat{R}_k}{\gamma_{k+1}} &= (1-\tau_k)\tau_k\|\bar{a}_k - \hat{a}_k\|^2 + \tfrac{\tau_k}{2}\|\hat{a}_k\|^2 - (1-\tau_k)\left(\tfrac{\gamma_k}{\gamma_{k+1}}-1\right)L_b\|\bar{a}_k\|^2 \\
&= \tau_k\left(\tfrac{3}{2}-\tau_k\right)\left\|\hat{a}_k - \tfrac{(1-\tau)}{(3/2-\tau_k)}\bar{a}_k\right\|^2 + (1-\tau_k)\left[\tfrac{\tau_k}{3-2\tau_k}+\left(1-\tfrac{\gamma_k}{\gamma_{k+1}}\right)L_b\right]\|\bar{a}\|^2.
\end{aligned}
$$

Since $\gamma_{k+1} \geq \left(\frac{3-2\tau_k}{3-(2-L_b^{-1})\tau_k}\right)\gamma_k$ due to (4.34), it is easy to show that $\hat{R}_k \geq 0$. In
addition, by (4.34), we also have $(1+2\tau_k)\eta_k - \frac{2\tau_k^2}{(1-\tau_k)\beta_k} \geq 0$. Using these conditions,
we can show from (4.82) that $\Delta G_k \geq -\frac{\eta_k\tau_k^2}{4}D_k^2 - \frac{\tau_k\rho_k}{2}\check{D}_k^2 \geq -\left(\frac{\tau_k^2\eta_k}{4}+\frac{\tau_k\rho_k}{2}\right)D_f^2$,
which is indeed the gap reduction condition (4.35). $\square$

Proof of Lemma 7: Parameter Updates

Similar to the proof of Lemma 5, we can show that the optimal rate of $\{\tau_k\}$ is
$\mathcal{O}(1/k)$. From the conditions (4.34), it is clear that if we choose $\tau_k := \frac{3}{k+4}$ then
$0 < \tau_k \leq \frac{3}{4} < 1$ for $k \geq 0$. Next, we choose $\gamma_{k+1} := \left(\frac{3-2\tau_k}{3-\tau_k}\right)\gamma_k$. Then γ_k

satisfies (4.34). Substituting $\tau_k = \frac{3}{k+4}$ into this formula we have $\gamma_{k+1} = \left(\frac{k+2}{k+3}\right)\gamma_k$. By induction, we obtain $\gamma_{k+1} = \frac{3\gamma_1}{k+3}$. Now, we choose $\eta_k := \frac{\gamma_{k+1}}{2\|A\|^2} = \frac{3\gamma_1}{2\|A\|^2(k+3)}$. Then, from the last condition of (4.34), we choose $\rho_k := \frac{\tau_k \gamma_{k+1}}{2\|A\|^2} = \frac{9\gamma_1}{2\|A\|^2(k+3)(k+4)}$.

To derive an update for β_k, from the third condition of (4.34) with equality, we can derive $\beta_k = \frac{2\tau_k^2}{(1-\tau_k)(1+2\tau_k)\eta_k} = \frac{6\|A\|^2(k+3)}{\gamma_1(k+1)(k+10)} < \frac{9\|A\|^2}{5\gamma_1(k+1)}$. We need to check the second condition $\beta_{k+1} \geq (1 - \tau_k)\beta_k$ in (4.34). Indeed, we have $\beta_{k+1} = \frac{6\|A\|^2(k+4)}{\gamma_1(k+2)(k+11)} \geq (1-\tau_k)\beta_k = \frac{6\|A\|^2(k+3)}{\gamma_1(k+1)(k+10)}$, which is true for all $k \geq 0$. Hence, the second condition of (4.34) holds. $\qquad\square$

Proof of Theorem 2: Convergence of Algorithm 2

First, we check the conditions of Lemma 4. From the update rule (4.36), we have $\eta_0 = \frac{\gamma_1}{2\|A\|^2}$ and $\beta_1 = \frac{12\|A\|^2}{11\gamma_1}$. Hence, $5\gamma_1 = 10\|A\|^2\eta_0 > 2\|A\|^2\eta_0$, which satisfies the first condition of Lemma 4. Now, $\frac{2\gamma_1}{(5\gamma_1 - 2\eta_0\|A\|^2)\eta_0} = \frac{\|A\|^2}{\gamma_1} < \frac{12\|A\|^2}{11\gamma_1} = \beta_1$. Hence, the second condition of Lemma 4 holds.

Next, since $\tau_k = \frac{3}{k+4}$, $\rho_k = \frac{9\gamma_1}{2\|A\|^2(k+3)(k+4)}$ and $\eta_k = \frac{3\gamma_1}{2\|A\|^2(k+3)}$, we can derive

$$\frac{\tau_k^2 \eta_k}{4} + \frac{\tau_k \rho_k}{2} = \frac{81\gamma_1}{8\|A\|^2(k+3)(k+4)^2}$$
$$< \frac{81\gamma_1}{8\|A\|^2(k+3)(k+4)} - (1 - \tau_k)\frac{81\gamma_1}{8\|A\|^2(k+2)(k+3)}.$$

Substituting this inequality into (4.35) and rearrange the result we obtain

$$G_{k+1}(\bar{w}^{k+1}) - \frac{81\gamma_1 D_f^2}{8\|A\|^2(k+3)(k+4)} \leq (1 - \tau_k)\left[G_k(\bar{w}^k) - \frac{81\gamma_1 D_f^2}{8\|A\|^2(k+2)(k+3)}\right].$$

By induction, we obtain $G_k(\bar{w}^k) - \frac{81\gamma_1 D_f^2}{8\|A\|^2(k+2)(k+3)} \leq \omega_k\left[G_0(\bar{w}^0) - \frac{27\gamma_1 D_f^2}{16\|A\|^2}\right] \leq 0$ as long as $G_0(\bar{w}^0) \leq \frac{27\gamma_1 D_f^2}{16\|A\|^2}$. Now using Lemma 4, we have $G_0(\bar{w}^0) \leq \frac{\eta_0}{4}D_f^2 = \frac{\gamma_1}{8\|A\|^2}D_f^2 < \frac{27\gamma_1 D_f^2}{16\|A\|^2}$. Hence, $G_k(\bar{w}^k) \leq \frac{27\gamma_1 D_f^2}{16\|A\|^2(k+2)(k+3)}$.

Finally, by using Lemma 2 with $\beta_k := \frac{6\|A\|^2(k+3)}{\gamma_1(k+1)(k+10)}$ and $\beta_k \leq \frac{9\|A\|^2}{5\gamma_1(k+1)}$, and simplifying the results, we obtain the bounds in (4.37). If we choose $\gamma_1 := \|A\|$ then, we obtain the worst-case iteration-complexity of Algorithm 2 is $\mathcal{O}(\varepsilon^{-1})$. $\qquad\square$

References

1. A. Alotaibi, P.L. Combettes, N. Shahzad, Best approximation from the Kuhn-Tucker set of composite monotone inclusions. Numer. Funct. Anal. Optim. **36**(12), 1513–1532 (2015)
2. H.H. Bauschke, P.L. Combettes, *Convex Analysis and Monotone Operator Theory in Hilbert Spaces* (Springer, Berlin, 2011)
3. A. Beck, M. Teboulle, A fast dual proximal gradient algorithm for convex minimization and applications. Oper. Res. Lett. **42**(1), 1–6 (2014)
4. J. Bolte, S. Sabach, M. Teboulle, Proximal alternating linearized minimization for non-convex and nonsmooth problems. Math. Program. **146**(1–2), 459–494 (2014)
5. S. Boyd, N. Parikh, E. Chu, B. Peleato, J. Eckstein, Distributed optimization and statistical learning via the alternating direction method of multipliers. Found. Trends Mach. Learn. **3**(1), 1–122 (2011)
6. R.S. Burachik, V. Martín-Márquez, An approach for the convex feasibility problem via monotropic programming. J. Math. Anal. Appl. **453**(2), 746–760 (2017)
7. X. Cai, D. Han, X. Yuan, On the convergence of the direct extension of ADMM for three-block separable convex minimization models with one strongly convex function. Comput. Optim. Appl. **66**(1), 39–73 (2017)
8. E. Candès, B. Recht, Exact matrix completion via convex optimization. Commun. ACM **55**(6), 111–119 (2012)
9. V. Cevher, S. Becker, M. Schmidt, Convex optimization for big data: scalable, randomized, and parallel algorithms for big data analytics. IEEE Signal Process. Mag. **31**(5), 32–43 (2014)
10. A. Chambolle, T. Pock, A first-order primal-dual algorithm for convex problems with applications to imaging. J. Math. Imaging Vis. **40**(1), 120–145 (2011)
11. D. Davis, W. Yin, Convergence rate analysis of several splitting schemes, in *Splitting Methods in Communication, Imaging, Science, and Engineering* (Springer, Cham, 2016), pp. 115–163
12. D. Davis, W. Yin, Faster convergence rates of relaxed Peaceman-Rachford and ADMM under regularity assumptions. Math. Oper. Res. **42**(3), 783–805 (2017)
13. D. Davis, W. Yin, A three-operator splitting scheme and its optimization applications. Tech. Report. (2015)
14. W. Deng, W. Yin, On the global and linear convergence of the generalized alternating direction method of multipliers. J. Sci. Comput. **66**(3), 889–916 (2016)
15. J. Eckstein, D. Bertsekas, On the Douglas Rachford splitting method and the proximal point algorithm for maximal monotone operators. Math. Program. **55**, 293–318 (1992)
16. E. Ghadimi, A. Teixeira, I. Shames, M. Johansson, Optimal parameter selection for the alternating direction method of multipliers (ADMM): quadratic problems. IEEE Trans. Autom. Control **60**(3), 644–658 (2015)
17. T. Goldstein, B. ODonoghue, S. Setzer, Fast alternating direction optimization methods. SIAM J. Imaging Sci. **7**(3), 1588–1623 (2012)
18. B. He, X. Yuan, On non-ergodic convergence rate of Douglas–Rachford alternating direction method of multipliers. Numer. Math. **130**(3), 567–577 (2012)
19. B. He, X. Yuan, On the $O(1/n)$ convergence rate of the Douglas-Rachford alternating direction method. SIAM J. Numer. Anal. **50**, 700–709 (2012)
20. T. Lin, S. Ma, S. Zhang, On the global linear convergence of the ADMM with multi- block variables. SIAM J. Optim. **25**(3), 1478–1497 (2015)
21. T. Lin, S. Ma, S. Zhang, Iteration complexity analysis of multi-block ADMM for a family of convex minimization without strong convexity. J. Sci. Comput. **69**(1), 52–81 (2016)
22. T. Lin, S. Ma, S. Zhang, An extragradient-based alternating direction method for convex minimization. Found. Comput. Math. **17**(1), 35–59 (2017)
23. I. Necoara, J. Suykens, Applications of a smoothing technique to decomposition in convex optimization. IEEE Trans. Autom. Control **53**(11), 2674–2679 (2008)

24. A. Nemirovskii, Prox-method with rate of convergence $O(1/t)$ for variational inequalities with Lipschitz continuous monotone operators and smooth convex-concave saddle point problems. SIAM J. Optim. **15**(1), 229–251 (2004)
25. A. Nemirovskii, D. Yudin, *Problem Complexity and Method Efficiency in Optimization* (Wiley Interscience, New York, 1983)
26. Y. Nesterov, *Introductory Lectures on Convex Optimization: A Basic Course*. Applied Optimization, vol. 87 (Kluwer Academic Publishers, Norwell, 2004)
27. Y. Nesterov, Smooth minimization of non-smooth functions. Math. Program. **103**(1), 127–152 (2005)
28. Y. Ouyang, Y. Chen, G. Lan, E.J. Pasiliao, An accelerated linearized alternating direction method of multiplier. SIAM J. Imaging Sci. **8**(1), 644–681 (2015)
29. N. Parikh, S. Boyd, Proximal algorithms. Found. Trends Optim. **1**(3), 123–231 (2013)
30. R.T. Rockafellar, *Convex Analysis*. Princeton Mathematics Series, vol. 28 (Princeton University Press, Princeton, 1970)
31. R. Shefi, M. Teboulle, Rate of convergence analysis of decomposition methods based on the proximal method of multipliers for convex minimization. SIAM J. Optim. **24**(1), 269–297 (2014)
32. R. Shefi, M. Teboulle, On the rate of convergence of the proximal alternating linearized minimization algorithm for convex problems. EURO J. Comput. Optim. **4**(1), 27–46 (2016)
33. F. Simon, R. Holger, *A Mathematical Introduction to Compressive Sensing* (Springer, New York, 2013)
34. M. Tao, X. Yuan, On the $O(1/t)$-convergence rate of alternating direction method with logarithmic-quadratic proximal regularization. SIAM J. Optim. **22**(4), 1431–1448 (2012)
35. Q. Tran-Dinh, V. Cevher, Constrained convex minimization via model-based excessive gap, in *Proceedings of the Neural Information Processing Systems (NIPS)*, Montreal, vol. 27 Dec. 2014, pp. 721–729
36. Q. Tran-Dinh, O. Fercoq, V. Cevher, A smooth primal-dual optimization framework for nonsmooth composite convex minimization. SIAM J. Optim. **28**, 96–134 (2018)
37. P. Tseng, D. Bertsekas, Relaxation methods for problems with strictly convex cost and linear constraints. Math. Oper. Res. **16**(3), 462–481 (1991)
38. W. Wang, A. Banerjee, Bregman alternating direction method of multipliers, in *Advances in Neural Information Processing Systems* 27 (NIPS 2014), pp. 1–9
39. E. Wei, A. Ozdaglar, On the $O(1/k)$-convergence of asynchronous distributed alternating direction method of multipliers, in *Global Conference on Signal and Information Processing (GlobalSIP)* (IEEE, Piscataway, 2013), pp. 551–554

Chapter 5
Primal-Dual Proximal Algorithms for Structured Convex Optimization: A Unifying Framework

Puya Latafat and Panagiotis Patrinos

Abstract We present a simple primal-dual framework for solving structured convex optimization problems involving the sum of a Lipschitz-differentiable function and two nonsmooth proximable functions, one of which is composed with a linear mapping. The framework is based on the recently proposed asymmetric forward-backward-adjoint three-term splitting (AFBA); depending on the value of two parameters, (extensions of) known algorithms as well as many new primal-dual schemes are obtained. This allows for a unified analysis that, among other things, establishes linear convergence under four different regularity assumptions for the cost functions. Most notably, linear convergence is established for the class of problems with piecewise linear-quadratic cost functions.

Keywords Convex optimization · Primal-dual algorithms · Operator splitting · Linear convergence

AMS Subject Classifications 90C25, 47H05, 65K05, 49M29

5.1 Introduction

In this chapter we consider structured convex optimization problems of the form

$$\underset{x \in \mathbb{R}^n}{\text{minimize}} \ f(x) + g(x) + h(Lx), \tag{5.1}$$

P. Latafat
KU Leuven, Department of Electrical Engineering (ESAT-STADIUS), Leuven, Belgium

IMT School for Advanced Studies Lucca, Lucca, Italy
e-mail: puya.latafat@kuleuven.be; puya.latafat@imtlucca.it

P. Patrinos (✉)
KU Leuven, Department of Electrical Engineering (ESAT-STADIUS), Leuven, Belgium
e-mail: panos.patrinos@esat.kuleuven.be

© Springer Nature Switzerland AG 2018
P. Giselsson, A. Rantzer (eds.), *Large-Scale and Distributed Optimization*, Lecture Notes in Mathematics 2227,
https://doi.org/10.1007/978-3-319-97478-1_5

where L is a linear mapping, g and h are proper closed convex (possibly) nonsmooth functions, and f is convex, continuously differentiable with Lipschitz continuous gradient. The working assumption throughout the chapter is that one can efficiently evaluate the gradient of f, the proximal mapping of the nonsmooth terms g and h, the linear mapping L and its adjoint.

This model is quite rich and captures a plethora of problems arising in machine learning, signal processing and control [13, 22, 31]. As a widely popular example consider consensus or sharing type problems:

$$\underset{x_1,\ldots,x_N}{\text{minimize}} \sum_{i=1}^{N} f_i(x_i) \tag{5.2a}$$

$$\textbf{subject to } x \in C, \tag{5.2b}$$

where $x = (x_1, \ldots, x_N)$. In the case of consensus $C = \{(x_1, \ldots, x_N) \mid x_1 = \ldots = x_N\}$, and for sharing $C = \{(x_1, \ldots, x_N) \mid \sum_{i=1}^{N} x_i = 0\}$. The two sets are orthogonal subspaces and indeed the sharing and consensus problems are dual to each other [7]. More generally, when C is a subspace problem (5.2) is referred to as *extended monotropic programming* [5]. This family of problems can be written in the form of (5.1) with $f(x) = \sum_{i=1}^{N} f_i(x_i)$, $h \equiv 0$, and $g = \iota_C$, where ι_X denotes the indicator of the set X.

A recent trend for solving problem (5.1), possibly with the smooth term $f \equiv 0$ or the nonsmooth term $g \equiv 0$, is to solve the monotone inclusion defined by the primal-dual optimality conditions [8–10, 14, 15, 18, 20, 23, 32]. The popularity of this approach is mainly due to the fact that it results in *fully split* primal-dual algorithms, in which the proximal mappings of g and h, the gradient of f, the linear mapping L and its adjoint are evaluated individually. In particular, there are no matrix inversions or inner loops involved.

Different convergence analysis techniques have been proposed in the literature for primal-dual algorithms. Some can be viewed as intelligent applications of classical splitting methods such as forward-backward splitting (FBS), Douglas-Rachford splitting (DRS) and forward-backward-forward splitting (FBFS). See for example [6, 8, 14, 15, 32], while others employ different tools to show convergence [10, 11, 18, 21]. Convergence rate of primal-dual schemes has also been analyzed using different approaches, see for example [11, 16, 26, 27].

Our approach here is a systematic one and relies on solving the primal-dual optimality conditions using a new three-term splitting, *asymmetric forward-backward-adjoint splitting* (AFBA) [23]. This splitting algorithm generalizes FBS to include a third linear monotone operator. Furthermore, it includes asymmetric preconditioning that is the key for developing our unifying primal-dual framework in which one can generate a wide range of algorithms by selecting different values for the scalar parameters θ and μ (cf. Algorithm 1). Many of the resulting algorithms are new, while some extend previously proposed algorithms and/or result in less conservative stepsize conditions. In short, this analysis provides us with a spectrum of primal-dual algorithms each of which may have an advantage over others depending on the application. For example, in [25] a special case was exploited to develop a randomized block coordinate variant for distributed applications where the stepsizes only depend on local information. The main idea was to exploit the fact that, for this particular primal-dual algorithm, the generated sequence is Fejér monotone with respect to $\| \cdot \|_S$, where S is a block diagonal positive definite matrix.

The convergence analysis of all the primal-dual algorithms is easily deduced from that of AFBA. It must be noted that this work complements the analysis in [23] where AFBA was introduced. The relationship between existing primal-dual algorithms is already documented in [23, Fig. 1]. In this work we simplify AFBA by considering a constant step in place of a dynamic one. This modification simplifies the analysis of the primal-dual framework. In addition, we provide a general and easy-to-check convergence condition for the stepsizes in Assumption II. Furthermore, we discuss four mild regularity assumptions on the functions involved in (5.1) that are sufficient for metric subregularity of the operator defining the primal-dual optimality conditions (cf. Lemmas 4.3 and 4.5). Linear convergence rate is then deduced based on the results developed for AFBA (cf. Theorem 3.6). These results do not impose additional restrictions on the stepsizes of the algorithms. It is important to note that the provided conditions are much weaker than strong convexity and in many cases do not imply a unique primal or dual solution.

It is worth mentioning that another class of primal-dual algorithms was introduced recently that rely on iterative projections onto half-spaces containing the set of solutions [1, 12]. This class of algorithms is not covered by the analysis in this work.

The paper is organized as follows. In Sect. 5.2 we present the new primal-dual framework and discuss several notable special cases. Section 5.3 is devoted to the introduction and analysis of a simplified variant of AFBA. In Sect. 5.4 we establish convergence for the proposed primal-dual framework based on the results developed in Sect. 5.3. In particular, linear convergence is established under four mild regularity assumptions for the cost functions.

Notation and Background

Throughout, $\mathbb{R}^n$ is the n-dimensional Euclidean space with inner product $\langle \cdot, \cdot \rangle$ and induced norm $\| \cdot \|$. The sets of symmetric, symmetric positive semi-definite and symmetric positive definite n by n matrices are denoted by $\mathcal{S}^n$, $\mathcal{S}^n_+$ and $\mathcal{S}^n_{++}$ respectively. We also write $P \succeq 0$ and $P \succ 0$ for $P \in \mathcal{S}^n_+$ and $P \in \mathcal{S}^n_{++}$ respectively. For $P \in \mathcal{S}^n_{++}$ we define the scalar product $\langle x, y \rangle_P = \langle x, Py \rangle$ and the induced norm $\|x\|_P = \sqrt{\langle x, x \rangle_P}$. For simplicity we use matrix notation for linear mappings when no ambiguity occurs.

An operator (or set-valued mapping) $A : \mathbb{R}^n \rightrightarrows \mathbb{R}^d$ maps each point $x \in \mathbb{R}^n$ to a subset Ax of $\mathbb{R}^d$. The graph of A is denoted by $\operatorname{gra} A = \{(x, y) \in \mathbb{R}^n \times \mathbb{R}^d \mid y \in Ax\}$ and the set of its zeros by $\operatorname{\mathbf{zer}} A = \{x \in \mathbb{R}^n \mid 0 \in Ax\}$. The mapping A is called monotone if $\langle x - x', y - y' \rangle \geq 0$ for all $(x, y), (x', y') \in \operatorname{gra} A$, and is said to be maximally monotone if its graph is not strictly contained in the graph of another monotone operator. The inverse of A is defined through its graph: $\operatorname{gra} A^{-1} := \{(y, x) \mid (x, y) \in \operatorname{gra} A\}$. The *resolvent* of A is defined by $J_A := (\operatorname{Id} + A)^{-1}$, where Id denotes the identity operator.

For an extended-real-valued function f, we use **dom** f to denote its domain. Let $f : \mathbb{R}^n \to \overline{\mathbb{R}} := \mathbb{R} \cup \{+\infty\}$ be a proper closed, convex function. Its subdifferential is the set-valued operator $\partial f : \mathbb{R}^n \rightrightarrows \mathbb{R}^n$

$$\partial f(x) = \{y \in \mathbb{R}^n \mid \forall z \in \mathbb{R}^n, \ \langle z - x, y \rangle + f(x) \leq f(z)\}.$$

The subdifferential is a maximally monotone operator. The resolvent of ∂f is called the *proximal operator* (or proximal mapping), and is single-valued. For a given $V \in \mathcal{S}_{++}^n$ the proximal mapping of f relative to $\| \cdot \|_V$ is uniquely determined by the resolvent of $V^{-1} \partial f$:

$$\mathbf{prox}_f^V(x) := (\operatorname{Id} + V^{-1} \partial f)^{-1} x = \underset{z \in \mathbb{R}^n}{\operatorname{argmin}} \{f(z) + \tfrac{1}{2}\|x - z\|_V^2\}.$$

The *Fenchel conjugate* of f, denoted f^*, is defined as

$$f^*(v) := \sup_{x \in \mathbb{R}^n} \{\langle v, x \rangle - f(x)\}.$$

The *infimal convolution* of two functions $f, g : \mathbb{R}^n \to \overline{\mathbb{R}}$ is defined as

$$(f \,\square\, g)(x) = \inf_{z \in \mathbb{R}^n} \{f(z) + g(x - z)\}.$$

Let X be a nonempty closed convex set and define the indicator of the set

$$\iota_X(x) := \begin{cases} 0 & x \in X \\ +\infty & x \notin X. \end{cases}$$

The distance to a set X with respect to $\| \cdot \|_V$ is given by $d_V(\cdot, X) = \iota_X \,\square\, \| \cdot \|_V$. We use $\Pi_X^V(\cdot)$ to denote the projection onto X with respect to $\| \cdot \|_V$.

The sequence $(x^k)_{k \in \mathbb{N}}$ is said to converge to $x^\star$ Q-linearly with Q-factor $\sigma \in (0, 1)$, if there exists $\bar{k} \in \mathbb{N}$ such that for all $k \geq \bar{k}$, $\|x^{k+1} - x^\star\| \leq \sigma \|x^k - x^\star\|$ holds. Furthermore, $(x^k)_{k \in \mathbb{N}}$ is said to converge to $x^\star$ R-linearly if there is a sequence of nonnegative scalars $(v_k)_{k \in \mathbb{N}}$ such that $\|x^k - x^\star\| \leq v^k$ and $(v_k)_{k \in \mathbb{N}}$ converges to zero Q-linearly.

5.2   A Simple Framework for Primal-Dual Algorithms

In this section we present a simple framework for primal-dual algorithms. For this purpose we consider the following extension of (5.1)

$$\underset{x \in \mathbb{R}^n}{\text{minimize}} \; f(x) + g(x) + (h \,\square\, l)(Lx), \tag{5.3}$$

where l is a strongly convex function. Notice that when $l = \iota_{\{0\}}$, the infimal convolution $h \,\square\, l$ reduces to h, and problem (5.1) is recovered.

Throughout this chapter the following assumptions hold for problem (5.3).

Assumption I

(i) $g : \mathbb{R}^n \to \overline{\mathbb{R}}$, $h : \mathbb{R}^r \to \overline{\mathbb{R}}$ *are proper closed convex functions, and* $L : \mathbb{R}^n \to \mathbb{R}^r$ *is a linear mapping.*

(ii) $f : \mathbb{R}^n \to \mathbb{R}$ *is convex, continuously differentiable, and for some* $\beta_f \in [0, \infty)$, ∇f *is* β_f-*Lipschitz continuous with respect to the metric induced by* $Q \succ 0$, *i.e., for all* $x, y \in \mathbb{R}^n$:

$$\|\nabla f(x) - \nabla f(y)\|_{Q^{-1}} \le \beta_f \|x - y\|_Q,$$

(iii) $l : \mathbb{R}^r \to \overline{\mathbb{R}}$ *is proper closed convex, its conjugate* l^* *is continuously differentiable, and for some* $\beta_l \in [0, \infty)$, ∇l^* *is* β_l-*Lipschitz continuous with respect to the metric induced by* $R \succ 0$.

(iv) *The set of solutions to* (5.3), *denoted by* $X^\star$, *is nonempty.*

(v) *(Constraint qualification) There exists* $x \in \operatorname{ri} \operatorname{\mathbf{dom}} g$ *such that* $Lx \in \operatorname{ri} \operatorname{\mathbf{dom}} h + \operatorname{ri} \operatorname{\mathbf{dom}} l$.

In Assumption I(*ii*) the constant β_f is not absorbed into the metric Q in order to be able to treat the case when ∇f is a constant in a uniform fashion by setting $\beta_f = 0$. The same reasoning applies to Assumption I(*iii*).

The dual problem is given by

$$\underset{u \in \mathbb{R}^r}{\text{minimize}} (g^* \,\square\, f^*)(-L^\top u) + h^*(u) + l^*(u). \tag{5.4}$$

Notice the similar structure of the dual problem in which l, f and h, g have swapped roles. A well-established approach for solving (5.3) is to consider the associated convex-concave saddle point problem given by

$$\underset{x \in \mathbb{R}^n}{\text{minimize}} \, \underset{u \in \mathbb{R}^r}{\text{maximize}} \, \mathcal{L}(x, u) := f(x) + g(x) + \langle Lx, u \rangle - h^*(u) - l^*(u). \tag{5.5}$$

The primal-dual optimality conditions are

$$\begin{cases} 0 \in \partial g(x) + \nabla f(x) + L^\top u, \\ 0 \in \partial h^*(u) + \nabla l^*(u) - Lx. \end{cases} \tag{5.6}$$

Under the constraint qualification condition, the set of solutions for the dual problem denoted by $U^\star$ is nonempty, a saddle point exists, and the duality gap is zero. In fact for any $x^\star \in X^\star$ and $u^\star \in U^\star$, the point $(x^\star, u^\star)$ is a primal-dual solution. See [28, Cor. 31.2.1] and [4, Thm. 19.1].

The right-hand side of the optimality conditions in (5.6) can be split as the sum of three operators:

$$\begin{pmatrix} 0 \\ 0 \end{pmatrix} \in \underbrace{\begin{pmatrix} \partial g(x) \\ \partial h^*(u) \end{pmatrix}}_{Az} + \underbrace{\begin{pmatrix} \nabla f(x) \\ \nabla l^*(u) \end{pmatrix}}_{Cz} + \underbrace{\begin{pmatrix} 0 & L^\top \\ -L & 0 \end{pmatrix}}_{M} \underbrace{\begin{pmatrix} x \\ u \end{pmatrix}}_{z}. \tag{5.7}$$

Operator A defined above, is maximally monotone [4, Thm. 21.2, Prop. 20.33], while operator C, being the gradient of $\tilde{f}(x, u) = f(x) + l^*(u)$, is cocoercive and M is skew-symmetric and as such monotone.

Throughout this section we shall use T to denote the operator above, i.e.,

$$0 \in Tz := Az + Cz + Mz. \tag{5.8}$$

Algorithm 1 describes the proposed primal-dual framework for solving (5.3). This framework is the result of solving the monotone inclusion (5.7) using the three-term splitting AFBA described in Sect. 5.3. We postpone the derivation and convergence analysis of Algorithm 1 to Sect. 5.4.

Notice that Algorithm 1 is not symmetric with respect to the primal and dual variables. Another variant may be obtained by switching their role. This would be equivalent to applying the algorithm to the dual problem (5.4).

The proposed framework involves two scalar parameters $\theta \in [0, \infty)$ and $\mu \in [0, 1]$. Different primal-dual algorithms correspond to different values for these parameters. The iterates in Algorithm 1 consist of two proximal updates followed by two correction steps that may or may not be performed depending on the parameters μ and θ. Below we discuss some values for these parameters that are most interesting.

Algorithm 1 A simple framework for primal-dual algorithms

Require: $x^0 \in \mathbb{R}^n$, $u^0 \in \mathbb{R}^r$, the algorithm parameters $\mu \in [0, 1]$, $\theta \in [0, \infty)$.
Initialize: Σ, Γ and λ based on Assumptions II(ii) and II(iii).
   for $k = 0, 1, \ldots$ **do**
      $\bar{x}^k = \mathbf{prox}_g^{\Gamma^{-1}}(x^k - \Gamma L^\top u^k - \Gamma \nabla f(x^k))$
      $\bar{u}^k = \mathbf{prox}_{h^*}^{\Sigma^{-1}}(u^k + \Sigma L((1 - \theta)x^k + \theta \bar{x}^k) - \Sigma \nabla l^*(u^k))$
      $\tilde{x}^k = \bar{x}^k - x^k$, $\tilde{u}^k = \bar{u}^k - u^k$
      $x^{k+1} = x^k + \lambda(\tilde{x}^k - \mu(2 - \theta)\Gamma L^\top \tilde{u}^k)$
      $u^{k+1} = u^k + \lambda(\tilde{u}^k + (1 - \mu)(2 - \theta)\Sigma L \tilde{x}^k)$
   end for

A variant of Algorithm 1 was introduced in [23, Alg. 3] that includes a dynamic stepsize, α_n, in the correction steps. In that work the connection between existing primal-dual algorithms was investigated by enforcing the dynamic stepsize $\alpha_n \equiv 1$. This approach results in cumbersome algebraic steps. By removing the dynamic stepsize we simplify the analysis substantially and provide a simple condition for convergence in Assumption II. Furthermore, in many distributed applications a dynamic stepsize is disadvantageous since it would entail global coordination. Moreover, in comparison to [23, Alg. 3] we generalize the algorithm by employing the matrices Σ and Γ as stepsizes in place of scalars and consider the Lipschitz continuity of ∇f and ∇l^* with respect to $\| \cdot \|_Q$ and $\| \cdot \|_R$.

In Sect. 5.4 we show that the sequence $(x^k, u^k)_{k\in\mathbb{N}}$ generated by Algorithm 1 converges to a primal-dual solution if Assumption II holds. Moreover, linear convergence rates are established if either one of four mild regularity assumptions holds for functions f, g, l and h (cf. Corollary 4.6).

Assumption II (Convergence Condition)

 (i) *(Algorithm parameters)* $\theta \in [0, \infty)$, $\mu \in [0, 1]$.
 (ii) *(Stepsizes)* $\Gamma \in \mathcal{S}^n_{++}$, $\Sigma \in \mathcal{S}^r_{++}$ *and (relaxation parameter)* $\lambda \in (0, 2)$
 (iii) *The following condition holds*

$$\begin{pmatrix} (\frac{2}{\lambda} - 1)\Gamma^{-1} - (1-\mu)(1-\theta)(2-\theta)L^{\top}\Sigma L - \frac{\beta_f}{2\lambda}Q & (\mu - (1-\mu)(1-\theta) - \frac{\theta}{\lambda})L^{\top} \\ (\mu - (1-\mu)(1-\theta) - \frac{\theta}{\lambda})L & (\frac{2}{\lambda} - 1)\Sigma^{-1} - \mu(2-\theta)L\Gamma L^{\top} - \frac{\beta_l}{2\lambda}R \end{pmatrix} \succ 0.$$
$$(5.9)$$

In Algorithm 1 the linear mappings L and $L^{\top}$ must be evaluated twice at every iteration. In the special cases when $\mu = 0, 1$ they may be evaluated only once per iteration by keeping track of the value computed in the previous iteration. As is evident from (5.9) the cases where both the algorithm and the convergence condition simplify are combinations of $\mu = 0, 1, \theta = 0, 1, 2$. Here we briefly discuss some of these special cases to demonstrate how (5.9) leads to simple conditions that are often less conservative than the conditions found in the literature. We have dubbed the algorithms based on whether the two proximal updates can be evaluated in parallel and if a primal or a dual correction step is performed.

The first algorithm is the result of setting $\theta = 2$ (regardless of μ) and leads to the algorithm of Condat and Vũ [15, 32] which itself is a generalization of the Chambolle-Pock algorithm [10]. With this choice of θ, the two proximal updates are performed sequentially while no correction step is required. In the box below a general condition is given for its convergence.

SNCA (Sequential No Corrector Algorithm): $\theta = 2$

Substituting $\theta = 2$ in (5.9) and dividing by $\frac{2}{\lambda} - 1$ yields

$$
\begin{pmatrix} \Gamma^{-1} - \frac{\beta_f}{2(2-\lambda)} Q & -L^\top \\ -L & \Sigma^{-1} - \frac{\beta_l}{2(2-\lambda)} R \end{pmatrix} \succ 0.
$$

If $l = \iota_{\{0\}}$, the infimal convolution $h \square l = h$, $l^* \equiv 0$ and $\beta_l = 0$. Using Schur complement yields the equivalent condition $\Gamma^{-1} - \frac{\beta_f}{2(2-\lambda)} Q - L^\top \Sigma L \succ 0$. If in addition $Q = \mathrm{Id}$ and $\Gamma = \gamma \mathrm{Id}$, $\Sigma = \sigma \mathrm{Id}$ for some scalars γ, σ, then the following sufficient condition may be used

$$
\sigma \gamma \|L\|^2 < 1 - \frac{\gamma \beta_f}{2(2-\lambda)}, \qquad \lambda \in (0, 2).
$$

Notice that this condition is less conservative than the condition of [15, Thm. 3.1] (see [23, Rem. 5.6]).

In the next algorithm proximal updates are evaluated sequentially, followed by a correction step for the primal variable, hence the name SPCA. Most notable property of this algorithm is that the generated sequence is Fejér monotone with respect to $\| \cdot \|_S$ where S is block diagonal. The algorithm introduced in [25] can be seen as an application of SPCA to the dual problem when the smooth term is zero.

SPCA (Sequential Primal Corrector Algorithm): $\theta = 1, \mu = 1, \lambda = 1$

In this case the left-hand side in (5.9) is block diagonal. Therefore the convergence condition simplifies to

$$
\Gamma^{-1} - \frac{\beta_f}{2} Q \succ 0, \qquad \Sigma^{-1} - L \Gamma L^\top - \frac{\beta_l}{2} R \succ 0.
$$

If $Q = \mathrm{Id}$, $R = \mathrm{Id}$, $\Gamma = \gamma \mathrm{Id}$, $\Sigma = \sigma \mathrm{Id}$ for some scalars γ, σ, then it is sufficient to have

$$
\gamma \beta_f < 2, \qquad \sigma \gamma \|L\|^2 < 1 - \frac{\sigma \beta_l}{2}. \tag{5.10}
$$

This special case generalizes the recent algorithm proposed in [18]. In particular we allow a third nonsmooth function g as well as the strongly convex function l. In addition to this improvement our convergence condition with $l^* \equiv 0$ (set $\beta_l = 0$ in (5.10)) is less restrictive and doubles the range of acceptable stepsize γ. The convergence condition in that work is given in our notation as $\gamma \beta_f < 1$ and $\sigma \gamma \|L\|^2 < 1$ [18, Cor. 3.2].

Next algorithm features sequential proximal updates that are followed by a correction step for the dual variable, for all values of $\theta \in (0, \infty)$. The parallel variant of this algorithm, referred to as PDCA, is discussed in a separate box below. We have observed that selecting θ so as to maximize the stepsizes, i.e., $\theta = 1.5$, leads to faster convergence [24]. Moreover, if we set $\Sigma = \Gamma^{-1}$, this choice of μ leads to a three-block ADMM or equivalently a generalization of DRS to include a third cocoercive operator (see [23, Sec. 5.4]).

SDCA (Sequential Dual Corrector Algorithm): $\theta \in (0, \infty), \mu = 0, \lambda = 1$
In this case the convergence condition simplifies to

$$\begin{pmatrix} \Gamma^{-1} - (1 - \theta)(2 - \theta)L^\top \Sigma L - \frac{\beta_f}{2}Q & -L^\top \\ -L & \Sigma^{-1} - \frac{\beta_l}{2}R \end{pmatrix} \succ 0.$$

If $l = \iota_{\{0\}}$ then $\beta_l = 0$ and using Schur complement we derive the following condition

$$\Gamma^{-1} - \left(\theta^2 - 3\theta + 3\right)L^\top \Sigma L - \frac{\beta_f}{2}Q \succ 0.$$

If in addition $Q = \mathrm{Id}$, and $\Gamma = \gamma\mathrm{Id}$, $\Sigma = \sigma\mathrm{Id}$ for some scalars γ, σ, then we have the following sufficient condition

$$(\theta^2 - 3\theta + 3)\sigma\gamma\|L\|^2 < 1 - \frac{\gamma\beta_f}{2}.$$

The next algorithm appears to be new and involves parallel proximal updates followed by a primal correction step.

PPCA (Parallel Primal Corrector Algorithm): $\theta = 0, \mu = 1$
The convergence condition is given by

$$\begin{pmatrix} (\frac{2}{\lambda} - 1)\Gamma^{-1} - \frac{\beta_f}{2\lambda}Q & L^\top \\ L & (\frac{2}{\lambda} - 1)\Sigma^{-1} - 2L\Gamma L^\top - \frac{\beta_l}{2\lambda}R \end{pmatrix} \succ 0.$$

If $f \equiv 0$ (set $\beta_f = 0$) and $\lambda = 1$, using Schur complement yields the following condition

$$\Sigma^{-1} - 3L\Gamma L^\top - \frac{\beta_l}{2}R \succ 0.$$

If in addition $R = \mathrm{Id}$ and $\Gamma = \gamma\mathrm{Id}$, $\Sigma = \sigma\mathrm{Id}$ for some scalars γ, σ, then the following sufficient condition may be used

$$3\sigma\gamma\|L\|^2 < 1 - \frac{\sigma\beta_l}{2}.$$

The parallel variant of SDCA is considered below. Interestingly, by switching the order of the proximal updates (since $\theta = 0$), PDCA may be seen as PPCA applied to the dual problem (5.4).

PDCA (Parallel Dual Corrector Algorithm): $\theta = 0$, $\mu = 0$

In this case the convergence condition simplifies to

$$\begin{pmatrix} (\frac{2}{\lambda} - 1)\Gamma^{-1} - 2L^\top \Sigma L - \frac{\beta_f}{2\lambda}Q & -L^\top \\ -L & (\frac{2}{\lambda} - 1)\Sigma^{-1} - \frac{\beta_l}{2\lambda}R \end{pmatrix} \succ 0.$$

If $l = \iota_{\{0\}}$ (set $\beta_l = 0$) and $\lambda = 1$, using Schur complement yields the following condition

$$\Gamma^{-1} - 3L^\top \Sigma L - \frac{\beta_f}{2}Q \succ 0.$$

If in addition $Q = \mathrm{Id}$, and $\Gamma = \gamma\mathrm{Id}$, $\Sigma = \sigma\mathrm{Id}$ for some scalars γ, σ, then the following sufficient condition may be used

$$3\sigma\gamma\|L\|^2 < 1 - \frac{\gamma\beta_f}{2}.$$

The last special case considered here involves sequential proximal updates followed by correction steps for the primal and dual variables. As noted before for this choice of μ, the linear mappings L and $L^\top$ must be evaluated twice at every iteration.

PPDCA (Parallel Primal and Dual Corrector Algorithm): $\theta = 0$, $\mu = 0.5$

In this case condition (5.9) reduces to:

$$\Gamma^{-1} - \frac{\lambda}{(2-\lambda)}L^\top \Sigma L - \frac{\beta_f}{2(2-\lambda)}Q \succ 0, \quad \Sigma^{-1} - \frac{\lambda}{(2-\lambda)}L\Gamma L^\top - \frac{\beta_l}{2(2-\lambda)}R \succ 0.$$

If $Q = \mathrm{Id}$, $R = \mathrm{Id}$, $\lambda = 1$ and $\Gamma = \gamma\mathrm{Id}$, $\Sigma = \sigma\mathrm{Id}$ for some scalars γ, σ, the following sufficient condition may be used

$$\sigma\gamma\|L\|^2 < \min\{1 - \frac{\gamma\beta_f}{2}, 1 - \frac{\sigma\beta_l}{2}\}.$$

This special case generalizes [8, Algorithm (4.8)] with the addition of the smooth function f and the strongly convex function l.

5.3 Simplified Asymmetric Forward-Backward-Adjoint Splitting

A new three-term splitting technique was introduced in [23] for the problem of finding $z \in \mathbb{R}^p$ such that

$$0 \in Tz := Az + Cz + Mz, \tag{5.11}$$

where A is maximally monotone, C is cocoercive and M is a monotone linear mapping. AFBA in its original form includes a dynamic stepsize, see [23, Alg. 1]. In this work we simplify the algorithm by considering a constant stepsize, see Algorithm 2. This variant of AFBA is particularly advantageous in distributed applications where global coordination may be infeasible. Furthermore, unlike [23, Alg. 1], cocoercivity of the operator C is considered with respect to some norm independent of the parameters of the algorithm, and the convergence condition is derived in terms of a matrix inequality. These changes simplify the analysis for the primal-dual algorithms discussed in Sect. 5.2. We remind the reader that Algorithm 1 is the result of solving the primal-dual optimality conditions using AFBA. We defer the derivation and convergence analysis of Algorithm 1 until Sect. 5.4.

Let us first recall the notion of cocoercivity.

Definition 3.1 (Cocoercivity) The operator $C : \mathbb{R}^p \to \mathbb{R}^p$ is said to be cocoercive with respect to $\| \cdot \|_U$ if for all $z, z' \in \mathbb{R}^p$

$$\langle Cz - Cz', z - z' \rangle \geq \|Cz - Cz'\|^2_{U^{-1}}. \tag{5.12}$$

A basic key inequality that we use is the following.

Lemma 3.2 (Three-Point Inequality) *Suppose that* $C : \mathbb{R}^p \to \mathbb{R}^p$ *is cocoercive with respect to* $\| \cdot \|_U$, *and let* $V : \mathbb{R}^p \to \mathbb{R}^p$ *be a linear mapping such that* $V \circ C = C$ *(identity is the trivial choice). Then for any three points* $z, z', z'' \in \mathbb{R}^p$ *we have*

$$\langle Cz - Cz', z' - z'' \rangle \leq \tfrac{1}{4}\|V^\top(z - z'')\|^2_U. \tag{5.13}$$

Proof Use the inequality, valid for any $a, b \in \mathbb{R}^p$,

$$\langle a, b \rangle = 2\langle \tfrac{1}{2}U^{\frac{1}{2}}a, U^{-\frac{1}{2}}b \rangle \leq \tfrac{1}{4}\|a\|^2_U + \|b\|^2_{U^{-1}}, \tag{5.14}$$

together with (5.12) and $V \circ C = C$ to derive

$$\langle Cz - Cz', z' - z'' \rangle = \langle V(Cz - Cz'), z - z'' \rangle + \langle Cz - Cz', z' - z \rangle$$

$$= \langle Cz - Cz', V^\top(z - z'') \rangle + \langle Cz - Cz', z' - z \rangle$$

$$\overset{(5.14)}{\leq} \frac{1}{4}\|V^\top(z - z'')\|_U^2 + \|Cz - Cz'\|_{U^{-1}}^2 + \langle Cz - Cz', z' - z\rangle$$

$$\overset{(5.12)}{\leq} \frac{1}{4}\|V^\top(z - z'')\|_U^2. \qquad\qquad\qquad \square$$

Our main motivation for considering V is to avoid conservative bounds in (5.13). For example, assume that the space is partitioned into two, $z = (z_1, z_2)$, and $Cz = (C_1 z_1, 0)$ where C_1 is cocoercive. Using inequality (5.14) without taking into account the structure of C, i.e., that $V \circ C = C$, would result in the whole vector appearing in the upper bound in (5.13).

Algorithm 2 involves two matrices H and S that are instrumental to its flexibility. In Sect. 5.4 we discuss a choice for H and S and demonstrate how Algorithm 1 is derived. Below we summarize the assumptions for the monotone inclusion (5.11) and the convergence conditions for Algorithm 2 (cf. Theorem 3.4).

Assumption III

  (i) *Assumptions for the monotone inclusion* (5.11):

   1. *The operator $A : \mathbb{R}^p \rightrightarrows \mathbb{R}^p$ is maximally monotone.*
   2. *The linear mapping $M : \mathbb{R}^p \to \mathbb{R}^p$ is monotone.*
   3. *The operator $C : \mathbb{R}^p \to \mathbb{R}^p$ is cocoercive with respect to $\|\cdot\|_U$. In addition, $V : \mathbb{R}^p \to \mathbb{R}^p$ is a linear mapping such that $V \circ C = C$ (identity is the trivial choice).*

  (ii) *Convergence conditions for Algorithm 2:*

   1. *The matrix $H := P + K$, where $P \in \mathcal{S}_{++}^p$ and K is a skew-symmetric matrix.*
   2. *The matrix $S \in \mathcal{S}_{++}^p$ and*

$$2P - \tfrac{1}{2}VUV^\top - D \succ 0, \qquad\qquad (5.15)$$

   where

$$D := (H + M^\top)^\top S^{-1}(H + M^\top). \qquad\qquad (5.16)$$

Algorithm 2 AFBA with constant stepsize

Require: $z^0 \in \mathbb{R}^p$
Initialize: set S and H according to Assumption III.
  for $k = 0, 1, \ldots$ **do**
   $\bar{z}^k = (H + A)^{-1}(H - M - C)z^k$
   $z^{k+1} = z^k + S^{-1}(H + M^\top)(\bar{z}^k - z^k)$
  end for

Lemma 3.3 *Let Assumption III hold. Consider the update for $\bar{z}$ in Algorithm 2*

$$\bar{z} = (H + A)^{-1}(H - M - C)z. \tag{5.17}$$

For all $z^\star \in \mathbf{zer}\, T$ the following holds

$$\langle z - z^\star, (H + M^\top)(\bar{z} - z)\rangle \le \tfrac{1}{4}\|V^\top(z - \bar{z})\|_U^2 - \|z - \bar{z}\|_P^2. \tag{5.18}$$

Proof Use (5.17) and the fact that $z^\star \in \mathbf{zer}\, T$, together with monotonicity of A at $z^\star$ and $\bar{z}$ to derive

$$0 \le \langle -Mz^\star - Cz^\star + Mz + Cz + H(\bar{z} - z), z^\star - \bar{z}\rangle. \tag{5.19}$$

In Lemma 3.2 set $z' = z^\star$ and $z'' = \bar{z}$

$$\langle Cz - Cz^\star, z^\star - \bar{z}\rangle \le \tfrac{1}{4}\|V^\top(z - \bar{z})\|_U^2. \tag{5.20}$$

For the remaining terms in (5.19) use skew-symmetry of K (twice) and monotonicity of M:

$$\begin{aligned}
\langle -Mz^\star &+ Mz + H(\bar{z} - z), z^\star - \bar{z}\rangle \\
&= \langle -Mz^\star + Mz + P(\bar{z} - z) + K(\bar{z} - z) + K(z^\star - \bar{z}), z^\star - \bar{z}\rangle \\
&= \langle (M - K)(z - z^\star) + P(\bar{z} - z), z^\star - \bar{z}\rangle \\
&= \langle (M - K)(z - z^\star) + P(\bar{z} - z), z^\star - z\rangle \\
&\quad + \langle (M - K)(z - z^\star) + P(\bar{z} - z), z - \bar{z}\rangle \\
&\le \langle P(\bar{z} - z), z^\star - z\rangle + \langle (M - K)(z - z^\star), z - \bar{z}\rangle - \|\bar{z} - z\|_P^2 \\
&\le \langle z - z^\star, (M^\top + H)(z - \bar{z})\rangle - \|\bar{z} - z\|_P^2.
\end{aligned}$$

Combining this with (5.19) and (5.20) completes the proof. □

Theorem 3.4 (Convergence) *Let Assumption III hold. Consider the sequence $(z^k)_{k\in\mathbb{N}}$ generated by Algorithm 2. Then the following inequality holds for all $k \in \mathbb{N}$*

$$\|z^{k+1} - z^\star\|_S^2 \le \|z^k - z^\star\|_S^2 - \|z^k - \bar{z}^k\|_{2P - \frac{1}{2}VUV^\top - D}^2, \tag{5.21}$$

and $(z^k)_{k\in\mathbb{N}}$ converges to a point $z^\star \in \mathbf{zer}\, T$.

Proof We show that the generated sequence is Fejér monotone with respect to **zer** T in the space equipped with the inner product $\langle \cdot, \cdot \rangle_S$. For any $z^\star \in$ **zer** T using the z^{k+1} update in Algorithm 2 we have

$$
\|z^{k+1} - z^\star\|_S^2 = \|z^k - z^\star\|_S^2 + \|S^{-1}(H + M^\top)(\bar{z}^k - z^k)\|_S^2
$$
$$
+ 2\langle z^k - z^\star, (H + M^\top)(\bar{z}^k - z^k)\rangle
$$
$$
\overset{(5.18)}{\leq} \|z^k - z^\star\|_S^2 + \|S^{-1}(H + M^\top)(\bar{z}^k - z^k)\|_S^2
$$
$$
+ \tfrac{1}{2}\|V^\top(z^k - \bar{z}^k)\|_U^2 - 2\|z^k - \bar{z}^k\|_P^2
$$
$$
\overset{(5.16)}{\leq} \|z^k - z^\star\|_S^2 - \|z^k - \bar{z}^k\|_{2P - \frac{1}{2}VUV^\top - D}^2.
$$

Therefore the sequence $(z^k - \bar{z}^k)_{k \in \mathbb{N}}$ converges to zero. Convergence of $(z^k)_{k \in \mathbb{N}}$ to a point in **zer** T follows by standard arguments; see the last part of the proof of [23, Thm. 3.1].   □

Our next goal is to establish linear convergence for Algorithm 2. Before continuing let us recall the notion of metric subregularity.

Definition 3.5 (Metric Subregularity) A mapping F is *metrically subregular* at $\bar{x}$ for $\bar{u}$ if $(\bar{x}, \bar{u}) \in$ gra F and there exists a positive constant κ, and neighborhoods $\mathcal{U}$ of $\bar{x}$ and $\mathcal{Y}$ of $\bar{u}$ such that

$$
d(x, F^{-1}\bar{u}) \leq \kappa d(\bar{u}, Fx \cap \mathcal{Y}), \quad \forall x \in \mathcal{U}.
$$

If in addition $\bar{x}$ is an isolated point of $F^{-1}\bar{u}$, i.e., $F^{-1}\bar{u} \cap \mathcal{U} = \{\bar{x}\}$, then F is said to be *strongly subregular* at $\bar{x}$ for $\bar{u}$. Otherwise stated:

$$
\|x - \bar{x}\| \leq \kappa d(\bar{u}, Fx \cap \mathcal{Y}), \quad \forall x \in \mathcal{U}.
$$

The neighborhood $\mathcal{Y}$ in the above definitions can be omitted [17, Ex. 3H.4].

Metric subregularity is a "one-point" version of metric regularity. We refer the interested reader to [17, Chap. 3] and [29, Chap. 9] for an extensive discussion.

The linear convergence for Algorithm 2 is established in Theorem 3.6 under two different assumptions: (1) when the operator T in (5.11) is metrically subregular at all $z^\star \in$ **zer** T for 0 or (2) when the operator Id $- T_{\mathrm{AFBA}}$ has this property, where T_{AFBA} denotes the operator that maps z^k to z^{k+1} in Algorithm 2. In Sect. 5.4 we exploit the first result, Theorem 3.6. The advantage of this result is that it is easier to characterize the metric subregularity of T in terms of its component. It is worth mentioning that for the preconditioned proximal point algorithm (a special case of Algorithm 2 with $C = 0$, $M = 0$, $K = 0$, $S = P$) the first assumption implies the second one [30, Prop. IV.2].

Theorem 3.6 (Linear Convergence) *Let Assumption III hold. Consider the sequence $(z^k)_{k \in \mathbb{N}}$ generated by Algorithm 2. Suppose that one of the following assumptions holds:*

(i) $T = A + M + C$ *is metrically subregular at all* $z^\star \in \mathbf{zer}\, T$ *for* 0.
(ii) Let T_{AFBA} *denote the operator that maps* z^k *to* z^{k+1} *in Algorithm 2, i.e.,* $z^{k+1} = T_{\mathrm{AFBA}}(z^k)$. *The operator* $\mathrm{Id} - T_{\mathrm{AFBA}}$ *is metrically subregular at all* $z^\star \in \mathbf{zer}\, T$ *for* 0.

Then $(z^k)_{k \in \mathbb{N}}$ *converges R-linearly to some* $z^\star \in \mathbf{zer}\, T$ *and* $(d_S(z^k, \mathbf{zer}\, T))_{k \in \mathbb{N}}$ *converges Q-linearly to zero.*

Proof The proof for Theorem 3.6*(i)* can be found in [23, Thm. 3.3]. The proof of the second part follows by noting that

$$d_S^2(z^{k+1}, \mathbf{zer}\, T) \leq \|z^{k+1} - \Pi^S_{\mathbf{zer}\, T}(z^k)\|_S^2$$

$$\overset{(5.21)}{\leq} \|z^k - \Pi^S_{\mathbf{zer}\, T}(z^k)\|_S^2 - \|z^k - \bar{z}^k\|^2_{2P - \frac{1}{2}VUV^\top - D} \qquad (5.22)$$

$$= d_S^2(z^k, \mathbf{zer}\, T) - \|z^k - z^{k+1}\|_W^2, \qquad (5.23)$$

where W is some symmetric positive definite matrix given by replacing $\bar{z}^k - z^k = (H + M^\top)^{-1} S(z^{k+1} - z^k)$ in (5.22). It remains to bound $d_S(z^k, \mathbf{zer}\, T)$ by $\|z^k - z^{k+1}\|_W$. By Theorem 3.4, z^k converges to some $z^\star \in \mathbf{zer}\, T$. Since $I - T_{\mathrm{AFBA}}$ is metrically subregular at $z^\star$ for 0, there exists $\bar{k} \in \mathbb{N}$, a positive constant κ, and a neighborhood $\mathcal{U}$ of $z^\star$ such that

$$d(z^k, \mathbf{zer}\, T) \leq \kappa \|z^k - z^{k+1}\|, \quad \forall k \geq \bar{k}, \qquad (5.24)$$

where we used the fact that the set of fixed points of T_{AFBA} is equal to $\mathbf{zer}\, T$. Therefore for $k \geq \bar{k}$ we have

$$d_S^2(z^k, \mathbf{zer}\, T) \leq \|z^k - \Pi_{\mathbf{zer}\, T}(z^k)\|_S^2 \leq \|S\| \|z^k - \Pi_{\mathbf{zer}\, T}(z^k)\|^2$$

$$= \|S\| d^2(z^k, \mathbf{zer}\, T) \overset{(5.24)}{\leq} \kappa^2 \|S\| \|z^k - z^{k+1}\|^2$$

$$\leq \kappa^2 \|S\| \|W^{-1}\| \|z^k - z^{k+1}\|_W^2 \qquad (5.25)$$

Combining (5.25) with (5.23) proves the Q-linear convergence rate for the distance from the set of solutions. Using this Q-linear convergence rate for the distance, it follows from (5.23) that $(\|z^k - z^{k+1}\|_W)_{k \in \mathbb{N}}$ converges R-linearly to zero and hence also $(z^k)_{k \in \mathbb{N}}$ converges R-linearly. $\qquad \square$

5.4 A Unified Convergence Analysis for Primal-Dual Algorithms

Our goal in this section is to describe how Algorithm 1 is derived and to establish its convergence. The idea is to write the primal-dual optimality conditions as a monotone inclusion involving the sum of a maximally monotone, a linear monotone and a cocoercive operator, cf. (5.7). This monotone inclusion is then solved using *asymmetric forward-backward-adjoint* splitting (AFBA) described in Sect. 5.3. In order to recover Algorithm 1 simply apply Algorithm 2 to this monotone inclusion with the following parameters: Let $\theta \in [0, \infty)$ and set $H = P + K$ with

$$P = \begin{pmatrix} \Gamma^{-1} & -\frac{\theta}{2}L^\top \\ -\frac{\theta}{2}L & \Sigma^{-1} \end{pmatrix}, \quad K = \begin{pmatrix} 0 & \frac{\theta}{2}L^\top \\ -\frac{\theta}{2}L & 0 \end{pmatrix}, \tag{5.26}$$

and $S = \left(\lambda \mu S_1^{-1} + \lambda(1-\mu)S_2^{-1}\right)^{-1}$ where $\mu \in [0, 1]$, $\lambda \in (0, 2)$ with

$$S_1 = \begin{pmatrix} \Gamma^{-1} & (1-\theta)L^\top \\ (1-\theta)L & \Sigma^{-1}+(1-\theta)(2-\theta)L\Gamma L^\top \end{pmatrix}, \quad S_2 = \begin{pmatrix} \Gamma^{-1}+(2-\theta)L^\top \Sigma L & -L^\top \\ -L & \Sigma^{-1} \end{pmatrix}.$$

Notice that with P and K set as in (5.26), H has a lower (block) triangular structure. Therefore the backward step $(H + A)^{-1}$ in Algorithm 2 can be carried out sequentially [23, Lem. 3.1]. Algorithm 1 is derived by noting this and substituting S and H defined above. We refer the reader to [23, Sec. 5] for a more detailed procedure.

   Our next goal is to verify that Assumptions I and II are sufficient for Assumption III to hold. As noted before, the operator A is maximally monotone [4, Thm. 21.2, Prop. 20.33], and the linear mapping M is skew-adjoint and as such monotone. The operator C is cocoercive with respect to the metric induced by $U = \text{blkdiag}(\beta_f Q, \beta_l R)$. In Assumption III($ii$) we use the linear mapping V in order to avoid conservative requirements. The special cases when $f \equiv 0$ (or $l^* \equiv 0$) are captured by setting $V = \text{blkdiag}(0_n, I_r)$ (or $V = \text{blkdiag}(I_n, 0_r)$) where I_s and 0_s denote the identity and zero matrices in $\mathbb{R}^{s \times s}$. It remains to verify Assumption III(ii). Evaluating D according to (5.16) yields the following $D = \lambda \mu D_1 + \lambda(1-\mu)D_2$ where

$$D_1 = \begin{pmatrix} \Gamma^{-1} & -L^\top \\ -L & \Sigma^{-1} + (2-\theta)L\Gamma L^\top \end{pmatrix},$$

$$D_2 = \begin{pmatrix} \Gamma^{-1} + (1-\theta)(2-\theta)L^\top \Sigma L & (1-\theta)L^\top \\ (1-\theta)L & \Sigma^{-1} \end{pmatrix}.$$

Noting that $\Gamma, \Sigma \succ 0$, and using Schur complement for D_1 and P defined in (5.26) we have

$$D_1 \succ 0 \quad \Leftrightarrow \quad \Sigma^{-1} + (1-\theta)L\Gamma L^\top \succ 0, \qquad \Sigma^{-1} - \tfrac{\theta^2}{4}L\Gamma L^\top \succ 0 \quad \Leftrightarrow \quad P \succ 0.$$

Thus, since $1 - \theta \geq -\tfrac{\theta^2}{4}$ for all θ, we have that $D_1 \succ 0$ if $P \succ 0$. It can be shown that the same argument applies for S_1, S_2 and D_2. The sum of two positive definite matrices is also positive definite, therefore $S, D \succ 0$ if $P \succ 0$. The matrix P is symmetric positive definite if (5.15) holds. The convergence conditions in Assumption II are simply the result of replacing D, P, U and V in (5.15).

We showed that Assumption III holds for the described choice of H and S. Algorithm 1 is a simple application of Algorithm 2 for solving the monotone inclusion (5.7). Therefore, the convergence of Algorithm 1 follows directly from that of AFBA (cf. Theorem 3.4). This result is summarized in the following theorem.

Theorem 4.1 (Convergence) *Let Assumptions I and II hold. The sequence $(z^k)_{k\in\mathbb{N}} = (x^k, u^k)_{k\in\mathbb{N}}$ generated by Algorithm 1 converges to a point $z^\star \in \mathbf{zer}\, T$.*

5.4.1 Linear Convergence

In this section we explore sufficient conditions for f, g, h and l under which Algorithm 1 achieves linear convergence. We saw that Algorithm 1 is an instance of AFBA, therefore its linear convergence may be established based on the results developed in Theorem 3.6(*i*). In Lemma 4.3 and 4.5 we provide four regularity assumptions under which T defining the primal-dual optimality conditions is metrically subregular at $z^\star \in \mathbf{zer}\, T$ for 0. The linear convergence for Algorithm 1 is then deduced in Corollary 4.6.

Let us first recall the notion of quadratic growth: a proper closed convex function g is said to have *quadratic growth* at $\bar{x}$ for 0 with $0 \in \partial g(\bar{x})$ if there exists a neighborhood $\mathcal{U}$ of $\bar{x}$ such that

$$g(x) \geq \inf g + cd^2(x, \partial g^{-1}(0)), \quad \forall x \in \mathcal{U}. \tag{5.27}$$

Metric subregularity of the subdifferential operator and the quadratic growth condition are known to be equivalent [3, 19]. In particular, ∂g is metrically subregular at $\bar{x}$ for $\bar{u}$ with $\bar{u} \in \partial g(\bar{x})$ if and only if the quadratic growth condition (5.27) holds for $g(\cdot) - \langle \bar{u}, \cdot \rangle$, i.e., there exists a positive constant c and a neighborhood $\mathcal{U}$ of $\bar{x}$ such that [3, Thm. 3.3]

$$g(x) \geq g(\bar{x}) + \langle \bar{u}, x - \bar{x} \rangle + cd^2(x, \partial g^{-1}(\bar{u})), \quad \forall x \in \mathcal{U}. \tag{5.28}$$

Strong subregularity has a similar characterization [3, Thm. 3.5]

$$g(x) \geq g(\bar{x}) + \langle \bar{u}, x - \bar{x} \rangle + c\|x - \bar{x}\|^2, \quad \forall x \in \mathcal{U}. \tag{5.29}$$

Next, let us define the following general growth condition.

Definition 4.2 (Quadratic Growth Relative to a Set) Consider a proper closed convex function g and a pair $(\bar{x}, \bar{u}) \in \text{gra}\,\partial g$. We say that g has quadratic growth at $\bar{x}$ for $\bar{u}$ relative to a nonempty closed convex set X containing $\bar{x}$, if there exists a positive constant c and a neighborhood $\mathcal{U}$ of $\bar{x}$ such that

$$g(x) \geq g(\bar{x}) + \langle \bar{u}, x - \bar{x} \rangle + cd^2(x, X), \quad \forall x \in \mathcal{U}. \tag{5.30}$$

From the above definition it is evident that metric subregularity and strong subregularity characterized in (5.28) and (5.29) are recovered when $X = \partial g^{-1}(\bar{u})$ and $X = \{\bar{x}\}$, respectively.

Another regularity assumption used in Lemma 4.3 is the notion of local strong convexity: a proper closed convex function g is said to be locally strongly convex in a neighborhood of $\bar{x}$, denoted by $\mathcal{U}$, if there exists a positive constant c such that

$$g(x') \geq g(x) + \langle v, x' - x \rangle + \tfrac{c}{2}\|x' - x\|^2, \quad \forall x, x' \in \mathcal{U}, \ \forall v \in \partial g(x).$$

Notice that local strong convexity in a neighborhood of $\bar{x}$ implies (5.29), but (5.29) is *much weaker* than local strong convexity since it holds only at $\bar{x}$ and only for $\bar{u} \in \partial g(\bar{x})$.

In the next lemma we provide three different regularity assumptions that are sufficient for metric subregularity of the operator defining the primal-dual optimality conditions. In Lemma 4.3(i) (or Lemma 4.3(ii)) we use local strong convexity, as well as the quadratic growth condition (5.30) relative to the set of primal solutions (or dual solutions). Interestingly, this regularity assumption does not entail a unique primal-dual solution.

Lemma 4.3 *Let Assumption I hold. The operator T defining the primal-dual optimality conditions, cf. (5.8), is metrically subregular at $z^\star = (x^\star, u^\star)$ for 0 with $0 \in Tz^\star$ if one of the following assumptions holds:*

(i) *$f + g$ has quadratic growth at $x^\star$ for $-L^\top u^\star$ relative to the set of primal solutions $X^\star$, and $h^* + l^*$ is locally strongly convex in a neighborhood of $u^\star$. In this case the set of dual solutions is a singleton, $U^\star = \{u^\star\}$.*

(ii) *$f + g$ is locally strongly convex in a neighborhood of $x^\star$, and $h^* + l^*$ has quadratic growth at $u^\star$ for $Lx^\star$ relative to the set of dual solutions $U^\star$. In this case the set of primal solutions is a singleton, $X^\star = \{x^\star\}$.*

(iii) *$\nabla f + \partial g$ is strongly subregular at $x^\star$ for $-L^\top u^\star$ and $\partial h^* + \nabla l^*$ is strongly subregular at $u^\star$ for $Lx^\star$. In this case the set of primal-dual solutions is a singleton, $\mathbf{zer}\, T = \{(x^\star, u^\star)\}$.*

Proof

4.3(*i*) Consider the point $z^\star = (x^\star, u^\star)$. By definition of quadratic growth there exists a neighborhood $\mathcal{U}_{x^\star}$ and a positive constant c_1 such that

$$(f+g)(x) \geq (f+g)(x^\star) + \langle -L^\top u^\star, x - x^\star \rangle + c_1 d^2(x, X^\star), \quad \forall x \in \mathcal{U}_{x^\star}.$$
(5.31)

Let $\mathcal{U}_{u^\star}$ denote the neighborhood of $u^\star$ in which the local strong convexity of $h^* + l^*$ holds. Fix a point $z = (x, u) \in \mathcal{Z}_{z^\star} := \mathcal{U}_{x^\star} \times \mathcal{U}_{u^\star}$. Now take $v = (v_1, v_2) \in Tz = Az + Mz + Cz$, i.e.,

$$\begin{cases} v_1 \in \partial g(x) + \nabla f(x) + L^\top u, \\ v_2 \in \partial h^*(u) + \nabla l^*(u) - Lx, \end{cases}$$
(5.32)

Let $z_0 = (x_0, u_0)$ denote the projection of z onto the set of solutions, **zer** T. The subgradient inequality for $f + g$ at x using (5.32) gives

$$\langle v_1, x - x_0 \rangle \geq (f+g)(x) - (f+g)(x_0) + \langle L^\top u, x - x_0 \rangle.$$
(5.33)

Noting that $0 \in Tz_0$, by the subgradient inequality for $h^* + l^*$ at u_0 we have

$$(h^* + l^*)(u) \geq (h^* + l^*)(u_0) + \langle Lx_0, u - u_0 \rangle.$$
(5.34)

Summing (5.33) and (5.34) yields

$$\langle v_1, x - x_0 \rangle \geq \mathcal{L}(x, u) - \mathcal{L}(x_0, u_0) = \mathcal{L}(x, u) - \mathcal{L}(x^\star, u^\star),$$
(5.35)

where $\mathcal{L}$ is the Lagrangian defined in (5.5). By local strong convexity of $h^* + l^*$ at $u \in \mathcal{U}_{u^\star}$ (for some strong convexity parameter c_2):

$$(h^* + l^*)(u^\star) \geq (h^* + l^*)(u) + \langle v_2 + Lx, u^\star - u \rangle + \tfrac{c_2}{2}\|u^\star - u\|^2.$$

Sum this inequality with (5.31) to derive

$$\mathcal{L}(x, u) - \mathcal{L}(x^\star, u^\star) \geq c_1 d^2(x, X^\star) + \langle v_2, u^\star - u \rangle + \tfrac{c_2}{2}\|u^\star - u\|^2, \quad z \in \mathcal{Z}_{z^\star}.$$
(5.36)

It follows from (5.35) and (5.36) that

$$\langle v_2, u - u^\star \rangle + \langle v_1, x - x_0 \rangle \geq \tfrac{c_2}{2}\|u^\star - u\|^2 + c_1 d^2(x, X^\star)$$

$$= \tfrac{c_2}{2}\|u^\star - u\|^2 + c_1 \|x - x_0\|^2$$

$$\geq c\left(\|u^\star - u\|^2 + \|x - x_0\|^2\right),$$
(5.37)

where $c = \mathbf{min}\{c_1, \frac{c_2}{2}\}$. By the Cauchy-Schwarz inequality

$$\langle v_1, x - x_0 \rangle + \langle v_2, u - u^\star \rangle \leq \|v\| \left(\|u - u^\star\|^2 + \|x - x_0\|^2 \right)^{\frac{1}{2}}. \quad (5.38)$$

Combining (5.37) and (5.38) yields

$$\|v\| \geq c \left(\|u - u^\star\|^2 + \|x - x_0\|^2 \right)^{\frac{1}{2}} \geq c\|z - z_0\| = cd(z, T^{-1}0).$$

Since $v \in Tz$ was selected arbitrarily we have that

$$d(z, T^{-1}0) \leq \tfrac{1}{c}d(Tz, 0), \quad \forall z \in \mathcal{Z}_{z^\star}.$$

This completes the first claim. Next, consider $\bar{z}^\star = (\bar{x}^\star, \bar{u}^\star) \in \mathbf{zer}\, T$ such that $\bar{z}^\star \in \mathcal{Z}_{z^\star}$. Setting $z = \bar{z}^\star$ in (5.36) yields

$$0 = \mathcal{L}(\bar{x}^\star, \bar{u}^\star) - \mathcal{L}(x^\star, u^\star) \geq \tfrac{c_2}{2}\|u^\star - \bar{u}^\star\|^2.$$

Therefore, $\bar{u}^\star = u^\star$ and since $\mathbf{zer}\, T$ is convex we conclude that $U^\star = \{u^\star\}$.

4.3(ii) The proof of the second part is similar to part 4.3(i). Therefore, we outline the proof. Let $\mathcal{U}_{x^\star}, \mathcal{U}_{u^\star}$ be the neighborhoods in the definition of local strong convexity of $f + g$ and quadratic growth of $h^* + l^*$, respectively. Fix $z \in \mathcal{Z}_{z^\star} = \mathcal{U}_{x^\star} \times \mathcal{U}_{u^\star}$. Take $v = (v_1, v_2)$ as in (5.32), and let z_0 denote the projection of z onto $\mathbf{zer}\, T$. In contrast to the previous part, sum the subgradient inequalities for $f + g$ at x_0 and for $h^* + l^*$ at u to derive

$$\langle v_2, u - u_0 \rangle \geq \mathcal{L}(x_0, u_0) - \mathcal{L}(x, u) = \mathcal{L}(x^\star, u^\star) - \mathcal{L}(x, u). \quad (5.39)$$

Use local strong convexity of $f + g$ at x, and the quadratic growth of $h^* + l^*$ at $u^\star$ for $Lx^\star$ relative to $U^\star$ to derive

$$\mathcal{L}(x^\star, u^\star) - \mathcal{L}(x, u) \geq c_1 d(u, U^\star) + \langle v_1, x^\star - x \rangle + \tfrac{c_2}{2}\|x - x^\star\|^2.$$

Combining this with (5.39) and arguing as in the previous part completes the proof.

4.3(iii) The proof of this part is slightly different. Let $\mathcal{U}_{x^\star}, \mathcal{U}_{u^\star}$ be the neighborhoods in the definitions of the two strong subregularity assumptions. Fix $z \in \mathcal{Z}_{z^\star} = \mathcal{U}_{x^\star} \times \mathcal{U}_{u^\star}$. Sum the subgradient inequality for $f + g$ at x and for $h^* + l^*$ at u to derive

$$\langle v, z - z^\star \rangle = \langle v_2, u - u^\star \rangle + \langle v_1, x - x^\star \rangle \geq \mathcal{L}(x, u^\star) - \mathcal{L}(x^\star, u) \quad (5.40)$$

On the other hand by [3, Thm. 3.5], $f + g$ has quadratic growth at x^* for $-L^\top u^*$ relative to $\{x^*\}$ and $h^* + l^*$ has quadratic growth at u^* for Lx^* relative to $\{u^*\}$. Summing the two yields

$$\mathcal{L}(x, u^*) - \mathcal{L}(x^*, u) \geq c_2\|u - u^*\|^2 + c_1\|x - x^*\|^2 \qquad (5.41)$$

Combining this inequality with (5.40) and using the Cauchy-Schwartz inequality as in previous parts completes the proof. Uniqueness of the solution follows from (5.41) by setting $z = \bar{z}^* \in \mathbf{zer}\, T$ such that $\bar{z}^* \in \mathcal{Z}_{z^*}$ and using the convexity of $\mathbf{zer}\, T$.   □

The assumptions of Lemma 4.3 are much weaker than strong convexity and do not always imply a unique primal-dual solution. Here we present two simple examples for demonstration. Notice that in the next example the assumption of Lemma 4.3 that $f + g$ has quadratic growth with respect to the set of primal solutions is equivalent to the metric subregularity assumption.

Example 1 Consider the problem

$$\underset{x \in \mathbb{R}}{\textbf{minimize}}\; g(x) + h(x) = \textbf{max}\{1 - x, 0\} + \tfrac{x}{2}\textbf{min}\{x, 0\}$$

The solution to this problem is not unique and any $x^* \in [1, \infty)$ solves this problem. The dual problem is given by

$$\underset{u \in \mathbb{R}}{\textbf{minimize}}\; g^*(-u) + h^*(u), \qquad (5.42)$$

where $g^*(u) = u + \iota_{[-1,0]}(u)$ and $h^*(u) = \tfrac{1}{2}u^2 + \iota_{\leq 0}(u)$. It is evident that the dual problem has the unique solution $u^* = 0$. It is easy to verify that g has quadratic growth at all the points $x^* \in [1, \infty)$ for 0 with respect to $X^* = [1, \infty)$. Moreover, we have $\partial g^{-1}(0) = [1, \infty)$, i.e., $X^* = \partial g^{-1}(0)$. In other words in this case the assumption of Lemma 4.3(i) for g is equivalent to the metric subregularity of ∂g at x^* for 0. Notice that ∂g is not strongly subregular at any point in $[1, \infty)$ for 0. Furthermore, h^* is globally strongly convex given that ∇h is Lipschitz. Therefore, according to Lemma 4.3(i) one would expect a unique dual solution but not necessarily a unique primal solution, which is indeed the case.

Example 2 Let $c \in [-1, 1]$ and consider

$$\underset{x \in \mathbb{R}}{\textbf{minimize}}\; g(x) + h(x) = |x| + cx.$$

When $c \in (-1, 1)$ the problem attains a unique minimum at $x^* = 0$. When $c = 1$ (or $c = -1$) all $x^* \in (-\infty, 0]$ (or $x^* \in [0, \infty)$) solves the problem. The dual problem is given by (5.42) with $g^*(u) = \iota_{[-1,1]}(u)$ and $h^*(u) = \iota_{\{c\}}(u)$. The unique dual solution is $u^* = c$. Furthermore, ∂h^* is strongly subregular at $u^* = c$ for all x^* given that $x^* \in \partial h^*(u^*)$. It is easy to verify that ∂g is metrically subregular at

$x^\star = 0$ for $u^\star \in [-1, 1]$ but is only strongly subregular at $x^\star = 0$ for $u^\star \in (-1, 1)$. Notice that $u^\star = c$, therefore by Lemma 4.3(iii) one would expect a unique primal-dual solution when $u^\star = c \in (-1, 1)$ which is indeed the case.

Another class of functions prevalent in optimization is the class of *piecewise linear-quadratic* (PLQ) functions, which is closed under scalar multiplication, addition, conjugation and Moreau envelope [29]. Recall the notion of piecewise linear-quadratic functions.

Definition 4.4 (Piecewise Linear-Quadratic Function) A function $f : \mathbb{R}^n \to \overline{\mathbb{R}}$ is called piecewise linear-quadratic (PLQ) if its domain can be represented as the union of finitely many polyhedral sets, and in each such set $f(x)$ is given by an expression of the form $\frac{1}{2}x^\top Q x + d^\top x + c$, for some $c \in \mathbb{R}$, $d \in \mathbb{R}^n$, and $Q \in \mathcal{S}^n$.

A wide range of functions used in optimization applications belong to this class, for example: affine functions, quadratic forms, indicators of polyhedral sets, polyhedral norms such as ℓ_1, and regularizers such as elastic net, Huber loss, hinge loss, and many more [2, 29]. For a proper closed convex PLQ function g, ∂g is piecewise polyhedral, and therefore metrically subregular at any z for any z' provided that $z' \in \partial g(z)$ [17, 29].

Lemma 4.5 *Let Assumption I hold. In addition, assume that f, g, l, and h are piecewise linear-quadratic. Then the operator T defining the primal-dual optimality conditions, cf. (5.8), is metrically subregular at any z for any z' provided that $z' \in Tz$.*

Proof The subdifferential ∂g, ∇f, ∂h^* and ∇l^* are piecewise polyhedral [29, Prop. 12.30, Thm. 11.14]. Therefore, A and C are piecewise polyhedral. Furthermore, the graph of M is polyhedral since M is linear. Therefore, the graph of $T = A + M + C$ is also piecewise polyhedral. The inverse of a piecewise polyhedral mapping is also piecewise polyhedral. Therefore by Dontchev and Rockafellar [17, Prop. 3H.1, 3H.3] the mapping T is metrically subregular at z for z' whenever $(z, z') \in \text{gra}\,T$.   □

The next corollary summarizes the linear convergence results based on the two previous lemmas and Theorem 3.6(i).

Corollary 4.6 (Linear Convergence for Algorithm 1) *Let Assumptions I and II hold. In addition, suppose that one of the following assumptions holds:*

(i) *f, g, l, and h are piecewise linear-quadratic.*
(ii) *f, g, l, and h satisfy at least one of the conditions of Lemma 4.3 at every $z^\star \in$ **zer** T (not necessarily the same condition at all the points).*

*Then the sequence $(z^k)_{k \in \mathbb{N}}$ generated by Algorithm 1 converges R-linearly to some $z^\star \in$ **zer** T, and $(d_S(z^k, \textbf{zer}\,T))_{k \in \mathbb{N}}$ converges Q-linearly to zero.*

Acknowledgments This work was supported by FWO PhD grant 1196818N; FWO research projects G086518N and G086318N: Fonds de la Recherche Scientifique – FNRS and the Fonds Wetenschappelijk Onderzoek – Vlaanderen under EOS Project no 30468160: SeLMA.

References

1. A. Alotaibi, P.L. Combettes, N. Shahzad, Solving coupled composite monotone inclusions by successive Fejér approximations of their Kuhn–Tucker set. SIAM J. Optim. **24**(4), 2076–2095 (2014)
2. A.Y. Aravkin, J.V. Burke, G. Pillonetto, Sparse/robust estimation and Kalman smoothing with nonsmooth log-concave densities: modeling, computation, and theory. J. Mach. Learn. Res. **14**, 2689–2728 (2013)
3. F.J.A. Artacho, M.H. Geoffroy, Characterization of metric regularity of subdifferentials. J. Convex Anal. **15**(2), 365–380 (2008)
4. H.H. Bauschke, P.L. Combettes, *Convex Analysis and Monotone Operator Theory in Hilbert Spaces* (Springer, Cham, 2011)
5. D.P. Bertsekas, Extended monotropic programming and duality. J. Optim. Theory Appl. **139**(2), 209–225 (2008)
6. R.I. Boţ, C. Hendrich, A Douglas-Rachford type primal-dual method for solving inclusions with mixtures of composite and parallel-sum type monotone operators. SIAM J. Optim. **23**(4), 2541–2565 (2013)
7. S. Boyd, N. Parikh, E. Chu, B. Peleato, J. Eckstein, Distributed optimization and statistical learning via the alternating direction method of multipliers. Found. Trends Mach. Learn. **3**(1), 1–122 (2011)
8. L.M. Briceño-Arias, P.L. Combettes, A monotone + skew splitting model for composite monotone inclusions in duality. SIAM J. Optim. **21**(4), 1230–1250 (2011)
9. L.M. Briceño-Arias, D. Davis, Forward-backward-half forward algorithm with non self- adjoint linear operators for solving monotone inclusions (2017). arXiv preprint arXiv:1703.03436
10. A. Chambolle, T. Pock, A first-order primal-dual algorithm for convex problems with applications to imaging. J. Math. Imaging Vision **40**(1), 120–145 (2011)
11. A. Chambolle, T. Pock, On the ergodic convergence rates of a first-order primal–dual algorithm. Math. Program. **159**(1), 253–287 (2016)
12. P.L. Combettes, J. Eckstein, Asynchronous block-iterative primal-dual decomposition methods for monotone inclusions. Math. Program. **168**, 645–672 (2016)
13. P.L. Combettes, I.-C. Pesquet, Proximal splitting methods in signal processing, in *Fixed-Point Algorithms for Inverse Problems in Science and Engineering* (Springer, New York, 2011), pp. 185–212
14. P.L. Combettes, J.-C. Pesquet, Primal-dual splitting algorithm for solving inclusions with mixtures of composite, Lipschitzian, and parallel-sum type monotone operators. Set- Valued Var. Anal. **20**(2), 307–330 (2012)
15. L. Condat, A primal-dual splitting method for convex optimization involving Lipschitzian, proximable and linear composite terms. J. Optim. Theory Appl. **158**(2), 460–479 (2013)
16. D. Davis, Convergence rate analysis of primal-dual splitting schemes. SIAM J. Optim. **25**(3), 1912–1943 (2015)
17. A.L. Dontchev, R.T. Rockafellar, *Implicit Functions and Solution Mappings*. Springer Monographs in Mathematics (Springer, New York, 2009), p. 208
18. Y. Drori, S. Sabach, M. Teboulle, A simple algorithm for a class of nonsmooth convex-concave saddle-point problems. Oper. Res. Lett. **43**(2), 209–214 (2015)
19. D. Drusvyatskiy, A.S. Lewis, Error bounds, quadratic growth, and linear convergence of proximal methods (2016). arXiv preprint arXiv:1602.06661

20. E. Esser, X. Zhang, T.F. Chan, A general framework for a class of first order primal-dual algorithms for convex optimization in imaging science. SIAM J. Imag. Sci. **3**(4), 1015–1046 (2010)
21. B. He, X. Yuan, Convergence analysis of primal-dual algorithms for a saddle-point problem: from contraction perspective. SIAM J. Imag. Sci. **5**(1), 119–149 (2012)
22. N. Komodakis, J.C. Pesquet, Playing with duality: an overview of recent primal-dual approaches for solving large-scale optimization problems. IEEE Signal Process. Mag. **32**(6), 31–54 (2015)
23. P. Latafat, P. Patrinos, Asymmetric forward–backward–adjoint splitting for solving monotone inclusions involving three operators. Comput. Optim. Appl. **68** 57–93 (2017)
24. P. Latafat, L. Stella, P. Patrinos, New primal-dual proximal algorithm for distributed optimization, in *55th IEEE Conference on Decision and Control (CDC)* (2016), pp. 1959–1964
25. P. Latafat, N.M. Freris, P. Patrinos, A new randomized block-coordinate primal-dual proximal algorithm for distributed optimization (2017). arXiv preprint arXiv:1706.02882
26. J. Liang, J. Fadili, G. Peyré, Convergence rates with inexact non-expansive operators. Math. Program. **159**(1), 403–434 (2016)
27. D.R. Luke, R. Shefi, A globally linearly convergent method for pointwise quadratically supportable convex–concave saddle point problems. J. Math. Anal. Appl. **457**(2), 1568–1590 (2018). Special Issue on Convex Analysis and Optimization: New Trends in Theory and Applications
28. R.T. Rockafellar, *Convex Analysis* (Princeton University Press, Princeton, 2015)
29. R.T. Rockafellar, R.J.-B. Wets, *Variational Analysis*, vol. 317 (Springer, Berlin, 2009)
30. P. Sopasakis, A. Themelis, J. Suykens, P. Patrinos, A primal-dual line search method and applications in image processing, in *25th European Signal Processing Conference (EUSIPCO)* (2017), pp. 1065–1069
31. S. Sra, S. Nowozin, S.J. Wright, *Optimization for Machine Learning* (MIT Press, Cambridge, 2011). ISBN: 026201646X, 9780262016469
32. B.C. Vũ, A splitting algorithm for dual monotone inclusions involving cocoercive operators. Adv. Comput. Math. **38**(3), 667–681 (2013)

Chapter 6
Block-Coordinate Primal-Dual Method for Nonsmooth Minimization over Linear Constraints

D. Russell Luke and Yura Malitsky

Abstract We consider the problem of minimizing a convex, separable, nonsmooth function subject to linear constraints. The numerical method we propose is a block-coordinate extension of the Chambolle-Pock primal-dual algorithm. We prove convergence of the method without resorting to assumptions like smoothness or strong convexity of the objective, full-rank condition on the matrix, strong duality or even consistency of the linear system. Freedom from imposing the latter assumption permits convergence guarantees for misspecified or noisy systems.

Keywords Saddle-point problems · First order algorithms · Primal-dual algorithms · Coordinate methods · Randomized methods

AMS Subject Classifications 49M29, 65K10, 65Y20, 90C25

6.1 Introduction

We propose a randomized coordinate primal-dual algorithm for convex optimization problems of the form

$$\min_x g(x) \quad \text{s.t.} \quad x \in \operatorname{argmin}_z \frac{1}{2}\|Az - b\|^2. \tag{6.1}$$

This is a generalization of the more commonly encountered linear constrained convex optimization problem

$$\min_x g(x) \quad \text{s.t.} \quad Ax = b. \tag{6.2}$$

D. R. Luke (✉) · Y. Malitsky
Institute for Numerical and Applied Mathematics, University of Göttingen, Göttingen, Germany
e-mail: r.luke@math.uni-goettingen.de; yurii.malitskyi@math.uni-goettingen.de

© Springer Nature Switzerland AG 2018
P. Giselsson, A. Rantzer (eds.), *Large-Scale and Distributed Optimization*, Lecture Notes in Mathematics 2227,
https://doi.org/10.1007/978-3-319-97478-1_6

121

When b is in the range of A problem (6.2) and (6.1) have the same optimal solutions; but (6.1) has the advantage of having solutions even when b is *not* in the range of A. Such problems will be called *inconsistent* in what follows. Of course, the solution set to (6.1) can be modeled by a problem with the format (6.2) via the normal equations. The main point of this note, however, is that the two models suggest very different algorithms with different behaviors, as explained in [28] and in Sect. 6.2 below.

We do not assume that g is smooth, but this is not our main concern. Our main focus in this note is the efficient use of problem structure. In particular, we assume throughout that the problem can be decomposed in the following manner. For $x \in \mathbb{R}^n$, $A \in \mathbb{R}^{m \times n}$

$$A = [A_1, \ldots, A_p] \quad \text{and} \quad g(x) = \sum_{i=1}^{p} g_i(x_i),$$

where $x_i \in \mathbb{R}^{n_i}$, $\sum_{i=1}^{p} n_i = n$, $A_i \in \mathbb{R}^{m \times n_i}$ and $g_i : \mathbb{R}^{n_i} \to (-\infty, +\infty]$ is proper, convex and lower semi-continuous (lsc). The coordinate primal-dual method we propose below allows one to achieve improved stepsize choice, tailored to the individual blocks of coordinates. To this, we add an intrinsic randomization to the algorithm which is particularly well suited for large-scale problems and distributed implementations. Another interesting property of the proposed method is that in the absence of the existence of Lagrange multipliers one can still obtain meaningful convergence results.

Randomization is currently the leading technique for handling extremely large-scale problems. Algorithms employing some sort of randomization have been around for more than 50 years, but they have only recently emerged as the preferred—indeed, sometimes the only feasible—numerical strategy for large-scale problems in machine learning and data analysis. The dominant randomization strategies can be divided roughly into two categories. To the first category belong stochastic methods, where in every iteration the full vector is updated, but only a fraction of the given data is used. The main motivation behind such methods is to generate descent directions cheaply. The prevalent methods SAGA [14] and SVR [23] belong to this group. Another category is coordinate-block methods. These methods update only one coordinate (or one block of coordinates) in every iteration. As with stochastic methods, the per iteration cost is very low since only a fraction of the data is used, but coordinate-block methods can also be accelerated by choosing larger step sizes. Popular methods that belong to this group are [18, 31]. A particular class of coordinate methods is alternating minimization methods, which appear to be a promising approach to solving nonconvex problems, see [1, 5, 21].

To keep the presentation simple, we eschew many possible generalizations and extensions of our proposed method. For concreteness we focus our attention on the primal-dual algorithm (PDA) of Chambolle-Pock [10]. The PDA is a well-known first-order method for solving saddle point problems with nonsmooth structure. It is based on the *proximal* mapping associated with a function g defined by $\text{prox}_{\tau g} = (\text{Id} + \tau \partial g)^{-1}$ where $\partial g(\bar{x})$ is the convex *subdifferential* of g at $\bar{x}$, defined as the set of all vectors v with

$$g(x) - g(\bar{x}) - \langle v, x - \bar{x} \rangle \geq 0 \quad \forall x. \tag{6.3}$$

The PDA applied to the Lagrangian of problem (6.2) generates two sequences (x^k), (y^k) via

$$\begin{aligned}
x^{k+1} &= \operatorname{prox}_{\tau g}(x^k - \tau A^T y^k) \\
y^{k+1} &= y^k + \sigma(A(2x^{k+1} - x^k) - b).
\end{aligned} \tag{6.4}$$

Alternative approaches such as the alternating direction method of multipliers [19] are also currently popular for large-scale problems and are based on the *augmented* Lagrangian of (6.2). The advantage of PDA over ADMM, however, is that it does not require one to invert the matrix A, and hence can be applied to very large-scale problems. The PDA is very similar, in fact, to a special case of the proximal point algorithm [20] and of the so-called *proximal* ADMM [2, 16, 39].

The procedure we study in this note is given in Algorithm 1. The cost per iteration is very low: it requires two dot products $A_i t_i$, $A_i^T y^k$; and the full vector-vector operation is needed only for the dual variables y^k, u^k in steps 6–7. The algorithm will therefore be the most efficient if $m \leq n$. If in particular all blocks are of size 1, that is $n_i = 1$, $n = p$ and A_i is just the i-th column of the matrix A, then $A_i t_i$ reduces to the vector-scalar multiplication and $A_i^T y^k$ to the vector-vector dot product. Moreover, if the matrix A is sparse and A_i is its i-th column, then step 7 requires an update of only those coordinates which are nonzero in A_i. The memory storage is also relatively small: we have to keep x^k, and two dual variables y^k, u^k. Another important feature of the proposed algorithm is that it is well suited for the distributed optimization: since there are no operations with a full primal vector, we can keep different blocks x_i on different machines which are coupled only over dual variables.

We want to highlight that with $p = 1$ the proposed algorithm indeed coincides with the primal-dual algorithm of Chambolle-Pock [10]. In fact, in this case it is not difficult to prove by induction that $u^k = \sigma(Ax^k - b)$ and hence, $y^{k+1} = y^k + \sigma(A(2x^{k+1} - x^k) - b)$.

Algorithm 1 Coordinate primal-dual algorithm

1: Choose $x^0 \in \mathbb{R}^n$, $\sigma > 0$, $\tau \in \mathbb{R}^p_{++}$ and set $y^0 = u^0 = \sigma(Ax^0 - b)$.
2: **for** $k = 0, 1, 2, \ldots$ **do**
3:     $x^{k+1} = x^k$
4:     Pick an index $i \in \{1, \ldots, p\}$ uniformly at random.
5:     $x_i^{k+1} = \operatorname{prox}_{\frac{\tau_i}{p} g_i}(x_i^k - \frac{\tau_i}{p} A_i^T y^k)$, $t_i = x_i^{k+1} - x_i^k$
6:     $y^{k+1} = y^k + u^k + \sigma(p+1)A_i t_i$
7:     $u^{k+1} = u^k + \sigma A_i t_i$
8: **end for**

To the best of our knowledge the first randomized extension of Chambolle-Pock algorithm can be found in [46]. This method was proposed for solving the empirical risk minimization problem and is very popular for such kinds of problems. However, it converges under quite restrictive assumptions that never hold for problem (6.2), see more details in Sect. 6.2. Another interesting coordinate extension of the primal-dual algorithm is studied in [17]. This does not require any special assumptions like smoothness or strong convexity, but, unfortunately, it requires an increase in the dimensionality of the problem, which for our targeted problems is counterproductive.

Recently there have also appeared coordinate methods for abstract fixed point algorithms [13, 22]. Although they provide a useful and general way for developing coordinate algorithms, they do not allow the use of larger stepsizes—one of the major advantages of coordinate methods.

There are also some randomized methods for a particular choice of g in (6.2). For $g \equiv 0$ paper [41] proposes a stochastic version of the Kaczmarz method, see also [25] and a nice review [43]. When g is the elastic net regularization (sum of l_1 and squared l_2 norms), papers [27, 38] provide a stochastic sparse Kaczmarz method. Those methods belong to the first category of randomized methods by our classification, i.e., in every iteration they require an update of the whole vector.

In [28] the connection between the primal-dual algorithm (6.4) and the Tseng proximal gradient method [42] was shown. On the other hand, paper [18] provides a coordinate extension of the latter method for a composite minimization problem. Based on these two results we propose a coordinate primal-dual method which is an extension of the original PDA. The key feature of Algorithm 1 is that it requires neither strong duality nor the consistency of the system $Ax = b$ to achieve good numerical performance with convergence guarantees. This allows us, for instance, to recover the signals from noisy measurements without any knowledge about the noise or without the need to tune some external parameters as one must for lasso or basis denoising problems [8, 12].

In the next section we provide possible applications and connections to other approaches in the literature. Section 6.3 is dedicated to the convergence analysis of our method. We also give an alternative form of the method and briefly discuss possible generalizations of the method. Finally, Sect. 6.4 details several numerical examples.

6.2 Applications

We briefly mention a few of the more prominent applications for Algorithm 1. We begin with the simplest of these.

Linear Programming The linear programming problem

$$\min_{x} \langle c, x \rangle \quad \text{s.t. } Ax = b, \quad x \geq 0 \tag{6.5}$$

is a special case of (6.2) with $g(x) = \langle c, x \rangle + \delta_{x \geq 0}(x)$. In this case g is fully separable. As a practical example, one can consider the optimal transport problem (see for instance [37]).

Composite Minimization   The composite minimization problem

$$\min_{v} f(Kv) + r(v), \tag{6.6}$$

where both f, r are convex functions, $v \in \mathbb{R}^q$, and $K \in \mathbb{R}^{m \times q}$ is a linear operator. For simplicity assume that both f and r are nonsmooth but allow for easy evaluation of the associated proximal mappings. In this case the PDA in [10] is a widely used method to solve such problems. It is applied to (6.6) written in the primal-dual form

$$\min_{v} \max_{w} r(v) + \langle Kv, w \rangle - f^*(w). \tag{6.7}$$

Alternatively, one can recast (6.6) in the form (6.2). To see this, let $x = (v, w)$, $g(x) = r(v) + f(w)$, and $Ax = Kv - w$. Then (6.6) is equivalent to

$$\min_{x} g(x) \quad \text{s.t.} \quad Ax = 0. \tag{6.8}$$

Such reformulation is typical for the augmented Lagrangian method or ADMM [6]. However, this is not very common to do for the primal-dual method, since problem (6.7) is already simpler than (6.8). Although the number of matrix-vector operations for both applications of the PDA remains the same, in (6.8) we have a larger problem: $x \in \mathbb{R}^{m+q}$, and the multiplier $y \in \mathbb{R}^m$ instead of $v \in \mathbb{R}^q$, $w \in \mathbb{R}^m$. However, the advantage for formulation (6.8) over (6.7) is that the updates of the most expensive variable $x = (v, w)$ can be done in parallel, in contrast to the sequential update of v and w when we apply PDA to (6.7).

Still, the main objection to (6.8) is that PDA treats the matrix A as a black box—it does not exploit its structure. In particular, we have $\lambda(A^T A) = \lambda(K^T K) + 1$ which globally defines the stepsizes. But in our case $A = [K \mid -I]$, whose structure is very tempting to use: for the block w in $x = (v, w)$ we would like to apply larger steps that come from $\lambda(I) = 1 \leq \lambda(K^T K)$ (the last inequality is typical). Fortunately, the proposed algorithm does exactly this: for each block the steps are defined only by the respective block of A. This is very similar in spirit to the proximal heterogeneous implicit-explicit method studied in [21]. Notice that the paper [35] provides only the possibility to use different weights for different blocks but it does not allow one to enlarge stepsizes. In [29] a linesearch strategy is introduced that in fact allows one to use larger steps, but still this enlargement is based on the inequalities for the whole matrix A and the vector x. Fercoq and Bianchi [17] propose an extension (coordinate) of the primal-dual method that takes into account the structure of the linear operator, but this modification requires to increase the dimension of the problem.

Algorithm 1 can be applied without any smoothing of the nonsmooth functions f and r. This is different from the approach of [32, 46], where the convergence is only shown when the fidelity term f is smooth and the regularizer r is strongly convex.

Distributed Optimization The aim of distributed optimization over a network is to minimize a global objective using only local computation and communication. Problems in contemporary statistics and machine learning might be so large that even storing the input data is impossible in one machine. For applications and more in-depth overview we refer the reader to [3, 6, 15, 30] and the references therein.

By construction, Algorithm 1 is distributed and hence ideally suited for these types of problems. Indeed, we can assign to the i-th node of our network the block of variables x^i and the respective block matrix A_i. The dual variables y and u reside on the central processor. In each iteration of the algorithm implemented with this architecture only one random node is activated, requiring communication of the current dual vector y^k from the central processor in order to compute the update to the block x_i. The node then returns $A_i t_i$ to the central processor.

The model problem (6.2) is also particularly well-suited for a distributed optimization. Suppose we want to minimize a convex function $f : \mathbb{R}^m \to (-\infty, +\infty]$. A common approach is to assign for each node a copy of x and solve a constrained problem (the *product space formulation*):

$$\min_{x_1,\dots,x_p} f(x_1) + \cdots + f(x_p) \quad \text{s.t.} \quad x_1 = \cdots = x_p. \tag{6.9}$$

In fact, instead of $x_1 = \cdots = x_p$ we can introduce a more general but equivalent[1] constraint $Ax = 0$, where $x = (x_1, \dots, x_p)$ and the matrix A describes the given network. By this, we arrive at the following problem

$$\min_x g(x) \quad \text{s.t.} \quad Ax = 0, \tag{6.10}$$

where $g(x) = f(x_1) + \cdots + f(x_p)$ is a separable convex function. Solving such problems stochastically, where in every iteration only one or few nodes are activated, has attracted a lot of attention recently, see [4, 17, 24].

Inverse Problems Linear systems remain a central occupation of inverse problems. The interesting problems are *ill-posed* and the observed data A, b is noisy, requiring an appropriate *regularization*. A standard way is to consider either[2]

$$\min_x g(x) + \frac{\delta}{2}\|Ax - b\|^2 \quad \text{or} \quad \min_x g(x) \quad \text{s.t.} \quad \|Ax - b\| \le \delta, \tag{6.11}$$

---

[1] This means $Ax = 0$ if and only if $x_1 = \cdots = x_p$.

[2] The left and right problems are also known as Tikhonov and Morozov regularization respectively.

where g is the regularizer which describes our a priori knowledge about a solution and $\delta > 0$ is some parameter. The issue of how to choose this parameter is a major concern of numerical inverse problems. For the right hand-side problem this is easier to do: usually δ corresponds to the noise level of the given problem, but the optimization problem itself is harder than the left one due to the nonlinear constraints. Nevertheless, it can be still efficiently solved via PDA. Let $y = Ax - b$ and h be the indicator function of the closed ball $B(0, \delta)$. Then the above problem can be expressed as

$$\min_{x,y} g(x) + h(y) \quad \text{s.t.} \quad Ax - y = b, \tag{6.12}$$

which is a particular case of (6.2). Again the coordinate primal-dual method, in contrast to the original PDA, has the advantage of exploiting the structure of the matrix $[\,A\,|{-}I\,]$ which allows one to use larger steps for faster convergence of the method.

There is another approach which we want to discuss. In many applications we do not need to solve a regularized problem exactly, the so-called early stopping may even help to obtain a better solution. Theoretically, one can consider another regularized problem: $\min_x g(x)$ such that $Ax = b$. Since during iteration we can easily control the feasibility gap $\|Ax^k - b\|$, we do not need to converge to a feasible solution x^*, where $Ax^* = b$, but stop somewhere earlier. The only issue with such approach is that the system $Ax = b$ might be inconsistent (and due to the noise this is often the case). Hence to be precise, we have to solve the following problem

$$\min_x g(x) \quad \text{s.t.} \quad x \in \text{argmin}_z \left\{ f(z) := \frac{1}{2}\|Az - b\|^2 \right\}. \tag{6.13}$$

Obviously, the above constraint is equivalent to the $A^T A x = A^T b$. Fortunately, we are able to show that our proposed method (without any modification) in fact solves (6.13), so it does not need to work with $A^T A$ that standard analysis of PDA requires. Notice that $A^T A$ is likely to be less sparse and more ill-conditioned than A.

6.3 Analysis

We first introduce some notation. For any vector $\omega = (\omega_1, \ldots, \omega_p) \in \mathbb{R}^p_{++}$ we define the weighted norm by $\|x\|^2_\omega := \sum \omega_i \|x_i\|^2$ and the weighted proximal operator prox^ω_g by

$$\text{prox}^\omega_g := (\text{Id} + \text{Diag}(\omega^{-1}) \circ \partial g)^{-1} = (\text{Diag}(\omega) + \partial g)^{-1} \circ \text{Diag}(\omega).$$

The weighted proximal operator has the following characteristic property (prox-inequality):

$$\bar{x} = \text{prox}^\omega_g z \quad \Leftrightarrow \quad \langle \text{Diag}(\omega)(\bar{x} - z), \, x - \bar{x} \rangle \geq g(\bar{x}) - g(x) \quad \forall x \in \mathbb{R}^n. \tag{6.14}$$

From this point onwards we will fix

$$f(x) := \frac{1}{2}\|Ax - b\|^2. \tag{6.15}$$

Then $\nabla f(x) = A^T(Ax - b)$ and its partial derivative corresponding to i-th block is $\nabla_i f(x) = A_i^T(Ax - b)$. Let $\lambda = (\lambda_1, \dots, \lambda_p)$, where λ_i is the largest eigenvalue of $A_i^T A_i$, that is $\lambda_i = \lambda_{\max}(A_i^T A_i)$. Then the Lipschitz constant of the i-th partial gradient is $\|A_i\|^2 = \lambda_i$. By $U_i \colon \mathbb{R}^n \to \mathbb{R}^n$ we denote the projection operator: $U_i x = (0, \dots, x_i, \dots, 0)$. Since f is quadratic, it follows that for any $x, y \in \mathbb{R}^n$

$$f(y) - f(x) - \langle \nabla f(x), y - x \rangle = \frac{1}{2}\|A(x - y)\|^2. \tag{6.16}$$

We also have that for any $x, t \in \mathbb{R}^n$

$$f(x + U_i t) = f(x) + \langle \nabla f(x), U_i t \rangle + \frac{1}{2}\|A_i t_i\|^2 \le f(x) + \langle \nabla f(x), U_i t \rangle + \frac{\lambda_i}{2}\|t_i\|^2. \tag{6.17}$$

Now since i is a uniformly distributed random number over $\{1, 2, \dots, p\}$, from the above it follows

$$\mathbb{E}[f(x + U_i t)] \le f(x) + \frac{1}{p}\langle \nabla f(x), t \rangle + \frac{1}{2p}\|t\|_\lambda^2. \tag{6.18}$$

For our analysis it is more convenient to work with Algorithm 2 given below. It is equivalent to Algorithm 1 when the random variable i for the blocks are the same, though this might be not obvious at first glance. We give a short proof of this fact. Notice also that Algorithm 2 is entirely primal. The formulation of Algorithm 2 with $p = 1$ (not random!) is related to the Tseng proximal gradient method [28, 42] and to its stochastic extension, the APPROX method [18]. The proposed method requires stepsizes: $\sigma > 0$ and $\tau \in \mathbb{R}_{++}^p$. The necessary condition for convergence of Algorithms 1 and 2 is, as we will see, $\tau_i \sigma \|A_i\|^2 < 1$. We have strict inequality for the same reason that one needs $\tau \sigma \|A\|^2 < 1$ in the original PDA.

Algorithm 2 Coordinate primal-dual algorithm (equivalent form)

1: Choose $x^0 \in \mathbb{R}^n, \sigma > 0, \tau \in \mathbb{R}_{++}^p$ and set $s^0 = x^0, \theta_0 = 1$.
2: **for** $k = 0, 1, 2, \dots$ **do**
3:    $z^k = \theta_k x^k + (1 - \theta_k)s^k$
4:    $x^{k+1} = x^k, s^{k+1} = z^k$
5:    Pick an index $i \in \{1, \dots, p\}$ uniformly at random.
6:    $x_i^{k+1} = \text{prox}_{\frac{\tau_i}{p} g_i}(x_i^k - \frac{\tau_i \sigma}{p \theta_k}\nabla_i f(z^k))$
7:    $s_i^{k+1} = z_i^k + p\theta_k(x_i^{k+1} - x_i^k)$
8:    $\theta_{k+1} = \frac{1}{k+2}$
9: **end for**

Let $f_* = \min f$ and S be the solution set of (6.1). Observe that if the linear system $Ax = b$ is consistent, then $f_* = 0$ and $\operatorname{argmin} f = \{x : Ax = b\}$. Otherwise, $f_* > 0$ and $\operatorname{argmin} f = \{x : A^T Ax = A^T b\}$. We will often use an important simple identity:

$$f(x) - f(\bar{x}) = \frac{1}{2}\|A(x - \bar{x})\|^2 \qquad (\forall x \in \mathbb{R}^n)(\forall \bar{x} \in S). \qquad (6.19)$$

Proposition 1 *If the index i selected at iteration k in step 4 of Algorithm 1 is identical to the index i selected at iteration k in step 5 of Algorithm 2, then both algorithms with the same starting point x^0 generate the same sequence (x^k).*

Proof We show that from Algorithm 2 one can recover all iterates of Algorithm 1 by setting $y^k = \frac{\sigma}{\theta_k}(Az^k - b)$, $u^k = \sigma(Ax^k - b)$. Then the proposition follows, since with $\nabla_i f(x) = A_i^T(Ax - b)$ we have

$$x_i^{k+1} = \operatorname{prox}_{\frac{\tau_i}{p} g_i}(x_i^k - \frac{\tau_i \sigma}{p\theta_k} A_i^T (Az^k - b)) = \operatorname{prox}_{\frac{\tau_i}{p} g_i}(x_i^k - \frac{\tau_i}{p} A_i^T y^k).$$

Evidently, for $k = 0$, one has $y^0 = \sigma(Az^0 - b) = \sigma(Ax^0 - b) = u^0$. Assume it holds for some $k \geq 0$. By step 3 in Algorithm 2, we have

$$Az^{k+1} = \theta_{k+1} Ax^{k+1} + (1 - \theta_{k+1}) As^{k+1}$$

$$= \theta_{k+1}(Ax^k + A_i t_i) + (1 - \theta_{k+1})(Az^k + p\theta_k A_i t_i),$$

where we have used that $Ax^{k+1} = Ax^k + A_i t_i$. Hence,

$$\frac{\sigma}{\theta_{k+1}}(Az^{k+1} - b) = \sigma(Ax^k - b) + \frac{\sigma}{\theta_k}(Az^k - b) + \sigma(p+1)A_i t_i$$

$$= u^k + y^k + \sigma(p+1)A_i t_i,$$

thus $y^{k+1} = \frac{\sigma}{\theta_{k+1}}(Az^{k+1} - b)$. Finally, $\sigma(Ax^{k+1} - b) = u^k + \sigma A_i t_i = u^{k+1}$. $\qquad \square$

We are now ready to state our main result. Since the iterates given by Algorithm 2 are random variables, our convergence result is of a probabilistic nature. Notice also that equalities and inequalities involving random variables should be always understood to hold almost surely, even if the latter is not explicitly stated.

Theorem 1 *Let $(x^k), (s^k)$ be given by Algorithm 2, $\tau_i \sigma \|A_i\|^2 < 1$ for all $i = 1, \ldots, p$, and S be the solution set of (6.1). Then*

(i) *If there exists a Lagrange multiplier for problem (6.1), then (x^k) and (s^k) converge a.s. to a solution of (6.1) and $f(x^k) - f_* = o(1/k)$, $f(s^k) - f_* = O(1/k^2)$ a.s. for the feasibility residual (6.15).*

(ii) *If S is a bounded set and g is bounded from below, then almost surely all limit points of (s^k) belong to S and $f(s^k) - f_* = o(1/k)$.*

The proof of Theorem 1 is based on several simple lemmas which we establish first. The next lemma uses the following notation for the full update, $\hat{x}^{k+1}$, of x^k in the k-th iteration of Algorithm 2, namely

$$\hat{x}^{k+1} = \text{prox}_{g/p}^{\tau^{-1}} \left(x^k - \frac{k+1}{p} \sigma \, \text{Diag}(\tau) \nabla f(z^k) \right). \tag{6.20}$$

Lemma 1 *For any fixed $x \in \mathbb{R}^n$ and any $k \in \mathbb{N}$ the following inequality holds:*

$$\frac{p}{2} \| \hat{x}^{k+1} - x^k \|_{\tau^{-1}}^2 + \frac{\sigma}{\theta_k} \langle \nabla f(z^k), \, \hat{x}^{k+1} - x \rangle + g(\hat{x}^{k+1}) - g(x)$$

$$\leq \frac{p}{2} \| x^k - x \|_{\tau^{-1}}^2 - \frac{p}{2} \| \hat{x}^{k+1} - x \|_{\tau^{-1}}^2. \tag{6.21}$$

Proof By the prox-inequality (6.14) with $\hat{x}^{k+1}$ given by (6.20) we have

$$\langle \text{Diag}(\tau^{-1})(\hat{x}^{k+1} - x^k), \, x - \hat{x}^{k+1} \rangle + \frac{\sigma}{p\theta_k} \langle \nabla f(z^k), \, x - \hat{x}^{k+1} \rangle \geq \frac{1}{p}(g(\hat{x}^{k+1}) - g(x)). \tag{6.22}$$

The statement then follows from the identity

$$\langle \text{Diag}(\tau^{-1})(\hat{x}^{k+1} - x^k), \, x - \hat{x}^{k+1} \rangle$$

$$= \frac{1}{2} \| x^k - x \|_{\tau^{-1}}^2 - \frac{1}{2} \| \hat{x}^{k+1} - x \|_{\tau^{-1}}^2 - \frac{1}{2} \| \hat{x}^{k+1} - x^k \|_{\tau^{-1}}^2. \tag{6.23}$$

$\square$

The next result provides a bound on the expectation of the residual at the $(k+1)$th iterate, conditioned on the kth iterate, which we denote by $\mathbb{E}_k$.

Lemma 2 *For any $x^* \in S$*

$$\frac{1}{\theta_k^2} \mathbb{E}_k[f(s^{k+1}) - f_*] \leq \frac{1}{\theta_{k-1}^2}(f(s^k) - f_*) - (f(x^k) - f_*) \tag{6.24}$$

$$+ \frac{1}{\theta_k} \langle \nabla f(z^k), \, \hat{x}^{k+1} - x^* \rangle$$

$$+ \frac{p}{2} \| \hat{x}^{k+1} - x^k \|_\lambda^2.$$

Proof First, by (6.18)

$$\mathbb{E}_k[f(s^{k+1})] \leq f(z^k) + \theta_k \langle \nabla f(z^k), \, \hat{x}^{k+1} - x^k \rangle + \theta_k^2 \frac{p}{2} \| \hat{x}^{k+1} - x^k \|_\lambda^2. \tag{6.25}$$

Since f is quadratic (6.15), by (6.16) we have

$$f(s^k) - f(z^k) = \langle \nabla f(z^k), s^k - z^k \rangle + \frac{1}{2} \|A(s^k - z^k)\|^2 \qquad (6.26)$$

$$= \frac{\theta_k}{1 - \theta_k} \langle \nabla f(z^k), z^k - x^k \rangle + \frac{1}{2} \|A(s^k - z^k)\|^2.$$

By (6.16) and $\|\alpha a + (1 - \alpha)b\|^2 = \alpha \|a\|^2 + (1 - \alpha)\|b\|^2 - \alpha(1 - \alpha)\|a - b\|^2$, we have

$$f_* - f(z^k) = f(x^*) - f(z^k) = \langle \nabla f(z^k), x^* - z^k \rangle + \frac{1}{2} \|A(z^k - x^*)\|^2 \qquad (6.27)$$

$$= \langle \nabla f(z^k), x^* - z^k \rangle + \theta_k(f(x^k) - f_*) + (1 - \theta_k)(f(s^k) - f_*)$$

$$- \frac{1}{2}\theta_k(1 - \theta_k)\|A(x^k - s^k)\|^2.$$

Notice that $x^k - s^k = \frac{1}{\theta_k}(s^k - z^k)$. Hence, summation of $\frac{1 - \theta_k}{\theta_k}$(6.26) and (6.27) yields

$$\frac{1 - \theta_k}{\theta_k} f(s^k) - \frac{1}{\theta_k} f(z^k) + f_* = \langle \nabla f(z^k), x^* - x^k \rangle + \theta_k(f(x^k) - f_*) \qquad (6.28)$$

$$+ (1 - \theta_k)(f(s^k) - f_*),$$

from which we conclude

$$\frac{1}{\theta_k^2} f(z^k) + \frac{1}{\theta_k} \langle \nabla f(z^k), x^* - x^k \rangle = \frac{(1 - \theta_k)^2}{\theta_k^2} f(s^k) + \frac{2 - \theta_k}{\theta_k} f_* - (f(x^k) - f_*).$$

$$(6.29)$$

Now summing up (6.25) multiplied by $\frac{1}{\theta_k^2}$ and (6.29), and using the identity $\frac{1 - \theta_k}{\theta_k} = \frac{1}{\theta_{k-1}}$, we obtain

$$\frac{1}{\theta_k^2} \mathbb{E}_k[f(s^{k+1}) - f_*] \leq \frac{1}{\theta_{k-1}^2}(f(s^k) - f_*) - (f(x^k) - f_*) \qquad (6.30)$$

$$+ \frac{1}{\theta_k} \langle \nabla f(z^k), \hat{x}^{k+1} - x^* \rangle$$

$$+ \frac{p}{2} \|\hat{x}^{k+1} - x^k\|_\lambda^2.$$

$\square$

The conditional expectations have the following useful representations. Expanding the expectation,

$$\mathbb{E}_k[\|x^{k+1} - x\|^2_{\tau-1}] = \sum_{i=1}^{p} \tau_i^{-1} \mathbb{E}_k[\|x_i^{k+1} - x_i\|^2]$$

$$= \sum_{i=1}^{p} \tau_i^{-1}\left(\frac{1}{p}\|\hat{x}_i^{k+1} - x\|^2 + \frac{p-1}{p}\|\hat{x}_i^k - x\|^2\right)$$

$$= \frac{1}{p}\|\hat{x}^{k+1} - x\|^2_{\tau-1} + \frac{p-1}{p}\|x^k - x\|^2_{\tau-1}.$$

This yields

$$\mathbb{E}_k\left[\frac{p^2}{2}\|x^{k+1} - x\|^2_{\tau-1}\right] = \frac{p}{2}\|\hat{x}^{k+1} - x\|^2_{\tau-1} + \frac{p^2 - p}{2}\|x^k - x\|^2_{\tau-1}. \qquad (6.31)$$

Another characterization we use follows similarly, namely

$$\mathbb{E}_k[g(x^{k+1})] = \sum_{i=1}^{p}\left(\frac{1}{p}g_i(\hat{x}_i^{k+1}) + \frac{p-1}{p}g_i(x_i^k)\right) = \frac{1}{p}g(\hat{x}^{k+1}) + \frac{p-1}{p}g(x^k)$$

which gives

$$\mathbb{E}_k[g(x^{k+1})] = \frac{1}{p}g(\hat{x}^{k+1}) + \left(1 - \frac{1}{p}\right)g(x^k). \qquad (6.32)$$

The next technical lemma is the last of the necessary preparations before proceeding to the proof of the main theorem.

Lemma 3 *The identity* $s^{k+1} = \sum_{j=0}^{k+1} \beta_{k+1}^j x^j$ *holds where the coefficients* $(\beta_{k+1}^j)_{j=0}^{k+1}$ *are nonnegative and sum to 1. In particular,*

$$\beta_{k+1}^j = \begin{cases} (1 - \theta_k)\beta_k^j, & j = 0, \ldots, k-1, \\ p\theta_{k-1}(1 - \theta_k) - (p-1)\theta_k, & j = k, \\ p\theta_k, & j = k+1. \end{cases} \qquad (6.33)$$

and $\beta_{k+1}^k + (p-1)\theta_k = (1 - \theta_k)\beta_k^k$.

Proof It is easy to prove by induction. For the reference see Lemma 2 in [18]. □

The proof of Theorem 1 consists of three parts. The first contains an important estimation, derived from previous lemmas. In the second and third parts we respectively and independently prove (i) and (ii). The condition in (ii) is not restrictive and often can be checked in advance without any effort. In contrast, verifying strong duality (existence of Lagrange multipliers) is usually very difficult.

Proof (Theorem 1) Since $\tau_i \sigma \|A_i\|^2 < 1$ for all $i = 1, \ldots, p$, there exists $\varepsilon > 0$ such that $\tau_i^{-1} - \sigma \|A_i\|^2 \geq \varepsilon$. This yields for all $x \in \mathbb{R}^n$, $\|x\|_{\tau^{-1}}^2 - \sigma \|x\|_\lambda^2 \geq \|x\|_\varepsilon^2$.

By convexity of g,

$$g(s^k) = g(\sum_{j=0}^{k} \beta_k^j x^j) \leq \sum_{j=0}^{k} \beta_k^j g(x^j) =: \hat{g}_k.$$

Let $\hat{F}_k := \hat{g}_k + \sigma k(f(s^k) - f_*) \geq g(s^k) + \sigma k(f(s^k) - f_*)$.

By Lemma 3 and (6.32), it follows

$$\mathbb{E}_k[\hat{g}_{k+1}] = \mathbb{E}_k[\beta_{k+1}^{k+1} g(\hat{x}^{k+1})] + \sum_{j=0}^{k} \beta_{k+1}^j g(x^j) \tag{6.34}$$

$$= \theta_k g(\hat{x}^{k+1}) + ((p-1)\theta_k + \beta_{k+1}^k)g(x^k) + \sum_{j=0}^{k-1} \beta_{k+1}^j g(x^j) \tag{6.35}$$

$$= \theta_k g(\hat{x}^{k+1}) + (1-\theta_k)\beta_k^k g(x^k) + (1-\theta_k)\sum_{j=0}^{k-1} \beta_k^j g(x^j) \tag{6.36}$$

$$= \theta_k g(\hat{x}^{k+1}) + (1-\theta_k)\sum_{j=0}^{k} \beta_k^j g(x^j) = \theta_k g(\hat{x}^{k+1}) + (1-\theta_k)\hat{g}_k. \tag{6.37}$$

Let $x^* \in S$. Setting $x = x^*$ in (6.21) and adding to σ times (6.24) yields

$$\frac{\sigma}{\theta_k^2}\mathbb{E}_k[f(s^{k+1})] - f_*] \leq \frac{\sigma}{\theta_{k-1}^2}(f(s^k) - f_*) - \sigma(f(x^k) - f_*) + \frac{p}{2}\|x^k - x^*\|_{\tau^{-1}}^2$$

$$- \frac{p}{2}\|\hat{x}^{k+1} - x^*\|_{\tau^{-1}}^2 - \frac{p}{2}\|\hat{x}^{k+1} - x^k\|_\varepsilon^2 - (g(\hat{x}^{k+1}) - g_*). \tag{6.38}$$

Summing up (6.31) with $x = x^*$, (6.37) multiplied by $1/\theta_k = (k+1)$, and (6.38), we obtain

$$\mathbb{E}_k\left[\frac{p^2}{2}\|x^{k+1} - x^*\|_{\tau^{-1}}^2 + (k+1)(\hat{F}_{k+1} - g_*)\right] \leq \frac{p^2}{2}\|x^k - x^*\|_{\tau^{-1}}^2 + k(\hat{F}_k - g_*) \tag{6.39}$$

$$- \sigma(f(x^k) - f_*)$$

$$- \frac{p}{2}\|\hat{x}^{k+1} - x^k\|_\varepsilon^2.$$

If the term inside the expectation in (6.39), $V_k(x^*) := \frac{p^2}{2}\|x^k - x^*\|^2_{\tau-1} + k(\hat{F}_k - g_*)$, were nonnegative, then one could apply the supermartingale theorem [3] to obtain almost sure convergence directly. In our case this term need not be nonnegative, however it suffices to have $V_k(x^*)$ bounded from below. With the assumption that there exists a Lagrange multiplier u^*, we shall show this.

(i) Let u^* be a Lagrange multiplier for a solution x^* to problem (6.1). Then by definition of the saddle point we have

$$g(s^k) - g_* = g(s^k) - g(x^*) \geq \langle A^T A u^*, s^k - x^* \rangle$$

$$\geq -\|Au^*\| \cdot \|A(s^k - x^*)\| = -\delta\sqrt{f(s^k) - f_*},$$
$$\text{(6.40)}$$

where we have used (6.19) and denoted $\delta := \sqrt{2}\|Au^*\|$. Our goal is to show that $k(\hat{F}_k - g_*)$ is bounded from below. Indeed

$$k(\hat{F}_k - g_*) \geq k(g(s^k) - g_*) + \sigma k^2(f(s^k) - f_*)$$

$$\geq \frac{\sigma k^2}{2}(f(s^k) - f_*) + \left(\frac{\sigma k^2}{2}(f(s^k) - f_*) - \delta k\sqrt{f(s^k) - f_*}\right).$$
$$\text{(6.41)}$$

The real-valued function $t \mapsto \frac{\sigma k^2}{2}t^2 - \delta k t$ attains its smallest value $-\frac{\delta^2}{2\sigma}$ at $t = \frac{\delta}{\sigma k}$. Hence, we can deduce in (6.41) that

$$k(\hat{F}_k - g_*) \geq \frac{\sigma k^2}{2}(f(s^k) - f_*) - \frac{\delta^2}{2\sigma} \geq -\frac{\delta^2}{2\sigma}.$$
$$\text{(6.42)}$$

We can then apply the supermartingale theorem to the shifted random variable $V_k(x^*) + \delta^2/(2\sigma)$ to conclude that

$$\sum_{k=0}^{\infty} \sigma(f(x^k) - f_*) + \frac{p}{2}\|\hat{x}^{k+1} - x^k\|^2_\varepsilon < \infty \quad a.s.$$

and the sequence $V_k(x^*)$ converges almost surely to a random variable with distribution bounded below by $-\frac{\delta^2}{2\sigma}$. Thus $(f(x^k) - f_*) = o(1/k)$, $\lim_{k\to\infty} \|\hat{x}^{k+1} - x^k\|^2_\varepsilon = 0$ and, by the definition of $V_k(x^*)$, the sequence (x^k) is pointwise a.s. bounded thus $k(\hat{F}_k - g_*)$ is bounded above and hence, by (6.42), $f(s^k) - f_* = O(1/k^2)$ a.s.

By the definition of z^k in step 3 of Algorithm 2 with $\theta_k = \frac{1}{k+1}$, we conclude that $f(z^k) - f_* = O(1/k^2)$ a.s. This yields a useful estimate:

$$\langle \nabla f(z^k), \hat{x}^{k+1} - x^* \rangle = \langle A^T(Az^k - b), \hat{x}^{k+1} - x^* \rangle = \langle A^T A(z^k - x^*), \hat{x}^{k+1} - x^* \rangle$$

$$\leq \|A(z^k - x^*)\| \|A(\hat{x}^{k+1} - x^*)\|$$

$$= \sqrt{(f(z^k) - f_*)(f(\hat{x}^{k+1}) - f_*)}. \tag{6.43}$$

Since $f(\hat{x}^{k+1}) - f_* \to 0$ a.s., we have $\frac{\sigma}{\theta_k} \langle \nabla f(z^k), \hat{x}^{k+1} - x^* \rangle \to 0$.

Now consider the pointwise sequence of realizations of the random variables x^k over $\omega \in \Omega$ for which the sequence is bounded. Since these realizations of sequences are bounded, they possess cluster points. Let x' be one such cluster point. This point is feasible since in the limit $f(x') - f_* = 0$. Writing down inequality (6.22) for the convergent subsequence with $x = x^*$ and taking the limit we obtain $g(x') \leq g(x^*)$. Thus the pointwise cluster points $x' \in S$ a.s.

We show next that the cluster points are unique. Again, for the same pointwise realization over $\omega \in \Omega$ as above, suppose there is another cluster point x''. By the argument above, $x'' \in S$. The point $x^* \in S$ was arbitrary and so we can replace this by x'. To reduce clutter, denote $\alpha_k = k(\hat{F}_k - g_*)$ and denote the subsequence of (x^k) converging to x' by (x^{k_i}) for $i \in \mathbb{N}$ and the subsequence converging to x'' by (x^{k_j}) for $j \in \mathbb{N}$. We thus have

$$\lim_{k \to \infty} V_k(x') = \lim_{i \to \infty} V_{k_i}(x') = \lim_{i \to \infty} \left(\frac{p^2}{2} \|x^{k_i} - x'\|_{\tau^{-1}}^2 + \alpha_{k_i} \right) = \lim_{i \to \infty} \alpha_{k_i}$$

$$\tag{6.44}$$

$$= \lim_{j \to \infty} V_{k_j}(x') = \lim_{j \to \infty} \left(\frac{p^2}{2} \|x^{k_j} - x'\|_{\tau^{-1}}^2 + \alpha_{k_j} \right) = \|x'' - \tilde{x}_1\|_{\tau^{-1}}^2$$

$$+ \lim_{j \to \infty} \alpha_{k_j}, \tag{6.45}$$

from which $\lim_{i \to \infty} \alpha_{k_i} = \frac{p^2}{2} \|x'' - x'\|_{\tau^{-1}}^2 + \lim_{j \to \infty} \alpha_{k_j}$ follows. Similarly, replacing $\tilde{x}$ with x'', we derive $\lim_{j \to \infty} \alpha_{k_j} = \frac{p^2}{2} \|\tilde{x}_1 - x''\|_{\tau^{-1}}^2 + \lim_{i \to \infty} \alpha_{k_i}$, from which we conclude that $x' = x''$. Therefore, the whole sequence (x^k) converges pointwise almost surely to a solution.

Convergence of (s^k) First, recall from Lemma 3 that s^k is a convex combination of x^j for $j = 0, \ldots, k$. Second, notice that for all j, $\beta_{k+1}^j \to 0$ as $k \to \infty$. Hence, by the Toeplitz theorem (Exercise 66 in [36]) we conclude that (s^k) converges pointwise almost surely to the same solution as (x^k). By this, (i) is proved.

(ii) Taking the total expectation $\mathbb{E}$ of both sides of (6.39), we obtain

$$\mathbb{E}\left[V_{k+1}(x^*) + \frac{p}{2} \|\hat{x}^{k+1} - x^k\|_\varepsilon^2\right] \leq \mathbb{E}[V_k(x^*)].$$

Iterating the above,

$$\mathbb{E}\left[\frac{p^2}{2}\|x^k - x^*\|_{\tau-1}^2 + k(\hat{F}_k - g_*) + \sum_{i=1}^{k} \frac{p}{2}\|\hat{x}^i - x^{i-1}\|_{\varepsilon}^2\right] \le \frac{p^2}{2}\|x^0 - x^*\|_{\tau-1}^2 =: C.$$

(6.46)

Since $g(s^k) \le \hat{g}_k$, one has

$$\mathbb{E}\left[\frac{p^2}{2}\|x^k - x^*\|_{\tau-1}^2 + \sigma k^2(f(s^k) - f_*) + \sum_{i=1}^{k} \frac{p}{2}\|\hat{x}^i - x^{i-1}\|_{\varepsilon}^2\right] \le C + \mathbb{E}[k(g_* - g(s^k))].$$

(6.47)

From the above equation it follows that

$$\mathbb{E}\left[\frac{p^2}{2k}\|x^k - x^*\|_{\tau-1}^2 + \sigma k(f(s^k) - f_*)\right] \le \frac{C}{k} + \mathbb{E}[g_* - g(s^k)].$$ (6.48)

Recall that by our assumption, g is bounded from below: let $g(x) \ge -l$ for some $l > 0$. Thus, from (6.48) one can conclude that

$$\mathbb{E}[f(s^k) - f_*] \le \frac{C}{\sigma k^2} + \frac{g_* + l}{\sigma k}.$$

This means that almost surely $f(s^k) - f_* = O(1/k)$. Later we will improve this estimation. Let Ω_1 be an event of probability one such that for every $\omega \in \Omega_1$, $f(s^k(\omega)) - f_* = O(1/k)$.

Boundedness of (s^k) First, we show that $(s^i(\omega))_{i \in \mathcal{I}}$ with $\mathcal{I} = \{i : g(s^i) < g_*\}$ is a bounded sequence for every $\omega \in \Omega_1$. For this we revisit the arguments from [28, 40]. By assumption the set $S = \{x : g(x) \le g_*, f(x) \le f_*\}$ is nonempty and bounded. Consider the convex function $\varphi(x) = \max\{g(x) - g_*, f(x) - f_*\}$. Notice that S coincides with the level set $\mathcal{L}(0)$ of φ:

$$S = \mathcal{L}(0) = \{x : \varphi(x) \le 0\}.$$

Since $\mathcal{L}(0)$ is bounded, the set $\mathcal{L}(c) = \{x : \varphi(x) \le c\}$ is also bounded for any $c \in \mathbb{R}$. Fix any $c \ge 0$ such that $f(s^k(\omega)) \le c$ for all k and $\omega \in \Omega_1$. Since $g(s^i) - g_* < 0 \le c$ for $i \in \mathcal{I}$, we have that for every $\omega \in \Omega_1$, $s^i(\omega) \in \mathcal{L}(c)$, which is a bounded set. Hence, $(s^i(\omega))_{i \in \mathcal{I}}$ is bounded for every $\omega \in \Omega_1$.

Note that from (6.47) we have a useful consequence: both sequences (x^j), $(\hat{x}^j - x^{j-1})$ with $j \notin \mathcal{I}$ are pointwise a.s. bounded. From this it follows immediately that for $j \notin \mathcal{I}$, $(x^j - x^{j-1})$ is pointwise a.s. bounded as well. Hence, there exists an event Ω_2 of probability one such that $x^j(\omega)$, $(x^j - x^{j-1})(\omega)$ are bounded for each $\omega \in \Omega_2$. We now restrict ourselves to points $\omega \in \Omega_1 \cap \Omega_2$. Clearly, the latter set has a probability one. Let M_ω be such a constant that, for $\omega \in \Omega_1 \cap \Omega_2$

$$\|s^i(\omega)\| \leq M_\omega \qquad\qquad \forall i \in \mathcal{I}$$
$$\|(x^j + (n-1)(x^j - x^{j-1}))(\omega)\| \leq M_\omega \qquad \forall j \notin \mathcal{I}. \qquad (6.49)$$

We prove by induction that the sequence $(s^k(\omega))$ is bounded, hence (s^k) is pointwise a.s. bounded. Suppose that for index k, $\|s^k(\omega)\| \leq M_\omega$. If for index $k+1$, $g(s^{k+1}(\omega)) < g_*$ then we are done: $k+1 \in \mathcal{I}$ and hence, $\|s^{k+1}(\omega)\| \leq M_\omega$. If $g(s^{k+1}(\omega)) \geq g_*$, then $(k+1) \notin \mathcal{I}$. By the definition of s^k in Algorithm 2, $s^{k+1} = \theta_k(nx^{k+1} - (n-1)x^k) + (1 - \theta_k)s^k$, and hence

$$\|s^{k+1}\| \leq \theta_k \|x^{k+1} + (n-1)(x^{k+1} - x^k)\| + (1 - \theta_k)\|s^k\|,$$

which shows that $\|s^{k+1}(\omega)\| \leq M_\omega$. This completes the proof that $s^k(\omega)$ is bounded and hence (s^k) is pointwise a.s. bounded.

Convergence of (s^k) Recall that a.s. $f(s^k) - f_* = O(1/k)$. Hence, all limit points of $s^k(\omega)$ are feasible for any $\omega \in \Omega_1 \cap \Omega_2$. This means that for every $\omega \in \Omega_1 \cap \Omega_2$, $\liminf_{k\to\infty} g(s^k(\omega)) \geq g_*$. If one assumes that there exist an event Ω_3 of non-zero probability such that for every $\omega \in \bigcap_{i=1}^3 \Omega_i$, $\limsup_{k\to\infty} g(s^k(\omega)) > g_*$, then taking the limit superior in (6.48), one obtains a contradiction. This yields that almost surely $\lim_{k\to\infty} g(s^k) = g_*$ and hence, almost surely all limit points of (s^k) belong to S. Using the obtained result, we can improve the estimation for $f(s^k) - f_*$. Now from (6.48) it follows that almost surely $f(s^k) - f_* = o(1/k)$. $\qquad\square$

6.3.1 Possible Generalizations

Strong Convexity When g is strongly convex in (6.2), the estimates in Theorem 1 can be improved and in case (ii) the convergence of (s^k) to the solution of (6.2) can be proved. For this one needs to combine the proposed analysis and the analysis from [28].

Adding the Smooth Term and ESO Instead of solving (6.2), one can impose more structure for this problem:

$$\min_x g(x) + h(x) \quad \text{s.t.} \quad Ax = b, \qquad (6.50)$$

where in addition to the assumptions in Sect. 6.1, we assume that $h: \mathbb{R}^N \to \mathbb{R}$ is a smooth convex function with a Lipschitz-continuous gradient ∇h. In this case the new coordinate algorithm can be obtained by combining analysis from [18] and [28]. The expected separable overapproximation (ESO), proposed in [18], can help here much more, due to the new additional term h.

Parallelization For simplicity, our analysis has focused only on the update of one block in every iteration. Alternatively, one could choose a random subset of blocks and update all of them more or less by the same logic as in Algorithm 1. This is one of the most important aspects of the algorithm that provides for fast implementations. The analysis of such extension will involve only a bit more tedious expressions with expectation, for more details we refer again to [18], where this was done for the unconstrained case: $\min_x g(x) + h(x)$. More challenging is to develop an asynchronous parallel extension of the proposed method in the spirit of [34], but with the possibility of taking larger steps.

6.4 Numerical Experiments

This section collects several numerical tests to illustrate the performance of the proposed methods. Computations[3] were performed using Python 3.5 on a standard laptop running 64-bit Debian GNU/Linux.

6.4.1 Basis Pursuit

For a given matrix $A \in \mathbb{R}^{m \times n}$ and observation vector $b \in \mathbb{R}^m$, we want to find a sparse vector $x^\dagger \in \mathbb{R}^n$ such that $Ax^\dagger = b$. In standard circumstances we have $m \ll n$, so the linear system is underdetermined. Basis pursuit, proposed by Chen and Donoho [11, 12], involves solving

$$\min_x \|x\|_1 \quad \text{s.t.} \quad Ax = b. \tag{6.51}$$

In many cases it can be shown (see [7] and references therein) that basis pursuit is able to recover the true signal $x^\dagger$ even when $m \ll n$.

A simple observation says that problem (6.51) fits well in our framework. For given m, n we generate the matrix $A \in \mathbb{R}^{m \times n}$ in two ways:

- Experiment 1: A is a random Gaussian matrix, i.e., every entry of A drawn from the normal distribution $\mathcal{N}(0, 1)$. A sparse vector $x^\dagger \in \mathbb{R}^n$ is constructed by choosing at random 5% of its entries independently and uniformly from $(-10, 10)$.
- Experiment 2: $A = R\Phi$, where $\Phi \in \mathbb{R}^{n \times n}$ is the discrete cosine transform and $R \in \mathbb{R}^{m \times n}$ is the random projection matrix: Rx randomly extracts m coordinates from the given x. A sparse vector $x^\dagger \in \mathbb{R}^n$ is constructed by choosing at random

---

[3] All codes can be found on https://gitlab.gwdg.de/malitskyi/coo-pd.git.

50 of its entries among the first 100 coordinates independently from the normal distribution $\mathcal{N}(0, 1)$. The remaining entries of $x^\dagger$ are set to 0.

For both experiments we generate $b = Ax^\dagger$. The starting point for all algorithms is $x^0 = 0$. For all methods we use the same stopping criteria:

$$\|Ax^k - b\|_\infty \leq 10^{-6} \quad \text{and} \quad \text{dist}(-A^T y^k, \partial_{\|\cdot\|_1}(x^k))_\infty \leq 10^{-6}.$$

For given A and b we compare the performance of PDA and the proposed methods: Block-PDA with $n_{\text{block}} = n/w$ blocks of the width w each and Coo-PDA with every single coordinate as a block. In the first and the second experiments the width of blocks for the Block-PDA is $w = 50$.

The numerical behavior of all three methods depends strongly on the choice of the stepsizes. It is easy to artificially handicap PDA in comparisons with almost any algorithm: just take very bad steps for PDA. To be fair, for every test problem we present the best results of PDA among all with steps $\sigma = \dfrac{1}{2^j \|A\|}, \tau = \dfrac{2^j}{\|A\|}$, for $j = -15, -14, \ldots, 15$. Instead, for the proposed methods we always take the same step $\sigma = \dfrac{1}{2^j n_{\text{block}}}$, where we set $j = 11$ for the first experiment and $j = 8$ for the second one.

Tables 6.1 and 6.2 collect information of how many epochs and how much CPU time (in seconds) is needed for each method and every test problem to reach the 10^{-6} accuracy. The term "epoch" means one iteration of the PDA or n_{block} iterations of the coordinate PDA. By this, after k epochs the i-th coordinate of x will be updated on average the same number of times for the PDA and our proposed method. The CPU time might be not a good indicator, since it depends on the platform and the implementation. However, the number of epochs gives an exact number of arithmetic operations. This is a fair benchmark characteristic.

Table 6.1 Comparison of PDA, Block-PDA and Coo-PDA for problem (6.51)—Experiment 1

Algorithm	$m = 1000, n = 4000$		$m = 2000, n = 8000$		$m = 4000, n = 16,000$	
	Epoch	CPU	Epoch	CPU	Epoch	CPU
PDA	777	24	815	89	829	333
Block-PDA	108	4	103	12	107	51
Coo-PDA	79	2	73	7	94	34

Table 6.2 Comparison of PDA, Block-PDA and Coo-PDA for problem (6.51)—Experiment 2

Algorithm	$m = 1000, n = 4000$		$m = 2000, n = 8000$		$m = 4000, n = 16,000$	
	Epoch	CPU	Epoch	CPU	Epoch	CPU
PDA	303	9	284	29	286	122
Block-PDA	41	2	40	5	36	18
Coo-PDA	27	1	23	3	24	10

Concerning the stepsizes, we reiterate that the PDA was always taken with the best steps for every particular problem: for the first experiment (when A is Gaussian), the parameter j was usually among 4, 5, 6, 7. For the second experiment, the best values of j were among $0, \ldots, 6$.

6.4.2 Basis Pursuit with Noise

In many cases, a more realistic scenario is when our measurements are noisy. Now we wish to recover a sparse vector $x^\dagger \in \mathbb{R}^n$ such that $Ax^\dagger + \varepsilon = b$, where $\varepsilon \in \mathbb{R}^m$ is the unknown noise. Two standard approaches [8, 12] involve solving either the lasso problem or basis pursuit denoising:

$$\min_x \|x\|_1 + \frac{\delta}{2}\|Ax - b\|^2 \quad \text{or} \quad \min_x \|x\|_1 \quad \text{s.t.} \quad \|Ax - b\| \le \delta. \quad (6.52)$$

In order to recover $x^\dagger$, both problems require a delicate tuning of the parameter δ. For this one usually requires some a priori knowledge about the noise.

Even with noise, we still apply our method to the plain basis pursuit problem

$$\min_x \|x\|_1 \quad \text{s.t.} \quad Ax = b. \quad (6.53)$$

We know that the proposed method converges even when the linear system is inconsistent (and this is the case when A is rank-deficient). First, we may hope that the actual solution x^* of (6.53) to which the method converges is not far from the true $x^\dagger$. In fact, if the system $Ax = b$ is consistent, we have $\|A(x^* - x^\dagger)\| = \varepsilon$. Similarly, if that system is inconsistent, by (6.19) we have $f(x^\dagger) - f(x^*) = \frac{1}{2}\|A(x^\dagger - x^*)\|^2$. And hence, again we obtain $\|A(x^\dagger - x^*)\|^2 \le 2f(x^\dagger) = \varepsilon^2$. Second, it might happen that the trajectory of (x^k) at some point is even closer to $x^\dagger$; in this case the early stopping can help to recover $x^\dagger$. In fact, as our simulations show, the convergence of the coordinate PDA to the actual solution x^* is very slow, though, it is unusually fast in the beginning.

For given $m = 1000$, $n = 4000$ we generate two scenarios:

- $A \in \mathbb{R}^{m \times n}$ is a random Gaussian matrix.
- $A = A_L A_R$, where $A_L \in \mathbb{R}^{m \times m/2}$ and $A_R \in \mathbb{R}^{m/2 \times n}$ are two random Gaussian matrices.

The latter condition guarantees that the rank of A is at most $m/2$. Hence, with a high probability the system $Ax = b$ will be inconsistent due to the noise.

We generate a sparse vector $x^\dagger \in \mathbb{R}^n$ choosing at random 50 of its elements independently and uniformly from $(-10, 10)$. Then we define $b \in \mathbb{R}^m$ in two ways: either $b = Ax^\dagger + \varepsilon$, where $\varepsilon \in \mathbb{R}^m$ is a random Gaussian vector or b is obtained by rounding off every coordinate of $Ax^\dagger$ to the nearest integer.

For simplicity we run only the block PDA with blocks of width 50, thus we have $n_{\text{block}} = n/50$. The parameter σ is chosen as $\sigma = \frac{1}{2^{25}n_{\text{block}}}$. After every k-th epoch we compute the signal error $\frac{\|x^k - x^\dagger\|}{\|x^\dagger\|}$ and the feasibility gap $\|A^T(Ax^k - b)\|$. The results are presented in Figs. 6.1 and 6.2.

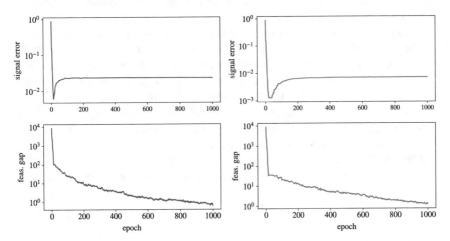

Fig. 6.1 Convergence of block-PDA for noisy basis pursuit. A is a random Gaussian matrix. Left (blue): problem with $b = Ax^\dagger + \varepsilon$, where ε is a random Gaussian vector. Right (green): problem with b obtained by rounding off $Ax^\dagger$ to the nearest integer. Top: signal error. Bottom: feasibility gap

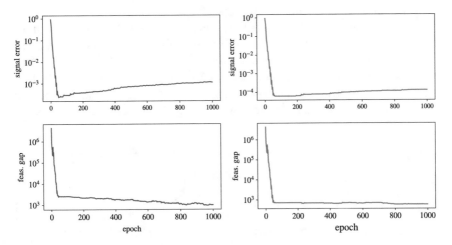

Fig. 6.2 Convergence of block-PDA for noisy basis pursuit. A is a low-rank matrix $A = A_L A_R$. Left (blue): problem with $b = Ax^\dagger + \varepsilon$, where ε is a random Gaussian vector. Right (green): problem with b obtained by rounding off $Ax^\dagger$ to the nearest integer. Top: signal error. Bottom: feasibility gap

As one can see from the plots, the convergence of (x^k) to the best approximation of $x^\dagger$ takes place just after few epochs (less than 100). This convergence is very fast for both the signal error and the feasibility gap. After that, both lines switch to the slow regime: the signal error slightly increases and stabilizes after that; the feasibility gap decreases relatively slowly. Although in practice we do not know $x^\dagger$, we still can use early stopping, since both lines change their behavior approximately at the same time and it is easy to control the feasibility gap.

Finally, we would like to illustrate the proposed approach on a realistic non-sparse signal. We set $m = 1000, n = 4000$ and generate a random signal $w \in \mathbb{R}^n$ that has a sparse representation in the dictionary of the discrete cosine transform, that is $w = \Phi x^\dagger$ with the matrix $\Phi \in \mathbb{R}^{n \times n}$ is the discrete cosine transform and $x^\dagger \in \mathbb{R}^n$ is the sparse vector with only 50 non-zero coordinates drawn from $\mathcal{N}(0, 1)$. The measurements are modeled by a random Gaussian matrix $M \in \mathbb{R}^{m \times n}$. The observed data is corrupted by noise: $b = Mw + \varepsilon$, where $\varepsilon \in \mathbb{R}^m$ is a random vector, whose entries are drawn from $\mathcal{N}(0, 10)$.

Obviously, we can rewrite the above equation as $b = Ax + \varepsilon$, where $A = M\Phi \in \mathbb{R}^{m \times n}$. We apply the proposed block-coordinate primal-dual method to the problem

$$\min_x \|x\|_1 \quad \text{s.t.} \quad Ax = b \tag{6.54}$$

with $\sigma = \frac{1}{2^{22} n_{\text{block}}}$ and the width 50 of each block. The behavior of the iterates x^k is depicted in Fig. 6.3. Figure 6.4 shows the true signal w and the reconstructed signal $\hat{w} = \Phi x^{30}$ after 30 epochs of the method. The signal error in this case is $\frac{\|\hat{w}-w\|}{\|w\|} = 0.0036$. Interestingly, after 1000 epochs we obtain the reconstructed signal $\hat{w}^{1000}$ for which the error is also quite reasonable: $\frac{\|\hat{w}^{1000}-w\|}{\|w\|} = 0.016$.

Notice that we have reconstructed the signal without any knowledge about the noise. The only parameter which requires some tuning from our side was the stepsize σ. However, the method is not very sensitive to the choice of σ. In fact, we were able to reconstruct the signal at least for any $\sigma = \frac{1}{2^j n_{\text{block}}}$ for j from the range $15, \ldots, 30$. Only the number of iterations of the method and the obtained accuracy changed, though still it was always enough for a good reconstruction.

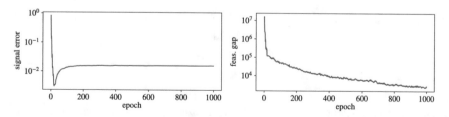

Fig. 6.3 Convergence of block-PDA for noisy basis pursuit with a non-sparse signal. $A = M\Phi$. Left: signal error. Right: feasibility gap

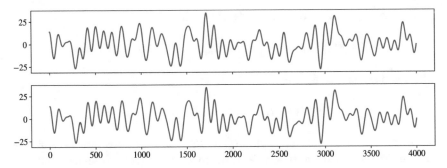

Fig. 6.4 Top: original signal w. Bottom: reconstructed signal $\hat{w}$ after 30 epochs. The signal error is 0.0036

6.4.3 Robust Principal Component Analysis

Given the observed data matrix $M \in \mathbb{R}^{n_1 \times n_2}$, robust principal component analysis aims to find a low rank matrix L and a sparse matrix S such that $M = L + S$. Problems which involve so many unknowns and the "low-rank" component should be difficult (in fact it is NP-hard), however, they can be successfully handled via robust principal component analysis, modeled by the following convex optimization problem

$$\min_{L,S} \|L\|_* + \lambda \|S\|_1 \quad \text{s.t.} \quad L + S = M. \tag{6.55}$$

Here $\|L\|_* = \sum_i \sigma_i(L)$ is the nuclear norm of L, $\|S\|_1 = \sum_{i,j} |S_{ij}|$, and $\lambda > 0$. In [9, 44] it was shown that under some mild assumptions problem (6.55) indeed recovers the true matrices L and S. The RPCA has many applications, see [9] and the references therein.

Problem (6.55) is well-suited for both ADMM [26, 45] and PDA. In fact, since the linear operator A that defines the linear constraint has a simple structure $A = [I \,|\, I]$, those two methods almost coincide. Obviously, the bottleneck for both methods is computing the prox-operator with respect to the variable L, which involves computing a singular value decomposition (SVD). As (6.55) is a particular case of (6.2) with two blocks, one can apply the proposed coordinate PDA. In this case, in every iteration one should update either the L block or the S block. Hence, on average k iterations of the method require only $k/2$ SVD. Keeping this in mind, we can hope for a faster convergence of our method compared to the original PDA.

For the experiment we use the settings from [44]. For given n_1, n_2 and rank r we set $L^\dagger = Q_1 Q_2$ with $Q_1 \in \mathbb{R}^{n_1 \times r}$ and $Q_2 \in \mathbb{R}^{r \times n_2}$, whose elements are i.i.d. $\mathcal{N}(0, 1)$ random variables. Then we generate $S^\dagger \in \mathbb{R}^{n_1 \times n_2}$ as a sparse matrix whose 5% non-zero elements are chosen uniformly at random from the range $[-500, 500]$. We set $M = L^\dagger + S^\dagger$ and $\lambda = \frac{1}{\sqrt{n_1}}$.

The KKT optimality conditions for (6.55) yield

$$Y \in \partial(\lambda\|S\|_1), \qquad Y \in \partial(\|L\|_*), \qquad S + L - M = 0,$$

which implies that (L, S) is optimal whenever

$$\partial(\lambda\|S\|_1) \cap \partial(\|L\|_*) \neq \varnothing \quad \text{and} \quad S + L - M = 0.$$

Since we do not want to compute $\partial(\|L\|_*)$ and $\partial(\|S\|_1)$, we just measure the distance between their two subgradients which we can compute from the iterates.

Given the current iterates (L^k, S^k, Y^k) for the PDA, we know that $\frac{1}{\tau}(L^{k-1} - L^k) - Y^{k-1} \in \partial(\|L^k\|_*)$ and $\frac{1}{\tau}(S^{k-1} - S^k) - Y^{k-1} \in \partial(\|S^k\|_1)$. Hence, we can terminate algorithm whenever

$$\frac{\|L^{k-1} - L^k + S^k - S^{k-1}\|}{\tau\|M\|} \leq \varepsilon \quad \text{and} \quad \frac{\|S^k + L^k - M\|_\infty}{\|M\|} \leq \varepsilon$$

where $\varepsilon > 0$ is the desired accuracy. A similar criteria was used for the Coo-PDA termination.

We run several instances of PDA with stepsizes $\tau = \frac{2^j}{L}$ and $\sigma = \frac{1}{2^j L}$ for $j = -6, -5, \ldots, 12$, where $L = \sqrt{2}$ is the norm of the operator $A = [\,I\,|\,I\,]$. Similarly, we run several instances of Coo-PDA with $\tau = (1, 1)$ and $\sigma = \frac{1}{2^j}$ for the same range of indices j. We show only the performance of the best instances for both methods, and it was always the case that $j \in \{6, 7, 8\}$. The accuracy ε was set to 10^{-6}. We also compare PDA-r and Coo-PDA-r, where instead of an exact SVD a fast randomized SVD solver was used. We chose the one from the scikit-learn library [33].

In Table 6.3 we show the benchmark for PDA and Coo-PDA with an exact evaluation of SVD and for PDA-r and Coo-PDA-r with randomized SVD. For different input data, the table collects the total number of iterations and the CPU time in seconds for all methods. Notice that for the Coo-PDA the number of evaluation of SVD is approximately half of the iterations, this is why it terminates faster.

Table 6.3 Comparison of PDA and Coo-PDA for problem (6.55) with exact and approximate evaluation of SVD

n_1	n_2	r	PDA		Coo-PDA		PDA-r		Coo-PDA-r	
			Iter	CPU	Iter	CPU	Iter	CPU	Iter	CPU
1000	500	20	161	173	219	121	104	39	159	29
1500	500	20	150	211	200	142	139	71	182	47
2000	500	50	154	279	205	185	130	87	180	61
1000	1000	50	188	1004	251	678	124	111	174	76
2000	1000	50	160	1355	200	879	91	146	120	95

We want to highlight that in all experiments the unknown matrix $L^\dagger$ was indeed recovered. And, as one can see from Table 6.3, in all experiments the coordinate primal-dual algorithms performed better than the standard ones.

Acknowledgements This research was supported by the German Research Foundation grant SFB755-A4.

References

1. H. Attouch, J. Bolte, P. Redont, A. Soubeyran, Proximal alternating minimization and projection methods for nonconvex problems: an approach based on the Kurdyka-Lojasiewicz inequality. Math. Oper. Res. **35**(2), 438–457 (2010)
2. S. Banert, R.I. Bot, E.R. Csetnek, Fixing and extending some recent results on the ADMM algorithm. arXiv:1612.05057 (2016, Preprint)
3. D.P. Bertsekas, J.N. Tsitsiklis, *Parallel and Distributed Computation: Numerical Methods* (Prentice-Hall, Upper Saddle River, 1989). ISBN: 0-13-648700-9
4. P. Bianchi, W. Hachem, I. Franck, A stochastic coordinate descent primal-dual algorithm and applications, in *2014 IEEE International Workshop on Machine Learning for Signal Processing (MLSP)* (IEEE, Piscataway, 2014), pp. 1–6
5. J. Bolte, S. Sabach, M. Teboulle, Proximal alternating linearized minimization for non-convex and nonsmooth problems. Math. Program. **146**(1–2), 459–494 (2014)
6. S. Boyd, N. Parikh, E. Chu, B. Peleato, J. Eckstein, Distributed optimization and statistical learning via the alternating direction method of multipliers. Found. Trends Mach. Learn. **3**(1), 1–122 (2011)
7. E.J. Candès, M.B. Wakin, An introduction to compressive sampling. IEEE Signal Process. Mag. **25**(2), 21–30 (2008)
8. E.J. Candès, J.K. Romberg, T. Tao, Stable signal recovery from incomplete and inaccurate measurements. Commun. Pure Appl. Math. **59**(8), 1207–1223 (2006)
9. E.J. Candès, X. Li, Y Ma, J. Wright, Robust principal component analysis? J. ACM **58**(3), 11 (2011)
10. A. Chambolle, T. Pock, A first-order primal-dual algorithm for convex problems with applications to imaging. J. Math. Imag. Vis. **40**(1), 120–145 (2011)
11. S.S. Chen, Basis pursuit, Ph.D. thesis, Department of Statistics, Stanford University Stanford, 1995
12. S.S. Chen, D.L. Donoho, M.A. Saunders, Atomic decomposition by basis pursuit. SIAM Rev. **43**(1), 129–159 (2001)
13. P.L. Combettes, J.-C. Pesquet, Stochastic quasi-Fejér block-coordinate fixed point iterations with random sweeping. SIAM J. Optim. **25**(2), 1221–1248 (2015)
14. A. Defazio, F. Bach, S. Lacoste-Julien, SAGA: a fast incremental gradient method with support for non-strongly convex composite objectives, in *Advances in Neural Information Processing Systems* (2014), pp. 1646–1654
15. J.C. Duchi, A. Agarwal, M.J. Wainwright, Dual averaging for distributed optimization: convergence analysis and network scaling. IEEE Trans. Autom. Control **57**(3), 592–606 (2012)
16. J. Eckstein, Some saddle-function splitting methods for convex programming. Optim. Methods Softw. **4**(1), 75–83 (1994)
17. O. Fercoq, P. Bianchi, A coordinate descent primal-dual algorithm with large step size and possibly non separable functions. arXiv:1508.04625 (2015, Preprint)
18. O. Fercoq, P. Richtaórik, Accelerated, parallel, and proximal coordinate descent. SIAM J. Optim. **25**(4), 1997–2023 (2015)

19. R. Glowinski, A. Marroco, Sur l'approximation, par elements finis d'ordre un, et las resolution, par penalisation-dualite' d'une classe de problemes de dirichlet non lineares. Revue Francais d'Automatique Informatique et Recherche Opeórationelle **9**(R-2), 41–76 (1975)
20. B. He, X. Yuan, Convergence analysis of primal-dual algorithms for a saddle-point problem: from contraction perspective. SIAM J. Imag. Sci. **5**(1), 119–149 (2012)
21. R. Hesse, D.R. Luke, S. Sabach, M.K. Tam, Proximal heterogeneous block implicit-explicit method and application to blind ptychographic diffraction imaging. SIAM J. Imag. Sci. **8**(1), 426–457 (2015)
22. F. Iutzeler, P. Bianchi, P. Ciblat, W. Hachem, Asynchronous distributed optimization using a randomized alternating direction method of multipliers, in *2013 IEEE 52nd Annual Conference on Decision and Control (CDC)* (IEEE, Piscataway, 2013), pp. 3671–3676
23. R. Johnson, T. Zhang, Accelerating stochastic gradient descent using predictive variance reduction, in *Advances in Neural Information Processing Systems* (2013), pp. 315–323
24. P. Latafat, N.M. Freris, P. Patrinos, A new randomized block-coordinate primal-dual proximal algorithm for distributed optimization. arXiv:1706.02882 (2017, Preprint)
25. D. Leventhal, A.S. Lewis, Randomized methods for linear constraints: convergence rates and conditioning. Math. Oper. Res. **35**(3), 641–654 (2010)
26. Z. Lin, M. Chen, Y. Ma, The augmented Lagrange multiplier method for exact recovery of corrupted low-rank matrices. arXiv:1009.5055 (2010, Preprint)
27. D.A. Lorenz, F. Schoüpfer, S. Wenger, The linearized Bregman method via split feasibility problems: analysis and generalizations. SIAM J. Imag. Sci. **7**(2), 1237–1262 (2014)
28. Y. Malitsky, The primal-dual hybrid gradient method reduces to a primal method for linearly constrained optimization problems. arXiv:1706.02602 (2017, Preprint)
29. Y. Malitsky, T. Pock, A first-order primal-dual algorithm with linesearch. SIAM J. Optim. **28**(1), 411–432 (2018)
30. A. Nedic, A. Ozdaglar, P.A. Parrilo, Constrained consensus and optimization in multi-agent networks. IEEE Trans. Autom. Control **55**(4), 922–938 (2010)
31. Y. Nesterov, Efficiency of coordinate descent methods on huge-scale optimization problems. SIAM J. Optim. **22**(2), 341–362 (2012)
32. B. Palaniappan, F. Bach, Stochastic variance reduction methods for saddle-point problems, in *Advances in Neural Information Processing Systems* (2016), pp. 1416–1424
33. F. Pedregosa, G. Varoquaux, A. Gramfort, V. Michel, B. Thirion, O. Grisel, M. Blondel, P. Prettenhofer, R. Weiss, V. Dubourg, J. Vanderplas, A. Passos, D. Cournapeau, M. Brucher, M. Perrot, E. Duchesnay, Scikit-learn: machine learning in Python. J. Mach. Learn. Res. **12**, 2825–2830 (2011)
34. Z. Peng, Y. Xu, M. Yan, W. Yin, Arock: an algorithmic framework for asynchronous parallel coordinate updates. SIAM J. Sci. Comput. **38**(5), A2851–A2879 (2016)
35. T. Pock, A. Chambolle, Diagonal preconditioning for first order primal-dual algorithms in convex optimization, in *2011 IEEE International Conference on Computer Vision (ICCV)* (IEEE, Piscataway, 2011), pp. 1762–1769
36. G. Poólya, G. Szegoü, *Problems and Theorems in Analysis I.* : Series. Integral Calculus Theory of Functions (Springer, Berlin, 1978)
37. F. Santambrogio, *Optimal Transport for Applied Mathematicians* (Birkaüuser, New York, 2015)
38. F. Schoüpfer, D.A. Lorenz, Linear convergence of the randomized sparse Kaczmarz method. arXiv:1610.02889 (2016, Preprint)
39. R. Shefi, M. Teboulle, Rate of convergence analysis of decomposition methods based on the proximal method of multipliers for convex minimization. SIAM J. Optim. **24**(1), 269–297 (2014)
40. M. Solodov, An explicit descent method for bilevel convex optimization. J. Convex Anal. **14**(2), 227 (2007)
41. T. Strohmer, R. Vershynin, A randomized Kaczmarz algorithm with exponential convergence. J. Fourier Anal. Appl. **15**(2), 262–278 (2009)

42. P. Tseng, On accelerated proximal gradient methods for convex-concave optimization (2008), http://www.mit.edu/dimitrib/PTseng/papers/apgm.pdf
43. S.J. Wright, Coordinate descent algorithms. Math. Program. **151**(1), 3–34 (2015)
44. J. Wright, A. Ganesh, S. Rao, Y. Peng, Y. Ma, Robust principal component analysis: exact recovery of corrupted low-rank matrices via convex optimization, in *Advances in Neural Information Processing Systems* (2009)
45. X. Yuan, J. Yang, Sparse and low-rank matrix decomposition via alternating direction methods. Pac. J. Optim. **9**, 167–180 (2013)
46. Y. Zhang, X. Lin, Stochastic primal-dual coordinate method for regularized empirical risk minimization, in *Proceedings of the 32nd International Conference on Machine Learning (ICML-15)* (2015), pp. 353–361

Chapter 7
Stochastic Forward Douglas-Rachford Splitting Method for Monotone Inclusions

Volkan Cevher, Bằng Công Vũ, and Alp Yurtsever

Abstract We propose a stochastic Forward-Douglas-Rachford Splitting framework for finding a zero point of the sum of three maximally monotone operators, one of which is cocoercive, in a real separable Hilbert space. We characterize the rate of convergence in expectation for strongly monotone operators. We further provide guidance on step-size sequence selection that achieve this rate, even when the strong convexity parameter is unknown.

Keywords Monotone inclusion · Monotone operator · Operator splitting · Cocoercive · Forward backward algorithm · Composite operator · Duality · Primal-dual algorithm

AMS Subject Classifications 47H05, 74S60, 49M29, 49M27, 90C25

7.1 Introduction

Forward-backward and Douglas-Rachford splitting methods are two fundamental algorithms for solving monotone inclusion problems in real Hilbert spaces [4]. These methods have also been successfully applied for solving various convex optimization problems from a vast set of applications; see [2, 7, 10, 14–16, 37] and the references therein.

Recently, a new splitting method appeared in the literature, unifying these two fundamental methods into the generalized forward-backward splitting method [31]. However, it is shown in [8] that the generalized forward-backward splitting is in fact a special instance of a more general splitting framework, namely the forward-Douglas-Rachford. Towards this goal, [17] extends the forward-Douglas-Rachford to the three-operator splitting method.

V. Cevher · B. C. Vũ · A. Yurtsever (✉)
École Polytechnique Fédérale de Lausanne (EPFL), Lausanne, Switzerland
e-mail: volkan.cevher@epfl.ch; bang.vu@epfl.ch; alp.yurtsever@epfl.ch

© Springer Nature Switzerland AG 2018
P. Giselsson, A. Rantzer (eds.), *Large-Scale and Distributed Optimization*, Lecture Notes in Mathematics 2227,
https://doi.org/10.1007/978-3-319-97478-1_7

149

These key results so far remained within the so-called deterministic setting. However, solving monotone inclusions in the stochastic setting is of great interest [12, 13, 30, 33–35, 39]. The main advantage of the stochastic approach against the deterministic setting is the reduced computational cost in large scale problems. In parallel to their deterministic counterparts, stochastic forward-backward splitting methods are proposed in [12, 13, 33]. A stochastic version of the Douglas-Rachford splitting can be found in [13]. Based on a product space reformulation technique and the work in [16], additional stochastic primal-dual methods are introduced in [12, 30, 34] for solving monotone inclusions involving cocoercive operators.

The objective of this work is to extend the recent advances in the deterministic methods involving three operators [8, 17, 31] to the stochastic setting. To this end, we focus on the stochastic version of the three-operator splitting method for solving the following problem template:

Problem 1 Let β be a strictly positive number, $(\mathcal{H}, \langle \cdot \mid \cdot \rangle)$ be a real separable Hilbert space, $\mathbf{A} \colon \mathcal{H} \to 2^{\mathcal{H}}$ and $\mathbf{B} \colon \mathcal{H} \to 2^{\mathcal{H}}$ be maximally monotone operators, $\mathbf{U}$ be self adjoint and positive definite, and let $\mathbf{Q} \colon \mathcal{H} \to \mathcal{H}$ satisfy

$$(\forall \mathbf{x} \in \mathcal{H})(\forall \mathbf{y} \in \mathcal{H}) \quad \langle \mathbf{x} - \mathbf{y} \mid \mathbf{Qx} - \mathbf{Qy} \rangle \geq \beta \langle \mathbf{Qx} - \mathbf{Qy} \mid \mathbf{U}(\mathbf{Qx} - \mathbf{Qy}) \rangle.$$

Define P as the set of points $\mathbf{x}$ in $\mathcal{H}$ such that

$$0 \in \mathbf{Ax} + \mathbf{Bx} + \mathbf{Qx}. \tag{7.1}$$

Our aim is to find a random vector that is P-valued almost surely (i.e., a random vector $\mathbf{x}$ that is in P almost surely).

Problem 1 covers a wide class of primal monotone inclusions, primal-dual monotone inclusions in product space, convex optimization, stochastic optimization, split feasibility and variational inequality problems [3–5, 14, 15, 22, 38].

The recent methods in [12, 35] are proposed for solving the special case of Problem 1 with $\mathbf{B} = 0$. A direct application of these methods in our template requires the computation the resolvent of the sum $\mathbf{A} + \mathbf{B}$, which is typically very expensive to evaluate in practice. Based on the framework in [17], we propose a splitting method which uses the stochastic estimate of the cocoercive operator $\mathbf{Q}$ and the resolvents of $\mathbf{A}$ and $\mathbf{B}$ separately. To the best of our knowledge, this is the first purely primal stochastic splitting method for solving Problem 1. Our proof technique is based on the stochastic Fejér monotone sequence [12], applied to the three operator splitting setting. Numerical experiments comparing the proposed method against the deterministic counterpart in some convex optimization applications can be found in [40].

The rest of this manuscript is organized as follows: Sect. 7.2 introduces the notation and it recalls some basic notions from the monotone operator theory. Then, Sect. 7.3 presents the main algorithm and proves the weak almost sure convergence. Finally, Sect. 7.4 establishes the rate of convergence in expectation under strong monotonicity assumption.

7.2 Notation, Background and Preliminary Results

Throughout the paper, $\mathcal{H}$ is a real separable Hilbert space. $\langle \cdot \mid \cdot \rangle$ and $\|\cdot\|$ denote the scalar product and its associated norm in $\mathcal{H}$. The symbols $\rightharpoonup$ and $\rightarrow$ denote weak and strong convergence respectively. We denote by $\ell_+^1(\mathbb{N})$ the set of summable sequences in $[0, +\infty[$, and by $\mathcal{B}(\mathcal{H})$ the space of linear operators from $\mathcal{H}$ to $\mathcal{H}$.

Let $U \in \mathcal{B}(\mathcal{H})$ be a self-adjoint and positive definite operator. Then, we define

$$\langle x \mid y \rangle_U = \langle Ux \mid y \rangle, \quad \text{and} \quad \|x\|_U = \sqrt{\langle Ux \mid x \rangle}.$$

The set of all fixed points of $T : \mathcal{H} \to \mathcal{H}$ is

$$\mathrm{Fix}(T) = \{x \in \mathcal{H} \mid Tx = x\}.$$

Let $A : \mathcal{H} \to 2^{\mathcal{H}}$ be a set-valued operator. The domain and the graph of A are

$$\mathrm{dom}\, A = \{x \in \mathcal{H} \mid Ax \neq \varnothing\}, \quad \text{and} \quad \mathrm{gra}\, A = \{(x, u) \in \mathcal{H} \times \mathcal{H} \mid u \in Ax\}.$$

The set of zeros and the range of A are

$$\mathrm{zer}\, A = \{x \in \mathcal{H} \mid 0 \in Ax\}, \quad \text{and} \quad \mathrm{ran}\, A = \{u \in \mathcal{H} \mid \exists x \in \mathcal{H}, \, u \in Ax\}.$$

The inverse of A is

$$A^{-1} : \mathcal{H} \to 2^{\mathcal{H}} : u \mapsto \{x \in \mathcal{H} \mid u \in Ax\}.$$

We denote the identity operator of $\mathcal{H}$ by Id. Then the resolvent of A is defined as

$$J_A = (\mathrm{Id} + A)^{-1}.$$

The parallel sum of $A : \mathcal{H} \to 2^{\mathcal{H}}$ and $B : \mathcal{H} \to 2^{\mathcal{H}}$ is defined as

$$A \,\square\, B = (A^{-1} + B^{-1})^{-1}.$$

A is a monotone operator if it satisfies

$$\langle x - y \mid u - v \rangle \geq 0, \quad \forall (x, u) \in \mathrm{gra}\, A, \ \forall (y, v) \in \mathrm{gra}\, A.$$

Moreover, it is maximally monotone if there exists no monotone operator $\widetilde{A} : \mathcal{H} \to \mathcal{H}$ such that $\mathrm{gra}\, A \subset \mathrm{gra}\, \widetilde{A} \neq \mathrm{gra}\, A$.

Let $\Gamma_0(\mathcal{H})$ be the class of proper lower semicontinuous convex functions from $\mathcal{H}$ to $]-\infty, +\infty]$. For any self-adjoint strongly positive operator $U \in \mathcal{B}(\mathcal{H})$ and

$f \in \Gamma_0(\mathcal{H})$, we define the proximal operator as

$$\mathrm{prox}_f^U : \mathcal{H} \to \mathcal{H} : x \mapsto \underset{y \in \mathcal{H}}{\mathrm{argmin}} \left(f(y) + \frac{1}{2}\|x - y\|_U^2 \right).$$

We use the convention $\mathrm{prox}_f = \mathrm{prox}_f^{\mathrm{Id}}$ for simplicity. Note that $\mathrm{prox}_f^U = J_{U^{-1}\partial f}$.
The conjugate function of f is

$$f^* : a \mapsto \sup_{x \in \mathcal{H}} \left(\langle a \mid x \rangle - f(x) \right).$$

Note that $\forall f \in \Gamma_0(\mathcal{H})$ and $\forall x \in \mathrm{dom}\,\partial f$,

$$y \in \partial f(x) \Leftrightarrow x \in \partial f^*(y),$$

or equivalently, $(\partial f)^{-1} = \partial f^*$.

The infimal convolution of the two functions f and g from $\mathcal{H}$ to $]-\infty, +\infty]$ is defined as

$$f \,\square\, g : x \mapsto \inf_{y \in \mathcal{H}} \left(f(y) + g(x - y) \right).$$

The relation between the infimal convolution and the parallel sum can be found in [4, Proposition 25.32].

The strong relative interior of a subset C of $\mathcal{H}$ is the set of points $x \in C$, such that the cone generated by $-x + C$ is a closed vector subspace of $\mathcal{H}$.

We refer to [4] for an account of the main results of convex analysis, monotone operator theory, and the theory of nonexpansive operators in the context of Hilbert spaces.

We will use a family of functions $(\varphi_c)_{c \in \mathbb{R}}$, where φ_c for any $c \in \mathbb{R}$ is defined as

$$\varphi_c : \,]0, +\infty[\,\to \mathbb{R} : t \mapsto \begin{cases} (t^c - 1)/c & \text{if } c \neq 0 \\ \log t & \text{if } c = 0. \end{cases}$$

Let (Ω, F, P) be a probability space. An $\mathcal{H}$-valued random variable is a measurable function $x : \Omega \to \mathcal{H}$, where $\mathcal{H}$ is endowed with the Borel σ-algebra. We denote by $\sigma(x)$ the σ-field generated by x. The expectation of a random variable x is denoted by $\mathbb{E}[x]$. The conditional expectation of x given a σ-field $\mathcal{A} \subset F$ is denoted by $\mathbb{E}[x|\mathcal{A}]$. Given a random variable $y : \Omega \to \mathcal{H}$, the conditional expectation of x given y is denoted by $\mathbb{E}[x|y]$. Throughout the text and in the algorithms, we use the letter $\mathbf{r}$ to denote an unbiased estimate. An $\mathcal{H}$-valued random process is a sequence $(x_n)_{n \in \mathbb{N}}$ of $\mathcal{H}$-valued random variables. See [24] for more details on probability theory in Hilbert spaces. The following lemma is a special instance of [12, Proposition 2.3].

Lemma 1 *Let C be a non-empty closed subset of $\mathcal{H}$ and let $(x_n)_{n\in\mathbb{N}}$ be an $\mathcal{H}$-valued random process. For every $n \in \mathbb{N}$, set $\mathcal{F}_n = \sigma(x_0, \dots, x_n)$. Suppose that, for every $x \in C$, there exist $[0, +\infty[$-valued random sequences $(\xi_n(x))_{n\in\mathbb{N}}$, $(\zeta_n(x))_{n\in\mathbb{N}}$ and $(t_n(x))_{n\in\mathbb{N}}$ such that $\xi_n(x)$, $\zeta_n(x)$ and $t_n(x)$ are $\mathcal{F}_n$-measurable, $(\zeta_n(x))_{n\in\mathbb{N}}$ and $(t_n(x))_{n\in\mathbb{N}}$ are summable P-almost surely, and*

$$\mathbb{E}[\|x_{n+1} - x\|^2 | \mathcal{F}_n] \le (1 + t_n(x))\|x_n - x\|^2 + \zeta_n(x) - \xi_n(x) \quad \mathsf{P}\text{-almost surely.}$$

Then, the followings hold:

(a) *$(x_n)_{n\in\mathbb{N}}$ is bounded P-almost surely, and $\sum_{n\in\mathbb{N}} \xi_n(x) < +\infty$ for every $x \in C$ P-almost surely.*

(b) *There exists $\widetilde{\Omega} \subset \Omega$ such that $\mathsf{P}(\widetilde{\Omega}) = 1$, and $(\|x_n(\omega) - x\|)_{n\in\mathbb{N}}$ converges for every $\omega \in \widetilde{\Omega}$ and every $x \in C$.*

(c) *Suppose that the set of weak cluster points of $(x_n)_{n\in\mathbb{N}}$ is a subset of C P-almost surely. Then, $(x_n)_{n\in\mathbb{N}}$ converges weakly to a C-valued random vector P-almost surely.*

Lemma 2 ([14, Lemma 3.7]) *Let $A : \mathcal{H} \to 2^{\mathcal{H}}$ be a maximally monotone operator, let $\mathbf{U} \in \mathcal{B}(\mathcal{H})$ be self-adjoint and strongly positive, and let $\mathcal{G}$ be the real Hilbert space obtained by endowing $\mathcal{H}$ with the scalar product $(x, y) \mapsto \langle x \mid y \rangle_{\mathbf{U}^{-1}}$. Then, the followings hold:*

(a) *$\mathbf{U}A : \mathcal{G} \to 2^{\mathcal{G}}$ is maximally monotone.*

(b) *$J_{\mathbf{U}A} : \mathcal{G} \to \mathcal{G}$ is firmly nonexpansive.*

Lemma 3 ([17, Lemma 2.2]) *Suppose that $\mathbf{U}$ is self-adjoint and positive definite, and let γ be a strictly positive number. Let A, B and $\mathbf{Q}$ be defined as in Problem 1. Set $T_2 = J_{\gamma \mathbf{U}B}$ and $T_1 = J_{\gamma \mathbf{U}A}$, and $\mathbf{T} = \mathrm{Id} - T_2 + T_1 \circ (2T_2 - \mathrm{Id} - \gamma \mathbf{U}\mathbf{Q} \circ T_2)$. Then, $\mathrm{Fix}(\mathbf{T}) \ne \varnothing$ whenever $\mathrm{zer}(A + B + \mathbf{Q}) \ne \varnothing$. Furthermore,*

$$\mathrm{zer}(A + B + \mathbf{Q}) = J_{\gamma \mathbf{U}B}(\mathrm{Fix}(\mathbf{T})).$$

7.3   Algorithm and Weak Almost Sure Convergence

In this section, we introduce the stochastic forward-Douglas-Rachford splitting method and provide the weak almost sure convergence analysis. Then, we present some variants of the proposed method for solving some important special cases of Problem 1, along with their convergence guarantees.

7.3.1   Stochastic Forward-Douglas-Rachford Splitting Method

Let γ and $(\lambda_n)_{n\in\mathbb{N}}$ be strictly positive numbers, and let $(\mathbf{r}_n)_{n\in\mathbb{N}}$ and $\overline{\mathbf{x}}_0$ be square integrable $\mathcal{H}$-valued random vectors. Then, stochastic forward-Douglas-Rachford

---

Algorithm 1 Stochastic forward-Douglas-Rachford splitting method (SFDR)

For $n = 0, 1, \ldots$
  1: $\mathbf{x}_n = J_{\gamma \mathbf{UB}} \overline{\mathbf{x}}_n$
  2: $\overline{\mathbf{x}}_{n+1} = \overline{\mathbf{x}}_n + \lambda_n \big(J_{\gamma \mathbf{UA}} (2\mathbf{x}_n - \overline{\mathbf{x}}_n - \gamma \mathbf{U} \mathbf{r}_n) - \mathbf{x}_n \big)$.
End for

---

splitting method (SFDR) applies to Problem 1 as described in Algorithm 1. Theorem 1 proves the weak almost sure convergence of SFDR.

Theorem 1 *Let* $(\varepsilon, \alpha) \in \,]0, 1[^2$ *satisfy* $\varepsilon \leq \lambda_n \leq 2 - \varepsilon - \alpha$ *and* $\varepsilon \leq \gamma \leq \alpha(2\beta - \varepsilon)$, *and suppose that*

$$\mathrm{zer}(A + B + Q) \neq \varnothing. \tag{7.2}$$

Also assume that the following conditions are satisfied for every $n \in \mathbb{N}$, *with* $F_n = \sigma(\overline{\mathbf{x}}_0, \ldots, \overline{\mathbf{x}}_n)$:

(i) $\mathbb{E}[\mathbf{r}_n | F_n] = \mathbf{Q}\mathbf{x}_n$ *P-almost surely.*
(ii) $\sum_{n \in \mathbb{N}} \mathbb{E}[\|\mathbf{r}_n - \mathbf{Q}\mathbf{x}_n\|^2 | F_n] < +\infty$ *P-almost surely.*

Then, the followings hold for some random vectors $\mathbf{x}$ *and* $\overline{\mathbf{y}}$, *which are* **P***-valued and* $\mathrm{Fix}(\mathbf{T})$*-valued respectively:*

(i) $\overline{\mathbf{x}}_n \rightharpoonup \overline{\mathbf{y}}$ *P-almost surely.*
(ii) $\mathbf{x}_n \rightharpoonup \mathbf{x}$ *P-almost surely.*
(iii) $\mathbf{x}_n \to \mathbf{x}$ *P-almost surely, if one of the following conditions is satisfied for some* $\widetilde{\boldsymbol{\Omega}} \subset \boldsymbol{\Omega}$ *with* $\mathsf{P}(\widetilde{\boldsymbol{\Omega}}) = 1$:

    (a) $\mathbf{Q}$ *is demiregular at* $\mathbf{x}(\omega)$, *for every* $\omega \in \widetilde{\boldsymbol{\Omega}}$.
    (b) A *is uniformly monotone at* $\mathbf{x}(\omega)$, *for every* $\omega \in \widetilde{\boldsymbol{\Omega}}$.
    (c) B *is uniformly monotone at* $\mathbf{x}(\omega)$, *for every* $\omega \in \widetilde{\boldsymbol{\Omega}}$.

Proof Define T_1 and T_2 as in Lemma 3. Set $P = \mathrm{Id} - T_2$, $R = 2T_2 - \mathrm{Id} - \gamma \mathbf{U} \mathbf{Q} \circ T_2$, and $Y = \mathrm{Id} - \gamma \mathbf{U} \mathbf{Q} \circ T_2$. Then we get $\mathbf{T} = P + T_1 \circ R$. Set $R_n = 2T_2 - \mathrm{Id} - \gamma \mathbf{U} \mathbf{r}_n$, $Y_n = \mathrm{Id} - \gamma \mathbf{U} \mathbf{r}_n$, and $\mathbf{T}_n = P + T_1 \circ R_n$. Then, $2P + R_n = Y_n$. Hence, Algorithm 1 yields

$$(\forall n \in \mathbb{N}) \quad \overline{\mathbf{x}}_{n+1} = \overline{\mathbf{x}}_n + \lambda_n (\mathbf{T}_n \overline{\mathbf{x}}_n - \overline{\mathbf{x}}_n).$$

Let $\overline{\mathbf{x}} \in \mathrm{Fix}(\mathbf{T})$. Then, upon setting $\mathbf{V} = \mathbf{U}^{-1}$, using [4, Corollary 2.3], we obtain

$$\|\overline{\mathbf{x}}_{n+1} - \overline{\mathbf{x}}\|_{\mathbf{V}}^2 = (1 - \lambda_n)\|\overline{\mathbf{x}}_n - \overline{\mathbf{x}}\|_{\mathbf{V}}^2 + \lambda_n \|\mathbf{T}_n \overline{\mathbf{x}}_n - \overline{\mathbf{x}}\|_{\mathbf{V}}^2 - \lambda_n (1 - \lambda_n) \|\mathbf{T}_n \overline{\mathbf{x}}_n - \overline{\mathbf{x}}_n\|_{\mathbf{V}}^2. \tag{7.3}$$

Moreover, by setting

$$(\forall n \in \mathbb{N}) \quad \xi_n = \langle -T_1 \circ R_n \bar{\mathbf{x}}_n + T_1 \circ R\bar{\mathbf{x}} \mid \gamma \mathbf{r}_n - \gamma \mathbf{Q} \circ T_2 \bar{\mathbf{x}} \rangle,$$

and using the firm expansiveness of T_1 and P with respect to the scalar product $\langle \cdot \mid \cdot \rangle_{\mathbf{V}}$, we get

$$
\begin{aligned}
\|\mathbf{T}_n \bar{\mathbf{x}}_n - \bar{\mathbf{x}}\|_{\mathbf{V}}^2 &= \|\mathbf{T}_n \bar{\mathbf{x}}_n - \mathbf{T}\bar{\mathbf{x}}\|_{\mathbf{V}}^2 \\
&= \|P\bar{\mathbf{x}}_n - P\bar{\mathbf{x}}\|_{\mathbf{V}}^2 + 2 \langle P\bar{\mathbf{x}}_n - P\bar{\mathbf{x}} \mid T_1 \circ R_n \bar{\mathbf{x}}_n - T_1 \circ R\bar{\mathbf{x}} \rangle_{\mathbf{V}} \\
&\quad + \|T_1 \circ R_n \bar{\mathbf{x}}_n - T_1 \circ R\bar{\mathbf{x}}\|_{\mathbf{V}}^2 \\
&\leq \langle P\bar{\mathbf{x}}_n - P\bar{\mathbf{x}} \mid \bar{\mathbf{x}}_n - \bar{\mathbf{x}} \rangle_{\mathbf{V}} + \langle T_1 \circ R_n \bar{\mathbf{x}}_n - T_1 \circ R\bar{\mathbf{x}} \mid R_n \bar{\mathbf{x}}_n - R\bar{\mathbf{x}} \rangle_{\mathbf{V}} \\
&\quad + 2 \langle P\bar{\mathbf{x}}_n - P\bar{\mathbf{x}} \mid T_1 \circ R_n \bar{\mathbf{x}}_n - T_1 \circ R\bar{\mathbf{x}} \rangle_{\mathbf{V}} \\
&\leq \langle P\bar{\mathbf{x}}_n - P\bar{\mathbf{x}} \mid \bar{\mathbf{x}}_n - \bar{\mathbf{x}} \rangle_{\mathbf{V}} + \langle T_1 \circ R_n \bar{\mathbf{x}}_n - T_1 \circ R\bar{\mathbf{x}} \mid Y_n \bar{\mathbf{x}}_n - Y\bar{\mathbf{x}} \rangle_{\mathbf{V}} \\
&= \langle \mathbf{T}_n \bar{\mathbf{x}}_n - \mathbf{T}\bar{\mathbf{x}} \mid \bar{\mathbf{x}}_n - \bar{\mathbf{x}} \rangle_{\mathbf{V}} + \xi_n.
\end{aligned}
$$

Now, by using the following equation,

$$2 \langle \mathbf{T}_n \bar{\mathbf{x}}_n - \mathbf{T}\bar{\mathbf{x}} \mid \bar{\mathbf{x}}_n - \bar{\mathbf{x}} \rangle_{\mathbf{V}} = \|\mathbf{T}_n \bar{\mathbf{x}}_n - \bar{\mathbf{x}}\|_{\mathbf{V}}^2 + \|\bar{\mathbf{x}}_n - \bar{\mathbf{x}}\|_{\mathbf{V}}^2 - \|(\mathrm{Id} - \mathbf{T}_n)\bar{\mathbf{x}}_n\|_{\mathbf{V}}^2,$$

we derive the following bound:

$$\|\mathbf{T}_n \bar{\mathbf{x}}_n - \bar{\mathbf{x}}\|_{\mathbf{V}}^2 \leq \|\bar{\mathbf{x}}_n - \bar{\mathbf{x}}\|_{\mathbf{V}}^2 - \|(\mathrm{Id} - \mathbf{T}_n)\bar{\mathbf{x}}_n\|_{\mathbf{V}}^2 + 2\xi_n. \tag{7.4}$$

Let us estimate ξ_n. We have $-T_1 \circ R_n = \mathrm{Id} - \mathbf{T}_n - T_2$, and $-T_1 \circ R = \mathrm{Id} - \mathbf{T} - T_2$. Then, it follows as

$$
\begin{aligned}
\xi_n &= \langle (\mathrm{Id} - \mathbf{T}_n)\bar{\mathbf{x}}_n - (\mathrm{Id} - \mathbf{T})\bar{\mathbf{x}} \mid \gamma \mathbf{r}_n - \gamma \mathbf{Q} \circ T_2 \bar{\mathbf{x}} \rangle - \gamma \langle T_2 \bar{\mathbf{x}}_n - T_2 \bar{\mathbf{x}} \mid \mathbf{r}_n - \mathbf{Q} \circ T_2 \bar{\mathbf{x}} \rangle \\
&= \gamma \langle (\mathrm{Id} - \mathbf{T}_n)\bar{\mathbf{x}}_n \mid \mathbf{Q} \circ T_2 \bar{\mathbf{x}}_n - \mathbf{Q} \circ T_2 \bar{\mathbf{x}} \rangle + \gamma \langle (\mathrm{Id} - \mathbf{T}_n)\bar{\mathbf{x}}_n \mid \mathbf{r}_n - \mathbf{Q} \circ T_2 \bar{\mathbf{x}}_n \rangle \\
&\quad - \gamma \langle T_2 \bar{\mathbf{x}}_n - T_2 \bar{\mathbf{x}} \mid \mathbf{r}_n - \mathbf{Q} \circ T_2 \bar{\mathbf{x}} \rangle \\
&= \gamma \langle (\mathrm{Id} - \mathbf{T}_n)\bar{\mathbf{x}}_n \mid \mathbf{Q} \circ T_2 \bar{\mathbf{x}}_n - \mathbf{Q} \circ T_2 \bar{\mathbf{x}} \rangle + \gamma \langle (\mathbf{T} - \mathbf{T}_n)\bar{\mathbf{x}}_n \mid \mathbf{r}_n - \mathbf{Q} \circ T_2 \bar{\mathbf{x}}_n \rangle \\
&\quad + \gamma \langle (\mathrm{Id} - \mathbf{T})\bar{\mathbf{x}}_n \mid \mathbf{r}_n - \mathbf{Q} \circ T_2 \bar{\mathbf{x}}_n \rangle - \gamma \langle T_2 \bar{\mathbf{x}}_n - T_2 \bar{\mathbf{x}} \mid \mathbf{r}_n - \mathbf{Q} \circ T_2 \bar{\mathbf{x}} \rangle.
\end{aligned}
$$

Now, we obtain the following estimates:

$$
\begin{aligned}
2\gamma \langle (\mathrm{Id} - \mathbf{T}_n)\bar{\mathbf{x}}_n \mid \mathbf{Q} \circ T_2 \bar{\mathbf{x}}_n - \mathbf{Q} \circ T_2 \bar{\mathbf{x}} \rangle &\leq \alpha \|(\mathrm{Id} - \mathbf{T}_n)\bar{\mathbf{x}}_n\|_{\mathbf{V}}^2 \\
&\quad + (\gamma^2/\alpha) \|\mathbf{U}\mathbf{Q} \circ T_2 \bar{\mathbf{x}}_n - \mathbf{U}\mathbf{Q} \circ T_2 \bar{\mathbf{x}}\|_{\mathbf{V}}^2,
\end{aligned}
$$

and

$$\langle (\mathbf{T} - \mathbf{T}_n)\bar{\mathbf{x}}_n \mid \mathbf{r}_n - \mathbf{Q} \circ T_2\bar{\mathbf{x}}_n \rangle \leq \|T_1 \circ R\bar{\mathbf{x}}_n - T_1 \circ R_n\bar{\mathbf{x}}_n\|_{\mathbf{V}} \|\mathbf{U}\mathbf{r}_n - \mathbf{U}\mathbf{Q} \circ T_2\bar{\mathbf{x}}_n\|_{\mathbf{V}}$$

$$\leq \|R_1\bar{\mathbf{x}}_n - R_n\bar{\mathbf{x}}_n\|_{\mathbf{V}} \|\mathbf{U}\mathbf{r}_n - \mathbf{U}\mathbf{Q} \circ T_2\bar{\mathbf{x}}_n\|_{\mathbf{V}}$$

$$\leq \gamma \|\mathbf{U}\mathbf{r}_n - \mathbf{U}\mathbf{Q} \circ T_2\bar{\mathbf{x}}_n\|_{\mathbf{V}}^2$$

$$\leq \gamma \|\mathbf{r}_n - \mathbf{Q} \circ T_2\bar{\mathbf{x}}_n\|_{\mathbf{U}}^2$$

$$\leq \gamma \|\mathbf{U}\| \|\mathbf{r}_n - \mathbf{Q} \circ T_2\bar{\mathbf{x}}_n\|^2.$$

Therefore, (7.4) becomes

$$\|\mathbf{T}_n\bar{\mathbf{x}}_n - \bar{\mathbf{x}}\|_{\mathbf{V}}^2 \leq \|\bar{\mathbf{x}}_n - \bar{\mathbf{x}}\|_{\mathbf{V}}^2 - (1-\alpha)\|(\mathrm{Id} - \mathbf{T}_n)\bar{\mathbf{x}}_n\|_{\mathbf{V}}^2 + \zeta_n$$

$$+ (\gamma^2/\alpha)\|\mathbf{U}\mathbf{Q} \circ T_2\bar{\mathbf{x}}_n - \mathbf{U}\mathbf{Q} \circ T_2\bar{\mathbf{x}}\|_{\mathbf{V}}^2 + 2\gamma^2\|\mathbf{U}\| \|\mathbf{r}_n - \mathbf{Q} \circ T_2\bar{\mathbf{x}}_n\|^2, \qquad (7.5)$$

where we set

$$\zeta_n = 2\gamma \langle (\mathrm{Id} - \mathbf{T})\bar{\mathbf{x}}_n \mid \mathbf{r}_n - \mathbf{Q} \circ T_2\bar{\mathbf{x}}_n \rangle - 2\gamma \langle T_2\bar{\mathbf{x}}_n - T_2\bar{\mathbf{x}} \mid \mathbf{r}_n - \mathbf{Q} \circ T_2\bar{\mathbf{x}} \rangle.$$

Since $\mathbf{T}$ and T_2 are continuous, $\bar{\mathbf{x}}_n$ is F_n-measurable, and $(\mathrm{Id} - \mathbf{T})\bar{\mathbf{x}}_n$ and $T_2\bar{\mathbf{x}}_n - T_2\bar{\mathbf{x}}$ are also F_n-measurable, we have

$$\mathbb{E}[\zeta_n | F_n] = 2\gamma \mathbb{E}[\langle (\mathrm{Id} - \mathbf{T})\bar{\mathbf{x}}_n \mid \mathbf{r}_n - \mathbf{Q} \circ T_2\bar{\mathbf{x}}_n \rangle | F_n]$$

$$- 2\gamma \mathbb{E}[\langle T_2\bar{\mathbf{x}}_n - T_2\bar{\mathbf{x}} \mid \mathbf{r}_n - \mathbf{Q} \circ T_2\bar{\mathbf{x}} \rangle | F_n]$$

$$= 2\gamma \langle (\mathrm{Id} - \mathbf{T})\bar{\mathbf{x}}_n \mid \mathbb{E}[\mathbf{r}_n - \mathbf{Q} \circ T_2\bar{\mathbf{x}}_n | F_n] \rangle$$

$$- 2\gamma \langle T_2\bar{\mathbf{x}}_n - T_2\bar{\mathbf{x}} \mid \mathbb{E}[\mathbf{r}_n - \mathbf{Q} \circ T_2\bar{\mathbf{x}} | F_n] \rangle$$

$$= -2\gamma \langle T_2\bar{\mathbf{x}}_n - T_2\bar{\mathbf{x}} \mid \mathbf{Q} \circ T_2\bar{\mathbf{x}}_n - \mathbf{Q} \circ T_2\bar{\mathbf{x}} \rangle$$

$$\leq -2\beta\gamma \|\mathbf{U}\mathbf{Q} \circ T_2\bar{\mathbf{x}}_n - \mathbf{U}\mathbf{Q} \circ T_2\bar{\mathbf{x}}\|_{\mathbf{V}}^2.$$

Taking conditional expectation of both sides with respect to F_n in (7.5), we obtain,

$$\mathbb{E}[\|\mathbf{T}_n\bar{\mathbf{x}}_n - \bar{\mathbf{x}}\|_{\mathbf{V}}^2 | F_n] \leq \|\bar{\mathbf{x}}_n - \bar{\mathbf{x}}\|_{\mathbf{V}}^2 - (1-\alpha)\mathbb{E}[\|(\mathrm{Id} - \mathbf{T}_n)\bar{\mathbf{x}}_n\|_{\mathbf{V}}^2 | F_n]$$

$$- \gamma(2\beta - \gamma/\alpha)\|\mathbf{Q} \circ T_2\bar{\mathbf{x}}_n - \mathbf{Q} \circ T_2\bar{\mathbf{x}}\|_{\mathbf{U}}^2$$

$$+ 2\gamma^2 \mathbb{E}[\|\mathbf{U}\| \|\mathbf{r}_n - \mathbf{Q} \circ T_2\bar{\mathbf{x}}_n\|^2 | F_n].$$

Next, we set

$$(\forall n \in \mathbb{N}) \quad \tau_{1,n} = \lambda_n(1 - \lambda_n) + \lambda_n(1 - \alpha) \quad \text{and} \quad \tau_{2,n} = \lambda_n\gamma(2\beta - \gamma/\alpha).$$

Then, $\tau_{1,n} \geq \varepsilon^2$ and $\tau_{2,n} \geq \varepsilon^3$ for every $n \in \mathbb{N}$. Now inserting (7.5) into (7.3), we obtain,

$$\mathbb{E}[\|\bar{\mathbf{x}}_{n+1} - \bar{\mathbf{x}}\|_{\mathbf{V}}^2 | F_n] \leq \|\bar{\mathbf{x}}_n - \bar{\mathbf{x}}\|_{\mathbf{V}}^2 - \tau_{1,n}\mathbb{E}[\|(\mathrm{Id} - \mathbf{T}_n)\bar{\mathbf{x}}_n\|_{\mathbf{V}}^2 | F_n]$$

$$- \tau_{2,n}\|\mathbf{Q} \circ T_{2,n}\bar{\mathbf{x}}_n - \mathbf{Q} \circ T_{2,n}\bar{\mathbf{x}}\|_{\mathbf{V}}^2 + 2\gamma^2\mathbb{E}[\|\mathbf{U}\|\|\mathbf{r}_n - \mathbf{Q} \circ T_2\bar{\mathbf{x}}_n\|^2 | F_n]$$

$$\leq \|\bar{\mathbf{x}}_n - \bar{\mathbf{x}}\|_{\mathbf{V}}^2 - \varepsilon^3\|\mathbf{Q} \circ T_{2,n}\bar{\mathbf{x}}_n - \mathbf{Q} \circ T_{2,n}\bar{\mathbf{x}}\|_{\mathbf{U}}^2$$

$$- \varepsilon^2\mathbb{E}[\|(\mathrm{Id} - \mathbf{T}_n)\bar{\mathbf{x}}_n\|_{\mathbf{V}}^2 | F_n]\| + 2\gamma^2\mathbb{E}[\|\mathbf{U}\|\|\mathbf{r}_n - \mathbf{Q} \circ T_2\bar{\mathbf{x}}_n\|^2 | F_n].$$

Hence, the sequence $(\bar{\mathbf{x}}_n)_{n\in\mathbb{N}}$ is stochastic Fejér monotone with respect to the target set $\mathrm{Fix}(\mathbf{T})$, see [19] for details. Therefore, by Lemma 1, the following estimate holds:

$$(\forall n \in \mathbb{N}) \quad \begin{cases} (\bar{\mathbf{x}}_n)_{n\in\mathbb{N}} \text{ is bounded P-almost surely} \\ \mathbb{E}[\|(\mathrm{Id} - \mathbf{T}_n)\bar{\mathbf{x}}_n\|^2 F_n] = \lambda_n^{-1}\mathbb{E}[\|(\bar{\mathbf{x}}_{n+1} - \bar{\mathbf{x}}_n)\|^2 F_n] \to 0 \\ \mathbf{Q} \circ T_2\bar{\mathbf{x}}_n - \mathbf{Q} \circ T_2\bar{\mathbf{x}} \to 0. \end{cases} \quad (7.6)$$

(i) Since $(\bar{\mathbf{x}}_n)_{n\in\mathbb{N}}$ is bounded P-almost surely, its weak cluster points exist P-almost surely. Let $\bar{\mathbf{y}}$ be a weak cluster point of $(\bar{\mathbf{x}}_n)_{n\in\mathbb{N}}$, i.e., there exists a subsequence $(\bar{\mathbf{x}}_{k_n})_{n\in\mathbb{N}}$ such that $\bar{\mathbf{x}}_{k_n} \rightharpoonup \bar{\mathbf{y}}$ P-almost surely. Moreover, we also have

$$\|\bar{\mathbf{x}}_n - \mathbf{T}\bar{\mathbf{x}}_n\|_{\mathbf{V}}^2 = \mathbb{E}[\|\bar{\mathbf{x}}_n - \mathbf{T}\bar{\mathbf{x}}_n\|_{\mathbf{V}}^2 | F_n]$$

$$\leq 2\mathbb{E}[\|\bar{\mathbf{x}}_n - \mathbf{T}_n\bar{\mathbf{x}}_n\|_{\mathbf{V}}^2 | F_n] + 2\mathbb{E}[\|\mathbf{T}\bar{\mathbf{x}}_n - \mathbf{T}_n\bar{\mathbf{x}}_n\|_{\mathbf{V}}^2 | F_n]$$

$$\to 0,$$

which implies that $\bar{\mathbf{x}}_n - \mathbf{T}\bar{\mathbf{x}}_n \to 0$ P-almost surely. To sum up, we have

$$\bar{\mathbf{x}}_{k_n} \rightharpoonup \bar{\mathbf{y}} \quad \text{and} \quad \bar{\mathbf{x}}_{k_n} - \mathbf{T}\bar{\mathbf{x}}_{k_n} \to 0 \quad \text{P-almost surely.}$$

Therefore, by the demiclosedness principle, $\bar{\mathbf{y}} \in \mathrm{Fix}(\mathbf{T})$ P-almost surely. Now, using Lemma 1, $(\bar{\mathbf{x}}_n)_{n\in\mathbb{N}}$ converges weakly to a random vector $\bar{\mathbf{y}}$, that is $\mathrm{Fix}(\mathbf{T})$-valued P-almost surely.

(ii) Note that T_2 is nonexpansive and $(\bar{\mathbf{x}}_n)_{n\in\mathbb{N}}$ is bounded P-almost surely. Also note $\mathbf{x}_n = T_2\bar{\mathbf{x}}_n$, and hence $(\mathbf{x}_n)_{n\in\mathbb{N}}$ is also bounded P-almost surely. Let $\mathbf{z}$ be a weak cluster point of $(\mathbf{x}_n)_{n\in\mathbb{N}}$, i.e., there exists a subsequence $(\mathbf{x}_{k_n})_{n\in\mathbb{N}}$ that converges weakly to $\mathbf{z}$ P-almost surely. Since $\mathrm{gra}(\mathbf{Q})$ is weak-to-strong sequentially closed, and since $\mathbf{Q}\mathbf{x}_n \to \mathbf{Q} \circ T_2\bar{\mathbf{x}}$, we get $\mathbf{Q} \circ T_2\bar{\mathbf{x}} = \mathbf{Q}\mathbf{z}$ and

$\mathbf{Qx}_{k_n} \to \mathbf{Qz}$. Set $\mathbf{a}_n = T_1(2\mathbf{x}_n - \overline{\mathbf{x}}_n - \gamma \mathbf{UQx}_n)$. Then, we derive from (7.6) that $\mathbf{a}_n - \mathbf{x}_n \to 0$ P-almost surely. We have

$$
\begin{cases}
(\mathbf{x}_{k_n}, \mathbf{Qx}_{k_n}) \in \mathrm{gra}(\mathbf{Q}), \quad \mathbf{x}_{k_n} \rightharpoonup \mathbf{z}, \quad \mathbf{Qx}_{k_n} \to \mathbf{Qz}; \\
(\mathbf{x}_{k_n}, \gamma^{-1}\mathbf{V}(\overline{\mathbf{x}}_{k_n} - \mathbf{x}_{k,n})) \in \mathrm{gra}(\mathbf{B}), \quad \gamma^{-1}\mathbf{V}(\overline{\mathbf{x}}_{k_n} - \mathbf{x}_{k,n}) \rightharpoonup \gamma^{-1}\mathbf{V}(\overline{\mathbf{y}} - \mathbf{z}); \\
(\mathbf{a}_{k_n}, \gamma^{-1}\mathbf{V}(\mathbf{x}_{k_n} - \mathbf{a}_{k_n} + \mathbf{x}_{k_n} - \overline{\mathbf{x}}_{k_n}) - \mathbf{Qx}_{k_n}) \in \mathrm{gra}(\mathbf{A}); \\
\gamma^{-1}\mathbf{V}(\mathbf{x}_{k_n} - \mathbf{a}_{k_n} + \mathbf{x}_{k_n} - \overline{\mathbf{x}}_{k_n}) - \mathbf{Qx}_{k_n} \rightharpoonup \gamma^{-1}\mathbf{V}(\mathbf{z} - \overline{\mathbf{y}}) - \mathbf{Qz}.
\end{cases}
$$

Therefore, by Bauschke and Combettes [4, Proposition 25.5], $\mathbf{z} \in \mathrm{zer}(\mathbf{A} + \mathbf{B} + \mathbf{Q})$ and $\gamma^{-1}\mathbf{V}(\overline{\mathbf{y}} - \mathbf{z}) \in \mathbf{Bz} \Rightarrow \mathbf{z} = J_{\gamma \mathbf{UB}}\overline{\mathbf{y}}$. Since every subsequence of the bounded sequence $(\mathbf{x}_n)_{n \in \mathbb{N}}$ converges weakly to $J_{\gamma \mathbf{UB}}\overline{\mathbf{y}}$, we conclude that $\mathbf{x}_n \rightharpoonup J_{\gamma \mathbf{UB}}\overline{\mathbf{y}}$ P-almost surely.

(iii) Let $\mathbf{\Omega}_1$ be the set of all $\omega \in \mathbf{\Omega}$ such that $\mathbf{x}_n(\omega) \rightharpoonup \mathbf{x}(\omega)$ and $\mathbf{Qx}_n(\omega) \to \mathbf{Qx}(\omega)$, and $\mathbf{a}_n(\omega) - \mathbf{x}_n(\omega) \to 0$. Then, $\mathsf{P}(\mathbf{\Omega}_1) = 1$, and hence $\mathsf{P}(\mathbf{\Omega}_1 \cap \widetilde{\mathbf{\Omega}}) = 1$.

(iii)(a) For every $\omega \in \mathbf{\Omega}_1 \cap \mathbf{\Omega}$, we have $\mathbf{x}_n(\omega) \rightharpoonup \mathbf{x}(\omega)$ and $\mathbf{Qx}_n \to \mathbf{Qx}(\omega)$. Since $\mathbf{Q}$ is demiregular at $\mathbf{x}(\omega)$, by definition, we obtain $\mathbf{x}_n(\omega) \to \mathbf{x}(\omega)$.

(iii)(b) As in the proof of (i), we have

$$
\begin{cases}
(\mathbf{a}_n, \gamma^{-1}\mathbf{V}(\mathbf{x}_n - \mathbf{a}_n + \mathbf{x}_n - \overline{\mathbf{x}}_n) - \mathbf{Qx}_n) \in \mathrm{gra}(\mathbf{A}) \\
\gamma^{-1}\mathbf{V}(\mathbf{x} - \overline{\mathbf{y}}) - \mathbf{Qx} \in \mathbf{Ax}.
\end{cases}
\tag{7.7}
$$

Since $\mathbf{A}$ is uniformly monotone at $\mathbf{x}(\omega)$, there exists a nonnegative increasing function $\phi\colon [0, +\infty[\to [0, +\infty]$, vanishing only at 0, such that

$$
\phi(\|(\mathbf{a}_n - \mathbf{x})(\omega)\|)
$$
$$
\leq \left\langle (\mathbf{a}_n - \mathbf{x})(\omega) \,\middle|\, \left(\gamma^{-1}\mathbf{V}(\mathbf{x}_n - \mathbf{a}_n + \mathbf{x}_n - \overline{\mathbf{x}}_n) \right.\right.
$$
$$
\left.\left. - \mathbf{Qx}_n - \gamma^{-1}\mathbf{V}(\mathbf{x} - \overline{\mathbf{y}}) + \mathbf{Qx}\right)(\omega)\right\rangle
$$
$$
= \langle (\mathbf{a}_n - \mathbf{x})(\omega) \mid (\mathbf{Qx} - \mathbf{Qx}_n)(\omega)\rangle
$$
$$
+ \left\langle (\mathbf{a}_n - \mathbf{x})(\omega) \mid \gamma^{-1}\mathbf{V}(\mathbf{x}_n - \mathbf{a}_n + \mathbf{x}_n - \overline{\mathbf{x}}_n - \mathbf{x} + \overline{\mathbf{y}})(\omega)\right\rangle
$$
$$
= t_{1,n} + t_{2,n},
\tag{7.8}
$$

where we set

$$
\begin{cases}
t_{1,n} = \langle (\mathbf{a}_n - \mathbf{x})(\omega) \mid (\mathbf{Qx} - \mathbf{Qx}_n)(\omega)\rangle \\
t_{2,n} = \left\langle (\mathbf{a}_n - \mathbf{x})(\omega) \mid \gamma^{-1}\mathbf{V}(\mathbf{x}_n - \mathbf{a}_n + \mathbf{x}_n - \overline{\mathbf{x}}_n - \mathbf{x} + \overline{\mathbf{y}})(\omega)\right\rangle \\
t_{3,n} = \left\langle (\mathbf{a}_n - \mathbf{x})(\omega) \mid \gamma^{-1}\mathbf{V}(\mathbf{x}_n - \mathbf{a}_n)(\omega)\right\rangle \\
t_{4,n} = \left\langle (\mathbf{a}_n - \mathbf{x}_n)(\omega) \mid \gamma^{-1}\mathbf{V}(\mathbf{x}_n - \overline{\mathbf{x}}_n - \mathbf{x} + \overline{\mathbf{y}})(\omega)\right\rangle.
\end{cases}
$$

Let us estimate each term on the right hand side of (7.8). Since $\mathbf{a}_n(\omega) - \mathbf{x}(\omega)$ converges weakly to 0, it is bounded, and since $\mathbf{Q}\mathbf{x}_n(\omega) \rightarrow \mathbf{Q}\mathbf{x}(\omega)$, we have

$$
\begin{aligned}
t_{1,n} &= \langle (\mathbf{a}_n - \mathbf{x})(\omega) \mid (\mathbf{Q}\mathbf{x} - \mathbf{Q}\mathbf{x}_n)(\omega) \rangle \\
&\leq \|(\mathbf{a}_n - \mathbf{x})(\omega)\| \, \|\mathbf{Q}\mathbf{x} - \mathbf{Q}\mathbf{x}_n)(\omega)\| \rightarrow 0.
\end{aligned}
$$

Next, we consider the second term of (7.8). We have

$$
\begin{aligned}
t_{2,n} &= t_{3,n} + \left\langle (\mathbf{a}_n - \mathbf{x})(\omega) \mid \gamma^{-1}\mathbf{V}(\mathbf{x}_n - \overline{\mathbf{x}}_n - \mathbf{x} + \overline{\mathbf{y}})(\omega) \right\rangle \\
&= t_{3,n} + t_{4,n} + \left\langle (\mathbf{x}_n - \mathbf{x})(\omega) \mid \gamma^{-1}\mathbf{V}(\mathbf{x}_n - \overline{\mathbf{x}}_n - \mathbf{x} + \overline{\mathbf{y}})(\omega) \right\rangle
\end{aligned}
\tag{7.9}
$$

$$
\begin{aligned}
&\leq t_{3,n} + t_{4,n} \\
&\rightarrow 0,
\end{aligned}
$$

where the last inequality follows from

$$
\left\langle (\mathbf{x}_n - \mathbf{x})(\omega) \mid \gamma^{-1}\mathbf{V}(\mathbf{x}_n - \overline{\mathbf{x}}_n - \mathbf{x} + \overline{\mathbf{y}})(\omega) \right\rangle \leq 0,
$$

which holds because $\mathbf{B}$ is monotone. Therefore, $\phi(\|(\mathbf{a}_n - \mathbf{x})(\omega)\|) \rightarrow 0$, and hence $(\mathbf{a}_n - \mathbf{x})(\omega) \rightarrow 0$. Therefore, $\mathbf{x}_n \rightarrow \mathbf{x}$ P-almost surely.

(iii)(c) Using the strong monotonicity of $\mathbf{B}$ and (7.9), there exists an increasing function $\psi \colon [0, +\infty[\rightarrow [0, +\infty]$, vanishing only at 0, such that

$$
\begin{aligned}
\psi(\|(\mathbf{x}_n - \mathbf{x})(\omega)\|) &\leq \left\langle (\mathbf{x}_n - \mathbf{x})(\omega) \mid \gamma^{-1}\mathbf{V}(\overline{\mathbf{x}}_n - \mathbf{x}_n + \mathbf{x} - \overline{\mathbf{y}})(\omega) \right\rangle \\
&\leq 2|t_{3,n}| + 2|t_{4,n}| \\
&\rightarrow 0,
\end{aligned}
$$

which implies $(\mathbf{x}_n - \mathbf{x})(\omega) \rightarrow 0$, and hence $\mathbf{x}_n - \mathbf{x} \rightarrow 0$ P-almost surely. $\qquad\square$

Corollary 1 *Let* $f \in \Gamma_0(\mathcal{H})$, $g \in \Gamma_0(\mathcal{H})$, *and let* $h \colon \mathcal{H} \rightarrow \,]-\infty, +\infty]$ *be a convex differentiable function with a* β^{-1}*-Lipschitz gradient. Let* P *be the set of all solutions* $\mathbf{x}$ *to*

$$
\underset{\mathbf{x} \in \mathcal{H}}{minimize} \; f(\mathbf{x}) + g(\mathbf{x}) + h(\mathbf{x}),
\tag{7.10}
$$

under the condition

$$
\mathrm{zer}(\partial f + \partial g + \nabla h) \neq \varnothing.
\tag{7.11}
$$

Let $(\varepsilon, \alpha) \in]0, 1[^2$, let γ be a strictly positive number, and let $(\lambda_n)_{n \in \mathbb{N}}$ be a sequence of strictly positive numbers such that $\varepsilon \leq \lambda_n \leq 2 - \varepsilon - \alpha$ and $\varepsilon \leq \gamma \leq \alpha(2\beta \|U\|^{-1} - \varepsilon)$. Suppose that $(\mathbf{r}_n)_{n \in \mathbb{N}}$ is a sequence of square integrable $\mathcal{H}$-valued random vectors, and $\overline{\mathbf{x}}_0$ is a square integrable $\mathcal{H}$-valued random vector.

Algorithm 2 SFDR for sum of three functions

For $n = 0, 1, \ldots$
  1: $\mathbf{x}_n = \text{prox}_{\gamma f}^{U^{-1}} \overline{\mathbf{x}}_n$
  2: $\overline{\mathbf{x}}_{n+1} = \overline{\mathbf{x}}_n + \lambda_n \left(\text{prox}_{\gamma g}^{U^{-1}} (2\mathbf{x}_n - \overline{\mathbf{x}}_n - \gamma U \mathbf{r}_n) - \mathbf{x}_n \right)$
End for

Suppose that the following conditions are satisfied for every $n \in \mathbb{N}$, with $F_n = \sigma(\overline{\mathbf{x}}_0, \ldots, \overline{\mathbf{x}}_n)$:

(i) $\mathbb{E}[\mathbf{r}_n | F_n] = \nabla h(\mathbf{x}_n)$ P-almost surely.
(ii) $\sum_{n \in \mathbb{N}} \mathbb{E}[\|\mathbf{r}_n - Q\mathbf{x}_n\|^2 | F_n] < +\infty$ P-almost surely.

Then, the followings hold for some random vectors $\mathbf{x}$, P-valued, and $\overline{\mathbf{y}}$, $\text{Fix}(\mathbf{T})$-valued, where $\mathbf{T}$ is defined in Lemma 3 with $T_2 = \text{prox}_{\gamma f}^{U^{-1}}$ and $T_1 = \text{prox}_{\gamma g}^{U^{-1}}$:

(i) $\overline{\mathbf{x}}_n \rightharpoonup \overline{\mathbf{y}}$ P-almost surely.
(ii) $\mathbf{x}_n \rightharpoonup \mathbf{x}$ P-almost surely.
(iii) $\mathbf{x}_n \to \mathbf{x}$ P-almost surely, if one of the following conditions is satisfied for some $\widetilde{\Omega} \subset \Omega$ with $P(\widetilde{\Omega}) = 1$:

> *(a) h is uniformly convex at $\mathbf{x}(\omega)$ for every $\omega \in \widetilde{\Omega}$*
> *(b) f is uniformly convex at $\mathbf{x}(\omega)$ for every $\omega \in \widetilde{\Omega}$.*
> *(c) g is uniformly convex at $\mathbf{x}(\omega)$ for every $\omega \in \widetilde{\Omega}$.*

Proof Conclusions (i) and (ii) follow from Theorem 1 with $\mathbf{A} = \partial f$, $\mathbf{B} = \partial g$, $\mathbf{Q} = \nabla h$ along with [4, Theorem 16.2] and [4, Proposition 16.5]. The last assertion follows from the fact that if f is uniformly convex at a point in the domain of ∂f, then ∂f is uniformly monotone at that point; and hence, it is demiregular [2, Proposition 2.4(v)]. □

Notice that condition (7.11) holds, if (7.10) has a solution and $0 \in \text{sri}\left(\text{dom}(f) - \text{dom}(g)\right)$, where sri denotes strong relative interior.

As a final remark, the recent work in [36] introduces an alternative stochastic method which applies if h and g are some averaged functions, in the sense that we can write them as

$$h = \frac{1}{p} \sum_{k=1}^{p} h_i \quad \text{and} \quad g = \frac{1}{p} \sum_{k=1}^{p} g_i,$$

where h_i is a differentiable convex function with β_i^{-1}-Lipschitz gradient and $g_i \in \Gamma_0(\mathbb{R}^d)$ for each $i \in \{1, \ldots, p\}$. This method is fundamentally different than our framework.

Corollary 2 *Let m be a strictly positive integer and β be a strictly positive real number. For every $i \in \{1, \ldots, m\}$, let $(\mathcal{H}_i, \langle \cdot \mid \cdot \rangle)$ be a real Hilbert space, $A_i : \mathcal{H}_i \to 2^{\mathcal{H}_i}$ and $B_i : \mathcal{H}_i \to 2^{\mathcal{H}_i}$ be maximally monotone operators. Let $Q_i : \mathcal{H}_1 \times \ldots \times \mathcal{H}_m \to \mathcal{H}_i$ satisfy*

$$\sum_{i=1}^{m} \langle Q_i\mathbf{x} - Q_i\mathbf{y} \mid x_i - y_i \rangle \geq \beta \sum_{i=1}^{m} \|Q_i\mathbf{x} - Q_i\mathbf{y}\|_{U_i}^2, \tag{7.12}$$

for every $\mathbf{x} = (x_1, \ldots, x_m)$, every $\mathbf{y} = (y_1, \ldots, y_m)$ and some self adjoint positive definite operator U_i on $\mathcal{H}_i$. Assume that the set X of $\mathbf{x} = (x_1, \ldots, x_m)$ that satisfies the following coupled system of inclusions

$$\begin{cases} 0 \in A_1x_1 + B_1x_1 + Q_1\mathbf{x} \\ \quad \vdots \\ 0 \in A_mx_m + B_mx_m + Q_m\mathbf{x} \end{cases} \tag{7.13}$$

is non-empty. Let $\overline{x}_{i,0}$ be an $\mathcal{H}_i$-valued and square integrable vector, and $(r_{i,n})_{n\in\mathbb{N}}$ be an $\mathcal{H}_i$-valued and squared random process. Let $(\varepsilon, \alpha) \in]0, 1[^2$, let γ be a strictly positive number, and let $(\lambda_n)_{n\in\mathbb{N}}$ be a sequence of strictly positive numbers such that $\varepsilon \leq \lambda_n \leq 2 - \varepsilon - \alpha$ and $\varepsilon \leq \gamma \leq \alpha(2\beta - \varepsilon)$.

Suppose that the following conditions are satisfied $\forall i \in \{1, \ldots, m\}$ with $F_n = \sigma(\overline{x}_{i,0}, \ldots, \overline{x}_{i,n})_{1 \leq i \leq m}$

(i) $\mathbb{E}[r_{i,n} \mid F_n] = Q_i(x_{1,n}, \ldots, x_{m,n})$ P-almost surely,
(ii) $\sum_{n\in\mathbb{N}} \mathbb{E}[\|r_{i,n} - Q_i(x_{1,n}, \ldots, x_{m,n})\|^2 \mid F_n] < +\infty$ P-almost surely.

Then, the followings hold for some random vector $\mathbf{x}$, which is X-valued P-almost surely:

(i) $(\mathbf{x}_n)_{n\in\mathbb{N}}$ converges weakly to $\mathbf{x}$ P-almost surely.
(ii) Suppose that, for every $\omega \in \widetilde{\Omega} \subset \Omega$ with $\mathsf{P}(\widetilde{\Omega}) = 1$,

    (a) $(Q_1, \ldots, Q_m)$ is demiregular at $(x_1(\omega), \ldots, x_m(\omega))$. Then,

$$\left(\forall j \in \{1, \ldots, m\}\right) \quad x_{j,n} \to x_j \text{ P-almost surely.}$$

    (b) A_j or B_j is uniformly monotone at $x_j(\omega)$, for some $j \in \{1, \ldots, m\}$. Then, $x_{j,n} \to \overline{x}_j$ P-almost surely.

Proof Define $\mathcal{H} = \mathcal{H}_1 \oplus \ldots \oplus \mathcal{H}_m$ with the scalar product and the norm

$$\langle \mathbf{x} \mid \mathbf{y} \rangle = \sum_{i=1}^{m} \langle x_i \mid y_i \rangle \quad \text{and} \quad \|\mathbf{x}\| = \sqrt{\langle \mathbf{x} \mid \mathbf{x} \rangle}, \tag{7.14}$$

where $\mathbf{x} = (x_1, \ldots, x_m)$ and $\mathbf{y} = (y_1, \ldots, y_m)$ denote the generic elements in $\mathcal{H}$, and $\oplus$ denotes the Hilbert space direct sum. Now, we define

$$\begin{cases} \mathbf{U}: \mathcal{H} \to \mathcal{H}: \mathbf{x} \mapsto (U_1 x_1, \ldots, U_m x_m); \\ \mathbf{A}: \mathcal{H} \to 2^{\mathcal{H}}: \mathbf{x} \mapsto \times_{i=1}^{m} A_i x_i; \\ \mathbf{B}: \mathcal{H} \to 2^{\mathcal{H}}: \mathbf{x} \mapsto \times_{i=1}^{m} B_i x_i; \\ \mathbf{Q}: \mathcal{H} \to \mathcal{H}: \mathbf{x} \mapsto (Q_1 \mathbf{x}, \ldots, Q_m \mathbf{x}). \end{cases}$$

Then, $\mathbf{U}$ is self-adjoint and positive definite on $\mathcal{H}$, and $\mathbf{A}$ and $\mathbf{B}$ are maximally monotone on $\mathcal{H}$ [4, Proposition 20.23]. Therefore, in view of Lemma 3, $\mathbf{UA}$ and $\mathbf{UB}$ are maximally monotone with respect to $\langle \cdot \mid \cdot \rangle_{\mathbf{U}^{-1}}$. By Bauschke and Combettes [4, Propostion 23.16], we have

$$(\forall \mathbf{x} \in \mathcal{H})(\forall \gamma \in]0, +\infty[) \quad \begin{cases} J_{\gamma \mathbf{UA}} \mathbf{x} = (J_{\gamma U_i A_i} x_i)_{1 \le i \le m} \\ J_{\gamma \mathbf{UB}} \mathbf{x} = (J_{\gamma U_i B_i} x_i)_{1 \le i \le m}. \end{cases}$$

Moreover, in view of (7.12) and (7.14), $\mathbf{Q}$ is a β-cocoercive operator. Upon setting,

$$(\forall n \in \mathbb{N}) \quad \begin{cases} \mathbf{x}_n = (x_{1,n}, \ldots, x_{m,n}) \\ \overline{\mathbf{x}}_{1,n} = (\overline{x}_{1,n}, \ldots, \overline{x}_{m,n}) \\ \mathbf{r}_n = (r_{1,n}, \ldots, r_{m,n}), \end{cases}$$

Algorithm 3 reduces to a special case of Algorithm 1. Moreover, all of the conditions in Theorem 1 on the operators as well as the stepsize γ and the relaxation parameter $(\lambda_n)_{n \in \mathbb{N}}$ are satisfied.

(i) This assertion follows from Theorem 1 (i).

(ii)(a) Suppose that $\mathbf{Q}$ is demiregular at $\mathbf{x}(\omega)$. By Theorem 1 (iii), $\mathbf{x}_n \to \mathbf{x}$ P−almost surely. This is equivalent to $(\forall i \in \{1, \ldots, m\})$ $x_{i,n} \to x_i(\omega)$.

Algorithm 3 SFDR for multivariate monotone inclusions

For $n = 0, 1, \ldots$
For $i = 1, 2, \ldots, m$
1: $x_{i,n} = \ldots$
End for
For $i = 1, 2, \ldots, m$
2: $\overline{x}_{i,n+1} = \ldots$
End for
End for

(ii)(b) Now, suppose that A_j is uniformly monotone at $x_j(\omega)$, for every $\omega \in \widetilde{\Omega}$. Let Ω_2 be the set of all $\omega \in \Omega$ such that (7.7) holds. Set $\Omega^* = \Omega_1 \cap \Omega_2 \cap \widetilde{\Omega}$. Then, $P(\Omega^*) = 1$. Fix $\omega \in \Omega$. We can rewrite (7.7), with $\mathbf{a}_n = (a_{i,n})_{1 \le i \le m}$, $\mathbf{x} = (x_1, \dots, x_m)$ and $\overline{\mathbf{y}} = (\overline{y}_1, \dots, \overline{y}_m)$, as follows:

$$\begin{cases} (a_{i,n}, \gamma^{-1} U_i^{-1}(x_{i,n} - a_{i,n} + x_{i,n} - \overline{x}_{i,n}) - Cx_{i,n}) \in \mathrm{gra}(A_i), \\ \gamma^{-1} U_i^{-1}(x_i - \overline{y}_i) - Cx_i \in A_i x_i^*. \end{cases}$$

Since A_j is uniformly monotone at $x_j(\omega)$, there exists an increasing function $\phi_j \colon [0, +\infty[\;\rightarrow\; [0, +\infty]$, vanishing only at 0, that satisfies the following inequality:

$$\phi_j(\|(a_{j,n} - x_j)(\omega)\|) \le \langle (a_{j,n} - x_j)(\omega) | (\gamma^{-1} U_j^{-1}(x_{j,n} - a_{j,n} + x_{j,n} - \overline{x}_{j,n})(\omega) \rangle$$
$$+ \langle (a_{j,n} - x_j)(\omega) | -Q_j x_n - \gamma^{-1} U_j^{-1}(x_j - \overline{y}_j) + Q_j x)(\omega) \rangle.$$

Now using the monotonicity of A_i, we get the following bound for $i \ne j$:

$$0 \le \left\langle (a_{i,n} - x_i)(\omega) \mid \left(\gamma^{-1} U_i^{-1}(x_{i,n} - a_{i,n} + x_{i,n} - \overline{x}_n) \right) \right\rangle$$
$$+ \left\langle (a_{i,n} - x_i)(\omega) \mid -Q_i x_{i,n} - \gamma^{-1} U_i^{-1}(x_i - \overline{y}_i) + Q x)(\omega) \right\rangle.$$

By adding the last two inequalities, we arrive at

$$\phi_j(\|(a_{j,n} - x_j)(\omega)\|) \le \left\langle (\mathbf{a}_n - \mathbf{x})(\omega) \mid \left(\gamma^{-1} V(\mathbf{x}_n - \mathbf{a}_n + \mathbf{x}_n - \overline{\mathbf{x}}_n) \right) \right\rangle$$
$$+ \left\langle (\mathbf{a}_n - \mathbf{x})(\omega) \mid -Q \mathbf{x}_n - \gamma^{-1} V(\mathbf{x} - \overline{\mathbf{y}}) + Q \mathbf{x})(\omega) \right\rangle$$
$$= \langle (\mathbf{a}_n - \mathbf{x})(\omega) \mid (Q \mathbf{x} - Q \mathbf{x}_n)(\omega) \rangle$$
$$+ \left\langle (\mathbf{a}_n - \mathbf{x})(\omega) \mid \gamma^{-1} V(\mathbf{x}_n - \mathbf{a}_n + \mathbf{x}_n - \overline{\mathbf{x}}_n - \mathbf{x} + \overline{\mathbf{y}})(\omega) \right\rangle$$
$$= t_{1,n} + t_{2,n}$$
$$\rightarrow 0.$$

Therefore, $a_{j,n}(\omega) \rightarrow x_j(\omega)$, and hence $x_{j,n} \rightarrow x_j$ P-almost surely.

In the case when B_j is uniformly monotone at $x_j(\omega)$, by using the same manner as above, there exists an increasing function $\phi_j \colon [0, +\infty[\rightarrow [0, +\infty]$ such that

$$\psi(\|(x_{j,n} - x_j)(\omega)\|) \le \left\langle (\mathbf{x}_n - \mathbf{x})(\omega) \mid \gamma^{-1} V(\overline{\mathbf{x}}_n - \mathbf{x}_n + \mathbf{x} - \overline{\mathbf{y}})(\omega) \right\rangle$$
$$\le 2|t_{3,n}| + 2|t_{4,n}|$$
$$\rightarrow 0.$$

This implies $x_{j,n}(\omega) \rightarrow x_j(\omega)$, and hence $x_{j,n} \rightarrow x_j$ P-almost surely. $\qquad\qquad \square$

Remark 1 Below, we highlight some important features of our framework, and provide some connections with the existing literature.

1. To the best of our knowledge, our framework is the first stochastic primal method for finding a zero point of sum of three operators. Some existing stochastic primal methods for finding a zero point of sum of two operators can be found in [12, 34, 35, 39].
2. We can cast our framework to deterministic setting by taking $\mathbf{U} = \mathrm{Id}$ and $\mathbf{r}_n = \mathbf{B}\mathbf{x}_n$. In this case, Algorithm 1 reduces to [17, Algorithm 1]. Further connections with the operator splitting methods with two operators can be found in [8, 25, 26]. In the case when $\mathbf{B}$ is a bounded linear monotone operator, an alternative framework can be found in [23].
3. When $\mathbf{B}$ is a normal cone to a closed vector subspace $\mathcal{W}$ and $\mathbf{U} = \mathrm{Id}$, we obtain a stochastic version of the algorithm presented in [8, Eq. (3.8)]. Furthermore, when $\mathcal{H}$ is a product space and $\mathcal{W}$ is its diagonal subspace, we obtain a stochastic version of the algorithm in [31].
4. When $\mathbf{B}$ is zero and $\mathbf{U} = \mathrm{Id}$, we get $\mathbf{x}_n = \bar{\mathbf{x}}_n$ for every $n \in \mathbb{N}$. In this case, Algorithm 1 reduces to stochastic forward-backward splitting which is recently investigated in [12, 13, 35].
5. When $(B_i)_{1 \le i \le m}$ are zero, (7.13) can be solved using the algorithm presented in [2] in the deterministic setting, and using the algorithm in [34] in the stochastic setting.
6. For convex minimization problems, almost sure convergence of projected stochastic gradient method was investigated in [3, 5, 27]. Similarly, almost sure convergence of proximal stochastic gradient method was investigated in [1, 33].
7. We present some applications of our framework in convex optimization in [40], along with some numerical experiments on Markowitz portfolio optimization and support vector machine classification.

7.3.2 Composite Monotone Inclusions Involving Cocoercive Operators

By using a simple product space technique, our result can be applied to a wide class of composite monotone inclusions involving cocoercive operators as in [37, 38]. In this section, we present an instance of our framework applied to the following generic problem template:

Problem 2 Let $\mathcal{H}$ and $\mathcal{G}$ be real Hilbert spaces and $L \in \mathcal{B}(\mathcal{H}, \mathcal{G})$. Let U and V be self adjoint positive definite operators on $\mathcal{H}$ and $\mathcal{G}$ respectively. Let $A: \mathcal{H} \to 2^{\mathcal{H}}$ and $B: \mathcal{G} \to 2^{\mathcal{G}}$ be maximally monotone operators. Let $C: \mathcal{H} \to \mathcal{H}$ satisfy

$$(\forall x \in \mathcal{H})(\forall y \in \mathcal{H}) \quad \langle x - y \mid Cx - Cy \rangle \ge \mu \|Cx - Cy\|_U^2$$

for some strictly positive number μ, and let $D \colon \mathcal{G} \to 2^{\mathcal{G}}$ be a maximally monotone operator such that

$$(\forall v \in \mathcal{G})(\forall w \in \mathcal{G}) \quad \left\langle v - w \mid D^{-1}v - D^{-1}w \right\rangle \geq v\|D^{-1}v - D^{-1}w\|_V^2$$

for some strictly positive number v. Let $z \in \mathcal{H}$ and $r \in \mathcal{G}$. Then, we define the primal problem as

$$\text{find } \overline{x} \in \mathcal{H} \quad \text{such that} \quad z \in A\overline{x} + L^*(B\square D)(L\overline{x} - r) + C\overline{x}, \tag{7.15}$$

and the dual problem as

$$\text{find } \overline{v}^* \in \mathcal{G} \quad \text{such that} \quad -r \in -L(A+C)^{-1}(-L^*\overline{v}^*) + B^{-1}\overline{v}^* + D^{-1}\overline{v}^*. \tag{7.16}$$

We denote the solution sets of (7.15) and (7.16) by $\mathcal{P}$ and $\mathcal{D}$ respectively.

Let γ and $(\lambda_n)_{n\in\mathbb{N}}$ be strictly positive numbers. Let $(r_{1,n})_{n\in\mathbb{N}}$ and $\overline{x}_0$ be square integrable $\mathcal{H}$-valued random vectors. Similarly, let $(r_{2,n})_{n\in\mathbb{N}}$ and $\overline{v}_0$ be square integrable $\mathcal{G}$-valued random vectors.

Corollary 3 *Suppose that we apply Algorithm 4 for solving Problem 2, which generates the sequence $(\overline{x}_n, \overline{v}_n)_{n\in\mathbb{N}}$. Set $\beta = \min\{\mu, v\}$ and suppose that*

$$z \in \mathrm{ran}(A + L^*(B \square D)(L \cdot -r) + C)$$

Let $(\varepsilon, \alpha) \in \,]0,1[^2$ satisfy $\varepsilon \leq \lambda_n \leq 2 - \varepsilon - \alpha$ and $\varepsilon \leq \gamma \leq \alpha(2\beta - \varepsilon)$. Suppose that the following conditions are satisfied for every $n \in \mathbb{N}$, with $F_n = \sigma(\overline{x}_0, \overline{v}_0, \ldots, \overline{x}_n, \overline{v}_n)$:

(i) $\mathbb{E}[r_{1,n}|F_n] = Cx_n$ and $\mathbb{E}[r_{2,n}|F_n] = D^{-1}v_n$ $\mathbb{P}$-almost surely.
(ii) $\sum_{n\in\mathbb{N}} \mathbb{E}[\|r_{1,n} - Cx_n\|^2|F_n] + \mathbb{E}[\|r_{2,n} - D^{-1}v_n\|^2|F_n] < +\infty$ $\mathbb{P}$-almost surely.

Then, the followings hold for some random vector (x^*, v^*), which is $\mathcal{P} \times \mathcal{D}$-valued $\mathbb{P}$-almost surely.

(i) $(x_n)_{n\in\mathbb{N}}$ converges weakly to x^* $\mathbb{P}$-almost surely.

Algorithm 4 Primal-dual SFDR

For $n = 0, 1, \ldots$
1: $x_n = \overline{x}_n - \gamma U L^* v_n$
2: $v_n = \overline{v}_n + \gamma V L x_n$
3: $\overline{x}_{n+1} = \overline{x}_n + \lambda_n\big(J_{\gamma U A}(2x_n - \overline{x}_n - \gamma U(r_{1,n} - z)) - x_n\big)$
4: $\overline{v}_{n+1} = \overline{v}_n + \lambda_n\big(J_{\gamma V B^{-1}}(2v_n - \overline{v}_n - \gamma V(r_{2,n} + r)) - v_n\big)$
End for

  (ii) $(v_n)_{n \in \mathbb{N}}$ converges weakly to v^* P-almost surely.
 (iii) Suppose that, for every $\omega \in \widetilde{\Omega} \subset \Omega$ with $\mathsf{P}(\widetilde{\Omega}) = 1$,

   (a) C is demiregular or A is uniformly monotone at $x^*(\omega)$. Then, $x_n \to x^*$
       P-almost surely.
   (b) D^{-1} is demiregular or B^{-1} is uniformly convex at $v^*(\omega)$. Then, $v_n \to v^*$
       P-almost surely.

Proof Define $\mathbf{U}: (x, v) \mapsto (Ux, Vv)$ and $\mathbf{S}: \mathcal{H} \to \mathcal{H}: (x, v) \mapsto (L^*v, -Lx)$.
Then $\mathbf{U}$ is self-adjoint and positive definite, and $\mathbf{S}$ is monotone and skew, and hence
it is maximally monotone [4, Example 20.30]. Denote $\mathbf{x}_n = (x_n, v_n)$ and $\overline{\mathbf{x}}_n = (\overline{x}_n, \overline{v}_n)$, then $(\forall n \in \mathbb{N})$ we have

$$\begin{cases} x_n = \overline{x}_n - \gamma UL^*v_n \\ v_n = \overline{v}_n + \gamma VLx_n \end{cases} \Leftrightarrow \begin{cases} \overline{x}_n - x_n = \gamma UL^*v_n \\ \overline{v}_n - v_n = -\gamma VLx_n \end{cases} \Leftrightarrow \quad \overline{\mathbf{x}}_n - \mathbf{x}_n \in \gamma \mathbf{USx}_n.$$

Therefore, for any $\gamma \in \,]0, +\infty[$,

$$\mathbf{x}_n = J_{\gamma \mathbf{US}}\overline{\mathbf{x}}_n.$$

Define $\mathbf{A}: \mathcal{H} \to \mathcal{H}: (x, v) \mapsto (-z+Ax) \times (r+B^{-1}v)$ and $\mathbf{Q}: \mathcal{H} \to \mathcal{H}: (x, v) \mapsto (Cx, D^{-1}v)$. Then, for every $\mathbf{x} = (x, v)$ and $\mathbf{y} = (y, w)$ in $\mathcal{H}$, we have

$$\langle \mathbf{x} - \mathbf{y} \mid \mathbf{Qx} - \mathbf{Qy} \rangle = \langle x - y \mid Cx - Cy \rangle + \left\langle v - w \mid D^{-1}v - D^{-1}w \right\rangle$$
$$\geq \mu \|Cx - Cy\|_U^2 + \nu \|D^{-1}v - D^{-1}w\|_V^2$$
$$\geq \beta \|\mathbf{Qx} - \mathbf{Qy}\|_U^2.$$

Moreover, it follows from [4, Proposition 20.23 and 23.16] that $\mathbf{A}$ is maximally monotone with the resolvent below:

$$(\forall \gamma \in \,]0, +\infty[)(\forall \mathbf{x} \in \mathcal{H}) \quad J_{\gamma \mathbf{UA}}\mathbf{x} = (J_{\gamma UA}x, J_{\gamma VB^{-1}}v).$$

Under condition (7.2), we know that $Z = \mathrm{zer}(\mathbf{A} + \mathbf{S} + \mathbf{Q}) \neq \varnothing$. Furthermore, as shown in [11],

$$(x, v) \in \mathrm{zer}(\mathbf{A} + \mathbf{S} + \mathbf{Q}) \Longrightarrow x \in \mathcal{P} \text{ and } v \in \mathcal{D}.$$

Upon setting $\mathbf{r}_n = (r_{1,n}, r_{2,n})$, the sequence $(\overline{x}_n, \overline{v}_n)_{n \in \mathbb{N}}$ generated by Algorithm 4 satisfies

$$(\forall n \in \mathbb{N}) \quad \left| \begin{array}{l} \mathbf{x}_n = J_{\gamma \mathbf{US}}\overline{\mathbf{x}}_n \\ \overline{\mathbf{x}}_{n+1} = \overline{\mathbf{x}}_n + \lambda_n \big(J_{\gamma \mathbf{UA}}(2\mathbf{x}_n - \overline{\mathbf{x}}_n - \gamma \mathbf{Ur}_n) - \mathbf{x}_n \big). \end{array} \right. \tag{7.17}$$

Note that (7.17) is a special case of Algorithm 1 with $\mathbf{B} = \mathbf{S}$. It is easy to verify that the conditions on the operators, on the relaxation parameter $(\lambda_n)_{n\in\mathbb{N}}$ and the step size γ, and on the stochastic gradients $(r_{1,n})_{n\in\mathbb{N}}$ and $(r_{2,n})_{n\in\mathbb{N}}$ ensures all assumptions of Theorem 1. Then, conclusions (i) and (ii) follow directly from Theorem 1.

(iii)(a) Suppose that C is demiregular or A is uniformly monotone at $x^*(\omega)$, for every $\omega \in \widetilde{\Omega}$. Let Ω_2 be the set of all $\omega \in \Omega$ such that (7.7) holds. Set $\Omega^* = \Omega_1 \cap \Omega_2 \cap \widetilde{\Omega}$. Then $\mathsf{P}(\Omega^*) = 1$. For every $\omega \in \Omega^*$, we have $\mathbf{Q}x_n(\omega) \to \mathbf{Q}x(\omega)$ and $Cx_n(\omega) \to Cx^*(\omega)$. Since $x_n(\omega) \rightharpoonup x(\omega)$, it follows from the definition of demiregular operators that $x_n(\omega) \to x(\omega)$. Next, suppose that A is uniformly monotone at $x^*(\omega)$. Then, by denoting $\mathbf{a}_n = (a_n, b_n)$, $\mathbf{x} = (x^*, v^*)$ and $\overline{\mathbf{y}} = (\overline{y}_1, \overline{y}_2)$, we can rewrite (7.7) as follows:

$$\begin{cases} (a_n, \gamma^{-1}U^{-1}(x_n - a_n + x_n - \overline{x}_n) - Cx_n) \in \operatorname{gra}(A) \\ \gamma^{-1}U^{-1}(x^* - \overline{y}_1) - Cx^* \in Ax^* \\ (b_n, \gamma^{-1}V^{-1}(v_n - b_n + v_n - \overline{v}_n) - D^{-1}v_n) \in \operatorname{gra}(B^{-1}) \\ \gamma^{-1}V^{-1}(v^* - \overline{y}_2) - D^{-1}v^* \in B^{-1}v^*. \end{cases}$$

Since A is uniformly monotone at $x^*(\omega)$, there exists a nonnegative increasing function $\phi \colon [0, +\infty[\to [0, +\infty]$, vanishing only at 0, that satisfies the following bound:

$$\phi(\|(a_n - x^*)(\omega)\|) \leq \Big\langle (a_n - x)(\omega) \mid \big(\gamma^{-1}U^{-1}(x_n - a_n + x_n - \overline{x}_n)\big) \Big\rangle$$
$$+ \Big\langle (a_n - x)(\omega) \mid -Cx_n - \gamma^{-1}U^{-1}(x^* - \overline{y}_1) + Cx^*)(\omega) \Big\rangle.$$

Now using the monotonicity of B^{-1}, we get

$$0 \leq \Big\langle (b_n - v^*)(\omega) \mid \big(\gamma^{-1}V^{-1}(v_n - b_n + v_n - \overline{v}_n)\big) \Big\rangle$$
$$+ \Big\langle (b_n - v^*)(\omega) \mid -D^{-1}v_n - \gamma^{-1}V^{-1}(v^* - \overline{y}_2) + D^{-1}v^*)(\omega) \Big\rangle.$$

By adding the last two inequalities with $\mathbf{V} = \mathbf{U}^{-1}$, we arrive at

$$\phi(\|(a_n - x)(\omega)\|) \leq \Big\langle (\mathbf{a}_n - \mathbf{x})(\omega) \mid \big(\gamma^{-1}\mathbf{V}(\mathbf{x}_n - \mathbf{a}_n + \mathbf{x}_n - \overline{\mathbf{x}}_n)\big) \Big\rangle$$
$$+ \Big\langle (\mathbf{a}_n - \mathbf{x})(\omega) \mid -\mathbf{Q}\mathbf{x}_n - \gamma^{-1}\mathbf{V}(\mathbf{x} - \overline{\mathbf{y}}) + \mathbf{Q}\mathbf{x})(\omega) \Big\rangle$$
$$= \langle (\mathbf{a}_n - \mathbf{x})(\omega) \mid (\mathbf{Q}\mathbf{x} - \mathbf{Q}\mathbf{x}_n)(\omega) \rangle$$
$$+ \Big\langle (\mathbf{a}_n - \mathbf{x})(\omega) \mid \gamma^{-1}\mathbf{V}(\mathbf{x}_n - \mathbf{a}_n + \mathbf{x}_n - \overline{\mathbf{x}}_n - \mathbf{x} + \overline{\mathbf{y}})(\omega) \Big\rangle$$
$$= t_{1,n} + t_{2,n}$$
$$\to 0.$$

Therefore, $a_n(\omega) \to x^*(\omega)$, and hence $x_n \to x^*$ P-almost surely.

(iii)(b) Follows by using the same arguments as the proof of Corollary 3 (iii)(b).

$\square$

Next, we present a direct consequence of the results above, where we apply our method to convex minimization problems.

Corollary 4 *Let $\mathcal{H}$ and $\mathcal{G}$ be real Hilbert spaces and $L \in \mathcal{B}(\mathcal{H}, \mathcal{G})$. Let $f \in \Gamma_0(\mathcal{H})$ and $g \in \Gamma_0(\mathcal{G})$, and let $h: \mathcal{H} \to \mathbb{R}$ be a convex differentiable function with μ-Lipschitz continuous gradient for some $\mu \in {]0, +\infty[}$. Suppose that ℓ is ν-strongly convex for some $\nu \in {]0, +\infty[}$. Let $z \in \mathcal{H}$ and $r \in \mathcal{G}$. Define the primal problem as*

$$\underset{x \in \mathcal{H}}{minimize}\ (f(x) - \langle z \mid x \rangle) + (\ell \ \square \ g)(Lx - r) + h(x),$$

and the dual problem as

$$\underset{v \in \mathcal{G}}{minimize}\ (f^* \ \square \ h^*)(z - L^*v) + g^*(v) + \ell^*(v) + \langle v \mid r \rangle.$$

We denote the solution sets of primal and dual problems by $\mathcal{P}_1$ and $\mathcal{D}_1$ respectively. Suppose that

$$z \in \operatorname{ran}(\partial f + L^*(\partial g \ \square \ \partial \ell)(L \cdot -r) + \nabla h)$$

Let γ and $(\lambda_n)_{n \in \mathbb{N}}$ be strictly positive, let U and V be self adjoint operators on $\mathcal{H}$ and $\mathcal{G}$ respectively. Let $\overline{x}_0$ be a square integrable $\mathcal{H}$-valued random vector, and $(r_{1,n})_{n \in \mathbb{N}}$ be a sequence of square integrable $\mathcal{H}$-valued random vectors. Similarly, let $\overline{v}_0$ be a square integrable $\mathcal{G}$-valued random vector, and $(r_{2,n})_{n \in \mathbb{N}}$ be a sequence of square integrable $\mathcal{G}$-valued random vectors.

Let $(\varepsilon, \alpha) \in {]0, 1[}^2$ and $\beta = \min\{\mu, \nu\}$ satisfy $\varepsilon \leq \lambda_n \leq 2 - \varepsilon - \alpha$ and $\varepsilon \leq \gamma \leq \alpha(2\beta \max\{\|U\|, \|V\|\}^{-1} - \varepsilon)$. Let $(\overline{x}_n, \overline{v}_n)_{n \in \mathbb{N}}$ be a sequence generated by Algorithm 5. Suppose that the following conditions are satisfied for every $n \in \mathbb{N}$, with $\boldsymbol{F}_n = \sigma(\overline{x}_0, \overline{v}_0, \ldots, \overline{x}_n, \overline{v}_n)$:

(i) $\mathbb{E}[r_{1,n} \mid \boldsymbol{F}_n] = \nabla h(x_n)$ and $\mathbb{E}[r_{2,n} \mid \boldsymbol{F}_n] = \nabla \ell^*(v_n)$ P-almost surely.

Algorithm 5 Primal-dual SFDR for composite minimization problem

---

For $n = 0, 1, \ldots$

1: $x_n = \overline{x}_n - \gamma U L^* v_n$

2: $v_n = \overline{v}_n + \gamma V L x_n$

3: $\overline{x}_{n+1} = \overline{x}_n + \lambda_n \left(\operatorname{prox}_{\gamma f}^{U^{-1}} (2x_n - \overline{x}_n - \gamma U(r_{1,n} - z)) - x_n \right)$

4: $\overline{v}_{n+1} = \overline{v}_n + \lambda_n \left(\operatorname{prox}_{\gamma g^*}^{V^{-1}} (2v_n - \overline{v}_n - \gamma V(r_{2,n} + r)) - v_n \right).$

End for

(ii) $\sum_{n\in\mathbb{N}} \mathbb{E}[\|r_{1,n} - \nabla h(x_n)\|^2 | F_n] + \mathbb{E}[\|r_{2,n} - \nabla\ell^*(v_n)\|^2 | F_n] < +\infty$ P-almost surely.

Then, the followings hold for some random vector (x^*, v^*), which is $\mathcal{P}_1 \times \mathcal{D}_1$-valued P-almost surely:

(i) $(x_n)_{n\in\mathbb{N}}$ converges weakly to x^*.

(ii) $(v_n)_{n\in\mathbb{N}}$ converges weakly to v^*.

(iii) Suppose that, for every $\omega \in \widetilde{\Omega} \subset \Omega$ with $\mathsf{P}(\widetilde{\Omega}) = 1$,

(a) h or f is uniformly convex at $x^*(\omega)$. Then, $x_n \to x^*$ P-almost surely.

(b) ℓ^* or g^* is uniformly convex at $v^*(\omega)$. Then, $v_n \to v^*$ P-almost surely.

Proof Follows from Corollary 3, using the same arguments as in the proof of [37, Corollary 4.2]. □

Remark 2 We highlight some important remarks about our framework below.

1. Our stochastic primal-dual splitting algorithms in this section appear to be novel even in the deterministic setting. In the deterministic setting, when $C = 0$ and $D^{-1} = 0$, by taking $r_{1,n} = 0$ and $r_{2,n} = 0$ for every $n \in \mathbb{N}$, and setting $U = \mathrm{Id}$, $V = \mathrm{Id}$, Algorithm (4) reduces to an error-free version of [9, Eq. (2.22)]. Another variant of this algorithm can be found in [7].
2. Our approach follows the reformulation technique in [11] and [37]. We reformulate the primal-dual inclusion in the form of (7.1), and then we apply Algorithm 1.
3. Our conditions on the stochastic gradient estimates $(r_{1,n})_{n\in\mathbb{N}}$ and $(r_{2,n})_{n\in\mathbb{N}}$ are the same as in [35], but different from the conditions in [12, 30, 39].
4. Our algorithms in this section require the inverse of $(\mathrm{Id} + \gamma^2 UL^*VL)$. This inversion can be computed efficiently in some specific applications, see [6, 7, 31].

7.4 Convergence Rate

In this section, we present a variant of Algorithm 1 that applies to a special case of Problem 1, where either $\mathbf{Q}$ or $\mathbf{B}$ is strongly monotone. Recall that this ensures uniqueness of solution $\mathbf{x}^*$. In this case, we the provide convergence rate in expectation of squared norm error, and our results match with the rates shown in [35] for strongly monotone inclusions.

Let $(\gamma_n)_{n\in\mathbb{N}}$ be a sequence of strictly positive numbers, and $(\mathbf{r}_n)_{n\in\mathbb{N}}$ be a sequence of square integrable $\mathcal{H}$-valued random vectors. Let $\bar{\mathbf{x}}_{A,0}$ be a square integrable $\mathcal{H}$-valued random vector, $\bar{\mathbf{x}}_{B,0} = J_{\gamma_0 \mathbf{UB}}\bar{\mathbf{x}}_{A,0}$ and $\mathbf{u}_{B,0} = (\mathbf{U}\gamma_0)^{-1}(\mathrm{Id} - J_{\gamma_0 \mathbf{UB}})\bar{\mathbf{x}}_{A,0}$.

Theorem 2 *Set* $\mathbf{V} = \mathbf{U}^{-1}$ *and suppose that there exist parameters* $\mu_Q \in$ $]0, +\infty[$, $\mu_B \in [0, +\infty[$, *and* $\beta \in [0, +\infty[$ *such that*

$$(\forall(\mathbf{x}, \mathbf{y}) \in \mathcal{H}^2)(\mathbf{w} \in \mathbf{Bx})(\mathbf{v} \in \mathbf{By}) \quad \langle \mathbf{x} - \mathbf{y} \mid \mathbf{w} - \mathbf{v} \rangle \geq \mu_B \|\mathbf{x} - \mathbf{y}\|_{\mathbf{V}}^2$$

Algorithm 6 SFDR for strongly monotone inclusions
___
For $n = 0, 1, \ldots$
  1: $\overline{\mathbf{x}}_{B,n+1} = J_{\gamma_n \mathbf{UB}}(\overline{\mathbf{x}}_{A,n} + \gamma_n \mathbf{u}_{B,n})$
  2: $\mathbf{u}_{B,n+1} = \gamma_n^{-1}(\overline{\mathbf{x}}_{A,n} - \overline{\mathbf{x}}_{B,n+1}) + \mathbf{u}_{B,n}$
  3: $\overline{\mathbf{x}}_{A,n+1} = J_{\gamma_{n+1} \mathbf{UA}}(\overline{\mathbf{x}}_{B,n+1} - \gamma_{n+1} \mathbf{u}_{B,n+1} - \gamma_{n+1} \mathbf{Ur}_{n+1}).$
End for

---

and

$$\langle \mathbf{x} - \mathbf{y} \mid \mathbf{Qx} - \mathbf{Qy} \rangle \geq \mu_Q \|\mathbf{x} - \mathbf{y}\|_\mathbf{V}^2. \tag{7.18}$$

Let $0 \leq \gamma_n < \min\{2(1 - \eta)\beta, (2\eta\mu_Q)^{-1}\}$ *for every* $n \in \mathbb{N}$ *and for some* $\eta \in]0, 1[$. *Furthermore, suppose that the following conditions are satisfied* $(\forall n \in \mathbb{N})$ *with* $F_n = \sigma(\overline{\mathbf{x}}_{A,k})_{0 \leq k \leq n}$:

  (i) $\mathbb{E}[\mathbf{r}_{n+1} | F_n] = \mathbf{Q}\overline{\mathbf{x}}_{B,n+1}$ P-*almost surely.*
  (ii) $(\exists c \in [0, +\infty[)(\exists t \in \mathbb{R})$ $\sum_{k=0}^n \mathbb{E}[\|\mathbf{r}_k - \mathbf{Q}\overline{\mathbf{x}}_{B,k}\|^2] \leq cn^t$.

Then, the followings hold:

  (i) For every $n \in \mathbb{N}$,

$$(1 + 2\gamma_n\mu_B)\mathbb{E}[\|\overline{\mathbf{x}}_{B,n+1} - \mathbf{x}^*\|_\mathbf{V}^2 | F_n] + \gamma_n^2 \mathbb{E}[\|\mathbf{u}_{B,n+1} - \mathbf{x}^*\|_\mathbf{V}^2 | F_n]$$
$$\leq (1 - 2\gamma_n\mu_Q\eta)\|\overline{\mathbf{x}}_{B,n} - \mathbf{x}^*\|_\mathbf{V}^2 + \gamma_n^2\|\mathbf{u}_{B,n} - \mathbf{x}^*\|_\mathbf{V}^2$$
$$+ 2\gamma_n^2 \mathbb{E}[\|\mathbf{r}_n - \mathbf{Q}\overline{\mathbf{x}}_{B,n}\|^2 | F_n] \tag{7.19}$$

  (ii) Assume that we generate $(\gamma_n)_{n \in \mathbb{N}}$ *by the recursion below:*

$$\gamma_{n+1} = \frac{-\gamma_n^2\mu_Q\eta + \sqrt{(\gamma_n^2\mu_Q\eta)^2 + (1 + 2\gamma_n\mu_B)\gamma_n^2}}{1 + 2\gamma_n\mu_B}.$$

Then, $\mathbb{E}[\|\overline{\mathbf{x}}_{B,n} - \mathbf{x}^*\|^2] = \mathcal{O}(1/n^2) + \mathcal{O}(1/n^{2-t})$.

  (iii) Let $\alpha \in]0, 1]$ *and* $(\tau_0, c) \in]0, +\infty[^2$. *Suppose that* $(2\mathbb{E}[\|\mathbf{r}_k - \mathbf{Q}\overline{\mathbf{x}}_{B,k}\|^2])_{n \in \mathbb{N}}$ *and* $(\mathbb{E}[\|\mathbf{u}_{B,n} - \mathbf{x}^*\|_\mathbf{V}^2])_{n \in \mathbb{N}}$ *are uniformly bounded by* $\tau_0(2\mu_Q c\eta)^2$ *(note that this ensures condition (ii) with* $t = 1$). *Let* n_0 *be a strictly positive integer such that* $c_0 = 2c\mu_Q\eta \leq n_0^\alpha$ *and* $c \leq \min\{2(1 - \eta)\beta, (2\eta\mu_Q)^{-1}\}n_0^\alpha$. *Set* $t_0 = 1 - 2^{\alpha-1} \geq 0$ *and* $s_n = \mathbb{E}[\|\overline{\mathbf{x}}_{B,n} - \mathbf{x}^*\|_\mathbf{V}^2]$ *for every* $n \in \mathbb{N}$. *Let us choose* $(\forall n \geq 2n_0)$ $\gamma_n = c.n^{-\alpha}$ *for some* $\alpha \in]0, 1]$. *Then, for every* $n \geq 2n_0$, *we have*

$$s_{n+1} \leq \left(\tau_0 c_0^2 \varphi_{1-2\alpha}(n) + s_{n_0} \exp\left(\frac{c_0 n_0^{1-\alpha}}{1-\alpha}\right)\right) \exp\left(\frac{-c_0 t_0(n+1)^{1-\alpha}}{1-\alpha}\right) + \frac{\tau_0 2^\alpha c_0}{(n-2)^\alpha}$$
$$= \mathcal{O}(1/n^\alpha)$$

if $\alpha \in \ ]0, 1[$ *and*

$$s_{n+1} \le s_{n_0} \Big(\frac{n_0}{n+1}\Big)^{c_0} + \frac{\tau_0 c_0^2}{(n+1)^{c_0}} \Big(1 + \frac{1}{n_0}\Big)^{c_0} \varphi_{c_0-1}(n) = \mathcal{O}(1/n)$$

if $\alpha = 1$ *and* $c_0 \ge 1$ *for the equality.*

Proof

(i) We have

$$(\forall n \in \mathbb{N}) \quad \begin{cases} \mathbf{u}_{A,n} = \gamma_n^{-1}(\overline{\mathbf{x}}_{B,n} - \overline{\mathbf{x}}_{A,n}) - (\mathbf{u}_{B,n} + \mathbf{U}\mathbf{r}_n) \in \mathbf{U}A\overline{\mathbf{x}}_{A,n} \\ \mathbf{u}_{B,n} \in \mathbf{U}B\overline{\mathbf{x}}_{B,n} \\ \gamma_n(\mathbf{u}_{B,n+1} - \mathbf{u}_{B,n}) = \overline{\mathbf{x}}_{A,n} - \overline{\mathbf{x}}_{B,n+1} \\ \gamma_n(\mathbf{u}_{A,n} + \mathbf{u}_{B,n} + \mathbf{U}\mathbf{r}_n) = \overline{\mathbf{x}}_{B,n} - \overline{\mathbf{x}}_{A,n} \\ \gamma_n(\mathbf{u}_{B,n+1} + \mathbf{u}_{A,n} + \mathbf{U}\mathbf{r}_n) = \overline{\mathbf{x}}_{B,n} - \overline{\mathbf{x}}_{B,n+1}. \end{cases}$$

Note that $\mathbf{V} = \mathbf{U}^{-1}$, and set

$$\begin{cases} \chi_n = 2\gamma_n\big(\langle\overline{\mathbf{x}}_{A,n} - \mathbf{x}^* \mid \mathbf{u}_{A,n} + \mathbf{U}\mathbf{r}_n\rangle_{\mathbf{V}} + \langle\overline{\mathbf{x}}_{B,n+1} - \mathbf{x}^* \mid \mathbf{u}_{B,n+1}\rangle_{\mathbf{V}}\big) \\ \chi_{1,n} = 2\langle\overline{\mathbf{x}}_{B,n+1} - \mathbf{x}^* \mid \overline{\mathbf{x}}_{B,n} - \overline{\mathbf{x}}_{B,n+1}\rangle_{\mathbf{V}} \\ \chi_{2,n} = 2\langle\overline{\mathbf{x}}_{A,n} - \overline{\mathbf{x}}_{B,n+1} \mid \overline{\mathbf{x}}_{B,n} - \overline{\mathbf{x}}_{A,n}\rangle_{\mathbf{V}} \\ \chi_{3,n} = 2\gamma_n\langle\overline{\mathbf{x}}_{B,n+1} - \overline{\mathbf{x}}_{A,n} \mid \mathbf{u}_{B,n} - \mathbf{u}_B^*\rangle_{\mathbf{V}} \\ \qquad = 2\gamma_n^2\langle\mathbf{u}_{B,n} - \mathbf{u}_{B,n+1} \mid \mathbf{u}_{B,n} - \mathbf{u}_B^*\rangle_{\mathbf{V}} \\ \chi_{4,n} = 2\gamma_n\langle\overline{\mathbf{x}}_{B,n+1} - \overline{\mathbf{x}}_{A,n} \mid \mathbf{u}_B^*\rangle_{\mathbf{V}}. \end{cases}$$

for every $n \in \mathbb{N}$. Then, simple calculations show that

$$\begin{cases} \chi_{2,n} = \|\overline{\mathbf{x}}_{B,n} - \overline{\mathbf{x}}_{B,n+1}\|_{\mathbf{V}}^2 - \|\overline{\mathbf{x}}_{A,n} - \overline{\mathbf{x}}_{B,n+1}\|_{\mathbf{V}}^2 - \|\overline{\mathbf{x}}_{B,n} - \overline{\mathbf{x}}_{A,n}\|_{\mathbf{V}}^2 \\ \chi_{1,n} = \|\overline{\mathbf{x}}_{B,n} - \mathbf{x}^*\|_{\mathbf{V}}^2 - \|\overline{\mathbf{x}}_{B,n+1} - \mathbf{x}^*\|_{\mathbf{V}}^2 - \|\overline{\mathbf{x}}_{B,n} - \overline{\mathbf{x}}_{B,n+1}\|_{\mathbf{V}}^2 \\ \chi_{3,n} = \gamma_n^2\big(\|\mathbf{u}_{B,n+1} - \mathbf{u}_{B,n}\|_{\mathbf{V}}^2 + \|\mathbf{u}_{B,n} - \mathbf{u}_B^*\|_{\mathbf{V}}^2 - \|\mathbf{u}_{B,n+1} - \mathbf{u}_B^*\|_{\mathbf{V}}^2\big) \\ \qquad = \|\overline{\mathbf{x}}_{A,n} - \overline{\mathbf{x}}_{B,n+1}\|_{\mathbf{V}}^2 + \gamma_n^2\big(\|\mathbf{u}_{B,n} - \mathbf{u}_B^*\|_{\mathbf{V}}^2 - \|\mathbf{u}_{B,n+1} - \mathbf{u}_B^*\|_{\mathbf{V}}^2\big). \end{cases} \tag{7.20}$$

Furthermore, for every $n \in \mathbb{N}$, we can express χ_n as follows:

$$\begin{aligned} \chi_n &= 2\gamma_n\big(\langle\overline{\mathbf{x}}_{A,n} - \overline{\mathbf{x}}_{B,n+1} \mid \mathbf{u}_{A,n} + \mathbf{U}\mathbf{r}_n\rangle_{\mathbf{V}} \\ &\qquad + \langle\overline{\mathbf{x}}_{B,n+1} - \mathbf{x}^* \mid \mathbf{u}_{B,n+1} + \mathbf{u}_{A,n} + \mathbf{U}\mathbf{r}_n\rangle_{\mathbf{V}}\big) \\ &= 2\gamma_n\langle\overline{\mathbf{x}}_{A,n} - \overline{\mathbf{x}}_{B,n+1} \mid \mathbf{u}_{A,n} + \mathbf{U}\mathbf{r}_n\rangle_{\mathbf{V}} + \chi_{1,n} \\ &= 2\gamma_n\big(\langle\overline{\mathbf{x}}_{A,n} - \overline{\mathbf{x}}_{B,n+1} \mid \mathbf{u}_{A,n} + \mathbf{U}\mathbf{r}_n + \mathbf{u}_{B,n}\rangle_{\mathbf{V}} \\ &\qquad - \langle\overline{\mathbf{x}}_{A,n} - \overline{\mathbf{x}}_{B,n+1} \mid \mathbf{u}_{B,n}\rangle_{\mathbf{V}}\big) + \chi_{1,n} \\ &= \chi_{2,n} + \chi_{1,n} + 2\gamma_n\langle\overline{\mathbf{x}}_{B,n+1} - \overline{\mathbf{x}}_{A,n} \mid \mathbf{u}_{B,n}\rangle_{\mathbf{V}} \end{aligned}$$

$$= \chi_{2,n} + \chi_{1,n} + 2\gamma_n \langle \overline{\mathbf{x}}_{B,n+1} - \overline{\mathbf{x}}_{A,n} \mid \mathbf{u}_{B,n} - \mathbf{u}_B^* \rangle_\mathbf{V}$$
$$+ 2\gamma_n \langle \overline{\mathbf{x}}_{B,n+1} - \overline{\mathbf{x}}_{A,n} \mid \mathbf{u}_B^* \rangle_\mathbf{V}$$
$$= \chi_{2,n} + \chi_{1,n} + \chi_{3,n} + \chi_{4,n}.$$

Now, summing the equalities in (7.20), we obtain,

$$\chi_n = \|\overline{\mathbf{x}}_{B,n} - \mathbf{x}^*\|_\mathbf{V}^2 - \|\overline{\mathbf{x}}_{B,n+1} - \mathbf{x}^*\|_\mathbf{V}^2 - \|\overline{\mathbf{x}}_{B,n} - \overline{\mathbf{x}}_{A,n}\|_\mathbf{V}^2$$
$$+ \gamma_n^2 \big(\|\mathbf{u}_{B,n} - \mathbf{u}_B^*\|_\mathbf{V}^2 - \|\mathbf{u}_{B,n+1} - \mathbf{u}_B^*\|_\mathbf{V}^2 \big) + \chi_{4,n}.$$

It follows from $\mathbf{u}_A^* \in \mathbf{U}\mathbf{A}\mathbf{x}^*$ and the monotonicity of $\mathbf{U}\mathbf{A}$ that

$$\langle \overline{\mathbf{x}}_{A,n} - \mathbf{x}^* \mid \mathbf{u}_{A,n} - \mathbf{u}_A^* \rangle_\mathbf{V} \geq 0,$$

and hence

$$\chi_n = 2\gamma_n \big(\langle \overline{\mathbf{x}}_{A,n} - \mathbf{x}^* \mid \mathbf{u}_{A,n} - \mathbf{u}_A^* \rangle_\mathbf{V} + \langle \overline{\mathbf{x}}_{A,n} - \mathbf{x}^* \mid \mathbf{u}_A^* + \mathbf{U}\mathbf{r}_n \rangle_\mathbf{V}$$
$$+ \langle \overline{\mathbf{x}}_{B,n+1} - \mathbf{x}^* \mid \mathbf{u}_{B,n+1} \rangle_\mathbf{V} \big)$$
$$\geq 2\gamma_n \big(\langle \overline{\mathbf{x}}_{A,n} - \mathbf{x}^* \mid \mathbf{u}_A^* + \mathbf{U}\mathbf{r}_n \rangle_\mathbf{V} + \langle \overline{\mathbf{x}}_{B,n+1} - \mathbf{x}^* \mid \mathbf{u}_{B,n+1} \rangle_\mathbf{V} \big)$$
$$= 2\gamma_n \big(\langle \overline{\mathbf{x}}_{A,n} - \mathbf{x}^* \mid \mathbf{u}_A^* + \mathbf{U}\mathbf{r}_n \rangle_\mathbf{V} + \langle \overline{\mathbf{x}}_{B,n+1} - \mathbf{x}^* \mid \mathbf{u}_{B,n+1} - \mathbf{u}_B^* \rangle_\mathbf{V}$$
$$+ \langle \overline{\mathbf{x}}_{B,n+1} - \mathbf{x}^* \mid \mathbf{u}_B^* \rangle_\mathbf{V} \big)$$
$$\geq 2\gamma_n \big(\langle \overline{\mathbf{x}}_{A,n} - \mathbf{x}^* \mid \mathbf{u}_A^* + \mathbf{U}\mathbf{r}_n \rangle_\mathbf{V} + \mu_B \|\overline{\mathbf{x}}_{B,n+1} - \mathbf{x}^*\|_\mathbf{V}^2 \qquad (7.21)$$
$$+ \langle \overline{\mathbf{x}}_{B,n+1} - \mathbf{x}^* \mid \mathbf{u}_B^* \rangle_\mathbf{V} \big),$$

where the last inequality follows from the assumption that $\mathbf{U}\mathbf{B}$ is μ_B-strongly monotone. Set, for every $n \in \mathbb{N}$,

$$\mathbf{x}_{A,n}^e = J_{\gamma_n \mathbf{U}\mathbf{A}} \big((\overline{\mathbf{x}}_{B,n} - \gamma_n \mathbf{U}\mathbf{u}_{B,n} - \gamma_n \mathbf{U}\mathbf{Q}\overline{\mathbf{x}}_{B,n}) \big).$$

Then using the firm non-expansiveness of $J_{\gamma_n \mathbf{U}\mathbf{A}}$ with respect to the norm $\|\cdot\|_\mathbf{V}$, we get

$$\|\mathbf{x}_{A,n}^e - \overline{\mathbf{x}}_{A,n}\|_\mathbf{V} \leq \gamma_n \|\mathbf{U}(\mathbf{Q}\overline{\mathbf{x}}_{B,n} - \mathbf{r}_n)\|_{\mathbf{U}^{-1}} = \gamma_n \|\mathbf{Q}\overline{\mathbf{x}}_{B,n} - \mathbf{r}_n\|_\mathbf{U}.$$

Let us set

$$\begin{cases} \chi_{5,n} = \langle \overline{\mathbf{x}}_{A,n} - \mathbf{x}_{A,n}^e \mid \mathbf{r}_n - \mathbf{Q}\overline{\mathbf{x}}_{B,n} \rangle \\ \chi_{6,n} = \langle \mathbf{x}_{A,n}^e - \mathbf{x}^* \mid \mathbf{r}_n - \mathbf{Q}\overline{\mathbf{x}}_{B,n} \rangle \\ \chi_{7,n} = \chi_{5,n} + \chi_{6,n} = \langle \overline{\mathbf{x}}_{A,n} - \mathbf{x}^* \mid \mathbf{r}_n - \mathbf{Q}\overline{\mathbf{x}}_{B,n} \rangle. \end{cases}$$

Then, we have

$$\chi_{5,n} = \langle \overline{\mathbf{x}}_{A,n} - \mathbf{x}^e_{A,n} \mid \mathbf{U}\mathbf{r}_n - \mathbf{U}\mathbf{Q}\overline{\mathbf{x}}_{B,n} \rangle_{\mathbf{V}}$$
$$\leq \|\overline{\mathbf{x}}_{A,n} - \mathbf{x}^e_{A,n}\|_{\mathbf{V}} \|\mathbf{U}\mathbf{r}_n - \mathbf{U}\mathbf{Q}\overline{\mathbf{x}}_{B,n}\|_{\mathbf{V}}$$
$$\leq \gamma_n \|\mathbf{r}_n - \mathbf{Q}\overline{\mathbf{x}}_{B,n}\|^2_{\mathbf{U}},$$

and since $\mathbf{x}^e_{A,n}$ is $\mathbf{F}_n$-measurable, we obtain

$$\mathbb{E}[\chi_{6,n}|\mathbf{F}_n] = \langle \mathbf{x}^e_{A,n} - \mathbf{x}^* \mid \mathbb{E}[\mathbf{r}_n - \mathbf{Q}\overline{\mathbf{x}}_{B,n}|\mathbf{F}_n] \rangle = 0.$$

Furthermore, for every $\eta \in \,]0, 1[$, since $\mathbf{U}\mathbf{Q}$ is β-cocoercive and μ_Q-strongly monotone, we have

$$2\langle \overline{\mathbf{x}}_{A,n} - \mathbf{x}^* \mid \mathbf{U}\mathbf{r}_n \rangle_{\mathbf{V}} = 2\langle \overline{\mathbf{x}}_{A,n} - \mathbf{x}^* \mid \mathbf{U}\mathbf{Q}\overline{\mathbf{x}}_{B,n} \rangle_{\mathbf{V}}$$
$$+ 2\langle \overline{\mathbf{x}}_{A,n} - \mathbf{x}^* \mid \mathbf{U}\mathbf{r}_n - \mathbf{U}\mathbf{Q}\overline{\mathbf{x}}_{B,n} \rangle_{\mathbf{V}}$$
$$= 2\langle \overline{\mathbf{x}}_{A,n} - \overline{\mathbf{x}}_{B,n} \mid \mathbf{U}\mathbf{Q}\overline{\mathbf{x}}_{B,n} - \mathbf{U}\mathbf{Q}\mathbf{x}^* \rangle_{\mathbf{V}} + 2\langle \overline{\mathbf{x}}_{A,n} - \mathbf{x}^* \mid \mathbf{Q}\mathbf{x}^* \rangle \quad (7.22)$$
$$+ 2\langle \overline{\mathbf{x}}_{B,n} - \mathbf{x}^* \mid \mathbf{U}\mathbf{Q}\overline{\mathbf{x}}_{B,n} - \mathbf{U}\mathbf{Q}\mathbf{x}^* \rangle_{\mathbf{V}} + 2\chi_{7,n}$$
$$\geq \frac{-1}{2\beta(1-\eta)} \|\overline{\mathbf{x}}_{A,n} - \overline{\mathbf{x}}_{B,n}\|^2_{\mathbf{V}} - 2\beta(1-\eta)\|\mathbf{Q}\overline{\mathbf{x}}_{B,n} - \mathbf{Q}\mathbf{x}^*\|^2_{\mathbf{U}}$$
$$+ 2\eta\mu_Q \|\overline{\mathbf{x}}_{B,n} - \mathbf{x}^*\|^2_{\mathbf{V}}$$
$$+ 2\beta(1-\eta)\|\mathbf{Q}\overline{\mathbf{x}}_{B,n} - \mathbf{Q}\mathbf{x}^*\|^2_{\mathbf{U}} + 2\langle \overline{\mathbf{x}}_{A,n} - \mathbf{x}^* \mid \mathbf{Q}\mathbf{x}^* \rangle + 2\chi_{7,n}$$
$$\geq \frac{-1}{2\beta(1-\eta)} \|\overline{\mathbf{x}}_{A,n} - \overline{\mathbf{x}}_{B,n}\|^2_{\mathbf{V}} + 2\eta\mu_Q \|\overline{\mathbf{x}}_{B,n} - \mathbf{x}^*\|^2_{\mathbf{V}} + 2\chi_{7,n}$$
$$+ 2\langle \overline{\mathbf{x}}_{A,n} - \mathbf{x}^* \mid \mathbf{Q}\mathbf{x}^* \rangle. \quad (7.23)$$

Now, inserting (7.23) into (7.21), we arrive at

$$\chi_n \geq 2\gamma_n \big(\langle \overline{\mathbf{x}}_{A,n} - \mathbf{x}^* \mid \mathbf{u}^*_A + \mathbf{U}\mathbf{Q}\mathbf{x}^* \rangle_{\mathbf{V}} + \langle \overline{\mathbf{x}}_{B,n+1} - \mathbf{x}^* \mid \mathbf{u}^*_B \rangle_{\mathbf{V}} \big) + 2\gamma_n\chi_{7,n}$$
$$\frac{-\gamma_n}{2\beta(1-\eta)} \|\overline{\mathbf{x}}_{A,n} - \overline{\mathbf{x}}_{B,n}\|^2_{\mathbf{V}} + 2\eta\mu_Q\gamma_n\|\overline{\mathbf{x}}_{B,n} - \mathbf{x}^*\|^2 + 2\mu_B\gamma_n\|\overline{\mathbf{x}}_{B,n+1} - \mathbf{x}^*\|^2_{\mathbf{V}},$$
$$(7.24)$$

since it follows that

$$2\gamma_n \big(\langle \overline{\mathbf{x}}_{A,n} - \mathbf{x}^* \mid \mathbf{u}^*_A + \mathbf{U}\mathbf{Q}\mathbf{x}^* \rangle_{\mathbf{V}} + \langle \overline{\mathbf{x}}_{B,n+1} - \mathbf{x}^* \mid \mathbf{u}^*_B \rangle_{\mathbf{V}} \big) - \chi_{4,n}$$
$$= 2\gamma_n(\langle \overline{\mathbf{x}}_{A,n} - \mathbf{x}^* \mid \mathbf{u}^*_A + \mathbf{u}^*_B + \mathbf{U}\mathbf{Q}\mathbf{x}^* \rangle_{\mathbf{V}}$$
$$= 0.$$

We derive from (7.24) and (7.21) that

$$(1 + 2\gamma_k\mu_B)\|\bar{\mathbf{x}}_{B,n+1} - \mathbf{x}^*\|_{\mathbf{V}}^2 + (1 - \frac{\gamma_n}{2(1-\eta)\beta})\|\bar{\mathbf{x}}_{A,n} - \bar{\mathbf{x}}_{B,n}\|_{\mathbf{V}}^2$$

$$+ \gamma_n^2\|\mathbf{u}_{B,n+1} - \mathbf{x}^*\|_{\mathbf{V}}^2$$

$$\leq (1 - 2\gamma_n\mu_Q\eta)\|\bar{\mathbf{x}}_{B,n} - \mathbf{x}^*\|_{\mathbf{V}}^2 + \gamma_n^2\|\mathbf{u}_{B,n} - \mathbf{x}^*\|_{\mathbf{V}}^2 - 2\gamma_n\chi_{7,n}.$$

By our assumptions on $(\gamma_n)_{n\in\mathbb{N}}$, we have $(1 - \frac{\gamma_n}{2(1-\eta)\beta}) \geq 0$. Then,

$$(1 + 2\gamma_k\mu_B)\|\bar{\mathbf{x}}_{B,n+1} - \mathbf{x}^*\|_{\mathbf{V}}^2 + \gamma_n^2\|\mathbf{u}_{B,n+1} - \mathbf{x}^*\|_{\mathbf{V}}^2 \leq (1 - 2\gamma_n\mu_Q\eta)\|\bar{\mathbf{x}}_{B,n} - \mathbf{x}^*\|_{\mathbf{V}}^2$$

$$+ \gamma_n^2\|\mathbf{u}_{B,n} - \mathbf{x}^*\|_{\mathbf{V}}^2 - 2\gamma_n\chi_{7,n}.$$

Now, taking the conditional expectation with respect to $\boldsymbol{F}_n$, we obtain

$$(1 + 2\gamma_n\mu_B)\mathbb{E}[\|\bar{\mathbf{x}}_{B,n+1} - \mathbf{x}^*\|_{\mathbf{V}}^2|\boldsymbol{F}_n] + \gamma_n^2\mathbb{E}[\|\mathbf{u}_{B,n+1} - \mathbf{x}^*\|_{\mathbf{V}}^2|\boldsymbol{F}_n]$$

$$\leq (1 - 2\gamma_n\mu_Q\eta)\|\bar{\mathbf{x}}_{B,n} - \mathbf{x}^*\|_{\mathbf{V}}^2 + \gamma_n^2\|\mathbf{u}_{B,n} - \mathbf{x}^*\|_{\mathbf{V}}^2 - 2\gamma_n\mathbb{E}[\chi_{7,n}|\boldsymbol{F}_n]$$

$$= (1 - 2\gamma_n\mu_Q\eta)\|\bar{\mathbf{x}}_{B,n} - \mathbf{x}^*\|_{\mathbf{V}}^2 + \gamma_n^2\|\mathbf{u}_{B,n} - \mathbf{x}^*\|_{\mathbf{V}}^2 - 2\gamma_n\mathbb{E}[\chi_{5,n}|\boldsymbol{F}_n]$$

$$\leq (1 - 2\gamma_n\mu_Q\eta)\|\bar{\mathbf{x}}_{B,n} - \mathbf{x}^*\|_{\mathbf{V}}^2 + \gamma_n^2\|\mathbf{u}_{B,n} - \mathbf{x}^*\|_{\mathbf{V}}^2$$

$$+ 2\gamma_n^2\mathbb{E}[\|\mathbf{r}_n - \mathbf{Q}\bar{\mathbf{x}}_{B,n}\|^2|\boldsymbol{F}_n],$$

which proves (7.19).

(ii) As indicated in the proof of [17], we have

$$\gamma_n^{-2}(1 + 2\gamma_n\mu_B) = \gamma_{n+1}^{-2}(1 - 2\gamma_{n+1}\mu_Q\eta),$$

and

$$\lim_{n\to\infty}(n + 1)\gamma_n = (\eta\mu_Q + \mu_B)^{-1}. \tag{7.25}$$

Therefore, by dividing both sides of (7.19) by γ_n^2, and taking the expectations, we obtain

$$\gamma_{n+1}^{-2}(1 - 2\gamma_{n+1}\mu_Q\eta)\mathbb{E}[\|\bar{\mathbf{x}}_{B,n+1} - \mathbf{x}^*\|_{\mathbf{V}}^2|] + \mathbb{E}[\|\mathbf{u}_{B,n+1} - \mathbf{x}^*\|_{\mathbf{V}}^2]$$

$$\leq \gamma_n^{-2}(1 - 2\gamma_n\mu_Q\eta)\mathbb{E}[\|\bar{\mathbf{x}}_{B,n} - \mathbf{x}^*\|_{\mathbf{V}}^2] + \mathbb{E}[\|\mathbf{u}_{B,n} - \mathbf{x}^*\|_{\mathbf{V}}^2]$$

$$+ 2\mathbb{E}[\|\mathbf{r}_n - \mathbf{Q}\bar{\mathbf{x}}_{B,n}\|^2].$$

Now, summing this inequality from $n = 0$ to $n = N$, we get

$$\gamma_{N+1}^{-2}(1 - 2\gamma_{N+1}\mu_Q\eta)\mathbb{E}[\|\bar{\mathbf{x}}_{B,N+1} - \mathbf{x}^*\|_\mathbf{V}^2]] \leq \gamma_0^{-2}(1 - 2\gamma_0\mu_Q\eta)\mathbb{E}\|\bar{\mathbf{x}}_{B,0} - \mathbf{x}^*\|_\mathbf{V}^2$$

$$+ \mathbb{E}[\|\mathbf{u}_{B,0} - \mathbf{x}^*\|^2] + \sum_{k=0}^{N} \mathbb{E}[\|\mathbf{r}_n - Q\bar{\mathbf{x}}_{B,n}\|^2]. \tag{7.26}$$

In view of (7.25), (iii) follows from (7.26).

(iii) Set $\theta_n = 2\gamma_n\mu_Q\eta = c_0 n^{-\alpha}$. Then, we have

$$s_{n+1} \leq (1 - \theta_n)s_n + \tau_0\theta_n^2.$$

Now, the proof follows from [33, Lemma 4.4]. □

Corollary 5 *Consider problem (7.10) described in Corollary 1 (with the same assumptions on f, g and h as in Corollary 1). Further assume that h is μ_h-strongly convex with some parameter $\mu_h \in \,]0, +\infty[$, and g is μ_g-strongly convex for some $\mu_g \in [0, +\infty[$. Let $(\gamma_n)_{n\in\mathbb{N}}$ satisfy $0 \leq \gamma_n < \min\{2(1 - \eta)\beta, (2\eta\mu_h)^{-1}\}$, for some $\eta \in \,]0, 1[$. Let $(\mathbf{r}_n)_{n\in\mathbb{N}}$ be a sequence of square integrable $\mathcal{H}$-valued random vectors, and let $\bar{\mathbf{x}}_{f,0}$ be a square integrable $\mathcal{H}$-valued random vector. Set $\bar{\mathbf{x}}_{g,0} = \text{prox}_{\gamma_0 g}\,\bar{\mathbf{x}}_{f,0}$ and $\mathbf{u}_{g,0} = (\gamma_0)^{-1}(\text{Id} - \text{prox}_{\gamma_0 g})\bar{\mathbf{x}}_{f,0}$.*

Algorithm 7 SFDR for strongly convex minimization problems

For $n = 0, 1, \ldots$
1: $\bar{\mathbf{x}}_{g,n+1} = \text{prox}_{\gamma_n g}(\bar{\mathbf{x}}_{f,n} + \gamma_n\mathbf{u}_{g,n})$
2: $\mathbf{u}_{g,n+1} = \gamma_n^{-1}(\bar{\mathbf{x}}_{f,n} - \bar{\mathbf{x}}_{g,n+1}) + \mathbf{u}_{g,n}$
3: $\bar{\mathbf{x}}_{f,n+1} = \text{prox}_{\gamma_{n+1} f}(\bar{\mathbf{x}}_{g,n+1} - \gamma_{n+1}\mathbf{u}_{g,n+1} - \gamma_{n+1}\mathbf{r}_{n+1})$
End for

Suppose that the following conditions hold $(\forall n \in \mathbb{N})$, *with* $F_n = \sigma(\bar{\mathbf{x}}_{f,k})_{0\leq k \leq n}$:

(i) $\mathbb{E}[\mathbf{r}_{n+1}|F_n] = \nabla h(\bar{\mathbf{x}}_{g,n+1})$ P$-almost surely.
(ii) $(\exists c \in [0, +\infty[)(\exists t \in \mathbb{R}) \sum_{k=0}^{n} \mathbb{E}[\|\mathbf{r}_k - \nabla h(\bar{\mathbf{x}}_{g,k})\|^2] \leq cn^t$.

Then, the followings hold:

(i) *For every* $n \in \mathbb{N}$,

$$(1 + 2\gamma_n\mu_g)\mathbb{E}[\|\bar{\mathbf{x}}_{g,n+1} - \mathbf{x}^*\|^2|F_n] + \gamma_n^2\mathbb{E}[\|\mathbf{u}_{g,n+1} - \mathbf{x}^*\|^2|F_n]$$

$$\leq (1 - 2\gamma_n\mu_h\eta)\|\bar{\mathbf{x}}_{g,n} - \mathbf{x}^*\|^2 + \gamma_n^2\|\mathbf{u}_{g,n} - \mathbf{x}^*\|^2$$

$$+ 2\gamma_n^2\mathbb{E}[\|\mathbf{r}_n - \nabla h(\bar{\mathbf{x}}_{g,n})\|^2|F_n].$$

(ii) *For every $n \in \mathbb{N}$, define*

$$\gamma_{n+1} = \frac{-\gamma_n^2 \mu_h \eta + \sqrt{(\gamma_n^2 \mu_h \eta)^2 + (1 + 2\gamma_n \mu_g)\gamma_n^2}}{1 + 2\gamma_n \mu_g}.$$

Then, $\mathbb{E}[\|\bar{\mathbf{x}}_{g,n} - \mathbf{x}^*\|^2] = \mathcal{O}(1/n^2) + \mathcal{O}(1/n^{2-t})$.

(iii) *Let $\alpha \in \,]0, 1]$ and $(\tau_0, c) \in \,]0, +\infty[^2$. Suppose that $(2\mathbb{E}[\|\mathbf{r}_k - \nabla h(\bar{\mathbf{x}}_{g,k})\|^2])_{n\in\mathbb{N}}$ and $(\mathbb{E}[\|\mathbf{u}_{g,n} - \mathbf{x}^*\|_{\mathbf{V}}^2])_{n\in\mathbb{N}}$ are uniformly bounded by $\tau_0 (2\mu_Q c\eta)^2$ (note that this ensures condition (ii) with $t = 1$). Let n_0 be a strictly positive integer such that $c_0 = 2c\mu_Q\eta \le n_0^\alpha$ and $c \le \min\{2(1 - \eta)\beta, (2\eta\mu_Q)^{-1}\}n_0^\alpha$. Set $s_n = \mathbb{E}[\|\bar{\mathbf{x}}_{g,n} - \mathbf{x}^*\|_{\mathbf{V}}^2]$ for every $n \in \mathbb{N}$. Let us choose $(\forall n \ge 2n_0)$ $\gamma_n = cn^{-\alpha}$ for some $\alpha \in \,]0, 1]$. Then, for every $n \ge 2n_0$, we have*

$$s_{n+1} \le \begin{cases} \mathcal{O}(1/n^\alpha) \text{ if } \alpha \in \,]0, 1[\\ \mathcal{O}(1/n) \ \text{ if } \alpha = 1 \text{ and } c_0 \ge 1. \end{cases}$$

Remark 3 If we assume that we know $(\forall n \in \mathbb{N})$ $\bar{\mathbf{x}}_{B,n} \in \mathcal{M} \ni \mathbf{x}^*$, then we can replace condition (7.18) by

$$(\forall (\mathbf{x}, \mathbf{y}) \in \mathcal{M}) \quad \langle \mathbf{x} - \mathbf{y} \mid \mathbf{Qx} - \mathbf{Qy} \rangle \ge \mu_Q \|\mathbf{x} - \mathbf{y}\|_{\mathbf{V}}^2.$$

Remark 4 If $\mathbf{Q}$ is monotone and $\mathbf{UQ}$ is Lipschitzian with respect to the norm $\| \cdot \|_{\mathbf{V}}$ with the Lipschitz constant β_0, and if $\mu_B > 0$, then we can use the following rule to update step size:

$$(\forall n \in \mathbb{N}) \quad \gamma_{n+1} = \gamma_n(1 + 2\gamma_n(\mu_B - \gamma_n\beta_0^2/2))^{-1/2}.$$

In this case, under the same assumptions on the stochastic estimate $(\mathbf{r}_n)_{n\in\mathbb{N}}$ as in Theorem 2, we get $\mathbb{E}[\|\bar{\mathbf{x}}_{B,n} - \mathbf{x}^*\|_{\mathbf{V}}^2] = \mathcal{O}(1/n^2) + \mathcal{O}(1/n^{2-t})$.

Proof Using (7.22), since $\mathbf{UQ}$ is monotone and Lipschitzian with respect to the norm $\| \cdot \|_{\mathbf{V}}$, we have

$$2\langle \bar{\mathbf{x}}_{A,n} - \mathbf{x}^* \mid \mathbf{Ur}_n \rangle_{\mathbf{V}}$$

$$= 2\langle \bar{\mathbf{x}}_{A,n} - \bar{\mathbf{x}}_{B,n} \mid \mathbf{UQ}\bar{\mathbf{x}}_{B,n} - \mathbf{UQx}^* \rangle_{\mathbf{V}}$$

$$+ 2\langle \bar{\mathbf{x}}_{A,n} - \mathbf{x}^* \mid \mathbf{Qx}^* \rangle + 2\langle \bar{\mathbf{x}}_{B,n} - \mathbf{x}^* \mid \mathbf{UQ}\bar{\mathbf{x}}_{B,n} - \mathbf{UQx}^* \rangle_{\mathbf{V}} + 2\chi_{7,n}$$

$$\ge 2\langle \bar{\mathbf{x}}_{A,n} - \bar{\mathbf{x}}_{B,n} \mid \mathbf{UQ}\bar{\mathbf{x}}_{B,n} - \mathbf{UQx}^* \rangle_{\mathbf{V}} + 2\langle \bar{\mathbf{x}}_{A,n} - \mathbf{x}^* \mid \mathbf{Qx}^* \rangle + 2\chi_{7,n}$$

$$\ge \frac{-1}{\gamma_n}\|\bar{\mathbf{x}}_{A,n} - \bar{\mathbf{x}}_{B,n}\|_{\mathbf{V}}^2 - \gamma_n\|\mathbf{Q}\bar{\mathbf{x}}_{B,n} - \mathbf{Qx}^*\|_{\mathbf{U}}^2 + 2\chi_{7,n} + 2\langle \bar{\mathbf{x}}_{A,n} - \mathbf{x}^* \mid \mathbf{Qx}^* \rangle$$

$$\ge \frac{-1}{\gamma_n}\|\bar{\mathbf{x}}_{A,n} - \bar{\mathbf{x}}_{B,n}\|_{\mathbf{V}}^2 - \gamma_n\beta_0^2\|\bar{\mathbf{x}}_{B,n} - \mathbf{x}^*\|_{\mathbf{V}}^2 + 2\chi_{7,n} + 2\langle \bar{\mathbf{x}}_{A,n} - \mathbf{x}^* \mid \mathbf{Qx}^* \rangle.$$

Now, using the same arguments as in the proof of Theorem 2, we can derive
$$\mathbb{E}[\|\bar{\mathbf{x}}_{B,n} - \mathbf{x}^*\|_{\mathbf{V}}^2] = \mathcal{O}(1/n^2) + \mathcal{O}(1/n^{2-t}).$$ □

Remark 5 Here are some highlights of our framework and the connections with the existing literature:

1. [35] investigates the rate of convergence in expectation for stochastic forward-backward splitting under strong monotonicity assumption, and obtains $\mathcal{O}(1/n)$ rate, which is the same rate as we get here for $t = 1$.
2. [29] obtains $\mathcal{O}(1/n)$ rate for variational inequalities involving Lipschitzian and monotone operators.
3. For the convex minimization setting, [33] provides further connections to existing works in [1, 18, 20–22, 28, 32].

Acknowledgments This project has received funding from the European Research Council (ERC) under the European Union's Horizon 2020 research and innovation programme (grant agreement n° 725594—time-data). This work was supported by the Swiss National Science Foundation (SNSF) under grant number 200021_178865/1.

References

1. Y.F. Atchadé, G. Fort, E. Moulines, On perturbed proximal gradient algorithms. J. Mach. Learn. Res. **18**(10), 1–33 (2017)
2. H. Attouch, L.M. Briceñõ-Arias, P.L. Combettes, A parallel splitting method for coupled monotone inclusions. SIAM J. Control Optim. **48**(5), 3246–3270 (2010)
3. K. Barty, J.-S. Roy, C. Strugarek, Hilbert-valued perturbed subgradient algorithms. Math. Oper. Res. **32**(3), 551–562 (2007)
4. H.H. Bauschke, P.L. Combettes, *Convex Analysis and Monotone Operator Theory in Hilbert Spaces* (Springer, New York, 2011)
5. A. Bennar, J.-M. Monnez, Almost sure convergence of a stochastic approximation process in a convex set. Int. J. Appl. Math. **20**(5), 713–722 (2007)
6. P. Bianchi, W. Hachem, F. Iutzeler, A coordinate descent primal-dual algorithm and application to distributed asynchronous optimization. IEEE Trans. Autom. Control **61**(10), 2947–2957 (2016)
7. K. Bredies, H. Sun, Preconditioned Douglas-Rachford splitting methods for convex-concave saddle-point problems. SIAM J. Numer. Anal. **53**(1), 421–444 (2015)
8. L.M. Briceñõ-Arias, Forward–Douglas–Rachford splitting and forward–partial inverse method for solving monotone inclusions. Optimization **64**(5), 1239–1261 (2015)
9. L.M. Briceñõ-Arias, P.L. Combettes, A monotone+skew splitting model for composite monotone inclusions in duality. SIAM J. Optim. **21**(4), 1230–1250 (2011)
10. P.L. Combettes, J.-C. Pesquet, A Douglas–Rachford splitting approach to nonsmooth convex variational signal recovery. IEEE J. Sel. Top. Sign. Proces. **1**(4), 564–574 (2007)
11. P.L. Combettes, J.-C. Pesquet, Primal–dual splitting algorithm for solving inclusions with mixtures of composite, Lipschitzian, and parallel–sum type monotone operators. Set-Valued Var. Anal. **20**(2), 307–330 (2012)
12. P.L. Combettes, J.-C. Pesquet, Stochastic quasi-Fejér block-coordinate fixed point iterations with random sweeping. SIAM J. Optim. **25**(2), 1221–1248 (2015)
13. P.L. Combettes, J.-C. Pesquet, Stochastic approximations and perturbations in forward–backward splitting for monotone operators. Pure Appl. Funct. Anal. **1**(1), 13–37 (2016)

14. P.L. Combettes, B.C. Vũ, Variable metric forward–backward splitting with applications to monotone inclusions in duality. Optimization **63**(9), 1289–1318 (2014)
15. P.L. Combettes, V.R. Wajs, Signal recovery by proximal forward-backward splitting. Multi-scale Model. Simul. **4**(4), 1168–1200 (2005)
16. P.L. Combettes, L. Condat, J.-C. Pesquet, B.C. Vũ, A forward-backward view of some primal-dual optimization methods in image recovery, in *IEEE International Conference on Image Processing (ICIP)* (2014), pp. 4141–4145
17. D. Davis, W. Yin, A three-operator splitting scheme and its optimization applications. Set-Valued Var. Anal. **25**(4), 829–858 (2017)
18. J. Duchi, Y. Singer, Efficient online and batch learning using forward backward splitting. J. Mach. Learn. Res. **10**(Dec), 2899–2934 (2009)
19. Y. Ermol'ev, A. Tuniev, Random Fejeór and quasi-Fejeór sequences. Theory Optim. Solut.–Akademiya Nauk Ukrainskoi SSR Kiev **2**, 76–83 (1968)
20. E. Ghadimi, A. Teixeira, I. Shames, M. Johansson, On the optimal step-size selection for the alternating direction method of multipliers, in *Proceedings of the IFAC Workshop on Estimation and Control of Networked Systems* (2012)
21. E. Hazan, S. Kale, Beyond the regret minimization barrier: optimal algorithms for stochastic strongly–convex optimization. J. Mach. Learn. Res. **15**(Jul), 2489–2512 (2014)
22. G. Lan, An optimal method for stochastic composite optimization. Math. Program. **133**(1–2), 365–397 (2012)
23. P. Latafat, P. Patrinos, Asymmetric forward-backward-adjoint splitting for solving monotone inclusions involving three operators. Comput. Optim. Appl. **68**(1), 57–93 (2017)
24. M. Ledoux, M. Talagrand, *Probability in Banach Spaces: Isoperimetry and Processes* (Springer, Berlin, 1991)
25. P.L. Lions, B. Mercier, Splitting algorithms for the sum of two nonlinear operators. SIAM J. Numer. Anal. **16**(6), 964–979 (1979)
26. B. Mercier, *Lectures on Topics in Finite Element Solution of Elliptic Problems*. Tata Institute Lectures on Mathematics and Physics (Springer, Berlin, 1979)
27. J.-M. Monnez, Almost sure convergence of stochastic gradient processes with matrix step sizes. Statist. Probab. Lett. **76**(5), 531–536 (2006)
28. E. Moulines, F.R. Bach, Non-asymptotic analysis of stochastic approximation algorithms for machine learning, in *Advances in Neural Information Processing Systems 24 (NIPS)* (2011), pp. 451–459
29. A. Nemirovski, Prox-method with rate of convergence $O(1/t)$ for variational inequalities with Lipschitz continuous monotone operators and smooth convex-concave saddle point problems. SIAM J. Optim. **15**(1), 229–251 (2005)
30. J.-C. Pesquet, A. Repetti, A class of randomized primal-dual algorithms for distributed optimization. J. Nonlinear Convex Anal. **16**(12), 2353–2490 (2015)
31. H. Raguet, J. Fadili, G. Peyré, A generalized forward-backward splitting. SIAM J. Imag. Sci. **6**(3), 1199–1226 (2013)
32. A. Rakhlin, O. Shamir, K. Sridharan, Making gradient descent optimal for strongly convex stochastic optimization, in *29th International Conference on Machine Learning (ICML)* (2012)
33. L. Rosasco, S. Villa, B.C. Vũ, Convergence of stochastic proximal gradient algorithm (2014), arXiv:1403.5074v3
34. L. Rosasco, S. Villa, B.C. Vũ, A stochastic inertial forward–backward splitting algorithm for multivariate monotone inclusions. Optimization **65**(6), 1293–1314 (2016)
35. L. Rosasco, S. Villa, B.C. Vũ, Stochastic forward–backward splitting for monotone inclusions. J. Optim. Theory Appl. **169**(2), 388–406 (2016)
36. E.K. Ryu, W. Yin, Proximal-proximal-gradient method (2017). arXiv:1708.06908v2
37. B.C. Vũ, A splitting algorithm for dual monotone inclusions involving cocoercive operators. Adv. Comput. Math. **38**(3), 667–681 (2013)
38. B.C. Vũ, A splitting algorithm for coupled system of primal–dual monotone inclusions. J Optim. Theory Appl. **164**(3), 993–1025 (2015)

39. B.C. Vũ, Almost sure convergence of the forward–backward–forward splitting algorithm. Optim. Lett. **10**(4), 781–803 (2016)
40. A. Yurtsever, B.C. Vũ, V. Cevher, Stochastic three–composite convex minimization, in *Advances in Neural Information Processing Systems 29 (NIPS)* (2016), pp. 4329–4337

Chapter 8
Mirror Descent and Convex Optimization Problems with Non-smooth Inequality Constraints

Anastasia Bayandina, Pavel Dvurechensky, Alexander Gasnikov, Fedor Stonyakin, and Alexander Titov

Abstract We consider the problem of minimization of a convex function on a simple set with convex non-smooth inequality constraint and describe first-order methods to solve such problems in different situations: smooth or non-smooth objective function; convex or strongly convex objective and constraint; deterministic or randomized information about the objective and constraint. Described methods are based on Mirror Descent algorithm and switching subgradient scheme. One of our focus is to propose, for the listed different settings, a Mirror Descent with adaptive stepsizes and adaptive stopping rule. We also construct Mirror Descent for problems with objective function, which is not Lipschitz, e.g., is a quadratic function. Besides that, we address the question of recovering the dual solution in the considered problem.

A. Bayandina
Moscow Institute of Physics and Technology, Dolgoprudny, Moscow Region, Russia

Skolkovo Institute of Science and Technology, Skolkovo Innovation Center, Moscow, Russia

P. Dvurechensky (✉)
Weierstrass Institute for Applied Analysis and Stochastics, Berlin, Germany

Institute for Information Transmission Problems RAS, Moscow, Russia
e-mail: pavel.dvurechensky@wias-berlin.de

A. Gasnikov
Moscow Institute of Physics and Technology, Dolgoprudny, Moscow Region, Russia

Institute for Information Transmission Problems RAS, Moscow, Russia
e-mail: gasnikov@yandex.ru

F. Stonyakin
V.I. Vernadsky Crimean Federal University, Simferopol, Russia

Moscow Institute of Physics and Technology, Dolgoprudny, Moscow Region, Russia
e-mail: fedyor@mail.ru

A. Titov
Moscow Institute of Physics and Technology, Dolgoprudny, Moscow Region, Russia
e-mail: a.a.titov@phystech.edu

© Springer Nature Switzerland AG 2018
P. Giselsson, A. Rantzer (eds.), *Large-Scale and Distributed Optimization*, Lecture Notes in Mathematics 2227,
https://doi.org/10.1007/978-3-319-97478-1_8

Keywords Constrained non-smooth convex optimization · Stochastic adaptive mirror descent · Primal-dual methods · Restarts

AMS Subject Classifications 90C25, 90C30, 90C06, 65Y20

8.1 Introduction

We consider the problem of minimization of a convex function on a simple set with convex non-smooth inequality constraint and describe first-order methods to solve such problems in different situations: smooth or non-smooth objective function; convex or strongly convex objective and constraint; deterministic or randomized information about the objective and constraint. The reason for considering first-order methods is potential large (more than 10^5) number of decision variables.

Because of the non-smoothness presented in the problem, we consider subgradient methods. These methods have a long history starting with the method for deterministic unconstrained problems and Euclidean setting in [28] and the generalization for constrained problems in [25], where the idea of steps switching between the direction of subgradient of the objective and the direction of subgradient of the constraint was suggested. Non-Euclidean extension, usually referred to as Mirror Descent, originated in [17, 19] and later analyzed in [5]. An extension for constrained problems was proposed in [19], see also recent version in [6]. Mirror Descent for unconstrained stochastic optimization problems was introduced in [16], see also [12, 15], and extended for stochastic optimization problems with expectation constraints in [14]. To prove faster convergence rate of Mirror Descent for strongly convex objective in unconstrained case, the restart technique [18–20] was used in [12]. An alternative approach for strongly convex stochastic optimization problems with strongly convex expectation constraints is used in [14].

Usually, the stepsize and stopping rule for Mirror Descent requires to know the Lipschitz constant of the objective function and constraint, if any. Adaptive stepsizes, which do not require this information, are considered in [7] for problems without inequality constraints, and in [6] for constrained problems. Nevertheless, the stopping criterion, expressed in the number of steps, still requires knowledge of Lipschitz constants. One of our focus in this chapter is to propose, for constrained problems, a Mirror Descent with adaptive stepsizes and adaptive stopping rule. We also adopt the ideas of [21, 24] to construct Mirror Descent for problems with objective function, which is not Lipschitz, e.g., a quadratic function. Another important issue, we address, is recovering the dual solution of the considered problem, which was considered in different contexts in [1, 4, 23].

Formally speaking, we consider the following convex constrained minimization problem

$$\min\{f(x): \quad x \in X \subset E, \quad g(x) \leq 0\}, \tag{8.1}$$

where X is a convex closed subset of a finite-dimensional real vector space E, $f :
X \to \mathbb{R}$, $g : E \to \mathbb{R}$ are convex functions.

We assume g to be a non-smooth Lipschitz-continuous function and the problem
(8.1) to be regular. The last means that there exists a point $\bar{x}$ in relative interior of
the set X, such that $g(\bar{x}) < 0$.

Note that, despite problem (8.1) contains only one inequality constraint, considered algorithms allow to solve more general problems with a number of constraints
given as $\{g_i(x) \leq 0, i = 1, \ldots, m\}$. The reason is that these constraints can be
aggregated and represented as an equivalent constraint given by $\{g(x) \leq 0\}$, where
$g(x) = \max_{i=1,\ldots,m} g_i(x)$.

The the rest of the chapter is divided in three parts. In Sect. 8.2, we describe
some basic facts about Mirror Descent, namely, we define the notion of proximal
setup, the Mirror Descent step, and provide the main lemma about the progress
on each iteration of this method. Section 8.3 is devoted to deterministic constrained
problems, among which we consider convex non-smooth problems, strongly convex
non-smooth problems and convex problems with smooth objective. The last,
Sect. 8.4, considers randomized setting with available stochastic subgradients for
the objective and constraint and possibility to calculate the constraint function. We
consider methods for convex and strongly convex problems and provide complexity
guarantees in terms of expectation of the objective residual and constraint infeasibility, as long as in terms of large deviation probability for these two quantities.

Notation Given a subset I of natural numbers, we denote $|I|$ the number of its
elements.

8.2 Mirror Descent Basics

We consider algorithms, which are based on Mirror Descent method. Thus, we start
with the description of proximal setup and basic properties of Mirror Descent step.
Let E be a finite-dimensional real vector space and E^* be its dual. We denote the
value of a linear function $g \in E^*$ at $x \in E$ by $\langle g, x \rangle$. Let $\| \cdot \|_E$ be some norm on E,
$\| \cdot \|_{E,*}$ be its dual, defined by $\|g\|_{E,*} = \max_x \{\langle g, x \rangle, \|x\|_E \leq 1\}$. We use $\nabla f(x)$
to denote any subgradient of a function f at a point $x \in \mathrm{dom} f$.

We choose a prox-function $d(x)$, which is continuous, convex on X and

1. admits a continuous in $x \in X^0$ selection of subgradients $\nabla d(x)$, where $X^0 \subseteq X$
   is the set of all x, where $\nabla d(x)$ exists;
2. $d(x)$ is 1-strongly convex on X with respect to $\| \cdot \|_E$, i.e., for any $x \in X^0$, $y \in X$
   $d(y) - d(x) - \langle \nabla d(x), y - x \rangle \geq \frac{1}{2}\|y - x\|_E^2$.

Without loss of generality, we assume that $\min_{x \in X} d(x) = 0$.

We define also the corresponding Bregman divergence $V[z](x) = d(x) - d(z) - \langle \nabla d(z), x - z \rangle$, $x \in X$, $z \in X^0$. Standard proximal setups, i.e., Euclidean, entropy, ℓ_1/ℓ_2, simplex, nuclear norm, spectahedron can be found in [8].

Given a vector $x \in X^0$, and a vector $p \in E^*$, the Mirror Descent step is defined as

$$x_+ = \text{Mirr}[x](p) := \arg\min_{u \in X} \{ \langle p, u \rangle + V[x](u) \}$$

$$= \arg\min_{u \in X} \{ \langle p, u \rangle + d(u) - \langle \nabla d(x), u \rangle \}. \tag{8.2}$$

We make the simplicity assumption, which means that $\text{Mirr}[x](p)$ is easily computable. The following lemma [7] describes the main property of the Mirror Descent step. We prove it here for the reader convenience and to make the chapter self-contained.

Lemma 1 *Let f be some convex function over a set X, $h > 0$ be a stepsize, $x \in X^0$. Let the point x_+ be defined by $x_+ = \text{Mirr}[x](h \cdot (\nabla f(x) + \Delta))$, where $\Delta \in E^*$. Then, for any $u \in X$,*

$$h \cdot \left(f(x) - f(u) + \langle \Delta, x - u \rangle \right) \le h \cdot \langle \nabla f(x) + \Delta, x - u \rangle$$

$$\le \frac{h^2}{2} \| \nabla f(x) + \Delta \|_{E,*}^2 + V[x](u) - V[x_+](u). \tag{8.3}$$

Proof By optimality condition in (8.2), we have that there exists a subgradient $\nabla d(x_+)$, such that, for all $u \in X$,

$$\langle h \cdot (\nabla f(x) + \Delta) + \nabla d(x_+) - \nabla d(x), u - x_+ \rangle \ge 0.$$

Hence, for all $u \in X$,

$$\langle h \cdot (\nabla f(x) + \Delta), x - u \rangle \tag{8.4}$$

$$\le \langle h \cdot (\nabla f(x) + \Delta), x - x_+ \rangle + \langle \nabla d(x_+) - \nabla d(x), u - x_+ \rangle$$

$$= \langle h \cdot (\nabla f(x) + \Delta), x - x_+ \rangle + (d(u) - d(x) - \langle \nabla d(x), u - x \rangle)$$

$$- (d(u) - d(x_+) - \langle \nabla d(x_+), u - x_+ \rangle)$$

$$- (d(x_+) - d(x) - \langle \nabla d(x), x_+ - x \rangle)$$

$$\le \langle h \cdot (\nabla f(x) + \Delta), x - x_+ \rangle + V[x](u) - V[x_+](u) - \frac{1}{2} \| x_+ - x \|_E^2$$

$$\le V[x](u) - V[x_+](u) + \frac{h^2}{2} \| (\nabla f(x) + \Delta) \|_{E,*}^2,$$

where we used the fact that, for any $g \in E^*$,

$$\max_{y \in E} \langle g, y \rangle - \frac{1}{2} \|y\|_E^2 = \frac{1}{2} \|g\|_{E,*}^2.$$

By convexity of f, we obtain the left inequality in (8.3). □

8.3 Deterministic Constrained Problems

In this section, we consider problem (8.1) in two different settings, namely, non-smooth Lipschitz-continuous objective function f and general objective function f, which is not necessarily Lipschitz-continuous, e.g., a quadratic function. In both cases, we assume that g is non-smooth and is Lipschitz-continuous

$$|g(x) - g(y)| \leq M_g \|x - y\|_E, \quad x, y \in X. \tag{8.5}$$

Let x_* be a solution to (8.1). We say that a point $\tilde{x} \in X$ is an ε-*solution* to (8.1) if

$$f(\tilde{x}) - f(x_*) \leq \varepsilon, \quad g(\tilde{x}) \leq \varepsilon. \tag{8.6}$$

The methods we describe are based on the of Polyak's switching subgradient method [25] for constrained convex problems, also analyzed in [21], and Mirror Descent method originated in [19]; see also [7].

8.3.1 Convex Non-smooth Objective Function

In this subsection, we assume that f is a non-smooth Lipschitz-continuous function

$$|f(x) - f(y)| \leq M_f \|x - y\|_E, \quad x, y \in X. \tag{8.7}$$

Let x_* be a solution to (8.1) and assume that we know a constant $\Theta_0 > 0$ such that

$$d(x_*) \leq \Theta_0^2. \tag{8.8}$$

For example, if X is a compact set, one can choose $\Theta_0^2 = \max_{x \in X} d(x)$. We further develop line of research [1, 4], but we should also mention close works [6, 23]. In comparison to known algorithms in the literature, the main advantage of our method for solving (8.1) is that the stopping criterion does not require the knowledge of constants M_f, M_g, and, in this sense, the method is adaptive. Mirror Descent with stepsizes not requiring knowledge of Lipschitz constants can be found, e.g., in [7] for problems without inequality constraints, and, for constrained problems, in

---

Algorithm 1 Adaptive mirror descent (non-smooth objective)

Input: accuracy $\varepsilon > 0$; Θ_0 s.t. $d(x_*) \leq \Theta_0^2$.

1: $x^0 = \arg\min_{x \in X} d(x)$.
2: Initialize the set I as empty set.
3: Set $k = 0$.
4: **repeat**
5:      **if** $g(x^k) \leq \varepsilon$ **then**
6:          $M_k = \|\nabla f(x^k)\|_{E,*}$,
7:          $h_k = \frac{\varepsilon}{M_k^2}$
8:          $x^{k+1} = \text{Mirr}[x^k](h_k \nabla f(x^k))$ ("productive step")
9:          Add k to I.
10:     **else**
11:         $M_k = \|\nabla g(x^k)\|_{E,*}$
12:         $h_k = \frac{\varepsilon}{M_k^2}$
13:         $x^{k+1} = \text{Mirr}[x^k](h_k \nabla g(x^k))$ ("non-productive step")
14:     **end if**
15:     Set $k = k + 1$.
16: **until** $\sum_{j=0}^{k-1} \frac{1}{M_j^2} \geq \frac{2\Theta_0^2}{\varepsilon^2}$

Output: $\bar{x}^k := \frac{\sum_{i \in I} h_i x^i}{\sum_{i \in I} h_i}$

---

[6].The algorithm is similar to the one in [2], but, for the sake of consistency with other parts of the chapter, we use slightly different proof.

Theorem 1 *Assume that inequalities* (8.5) *and* (8.7) *hold and a known constant* $\Theta_0 > 0$ *is such that* $d(x_*) \leq \Theta_0^2$. *Then, Algorithm 1 stops after not more than*

$$k = \left\lceil \frac{2\max\{M_f^2, M_g^2\}\Theta_0^2}{\varepsilon^2} \right\rceil \tag{8.9}$$

iterations and $\bar{x}^k$ *is an* ε-*solution to* (8.1) *in the sense of* (8.6).

Proof First, let us prove that the inequality in the stopping criterion holds for k defined in (8.9). By (8.5) and (8.7), we have that, for any $i \in \{0, \ldots, k-1\}$, $M_i \leq \max\{M_f, M_g\}$. Hence, by (8.9), $\sum_{j=0}^{k-1} \frac{1}{M_j^2} \geq \frac{k}{\max\{M_f^2, M_g^2\}} \geq \frac{2\Theta_0^2}{\varepsilon^2}$.

Denote $[k] = \{i \in \{0, \ldots, k-1\}\}$, $J = [k] \setminus I$. From Lemma 1 with $\Delta = 0$, we have, for all $i \in I$ and all $u \in X$,

$$h_i \cdot \left(f(x^i) - f(u)\right) \leq \frac{h_i^2}{2}\|\nabla f(x^i)\|_{E,*}^2 + V[x^i](u) - V[x^{i+1}](u)$$

and, for all $i \in J$ and all $u \in X$,

$$h_i \cdot \left(g(x^i) - g(u)\right) \leq \frac{h_i^2}{2} \|\nabla g(x^i)\|_{E,*}^2 + V[x^i](u) - V[x^{i+1}](u).$$

Summing up these inequalities for i from 0 to $k - 1$, using the definition of h_i, $i \in \{0, \ldots, k - 1\}$, and taking $u = x_*$, we obtain

$$\sum_{i \in I} h_i \left(f(x^i) - f(x_*)\right) + \sum_{i \in J} h_i \left(g(x^i) - g(x_*)\right)$$

$$\leq \sum_{i \in I} \frac{h_i^2 M_i^2}{2} + \sum_{i \in J} \frac{h_i^2 M_i^2}{2} + \sum_{i \in [k]} \left(V[x^i](x_*) - V[x^{i+1}](x_*)\right)$$

$$\leq \frac{\varepsilon}{2} \sum_{i \in [k]} h_i + \Theta_0^2. \tag{8.10}$$

We also used that, by definition of x^0 and (8.8),

$$V[x^0](x_*) = d(x_*) - d(x^0) - \langle \nabla d(x^0), x_* - x^0 \rangle \leq d(x_*) \leq \Theta_0^2.$$

Since, for $i \in J$, $g(x^i) - g(x_*) \geq g(x^i) > \varepsilon$, by convexity of f and the definition of $\bar{x}^k$, we have

$$\left(\sum_{i \in I} h_i\right) \left(f(\bar{x}^k) - f(x_*)\right) \leq \sum_{i \in I} h_i \left(f(x^i) - f(x_*)\right) < \frac{\varepsilon}{2} \sum_{i \in [k]} h_i - \varepsilon \sum_{i \in J} h_i + \Theta_0^2$$

$$= \varepsilon \sum_{i \in I} h_i - \frac{\varepsilon^2}{2} \sum_{i \in [k]} \frac{1}{M_i^2} + \Theta_0^2 \leq \varepsilon \sum_{i \in I} h_i, \tag{8.11}$$

where in the last inequality, the stopping criterion is used. As long as the inequality is strict, the case of the empty I is impossible. Thus, the point $\bar{x}^k$ is correctly defined. Dividing both parts of the inequality by $\sum_{i \in I} h_i$, we obtain the left inequality in (8.6).

For $i \in I$, it holds that $g(x^i) \leq \varepsilon$. Then, by the definition of $\bar{x}^k$ and the convexity of g,

$$g(\bar{x}^k) \leq \left(\sum_{i \in I} h_i\right)^{-1} \sum_{i \in I} h_i g(x^i) \leq \varepsilon.$$

$\square$

Let us now show that Algorithm 1 allows to reconstruct an approximate solution to the problem, which is dual to (8.1). We consider a special type of problem (8.1) with g given by

$$g(x) = \max_{i \in \{1,\ldots,m\}} \{g_i(x)\}. \tag{8.12}$$

Then, the dual problem to (8.1) is

$$\varphi(\lambda) = \min_{x \in X} \left\{ f(x) + \sum_{i=1}^{m} \lambda_i g_i(x) \right\} \to \max_{\lambda_i \geq 0, i=1,\ldots,m} \varphi(\lambda), \tag{8.13}$$

where $\lambda_i \geq 0$, $i = 1, \ldots, m$ are Lagrange multipliers.

We slightly modify the assumption (8.8) and assume that the set X is bounded and that we know a constant $\Theta_0 > 0$ such that

$$\max_{x \in X} d(x) \leq \Theta_0^2.$$

As before, denote $[k] = \{j \in \{0, \ldots, k-1\}\}$, $J = [k] \setminus I$. Let $j \in J$. Then a subgradient of $g(x)$ is used to make the j-th step of Algorithm 1. To find this subgradient, it is natural to find an active constraint $i \in 1, \ldots, m$ such that $g(x^j) = g_i(x^j)$ and use $\nabla g(x^j) = \nabla g_i(x^j)$ to make a step. Denote $i(j) \in 1, \ldots, m$ the number of active constraint, whose subgradient is used to make a non-productive step at iteration $j \in J$. In other words, $g(x^j) = g_{i(j)}(x^j)$ and $\nabla g(x^j) = \nabla g_{i(j)}(x^j)$. We define an approximate dual solution on a step $k \geq 0$ as

$$\bar{\lambda}_i^k = \frac{1}{\sum\limits_{j \in I} h_j} \sum_{j \in J, i(j)=i} h_j, \quad i \in \{1, \ldots, m\}. \tag{8.14}$$

and modify Algorithm 1 to return a pair $(\bar{x}^k, \bar{\lambda}^k)$.

Theorem 2 *Assume that the set X is bounded, the inequalities (8.5) and (8.7) hold and a known constant $\Theta_0 > 0$ is such that $d(x_*) \leq \Theta_0^2$. Then, modified Algorithm 1 stops after not more than*

$$k = \left\lceil \frac{2 \max\{M_f^2, M_g^2\} \Theta_0^2}{\varepsilon^2} \right\rceil$$

iterations and the pair $(\bar{x}^k, \bar{\lambda}^k)$ returned by this algorithm satisfies

$$f(\bar{x}^k) - \varphi(\bar{\lambda}^k) \leq \varepsilon, \quad g(\bar{x}^k) \leq \varepsilon. \tag{8.15}$$

Proof From Lemma 1 with $\Delta = 0$, we have, for all $j \in I$ and all $u \in X$,

$$h_j \big(f(x^j) - f(u) \big) \leq \frac{h_j^2}{2} \| \nabla f(x^j) \|_{E,*}^2 + V[x^j](u) - V[x^{j+1}](u)$$

and, for all $j \in J$ and all $u \in X$,

$$h_j\big(g_{i(j)}(x^j) - g_{i(j)}(u)\big) \leq h_j \langle \nabla g_{i(j)}(x^j), x^j - u \rangle$$

$$= h_j \langle \nabla g(x^j), x^j - u \rangle$$

$$\leq \frac{h_j^2}{2} \|\nabla g(x^j)\|_{E,*}^2 + V[x^j](u) - V[x^{j+1}](u).$$

Summing up these inequalities for j from 0 to $k - 1$, using the definition of h_j, $j \in \{0, \ldots, k-1\}$, we obtain, for all $u \in X$,

$$\sum_{j \in I} h_j\big(f(x^j) - f(u)\big) + \sum_{j \in J} h_j\big(g_{i(j)}(x^j) - g_{i(j)}(u)\big)$$

$$\leq \sum_{i \in I} \frac{h_j^2 M_j^2}{2} + \sum_{j \in J} \frac{h_j^2 M_j^2}{2} + \sum_{j \in [k]} \left(V[x^j](u) - V[x^{j+1}](u)\right)$$

$$\leq \frac{\varepsilon}{2} \sum_{j \in [k]} h_j + \Theta_0^2.$$

Since, for $j \in J$, $g_{i(j)}(x^j) = g(x^j) > \varepsilon$, by convexity of f and the definition of $\bar{x}^k$, we have, for all $u \in X$,

$$\left(\sum_{j \in I} h_j\right)(f(\bar{x}^k) - f(u)) \leq \sum_{j \in I} h_j\left(f(x^j) - f(u)\right)$$

$$\leq \frac{\varepsilon}{2} \sum_{j \in [k]} h_j + \Theta_0^2 - \sum_{j \in J} h_j\big(g_{i(j)}(x^j) - g_{i(j)}(u)\big)$$

$$< \frac{\varepsilon}{2} \sum_{j \in [k]} h_i + \Theta_0^2 - \varepsilon \sum_{j \in J} h_i + \sum_{j \in J} h_j g_{i(j)}(u)$$

$$= \varepsilon \sum_{j \in I} h_j - \frac{\varepsilon^2}{2} \sum_{j \in [k]} \frac{1}{M_j^2} + \Theta_0^2 + \sum_{j \in J} h_j g_{i(j)}(u)$$

$$\leq \varepsilon \sum_{j \in I} h_j + \sum_{j \in J} h_j g_{i(j)}(u), \tag{8.16}$$

where in the last inequality, the stopping criterion is used. At the same time, by (8.14), for all $u \in X$,

$$\sum_{j \in J} h_j g_{i(j)}(u) = \sum_{i=1}^{m} \sum_{j \in J, i(j)=i} h_j g_{i(j)}(u) = \left(\sum_{j \in I} h_j\right) \sum_{i=1}^{m} \bar{\lambda}_i^k g_i(u).$$

This and (8.16) give, for all $u \in X$,

$$\left(\sum_{j \in I} h_j\right) f(\bar{x}^k) < \left(\sum_{j \in I} h_j\right) \left(f(u) + \varepsilon + \sum_{i=1}^{m} \bar{\lambda}_i^k g_i(u)\right).$$

Since the inequality is strict and holds for all $u \in X$, we have $\left(\sum_{j \in I} h_j\right) \neq 0$ and

$$f(\bar{x}^k) < \varepsilon + \min_{u \in X} \left\{ f(u) + \sum_{i=1}^{m} \bar{\lambda}_i^k g_i(u) \right\}$$

$$= \varepsilon + \varphi(\bar{\lambda}^k). \tag{8.17}$$

Second inequality in (8.15) follows from Theorem 1. □

8.3.2 Strongly Convex Non-smooth Objective Function

In this subsection, we consider problem (8.1) with assumption (8.7) and additional assumption of strong convexity of f and g with the same parameter μ, i.e.,

$$f(y) \geq f(x) + \langle \nabla f(x), y - x \rangle + \frac{\mu}{2} \|y - x\|_E^2, \quad x, y \in X$$

and the same holds for g. For example, $f(x) = x^2 + |x|$ is a Lipschitz-continuous and strongly convex function on $X = [-1; 1] \subset \mathbb{R}$. We also slightly modify assumptions on prox-function $d(x)$. Namely, we assume that $0 = \arg\min_{x \in X} d(x)$ and that d is bounded on the unit ball in the chosen norm $\| \cdot \|_E$, that is

$$d(x) \leq \frac{\Omega}{2}, \quad \forall x \in X : \|x\|_E \leq 1, \tag{8.18}$$

where Ω is some known number. Finally, we assume that we are given a starting point $x_0 \in X$ and a number $R_0 > 0$ such that $\|x_0 - x_*\|_E^2 \leq R_0^2$.

To construct a method for solving problem (8.1) under stated assumptions, we use the idea of restarting Algorithm 1. The idea of restarting a method for convex problems to obtain faster rate of convergence for strongly convex problems dates back to 1980s, see [19, 20]. The algorithm is similar to the one in [2], but, for the sake of consistency with other parts of the chapter, we use slightly different proof. To show that restarting algorithm is also possible for problems with inequality constraints, we rely on the following lemma.

Lemma 2 *Let f and g be strongly convex functions with the same parameter μ and x_* be a solution of the problem (8.1). If, for some $\tilde{x} \in X$,*

$$f(\tilde{x}) - f(x_*) \leq \varepsilon, \quad g(\tilde{x}) \leq \varepsilon,$$

then

$$\frac{\mu}{2}\|\tilde{x} - x_*\|_E^2 \leq \varepsilon.$$

Proof Since problem (8.1) is regular, by necessary optimality condition [9] at the point x_*, there exist $\lambda_0, \lambda \geq 0$ not equal to 0 simultaneously, and subgradients $\nabla f(x_*), \nabla g(x_*)$, such that

$$\langle \lambda_0 \nabla f(x_*) + \lambda \nabla g(x_*), x - x_* \rangle \geq 0, \quad \forall x \in X, \quad \lambda g(x_*) = 0.$$

Since λ_0 and λ are not equal to 0 simultaneously, three cases are possible.

1. $\lambda_0 = 0$ and $\lambda > 0$. Then, by optimality conditions, $g(x_*) = 0$ and $\langle \lambda \nabla g(x_*), \tilde{x} - x_* \rangle \geq 0$. Thus, by the Lemma assumption and strong convexity,

$$\varepsilon \geq g(\tilde{x}) \geq g(x_*) + \langle \nabla g(x_*), \tilde{x} - x_* \rangle + \frac{\mu}{2}\|\tilde{x} - x_*\|_E^2 \geq \frac{\mu}{2}\|\tilde{x} - x_*\|_E^2.$$

2. $\lambda_0 > 0$ and $\lambda = 0$. Then, by optimality conditions, $\langle \lambda_0 \nabla f(x_*), \tilde{x} - x_* \rangle \geq 0$. Thus, by the Lemma assumption and strong convexity,

$$f(x_*) + \varepsilon \geq f(\tilde{x}) \geq f(x_*) + \langle \nabla f(x_*), \tilde{x} - x_* \rangle + \frac{\mu}{2}\|\tilde{x} - x_*\|_E^2 \geq f(x_*) + \frac{\mu}{2}\|\tilde{x} - x_*\|_E^2.$$

3. $\lambda_0 > 0, \lambda > 0$. Then, by optimality conditions, $g(x_*) = 0$ and $\langle \lambda_0 \nabla f(x_*) + \lambda \nabla g(x_*), \tilde{x} - x_* \rangle \geq 0$. Thus, either $\langle \nabla g(x_*), \tilde{x} - x_* \rangle \geq 0$ and the proof is the same as in item 1, or $\langle \nabla f(x_*), \tilde{x} - x_* \rangle \geq 0$ and the proof is the same as in the item 2. $\qquad\square$

Theorem 3 *Assume that inequalities (8.5) and (8.7) hold and f, g are strongly convex with the same parameter μ. Also assume that the prox function $d(x)$ satisfies (8.18) and the starting point $x_0 \in X$ and a number $R_0 > 0$ are such that $\|x_0 - x_*\|_E^2 \leq R_0^2$. Then, the point x_p returned by Algorithm 2 is an ε-solution to (8.1) in the sense of (8.6) and $\|x_p - x_*\|_E^2 \leq \frac{2\varepsilon}{\mu}$. At the same time, the total number of iterations of Algorithm 1 does not exceed*

$$\left\lceil \log_2 \frac{\mu R_0^2}{2\varepsilon} \right\rceil + \frac{32\Omega \max\{M_f^2, M_g^2\}}{\mu\varepsilon}. \tag{8.19}$$

Proof Observe that, for all $p \geq 0$, the function $d_p(x)$ defined in Algorithm 2 is 1-strongly convex w.r.t. the norm $\|\cdot\|_E/R_p$. The conjugate of this norm is $R_p\|\cdot\|_{E,*}$.

Algorithm 2 Adaptive mirror descent (non-smooth strongly convex objective)

Input: accuracy $\varepsilon > 0$; strong convexity parameter μ; Ω s.t. $d(x) \le \frac{\Omega}{2}$ $\quad \forall x \in X : \|x\|_E \le 1$;
   starting point x_0 and number R_0 s.t. $\|x_0 - x_*\|_E^2 \le R_0^2$.

1: Set $d_0(x) = d\left(\frac{x-x_0}{R_0}\right)$.
2: Set $p = 1$.
3: **repeat**
4:     Set $R_p^2 = R_0^2 \cdot 2^{-p}$.
5:     Set $\varepsilon_p = \frac{\mu R_p^2}{2}$.
6:     Set x_p as the output of Algorithm 1 with accuracy ε_p, prox-function $d_{p-1}(\cdot)$ and $\frac{\Omega}{2}$ as Θ_0^2.
7:     $d_p(x) \leftarrow d\left(\frac{x-x_p}{R_p}\right)$.
8:     Set $p = p + 1$.
9: **until** $p > \log_2 \frac{\mu R_0^2}{2\varepsilon}$.
Output: x_p.

This means that, at each step k of inner Algorithm 1, M_k changes to $M_k R_{p-1}$, where $p \ge 1$ is the number of outer iteration.

We show, by induction, that, for all $p \ge 0$, $\|x_p - x_*\|_E^2 \le R_p^2$. For $p = 0$ it holds by the assumption on x_0 and R_0. Let us assume that this inequality holds for some p and show that it holds for $p + 1$. By (8.18), we have $d_p(x_*) \le \frac{\Omega}{2}$. Thus, on the outer iteration $p + 1$, by Theorem 1 and (8.6), after at most

$$k_{p+1} = \left\lceil \frac{\Omega \max\{M_f^2, M_g^2\} R_p^2}{\varepsilon_{p+1}^2} \right\rceil \tag{8.20}$$

inner iterations, $x_{p+1} = \bar{x}^{k_{p+1}}$ satisfies

$$f(x_{p+1}) - f(x_*) \le \varepsilon_{p+1}, \quad g(x_{p+1}) \le \varepsilon_{p+1},$$

where $\varepsilon_{p+1} = \frac{\mu R_{p+1}^2}{2}$. Then, by Lemma 2,

$$\|x_{p+1} - x_*\|_E^2 \le \frac{2\varepsilon_{p+1}}{\mu} = R_{p+1}^2.$$

Thus, we proved that, for all $p \ge 0$, $\|x_p - x_*\|_E^2 \le R_p^2 = R_0^2 \cdot 2^{-p}$. At the same time, we have, for all $p \ge 1$,

$$f(x_p) - f(x_*) \le \frac{\mu R_0^2}{2} \cdot 2^{-p}, \quad g(x_p) \le \frac{\mu R_0^2}{2} \cdot 2^{-p}.$$

Thus, if $p > \log_2 \frac{\mu R_0^2}{2\varepsilon}$, x_p is an ε-solution to (8.1) in the sense of (8.6) and

$$\|x_p - x_*\|_E^2 \leq R_0^2 \cdot 2^{-p} \leq \frac{2\varepsilon}{\mu}.$$

Let us now estimate the total number N of inner iterations, i.e., the iterations of Algorithm 1. Let us denote $\hat{p} = \left\lceil \log_2 \frac{\mu R_0^2}{2\varepsilon} \right\rceil$. According to (8.20), we have

$$N = \sum_{p=1}^{\hat{p}} k_p \leq \sum_{p=1}^{\hat{p}} \left(1 + \frac{\Omega \max\{M_f^2, M_g^2\} R_p^2}{\varepsilon_{p+1}^2} \right)$$

$$= \sum_{p=1}^{\hat{p}} \left(1 + \frac{16\Omega \max\{M_f^2, M_g^2\} 2^p}{\mu^2 R_0^2} \right)$$

$$\leq \hat{p} + \frac{32\Omega \max\{M_f^2, M_g^2\} 2^{\hat{p}}}{\mu^2 R_0^2} \leq \hat{p} + \frac{32\Omega \max\{M_f^2, M_g^2\}}{\mu \varepsilon}.$$

□

Similarly to Sect. 8.3.1, let us consider a special type of problem (8.1) with strongly convex g given by

$$g(x) = \max_{i \in \{1,\dots,m\}} \{g_i(x)\}. \tag{8.21}$$

and corresponding dual problem

$$\varphi(\lambda) = \min_{x \in X} \left\{ f(x) + \sum_{i=1}^{m} \lambda_i g_i(x) \right\} \rightarrow \max_{\lambda_i \geq 0, i \in \{1,\dots,m\}} \varphi(\lambda).$$

On each outer iteration p of Algorithm 2, there is the last inner iteration k_p of Algorithm 1. We define approximate dual solution as $\lambda_p = \bar{\lambda}^{k_p}$, where $\bar{\lambda}^{k_p}$ is defined in (8.14). We modify Algorithm 2 to return a pair (x_p, λ_p).

Combining Theorems 2 and 3, we obtain the following result.

Theorem 4 *Assume that g is given by (8.21), inequalities (8.5) and (8.7) hold and f, g are strongly convex with the same parameter μ. Also assume that the prox function $d(x)$ satisfies (8.18) and the starting point $x_0 \in X$ and a number $R_0 > 0$ are such that $\|x_0 - x_*\|_E^2 \leq R_0^2$. Then, the pair (x_p, λ_p) returned by Algorithm 2 satisfies*

$$f(x_p) - \varphi(\lambda_p) \leq \varepsilon, \quad g(x_p) \leq \varepsilon.$$

and $\|x_p - x_*\|_E^2 \leq \frac{2\varepsilon}{\mu}$. At the same time, the total number of inner iterations of Algorithm 1 does not exceed

$$\left\lceil \log_2 \frac{\mu R_0^2}{2\varepsilon} \right\rceil + \frac{32\Omega \max\{M_f^2, M_g^2\}}{\mu\varepsilon}.$$

8.3.3 General Convex Objective Function

In this subsection, we assume that the objective function f in (8.1) might not satisfy (8.7) and, hence, its subgradients could be unbounded. One of the examples is a quadratic function. We also assume that inequality (8.8) holds.

We further develop ideas in [21, 24] and adapt them for problem (8.1), in a way that our algorithm allows to use non-Euclidean proximal setup, as does Mirror Descent, and does not require to know the constant M_g. Following [21], given a function f for each subgradient $\nabla f(x)$ at a point $y \in X$, we define

$$v_f[y](x) = \begin{cases} \left\langle \dfrac{\nabla f(x)}{\|\nabla f(x)\|_{E,*}}, x - y \right\rangle, & \nabla f(x) \neq 0 \\ 0 & \nabla f(x) = 0 \end{cases}, \quad x \in X. \qquad (8.22)$$

The following result gives complexity estimate for Algorithm 3 in terms of $v_f[x_*](x)$. Below we use this theorem to establish complexity result for smooth objective f.

Theorem 5 *Assume that inequality (8.5) holds and a known constant $\Theta_0 > 0$ is such that $d(x_*) \leq \Theta_0^2$. Then, Algorithm 3 stops after not more than*

$$k = \left\lceil \frac{2\max\{1, M_g^2\}\Theta_0^2}{\varepsilon^2} \right\rceil \qquad (8.23)$$

iterations and it holds that $\min_{i \in I} v_f[x_*](x^i) \leq \varepsilon$ and $g(\bar{x}^k) \leq \varepsilon$.

Proof First, let us prove that the inequality in the stopping criterion holds for k defined in (8.23). Denote $[k] = \{i \in \{0, \ldots, k-1\}\}$, $J = [k] \setminus I$. By (8.5), we have that, for any $j \in J$, $\|\nabla g(x^j)\|_{E,*} \leq M_g$. Hence, since $|I| + |J| = k$, by (8.23), we obtain

$$|I| + \sum_{j \in J} \frac{1}{\|\nabla g(x^j)\|_{E,*}^2} \geq |I| + \frac{|J|}{M_g^2} \geq \frac{k}{\max\{1, M_g^2\}} \geq \frac{2\Theta_0^2}{\varepsilon^2}.$$

Algorithm 3 Adaptive mirror descent (general convex objective)

Input: accuracy $\varepsilon > 0$; Θ_0 s.t. $d(x_*) \le \Theta_0^2$.

  1: $x^0 = \arg\min_{x \in X} d(x)$.

  2: Initialize the set I as empty set.

  3: Set $k = 0$.

  4: **repeat**

  5:     **if** $g(x^k) \le \varepsilon$ **then**

  6:         $h_k = \frac{\varepsilon}{\|\nabla f(x^k)\|_{E,*}}$

  7:         $x^{k+1} = \text{Mirr}[x^k](h_k \nabla f(x^k))$ ("productive step")

  8:         Add k to I.

  9:     **else**

10:         $h_k = \frac{\varepsilon}{\|\nabla g(x^k)\|_{E,*}^2}$

11:         $x^{k+1} = \text{Mirr}[x^k](h_k \nabla g(x^k))$ ("non-productive step")

12:     **end if**

13:     Set $k = k + 1$.

14: **until** $|I| + \sum_{j \in J} \frac{1}{\|\nabla g(x^j)\|_{E,*}^2} \ge \frac{2\Theta_0^2}{\varepsilon^2}$

Output: $\bar{x}^k := \arg\min_{x^j,\, j \in I} f(x^j)$

From Lemma 1 with $u = x_*$ and $\Delta = 0$, by the definition of h_i, $i \in I$, we have, for all $i \in I$,

$$\varepsilon v_f[x_*](x^i) = \varepsilon \left\langle \frac{\nabla f(x^i)}{\|\nabla f(x^i)\|_{E,*}}, x^i - x_* \right\rangle = h_i \langle \nabla f(x^i), x^i - x_* \rangle$$

$$\le \frac{h_i^2}{2} \|\nabla f(x^i)\|_{E,*}^2 + V[x^i](x_*) - V[x^{i+1}](x_*)$$

$$= \frac{\varepsilon^2}{2} + V[x^i](x_*) - V[x^{i+1}](x_*). \tag{8.24}$$

Similarly, by the definition of h_i, $i \in J$, we have, for all $i \in J$,

$$\frac{\varepsilon(g(x^i) - g(x_*))}{\|\nabla g(x^i)\|_{E,*}^2} = h_i(g(x^i) - g(x_*)) \le \frac{h_i^2}{2} \|\nabla g(x^i)\|_{E,*}^2 + V[x^i](x_*) - V[x^{i+1}](x_*)$$

$$= \frac{\varepsilon^2}{2\|\nabla g(x^i)\|_{E,*}^2} + V[x^i](x_*) - V[x^{i+1}](x_*).$$

Whence, using that, for all $i \in J$, $g(x^i) - g(x_*) \ge g(x^i) > \varepsilon$, we have

$$-\frac{\varepsilon^2}{2\|\nabla g(x^i)\|_{E,*}^2} + V[x^i](x_*) - V[x^{i+1}](x_*) > 0. \tag{8.25}$$

Summing up inequalities (8.24) for $i \in I$ and applying (8.25) for $i \in J$, we obtain

$$\varepsilon |I| \min_{i \in I} v_f[x_*](x^i) \le \varepsilon \sum_{i \in I} v_f[x_*](x^i) < \frac{\varepsilon^2}{2} \cdot |I| + \Theta_0^2 - \sum_{i \in J} \frac{\varepsilon^2}{2\|\nabla g(x^i)\|_{E,*}^2},$$

where we also used that, by definition of x^0 and (8.8),

$$V[x^0](x_*) = d(x_*) - d(x^0) - \langle \nabla d(x^0), x_* - x^0 \rangle \le d(x_*) \le \Theta_0^2.$$

If the stopping criterion in Algorithm 3 is fulfilled, we get

$$\varepsilon |I| \min_{i \in I} v_f[x_*](x^i) < \varepsilon^2 |I|.$$

Since the inequality is strict, the set I is not empty and the output point $\bar{x}^k$ is correctly defined. Dividing both sides of the last inequality by $\varepsilon|I|$, we obtain the first statement of the Theorem. By definition of $\bar{x}^k$, it is obvious that $g(\bar{x}^k) \le \varepsilon$. $\square$

To obtain the complexity of our algorithm in terms of the values of the objective function f, we define non-decreasing function

$$\omega(\tau) = \begin{cases} \max_{x \in X}\{f(x) - f(x_*) : \|x - x_*\|_E \le \tau\} & \tau \ge 0, \\ 0 & \tau < 0. \end{cases} \tag{8.26}$$

and use the following lemma from [21].

Lemma 3 *Assume that f is a convex function. Then, for any $x \in X$,*

$$f(x) - f(x_*) \le \omega(v_f[x_*](x)). \tag{8.27}$$

Corollary 1 *Assume that the objective function f in (8.1) is defined as $f(x) = \max_{i \in \{1,\dots,m\}} f_i(x)$, where f_i, $i = 1,\dots,m$ are differentiable with Lipschitz-continuous gradient*

$$\|\nabla f_i(x) - \nabla f_i(y)\|_{E,*} \le L_i \|x - y\|_E \quad \forall x, y \in X, \quad i \in \{1,\dots,m\}. \tag{8.28}$$

Then $\bar{x}^k$ is $\tilde{\varepsilon}$-solution to (8.1) in the sense of (8.6), where

$$\tilde{\varepsilon} = \max\{\varepsilon, \varepsilon \max_{i=1,\dots,m} \|\nabla f_i(x_*)\|_{E,*} + \varepsilon^2 \max_{i=1,\dots,m} L_i/2\}.$$

Proof As it was shown in Theorem 5, $g(\bar{x}^k) \le \varepsilon$. It follows from (8.28) that

$$f_i(x) \le f_i(x_*) + \langle \nabla f_i(x_*), x - x_* \rangle + \frac{1}{2} L_i \|x - x_*\|_E^2$$

$$\le f_i(x_*) + \|\nabla f_i(x_*)\|_{E,*} \|x - x_*\|_E + \frac{1}{2} L_i \|x - x_*\|_E^2, \quad i = 1,\dots,m.$$

Whence, $\omega(\tau) \leq \tau \max_{i=1,...,m} \|\nabla f_i(x_*)\|_{E,*} + \frac{\tau^2 \max_{i=1,...,m} L_i}{2}$. By Lemma 3, non-decreasing property of ω and Theorem 5, we obtain

$$f(\bar{x}^k) - f(x_*) = \min_{i \in I} f(x^i) - f(x_*) \leq \min_{i \in I} \omega(v_f[x_*](x^i))$$

$$\leq \omega(\min_{i \in I} v_f[x_*](x^i)) \leq \omega(\varepsilon)$$

$$\leq \varepsilon \max_{i=1,...,m} \|\nabla f_i(x_*)\|_{E,*} + \frac{\varepsilon^2 \max_{i=1,...,m} L_i}{2}.$$

$\square$

8.4 Randomization for Constrained Problems

In this section, we consider randomized version of problem (8.1). This means that we still can use the value of the function $g(x)$ in an algorithm, but, instead of subgradients of f and g, we use their stochastic approximations. We combine the idea of switching subgradient method [25] and Stochastic Mirror Descent method introduced in [16]. More general case of stochastic optimization problems with expectation constraints is studied in [14]. We consider convex problems as long as strongly convex and, for each case, we have two types of algorithms. The first one allows to control expectation of the objective residual $f(\tilde{x}) - f(x_*)$ and inequality infeasibility $g(\tilde{x})$, where $\tilde{x}$ is the output of the algorithm. The second one allows to control probability of large deviation for these two quantities.

We introduce the following new assumptions. Given a point $x \in X$, we can calculate stochastic subgradients $\nabla f(x, \xi)$, $\nabla g(x, \zeta)$, where ξ, ζ are random vectors. These stochastic subgradients satisfy

$$\mathbb{E}[\nabla f(x, \xi)] = \nabla f(x) \in \partial f(x), \quad \mathbb{E}[\nabla g(x, \zeta)] = \nabla g(x) \in \partial g(x), \qquad (8.29)$$

and

$$\|\nabla f(x, \xi)\|_{E,*} \leq M_f, \quad \|\nabla g(x, \zeta)\|_{E,*} \leq M_g, \quad \text{a.s. in } \xi, \zeta. \qquad (8.30)$$

To motivate these assumptions, we consider the following example.

Example 1 ([3]) Consider Problem (8.1) with

$$f(x) = \frac{1}{2}\langle Ax, x \rangle,$$

where A is given $n \times n$ matrix, $X = S(1)$ being standard unit simplex, i.e., $X = \{x \in \mathbb{R}^n_+ : \sum_{i=1}^n x_i = 1\}$, and

$$g(x) = \max_{i \in \{1,\dots,m\}} \{\langle c_i, x \rangle\},$$

where $\{c_i\}_{i=1}^m$ are given vectors in $\mathbb{R}^n$.

Even if the matrix A is sparse, the gradient $\nabla f(x) = Ax$ is usually not. The exact computation of the gradient takes $O(n^2)$ arithmetic operations, which is expensive when n is large. In this setting, it is natural to use randomization to construct a stochastic approximation for $\nabla f(x)$. Let ξ be a random variable taking its values in $\{1,\dots,n\}$ with probabilities $(x_1,\dots,x_n)$ respectively. Let $A^{\langle i \rangle}$ denote the i-th column of the matrix A. Since $x \in S_n(1)$,

$$\mathbb{E}[A^{\langle \xi \rangle}] = A^{\langle 1 \rangle} \underbrace{\mathbb{P}(\xi = 1)}_{x_1} + \cdots + A^{\langle n \rangle} \underbrace{\mathbb{P}(\xi = n)}_{x_n}$$

$$= A^{\langle 1 \rangle} x_1 + \cdots + A^{\langle n \rangle} x_n = Ax.$$

Thus, we can use $A^{\langle \xi \rangle}$ as stochastic subgradient, which can be calculated in $O(n)$ arithmetic operations.

8.4.1 Convex Objective Function, Control of Expectation

In this subsection, we consider convex optimization problem (8.1) in randomized setting described above. In this setting the output of the algorithm is random. Thus, we need to change the notion of approximate solution. Let x_* be a solution to (8.1). We say that a (random) point $\tilde{x} \in X$ is an *expected ε-solution* to (8.1) if

$$\mathbb{E} f(\tilde{x}) - f(x_*) \le \varepsilon, \quad \text{and} \quad g(\tilde{x}) \le \varepsilon \quad \text{a.s.} \tag{8.31}$$

We also introduce a stronger assumption than (8.8). Namely, we assume that we know a constant $\Theta_0 > 0$ such that

$$\sup_{x,y \in X} V[x](y) \le \Theta_0^2. \tag{8.32}$$

The main difference between the method, which we describe below, and the method in [14] is the adaptivity of our method both in terms of stepsize and stopping rule, which means that we do not need to know the constants M_f, M_g in advance. We assume that on each iteration of the algorithm independent realizations of ξ and ζ are generated. The algorithm is similar to the one in [3], but, for the sake of consistency with other parts of the chapter, we use slightly different proof.

Algorithm 4 Adaptive stochastic mirror descent

Input: accuracy $\varepsilon > 0$; Θ_0 s.t. $V[x](y) \leq \Theta_0^2$, $\forall x, y \in X$.
1: $x^0 = \arg\min_{x \in X} d(x)$.
2: Initialize the set I as empty set.
3: Set $k = 0$.
4: **repeat**
5:     **if** $g(x^k) \leq \varepsilon$. **then**
6:         $M_k = \|\nabla f(x^k, \xi^k)\|_{E,*}$.
7:         $h_k = \Theta_0 \left(\sum_{i=0}^{k} M_i^2 \right)^{-1/2}$.
8:         $x^{k+1} = \text{Mirr}[x^k](h_k \nabla f(x^k, \xi^k))$ ("productive step").
9:         Add k to I.
10:    **else**
11:        $M_k = \|\nabla g(x^k, \zeta^k)\|_{E,*}$.
12:        $h_k = \Theta_0 \left(\sum_{i=0}^{k} M_i^2 \right)^{-1/2}$.
13:        $x^{k+1} = \text{Mirr}[x^k](h_k \nabla g(x^k, \zeta^k))$ ("non-productive step").
14:    **end if**
15:    Set $k = k + 1$.
16: **until** $k \geq \frac{2\Theta_0}{\varepsilon} \left(\sum_{i=0}^{k-1} M_i^2 \right)^{1/2}$.
Output: $\bar{x}^k = \frac{1}{|I|} \sum_{k \in I} x^k$.

Theorem 6 *Let equalities* (8.29) *and inequalities* (8.30) *hold. Assume that a known constant* $\Theta_0 > 0$ *is such that* $V[x](y) \leq \Theta_0^2$, $\forall x, y \in X$. *Then, Algorithm 4 stops after not more than*

$$k = \left\lceil \frac{4 \max\{M_f^2, M_g^2\} \Theta_0^2}{\varepsilon^2} \right\rceil \qquad (8.33)$$

iterations and $\bar{x}^k$ *is an expected* ε*-solution to* (8.1) *in the sense of* (8.31).

Proof First, let us prove that the inequality in the stopping criterion holds for k defined in (8.33). By (8.30), we have that, for any $i \in \{0, \ldots, k-1\}$, $M_i \leq \max\{M_f, M_g\}$. Hence, by (8.33), $\frac{2\Theta_0}{\varepsilon} \left(\sum_{j=0}^{k-1} M_j^2 \right)^{1/2} \leq \frac{2\Theta_0}{\varepsilon} \max\{M_f, M_g\} \sqrt{k} \leq k$.

Denote $[k] = \{i \in \{0, \ldots, k-1\}\}$, $J = [k] \setminus I$ and

$$\delta_i = \begin{cases} \langle \nabla f(x^i, \xi^i) - \nabla f(x^i), x_* - x^i \rangle, & \text{if } i \in I, \\ \langle \nabla g(x^i, \zeta^i) - \nabla g(x^i), x_* - x^i \rangle, & \text{if } i \in J. \end{cases} \qquad (8.34)$$

From Lemma 1 with $u = x_*$ and $\Delta = \nabla f(x^i, \xi^i) - \nabla f(x^i)$, we have, for all $i \in I$,

$$h_i\big(f(x^i) - f(x_*)\big) \leq \frac{h_i^2}{2}\|\nabla f(x^i, \xi^i)\|_{E,*}^2 + V[x^i](x_*) - V[x^{i+1}](x_*) + h_i\delta_i$$

and, from Lemma 1 with $u = x_*$ and $\Delta = \nabla g(x^i, \zeta^i) - \nabla g(x^i)$, for all $i \in J$,

$$h_i\big(g(x^i) - g(x_*)\big) \leq \frac{h_i^2}{2}\|\nabla g(x^i, \zeta^i)\|_{E,*}^2 + V[x^i](x_*) - V[x^{i+1}](x_*) + h_i\delta_i.$$

Dividing each inequality by h_i and summing up these inequalities for i from 0 to $k - 1$, using the definition of $h_i, i \in \{0, \ldots, k - 1\}$, we obtain

$$\sum_{i \in I}\big(f(x^i) - f(x_*)\big) + \sum_{i \in J}\big(g(x^i) - g(x_*)\big)$$

$$\leq \sum_{i \in [k]}\frac{h_i M_i^2}{2} + \sum_{i \in [k]}\frac{1}{h_i}\big(V[x^i](x_*) - V[x^{i+1}](x_*)\big) + \sum_{i \in [k]}\delta_i \qquad (8.35)$$

Using (8.32), we get

$$\sum_{i=0}^{k-1}\frac{1}{h_i}\big(V[x^i](x_*) - V[x^{i+1}](x_*)\big)$$

$$= \frac{1}{h_0}V[x^0](x_*) + \sum_{i=0}^{k-2}\Big(\frac{1}{h_{i+1}} - \frac{1}{h_i}\Big)V[x^{i+1}](x_*) - \frac{1}{h_{k-1}}V[x^k](x_*)$$

$$\leq \frac{\Theta_0^2}{h_0} + \Theta_0^2\sum_{k=0}^{k-2}\Big(\frac{1}{h_{i+1}} - \frac{1}{h_i}\Big) = \frac{\Theta_0^2}{h_{k-1}}.$$

Whence, by the definition of stepsizes h_i,

$$\sum_{i \in I}\big(f(x^i) - f(x_*)\big) + \sum_{i \in J}\big(g(x^i) - g(x_*)\big) \leq \sum_{i \in [k]}\frac{h_i M_i^2}{2} + \frac{\Theta_0^2}{h_{k-1}} + \sum_{i \in [k]}\delta_i$$

$$\leq \sum_{i=0}^{k-1}\frac{\Theta_0}{2}\frac{M_i^2}{\big(\sum_{j=0}^i M_j^2\big)^{1/2}} + \Theta_0\Big(\sum_{i=0}^{k-1}M_i^2\Big)^{1/2} + \sum_{i \in [k]}\delta_i$$

$$\leq 2\Theta_0\Big(\sum_{i=0}^{k-1}M_i^2\Big)^{1/2} + \sum_{i \in [k]}\delta_i,$$

where we used inequality $\sum_{i=0}^{k-1} \frac{M_i^2}{\left(\sum_{j=0}^{i} M_j^2\right)^{1/2}} \leq 2\left(\sum_{i=0}^{k-1} M_i^2\right)^{1/2}$, which can be proved by induction. Since, for $i \in J$, $g(x^i) - g(x_*) \geq g(x^i) > \varepsilon$, by convexity of f, the definition of $\bar{x}^k$, and the stopping criterion, we get

$$|I|\left(f(\bar{x}^k) - f(x_*)\right) < \varepsilon|I| - \varepsilon k + 2\Theta_0\left(\sum_{i=0}^{k-1} M_i^2\right)^{1/2} + \sum_{i=0}^{k-1} \delta_i \leq \varepsilon|I| + \sum_{i=0}^{k-1} \delta_i.$$
(8.36)

Taking the expectation and using (8.29), as long as the inequality is strict and the case of $I = \emptyset$ is impossible, we obtain

$$\mathbb{E}f(\bar{x}^k) - f(x_*) \leq \varepsilon.$$
(8.37)

At the same time, for $i \in I$ it holds that $g(x^i) \leq \varepsilon$. Then, by the definition of $\bar{x}^k$ and the convexity of g,

$$g(\bar{x}^k) \leq \frac{1}{|I|} \sum_{i \in I} g(x^i) \leq \varepsilon.$$

$\square$

8.4.2 Convex Objective Function, Control of Large Deviation

In this subsection, we consider the same setting as in previous subsection, but change the notion of approximate solution. Let x_* be a solution to (8.1). Given $\varepsilon > 0$ and $\sigma \in (0, 1)$, we say that a point $\tilde{x} \in X$ is an (ε, σ)-solution to (8.1) if

$$\mathbb{P}\{f(\tilde{x}) - f(x_*) \leq \varepsilon, \quad g(\tilde{x}) \leq \varepsilon\} \geq 1 - \sigma.$$
(8.38)

As in the previous subsection, we use an assumption expressed by inequality (8.32). We assume additionally to (8.30) that inequalities (8.5) and (8.7) hold. Unfortunately, it is not clear, how to obtain large deviation guarantee for an adaptive method. Thus, in this section, we assume that the constants M_f, M_g are known and use a simplified algorithm. We assume that on each iteration of the algorithm independent realizations of ξ and ζ are generated.

To analyze Algorithm 5 in terms of large deviation bound, we need the following known result, see, e.g., [10].

Lemma 4 (Azuma-Hoeffding Inequality) *Let $\eta^1, \ldots, \eta^n$ be a sequence of independent random variables taking their values in some set Ξ, and let $Z =$*

Algorithm 5 Stochastic mirror descent

Input: accuracy $\varepsilon > 0$; maximum number of iterations N; M_f, M_g s.t. (8.5), (8.7), (8.30) hold.
1: $x^0 = \arg\min\limits_{x \in X} d(x)$.
2: Set $h = \dfrac{\varepsilon}{\max\{M_f^2, M_g^2\}}$.
3: Set $k = 0$.
4: **repeat**
5:     **if** $g(x^k) \le \varepsilon$. **then**
6:         $x^{k+1} = \text{Mirr}[x^k](h\nabla f(x^k, \xi^k))$ ("productive step").
7:         Add k to I.
8:     **else**
9:         $x^{k+1} = \text{Mirr}[x^k](h\nabla g(x^k, \zeta^k))$ ("non-productive step").
10:    **end if**
11:    Set $k = k + 1$.
12: **until** $k \ge N$.
Output: If $I \ne \emptyset$, then $\bar{x}^k = \frac{1}{|I|} \sum\limits_{k \in I} x^k$. Otherwise $\bar{x}^k = NULL$.

$\phi(\eta^1, \ldots, \eta^n)$ for some function $\phi : \Xi^n \to \mathbb{R}$. Suppose that a. s.

$$\left| \mathbb{E}[Z|\eta^1, \ldots, \eta^i] - \mathbb{E}[Z|\eta^1, \ldots, \eta^{i-1}] \right| \le c_i, \quad i = 1, \ldots, n,$$

where c_i, $i \in \{1, \ldots, n\}$ are deterministic. Then, for each $t \ge 0$

$$\mathbb{P}\left(Z - \mathbb{E}Z \ge t\right) \le \exp\left\{ -\frac{t^2}{2 \sum\limits_{i=1}^{n} c_i^2} \right\}.$$

Theorem 7 *Let equalities (8.29) and inequalities (8.5), (8.7), (8.30) hold. Assume that a known constant $\Theta_0 > 0$ is such that $V[x](y) \le \Theta_0^2$, $\forall x, y \in X$, and the confidence level satisfies $\sigma \in (0, 0.5)$. Then, if in Algorithm 5*

$$N = \left\lceil 70 \frac{\max\{M_f^2, M_g^2\}\Theta_0^2}{\varepsilon^2} \ln\frac{1}{\sigma} \right\rceil, \tag{8.39}$$

$\bar{x}^k$ *is an (ε, σ)-solution to (8.1) in the sense of (8.38).*

Proof Let us denote $M = \max\{M_f, M_g\}$. In the same way as we obtained (8.35) in the proof of Theorem 6, we obtain

$$h \sum_{i \in I} \left(f(x^i) - f(x_*) \right) + h \sum_{i \in J} \left(g(x^i) - g(x_*) \right)$$

$$\le \frac{h^2 M^2 k}{2} + V[x^0](x_*) + h \sum_{i=0}^{k-1} \delta_i,$$

where δ_i, $i = 0, \ldots, k - 1$ are defined in (8.34). Since, for $i \in J$, $g(x^i) - g(x_*) \geq g(x^i) > \varepsilon$, by convexity of f, the definition of $\bar{x}^k$ and h, we get

$$h|I|\left(f(\bar{x}^k) - f(x_*)\right) < \varepsilon h|I| - \frac{\varepsilon^2 k}{2M^2} + \Theta_0^2 + h \sum_{i=0}^{k-1} \delta_i. \tag{8.40}$$

Using Cauchy-Schwarz inequality, (8.5), (8.7), (8.30), (8.32), we have

$$h|\delta_i| \leq 2hM \|x^i - x^*\|$$

$$\leq 2hM \sqrt{2V[x^i](x^*)} \leq 2\sqrt{2}hM\Theta_0 = 2\sqrt{2}\frac{\varepsilon\Theta_0}{M}.$$

Now we use Lemma 4 with $Z = \sum_{i=0}^{k-1} h\delta_i$. Clearly, $\mathbb{E}Z = \mathbb{E}\left[\sum_{i=0}^{k-1} h\delta_i\right] = 0$ and we can take $c_i = 2\sqrt{2}\frac{\varepsilon\Theta_0}{M}$. Then, by Lemma 4, for each $t \geq 0$,

$$\mathbb{P}\left\{\sum_{i=0}^{k-1} h\delta_i \geq t\right\} \leq \exp\left(-\frac{t^2}{2\sum_{i=0}^{k-1} c_i^2}\right) = \exp\left(-\frac{t^2 M^2}{16\varepsilon^2\Theta_0^2 k}\right).$$

In other words, for each $\sigma \in (0, 1)$

$$\mathbb{P}\left\{\sum_{i=0}^{k-1} h\delta_i \geq \frac{4\varepsilon\Theta_0}{M}\sqrt{k \ln\left(\frac{1}{\sigma}\right)}\right\} \leq \sigma.$$

Applying this inequality to (8.40), we obtain, for any $\sigma \in (0, 1)$,

$$\mathbb{P}\left\{h|I|\left(f(\bar{x}^k) - f(x_*)\right) < \varepsilon h|I| - \frac{\varepsilon^2 k}{2M^2} + \Theta_0^2 + \frac{4\varepsilon\Theta_0}{M}\sqrt{k \ln\left(\frac{1}{\sigma}\right)}\right\} \geq 1 - \sigma.$$

Then, by (8.39), we have

$$-\frac{\varepsilon^2 k}{2M^2} + \Theta_0^2 + \frac{4\varepsilon\Theta_0}{M}\sqrt{k \ln\left(\frac{1}{\sigma}\right)} < \Theta_0^2\left(-\frac{71}{2}\ln\left(\frac{1}{\sigma}\right) + 1 + 4\ln\left(\frac{1}{\sigma}\right)\sqrt{71}\right)$$

$$< \Theta_0^2\left(-\frac{3}{2}\ln\left(\frac{1}{\sigma}\right) + 1\right). \tag{8.41}$$

Since $\sigma \leq 0.5 < \exp(-2/3)$, we have $-\frac{3}{2} \ln\left(\frac{1}{\sigma}\right) + 1 < 0$ and

$$\mathbb{P}\left\{h|I|\left(f(\bar{x}^k) - f(x^*)\right) < h|I|\varepsilon\right\} \geq 1 - \sigma.$$

Thus, with probability at least $1 - \sigma$, the inequality is strict, the case of $I = \emptyset$ is impossible, and $\bar{x}^k$ is correctly defined. Dividing the both sides of it by $h \cdot |I|$, we obtain that $\mathbb{P}\left\{f(\bar{x}^k) - f(x^*) \leq \varepsilon\right\} \geq 1 - \sigma$. At the same time, for $i \in I$ it holds that $g(x^i) \leq \varepsilon$. Then, by the definition of $\bar{x}^k$ and the convexity of g, again with probability at least $1 - \sigma$

$$g(\bar{x}^k) \leq \frac{1}{|I|} \sum_{i \in I} g(x^i) \leq \varepsilon.$$

Thus, $\bar{x}^k$ is an (ε, σ)-solution to (8.1) in the sense of (8.38).   □

8.4.3 Strongly Convex Objective Function, Control of Expectation

In this subsection, we consider the setting of Sect. 8.4.1, but, as in Sect. 8.3.2, make the following additional assumptions. First, we assume that functions f and g are strongly convex. Second, without loss of generality, we assume that $0 = \arg\min_{x \in X} d(x)$. Third, we assume that we are given a starting point $x_0 \in X$ and a number $R_0 > 0$ such that $\|x_0 - x_*\|_E^2 \leq R_0^2$. Finally, we make the following assumption (cf. (8.18)) that d is bounded in the following sense. Assume that x_* is some fixed point and x is a random point such that $\mathbb{E}_x\left[\|x - x_*\|_E^2\right] \leq R^2$, then

$$\mathbb{E}_x\left[d\left(\frac{x - x_*}{R}\right)\right] \leq \frac{\Omega}{2}, \tag{8.42}$$

where Ω is some known number and $\mathbb{E}_x$ denotes the expectation with respect to random vector x. For example, this assumption holds for Euclidean proximal setup. Unlike the method introduced in [14] for strongly convex problems, we present a method, which is based on the restart of Algorithm 5. Unfortunately, it is not clear, whether the restart technique can be combined with adaptivity to constants M_f, M_g. Thus, we assume that these constants are known.

The following lemma can be proved in the same way as Lemma 2.

Lemma 5 Let f and g be strongly convex functions with the same parameter μ and x_* be a solution of problem (8.1). Assume that, for some random $\tilde{x} \in X$,

$$\mathbb{E}f(\tilde{x}) - f(x_*) \leq \varepsilon, \quad g(\tilde{x}) \leq \varepsilon.$$

Then

$$\frac{\mu}{2}\mathbb{E}\|\tilde{x} - x_*\|_E^2 \le \varepsilon.$$

Theorem 8 *Let equalities* (8.29) *and inequalities* (8.30) *hold and* f, g *be strongly convex with the same parameter* μ. *Also assume that the prox function* $d(x)$ *satisfies* (8.42) *and the starting point* $x_0 \in X$ *and a number* $R_0 > 0$ *are such that* $\|x_0 - x_*\|_E^2 \le R_0^2$. *Then, the point* x_p *returned by Algorithm 6 is an expected* ε-*solution to* (8.1) *in the sense of* (8.31) *and* $\mathbb{E}\|x_p - x_*\|_E^2 \le \frac{2\varepsilon}{\mu}$. *At the same time, the total number of inner iterations of Algorithm 5 does not exceed*

$$\left\lceil \log_2 \frac{\mu R_0^2}{2\varepsilon} \right\rceil + \frac{32\Omega \max\{M_f^2, M_g^2\}}{\mu\varepsilon}. \tag{8.43}$$

Proof Let us denote $M = \max\{M_f, M_g\}$. Observe that, for all $p \ge 0$, the function $d_p(x)$ defined in Algorithm 6 is 1-strongly convex w.r.t. the norm $\|\cdot\|_E / R_p$. The conjugate of this norm is $R_p\|\cdot\|_{E,*}$. This means that, at each outer iteration p, M changes to MR_{p-1}, where p is the number of outer iteration. We show by induction that, for all $p \ge 0$, $\mathbb{E}\|x_p - x_*\|_E^2 \le R_p^2$. For $p = 0$ it holds by the definition of x_0 and R_0.

Let us assume that this inequality holds for some $p - 1$ and show that it holds for p. At iteration p, we start Algorithm 5 with starting point x_{p-1} and stepsize $h_p = \frac{\varepsilon_p}{M^2 R_{p-1}^2}$. Using the same steps as in the proof of Theorem 7, after N_p iterations

---

Algorithm 6 Stochastic mirror descent (strongly convex objective, expectation control)

Input: accuracy $\varepsilon > 0$; strong convexity parameter μ; Ω s.t. $\mathbb{E}_x\left[d\left(\frac{x-x_*}{R}\right)\right] \le \frac{\Omega}{2}$ if

1: $\mathbb{E}_x\left[\|x - x_*\|_E^2\right] \le R^2$; starting point x_0 and number R_0 s.t. $\|x_0 - x_*\|_E^2 \le R_0^2$.

2: Set $d_0(x) = d\left(\frac{x-x_0}{R_0}\right)$.

3: Set $p = 1$.

4: **repeat**

5:   Set $R_p^2 = R_0^2 \cdot 2^{-p}$.

6:   Set $\varepsilon_p = \frac{\mu R_p^2}{2}$.

7:   Set $N_p = \left\lceil \frac{\max\{M_f^2, M_g^2\}\Omega R_{p-1}^2}{\varepsilon_p^2} \right\rceil$

8:   Set x_p as the output of Algorithm 5 with accuracy ε_p, number of iterations N_p, prox-function $d_{p-1}(\cdot)$ and $\frac{\Omega}{2}$ as Θ_0^2.

9:   $d_p(x) \leftarrow d\left(\frac{x-x_p}{R_p}\right)$.

10:   Set $p = p + 1$.

11: **until** $p > \log_2 \frac{\mu R_0^2}{2\varepsilon}$.

Output: x_p.

---

of Algorithm 5 (see (8.40)), we obtain

$$h_p|I_p|(f(\bar{x}_p^k) - f(x_*)) < \varepsilon_p h_p |I_p| - \frac{\varepsilon_p^2 N_p}{2M^2 R_{p-1}^2} + V_{p-1}[x_{p-1}](x_*) + h_p \sum_{i=0}^{N_p-1} \delta_i, \tag{8.44}$$

where $V_{p-1}[z](x)$ is the Bregman divergence corresponding to $d_{p-1}(x)$ and I_p is the set of "productive steps". Using the definition of d_{p-1}, we have

$$V_{p-1}[x_{p-1}](x_*) = d_{p-1}(x_*) - d_{p-1}(x_{p-1}) - \langle \nabla d_{p-1}(x_{p-1}), x_* - x_{p-1} \rangle \leq d_{p-1}(x_*).$$

Taking expectation with respect to x_{p-1} in (8.44) and using inductive assumption $\mathbb{E}\|x_{p-1} - x_*\|_E^2 \leq R_{p-1}^2$ and (8.42), we obtain, substituting N_p,

$$h_p|I_p|(f(\bar{x}_p^k) - f(x_*)) < \varepsilon_p h_p |I_p| - \frac{\varepsilon_p^2 N_p}{2M^2 R_{p-1}^2} + \frac{\Omega}{2} + h_p \sum_{i=0}^{N_p-1} \delta_i$$

$$\leq \varepsilon_p h_p |I_p| + h_p \sum_{i=0}^{N_p-1} \delta_i. \tag{8.45}$$

Taking the expectation and using (8.29), as long as the inequality is strict and the case of $I_p = \emptyset$ is impossible, we obtain

$$\mathbb{E}f(\bar{x}_p^k) - f(x_*) \leq \varepsilon_p. \tag{8.46}$$

At the same time, for $i \in I_p$ it holds that $g(x^i) \leq \varepsilon_p$. Then, by the definition of $\bar{x}_p^k$ and the convexity of g,

$$g(\bar{x}_p^k) \leq \frac{1}{|I_p|} \sum_{i \in I_p} g(x^i) \leq \varepsilon_p.$$

Thus, we can apply Lemma 5 and obtain

$$\mathbb{E}\|x_p - x_*\|_E^2 \leq \frac{2\varepsilon_p}{\mu} = R_p^2.$$

Thus, we proved that, for all $p \geq 0$, $\mathbb{E}\|x_p - x_*\|_E^2 \leq R_p^2 = R_0^2 \cdot 2^{-p}$. At the same time, we have, for all $p \geq 1$,

$$\mathbb{E}f(x_p) - f(x_*) \leq \frac{\mu R_0^2}{2} \cdot 2^{-p}, \quad g(x_p) \leq \frac{\mu R_0^2}{2} \cdot 2^{-p}.$$

Thus, if $p > \log_2 \frac{\mu R_0^2}{2\varepsilon}$, x_p is an ε-solution to (8.1) in the sense of (8.31) and

$$\mathbb{E}\|x_p - x_*\|_E^2 \le R_0^2 \cdot 2^{-p} \le \frac{2\varepsilon}{\mu}.$$

Let us now estimate the total number N of inner iterations, i.e., the iterations of Algorithm 1. Let us denote $\hat{p} = \left\lceil \log_2 \frac{\mu R_0^2}{2\varepsilon} \right\rceil$. We have

$$N = \sum_{p=1}^{\hat{p}} N_p \le \sum_{p=1}^{\hat{p}} \left(1 + \frac{\Omega \max\{M_f^2, M_g^2\} R_{p-1}^2}{\varepsilon_p^2} \right)$$

$$= \sum_{p=1}^{\hat{p}} \left(1 + \frac{16\Omega \max\{M_f^2, M_g^2\} 2^p}{\mu^2 R_0^2} \right) \le \hat{p} + \frac{32\Omega \max\{M_f^2, M_g^2\} 2^{\hat{p}}}{\mu^2 R_0^2}$$

$$\le \hat{p} + \frac{32\Omega \max\{M_f^2, M_g^2\}}{\mu\varepsilon}.$$

$\square$

8.4.4 Strongly Convex Objective Function, Control of Large Deviation

In this subsection, we consider the setting of Sect. 8.4.2, but make the following additional assumptions. First, we assume that functions f and g are strongly convex. Second, without loss of generality, we assume that $0 = \arg\min_{x \in X} d(x)$. Third, we assume that we are given a starting point $x_0 \in X$ and a number $R_0 > 0$ such that $\|x_0 - x_*\|_E^2 \le R_0^2$. Finally, instead of (8.32), we assume that the Bregman divergence satisfies quadratic growth condition

$$V[z](x) \le \frac{\Omega}{2} \|x - z\|_E^2, \quad x, z \in X. \tag{8.47}$$

where Ω is some known number. For example, this assumption holds for Euclidean proximal setup. Unlike the method introduced in [14] for strongly convex problems, we present a method, which is based on the restart of Algorithm 5. Unfortunately, it is not clear, whether the restart technique can be combined with adaptivity to constants M_f, M_g. Thus, we assume that these constants are known.

Theorem 9 *Let equalities (8.29) and inequalities (8.5), (8.7), (8.30) hold. Let f, g be strongly convex with the same parameter μ. Also assume that the Bregman divergence $V[z](x)$ satisfies (8.47) and the starting point $x_0 \in X$ and a number*

$R_0 > 0$ *are such that* $\|x_0 - x_*\|_E^2 \le R_0^2$. *Then, the point* x_p *returned by Algorithm 7 is an* (ε, σ)-*solution to* (8.1) *in the sense of* (8.38) *and* $\|x_p - x_*\|_E^2 \le \frac{2\varepsilon}{\mu}$ *with probability at least* $1 - \sigma$. *At the same time, the total number of inner iterations of Algorithm 5 does not exceed*

$$\left\lceil \log_2 \frac{\mu R_0^2}{2\varepsilon} \right\rceil + \frac{2240\Omega \max\{M_f^2, M_g^2\}}{\mu\varepsilon}\left(\ln\frac{1}{\sigma} + \ln\log_2\frac{\mu R_0^2}{2\varepsilon}\right).$$

Proof Let us denote $M = \max\{M_f, M_g\}$. Observe that, for all $p \ge 0$, the function $d_p(x)$ defined in Algorithm 7 is 1-strongly convex w.r.t. the norm $\|\cdot\|_E / R_p$. The conjugate of this norm is $R_p\|\cdot\|_{E,*}$. This means that, at each outer iteration p, M changes to MR_{p-1}, where p is the number of outer iteration.

Let A_p, $p \ge 0$ be the event $A_p = \{\|x_p - x_*\|_E^2 \le R_p^2\}$ and $\bar{A}_p$ be its complement. Note that, by the definition of x_0 and R_0, A_0 holds with probability 1. Denote $\hat{p} = \left\lceil \log_2 \frac{\mu R_0^2}{2\varepsilon} \right\rceil$.

We now show by induction that, for all $p \ge 1$, $\mathbb{P}\{A_p | A_{p-1}\} \ge 1 - \frac{\sigma}{\hat{p}}$. By inductive assumption, A_{p-1} holds and we have $\|x_{p-1} - x_*\|_E^2 \le R_{p-1}^2$. At iteration p, we start Algorithm 5 with starting point x_{p-1}, feasible set X_p and Bregman

---

Algorithm 7 Stochastic mirror descent (strongly convex objective, control of large deviation)

Input: accuracy $\varepsilon > 0$; strong convexity parameter μ; Ω s.t. $V[x](y) \le \frac{\Omega}{2}\|x-y\|_E^2$, $x, y \in X$; starting point x_0 and number R_0 s.t. $\|x_0 - x_*\|_E^2 \le R_0^2$.

1: Set $d_0(x) = d\left(\frac{x-x_0}{R_0}\right)$.
2: Set $p = 1$.
3: **repeat**
4:   Set $R_p^2 = R_0^2 \cdot 2^{-p}$.
5:   Set $\varepsilon_p = \frac{\mu R_p^2}{2}$.
6:   Set $N_p = \left\lceil 70\frac{\max\{M_f^2, M_g^2\}\Omega R_{p-1}^2}{\varepsilon_p^2}\ln\left(\frac{1}{\sigma}\log_2\frac{\mu R_0^2}{2\varepsilon}\right)\right\rceil$.
7:   Set $X_p = \{x \in X : \|x - x_{p-1}\|_E^2 \le R_{p-1}^2\}$.
8:   Set x_p as the output of Algorithm 5 with accuracy ε_p, number of iteration N_p, prox-function $d_{p-1}(\cdot)$, Ω as Θ_0^2 and X_p as the feasible set.
9:   $d_p(x) \leftarrow d\left(\frac{x-x_p}{R_p}\right)$.
10:   Set $p = p + 1$.
11: **until** $p > \log_2\frac{\mu R_0^2}{2\varepsilon}$.
Output: x_p.

---

divergence $V_{p-1}[z](x)$ corresponding to $d_{p-1}(x)$. Thus, by (8.47), we have

$$
\max_{x,z \in X_p} V_{p-1}[z](x) = \max_{x,z \in X_p} d\left(\frac{x - x_{p-1}}{R_{p-1}}\right) - d\left(\frac{z - x_{p-1}}{R_{p-1}}\right)
$$
$$
- \left\langle \nabla d\left(\frac{z - x_{p-1}}{R_{p-1}}\right), \frac{x - x_{p-1}}{R_{p-1}} - \frac{z - x_{p-1}}{R_{p-1}} \right\rangle
$$
$$
= \max_{x,z \in X_p} V\left[\frac{z - x_{p-1}}{R_{p-1}}\right]\left(\frac{x - x_{p-1}}{R_{p-1}}\right)
$$
$$
\leq \max_{x,z \in X_p} \frac{\Omega \|x - z\|_E^2}{2R_{p-1}^2} \leq \Omega.
$$

Hence, by Theorem 7 with $\sigma_p = \frac{\sigma}{\hat{p}}$, after N_p iterations of Algorithm 5, we have

$$
\mathbb{P}\left\{f(x_p) - f(x_*) \leq \varepsilon_p, \quad g(x_p) \leq \varepsilon_p | A_{p-1}\right\} \geq 1 - \frac{\sigma}{\hat{p}}.
$$

Whence, by Lemma 2,

$$
\mathbb{P}\left\{A_p | A_{p-1}\right\} = \mathbb{P}\left\{\|x_p - x_*\|_E^2 \leq R_p^2 | A_{p-1}\right\} \geq 1 - \frac{\sigma}{\hat{p}},
$$

which finishes the induction proof.

At the same time,

$$
\mathbb{P}\left\{f(x_{\hat{p}}) - f(x_*) > \varepsilon_{\hat{p}} \quad \text{or} \quad g(x_{\hat{p}}) > \varepsilon_{\hat{p}}\right\}
$$
$$
= \mathbb{P}\left\{f(x_{\hat{p}}) - f(x_*) > \varepsilon_{\hat{p}} \quad \text{or} \quad g(x_{\hat{p}}) > \varepsilon_{\hat{p}} \,\middle|\, A_{\hat{p}-1} \cup \bar{A}_{\hat{p}-1}\right\}
$$
$$
= \mathbb{P}\left\{f(x_{\hat{p}}) - f(x_*) > \varepsilon_{\hat{p}} \quad \text{or} \quad g(x_{\hat{p}}) > \varepsilon_{\hat{p}} \,\middle|\, A_{\hat{p}-1}\right\} \mathbb{P}\{A_{\hat{p}-1}\}
$$
$$
+ \mathbb{P}\left\{f(x_{\hat{p}}) - f(x_*) > \varepsilon_{\hat{p}} \quad \text{or} \quad g(x_{\hat{p}}) > \varepsilon_{\hat{p}} \,\middle|\, \bar{A}_{\hat{p}-1}\right\} \mathbb{P}\{\bar{A}_{\hat{p}-1}\}
$$
$$
\leq \frac{\sigma}{\hat{p}} + \mathbb{P}\{\bar{A}_{\hat{p}-1}\} \overset{(*)}{\leq} \frac{\sigma}{\hat{p}} + \mathbb{P}\left\{f(x_{\hat{p}-1}) - f(x_*) > \varepsilon_{\hat{p}-1} \quad \text{or} \quad g(x_{\hat{p}-1}) > \varepsilon_{\hat{p}-1}\right\}
$$
$$
\leq 2 \cdot \frac{\sigma}{\hat{p}} + \mathbb{P}\{\bar{A}_{\hat{p}-2}\} \leq \ldots \leq \frac{\hat{p} - 1}{\hat{p}} \cdot \sigma + \mathbb{P}\{\bar{A}_1\}, \tag{8.48}
$$

where $(*)$ follows from Lemma 2. Using that $\mathbb{P}\{A_1\} = \mathbb{P}\{A_1 | A_0\} \geq 1 - \frac{\sigma}{\hat{p}}$ and, hence, $\mathbb{P}\{\bar{A}_1\} \leq \frac{\sigma}{\hat{p}}$, we obtain

$$
\mathbb{P}\left\{f(x_{\hat{p}}) - f(x_*) \leq \varepsilon, \quad g(x_{\hat{p}}) \leq \varepsilon\right\} \geq 1 - \sigma.
$$

Hence,

$$\mathbb{P}\left\{\|x_{\hat{p}} - x_*\|_E^2 \le \frac{2\varepsilon}{\mu}\right\} \ge 1 - \sigma.$$

Let us now estimate the total number N of inner iterations, i.e., the iterations of Algorithm 5. We have

$$N = \sum_{p=1}^{\hat{p}} N_p \le \sum_{p=1}^{\hat{p}} \left(1 + 70\frac{\Omega \max\{M_f^2, M_g^2\}R_{p-1}^2}{\varepsilon_p^2} \ln\left(\frac{1}{\sigma} \log_2 \frac{\mu R_0^2}{2\varepsilon}\right)\right)$$

$$= \sum_{p=1}^{\hat{p}} \left(1 + 1120\frac{\Omega \max\{M_f^2, M_g^2\}2^p}{\mu^2 R_0^2} \ln\left(\frac{1}{\sigma} \log_2 \frac{\mu R_0^2}{2\varepsilon}\right)\right)$$

$$\le \hat{p} + 2240\frac{\Omega \max\{M_f^2, M_g^2\}2^{\hat{p}}}{\mu^2 R_0^2} \ln\left(\frac{1}{\sigma} \log_2 \frac{\mu R_0^2}{2\varepsilon}\right)$$

$$\le \hat{p} + 2240\frac{\Omega \max\{M_f^2, M_g^2\}}{\mu\varepsilon} \left(\ln\frac{1}{\sigma} + \ln \log_2 \frac{\mu R_0^2}{2\varepsilon}\right).$$

$\square$

8.5 Discussion

We conclude with several remarks concerning possible extensions of the described results.

Obtained results can be easily extended for *composite optimization problems* of the form

$$\min\{f(x) + c(x) : x \in X \subset E, g(x) + c(x) \le 0\}, \tag{8.49}$$

where X is a convex closed subset of finite-dimensional real vector space E, $f : X \to \mathbb{R}, g : E \to \mathbb{R}, c : X \to \mathbb{R}$ are convex functions. Mirror Descent for unconstrained composite problems was proposed in [11], see also [29] for corresponding version of Dual Averaging [22]. To deal with composite problems (8.49), the Mirror Descent step should be changed to

$$x_+ = \text{Mirr}[x](p) = \arg\min_{u \in X} \left\{\langle p, u\rangle + d(u) + c(u) - \langle \nabla d(x), u\rangle\right\} \quad \forall x \in X^0,$$

where X^0 is defined in Sect. 8.2. The counterpart of Lemma 1 is as follows.

Lemma 6 *Let f be some convex function over a convex closed set X, $h > 0$ be a stepsize, $x \in X^0$. Let the point x_+ be defined by $x_+ = \text{Mirr}[x](h \cdot (\nabla f(x) + \Delta))$, where $\Delta \in E^*$. Then, for any $u \in X$,*

$$h \cdot \big(f(x) - f(u) + c(x_+) - c(u) + \langle \Delta, x - u \rangle\big)$$
$$\leq h \cdot \langle \nabla f(x) + \Delta, x - u \rangle - h \cdot \langle \nabla c(x_+), u - x_+ \rangle$$
$$\leq \frac{h^2}{2} \|\nabla f(x) + \Delta\|_{E,*}^2 + V[x](u) - V[x_+](u).$$

We considered restarting Mirror Descent only in the case of strongly convex functions. A possible extension can be in applying the restart technique to the case of uniformly convex functions f and g introduced in [26] and satisfying

$$f(y) \geq f(x) + \langle \nabla f(x), y - x \rangle + \frac{\mu}{2} \|y - x\|_E^\rho, \quad x, y \in X,$$

where $\rho \geq 2$, and the same holds for g. Restarting Dual Averaging [22] to obtain subgradient methods for minimizing such functions without functional constraints, both in deterministic and stochastic setting, was suggested in [13]. Another option is, as it was done in [27] for deterministic unconstrained problems, to use sharpness condition of f and g

$$\mu \left(\min_{x_* \in X_*} \|x - x_*\|_E \right)^\rho \leq f(x) - f_*, \quad \forall x \in X,$$

where f_* is the minimum value of f, X_* is the set of minimizers of f in Problem (8.1), and the same holds for g.

In stochastic setting, motivated by randomization for deterministic problems, we considered only problems with available values of g. As it was done in [14], one can consider more general problems of minimizing an expectation of a function under inequality constraint given by $\mathbb{E}G(x, \eta) \leq 0$, where η is random vector. In this setting one can deal only with stochastic approximation of this inequality constraint.

Acknowledgements The authors are very grateful to Anatoli Juditsky, Arkadi Nemirovski and Yurii Nesterov for fruitful discussions. The research by P. Dvurechensky and A. Gasnikov presented in Section 4 was conducted in IITP RAS and supported by the Russian Science Foundation grant (project 14-50-00150). The research by F. Stonyakin presented in subsection 3.3 was partially supported by the grant of the President of the Russian Federation for young candidates of sciences, project no. MK-176.2017.1.

References

1. A. Anikin, A. Gasnikov, A. Gornov, Randomization and sparsity in huge-scale optimization on an example of mirror descent. Proc. Moscow Inst. Phys. Technol. **8**(1), 11–24 (2016) (in Russian). arXiv:1602.00594

2. A. Bayandina, Adaptive mirror descent for constrained optimization, in *2017 Constructive Nonsmooth Analysis and Related Topics (Dedicated to the Memory of VF Demyanov) (CNSA)*, May 2017, pp. 1–4

3. A. Bayandina, Adaptive stochastic mirror descent for constrained optimization, in *2017 Constructive Nonsmooth Analysis and Related Topics (Dedicated to the Memory of VF Demyanov) (CNSA)*, May 2017, pp. 1–4

4. A. Bayandina, A.G.E. Gasnikova, S. Matsievsky, Primal-dual mirror descent for the stochastic programming problems with functional constraints. Comput. Math. Math. Phys. **58** (2018). arXiv:1604.08194

5. A. Beck, M. Teboulle, Mirror descent and nonlinear projected subgradient methods for convex optimization. Oper. Res. Lett. **31**(3), 167–175 (2003). ISSN: 0167-6377

6. A. Beck, A. Ben-Tal, N. Guttmann-Beck, L. Tetruashvili, The comirror algorithm for solving nonsmooth constrained convex problems. Oper. Res. Lett. **38**(6), 493–498 (2010). ISSN: 0167-6377

7. A. Ben-Tal, A. Nemirovski, *Lectures on Modern Convex Optimization* (Society for Industrial and Applied Mathematics, Philadelphia, 2001)

8. A. Ben-Tal, A. Nemirovski, *Lectures on Modern Convex Optimization (Lecture Notes)*. Personal web-page of A. Nemirovski (2015)

9. A. Birjukov, *Optimization Methods: Optimality Conditions in Extremal Problems*. Moscow Institute of Physics and Technology (2010, in Russian)

10. S. Boucheron, G. Lugosi, P. Massart, *Concentration Inequalities: A Nonasymptotic Theory of Independence* (Oxford University Press, Oxford, 2013)

11. J.C. Duchi, S. Shalev-Shwartz, Y. Singer, A. Tewari, Composite objective mirror descent, in *COLT 2010 the 23rd Conference on Learning Theory* (2010), pp. 14–26. ISBN: 9780982252925

12. A. Juditsky, A. Nemirovski, First order methods for non-smooth convex large-scale optimization, I: general purpose methods, in *Optimization for Machine Learning*, ed. by S.J. Wright, S. Sra, S. Nowozin (MIT Press, Cambridge, 2012), pp. 121–184

13. A. Juditsky, Y. Nesterov, Deterministic and stochastic primal-dual subgradient algorithms for uniformly convex minimization. Stochastic Syst. **4**(1), 44–80 (2014)

14. G. Lan, Z. Zhou, Algorithms for stochastic optimization with expectation constraints (2016). arXiv preprint arXiv:1604.03887

15. A. Nedic, S. Lee, On stochastic subgradient mirror-descent algorithm with weighted averaging. SIAM J. Optim. **24**(1), 84–107 (2014)

16. A. Nemirovski, A. Juditsky, G. Lan, A. Shapiro, Robust stochastic approximation approach to stochastic programming. SIAM J. Optim. **19**(4), 1574–1609 (2009)

17. A. Nemirovskii, Efficient methods for large-scale convex optimization problems. Ekonomika i Matematicheskie Metody **15** (1979, in Russian)

18. A. Nemirovskii, Y. Nesterov, Optimal methods of smooth convex minimization. USSR Comput. Math. Math. Phys. **25**(2), 21–30 (1985). ISSN: 0041-5553

19. A. Nemirovsky, D. Yudin, *Problem Complexity and Method Efficiency in Optimization* (Wiley, New York, 1983)

20. Y. Nesterov, A method of solving a convex programming problem with convergence rate $O(1/k^2)$. Sov. Math. Dokl. **27**(2), 372–376 (1983)

21. Y. Nesterov, *Introductory Lectures on Convex Optimization: A Basic Course* (Kluwer Academic Publishers, Norwell, 2004)

22. Y. Nesterov, Primal-dual subgradient methods for convex problems. Math. Program. **120**(1), 221–259 (2009). ISSN: 1436-4646. First appeared in 2005 as CORE discussion paper 2005/67

23. Y. Nesterov, New primal-dual subgradient methods for convex problems with functional constraints, 2015. http://lear.inrialpes.fr/workshop/osl2015/slides/osl2015_yurii.pdf

24. Y. Nesterov, Subgradient methods for convex functions with nonstandard growth properties, 2016. http://www.mathnet.ru:8080/PresentFiles/16179/growthbm_nesterov.pdf

25. B. Polyak, A general method of solving extremum problems. Sov. Math. Dokl. **8**(3), 593–597 (1967)

26. B. Polyak, Existence theorems and convergence of minimizing sequences in extremum problems with restrictions. Sov. Math. Dokl. **7**, 72–75 (1967)
27. V. Roulet, A. d'Aspremont, Sharpness, restart and acceleration (2017). arXiv preprint arXiv:1702.03828
28. N.Z. Shor, Generalized gradient descent with application to block programming. Kibernetika **3**(3), 53–55 (1967)
29. L. Xiao, Dual averaging methods for regularized stochastic learning and online optimization. J. Mach. Learn. Res. **11**, 2543–2596 (2010). ISSN: 1532-4435

Chapter 9
Frank-Wolfe Style Algorithms for Large Scale Optimization

Lijun Ding and Madeleine Udell

Abstract We introduce a few variants on Frank-Wolfe style algorithms suitable for large scale optimization. We show how to modify the standard Frank-Wolfe algorithm using stochastic gradients, approximate subproblem solutions, and sketched decision variables in order to scale to enormous problems while preserving (up to constants) the optimal convergence rate $\mathcal{O}\left(\frac{1}{k}\right)$.

Keywords Large scale optimization · Frank-Wolfe algorithm · Stochastic gradient · Low memory optimization · Matrix completion

AMS Subject Classifications 90C06, 90C25

9.1 Introduction

This chapter describes variants on Frank-Wolfe style algorithms suitable for large scale optimization. Frank-Wolfe style algorithms enforce constraints by solving a linear optimization problem over the constraint set at each iteration, while competing approaches, such as projected or proximal gradient algorithms, generally require projection onto the constraint set. For important classes of constraints, such as the unit norm ball of the ℓ_1 or nuclear norm, linear optimization over the constraint set is much faster than projection onto the set. This paper provides a gentle introduction to three ideas that can be used to further improve the performance of Frank-Wolfe style algorithms for large scale optimization: stochastic gradients, approximate subproblem solutions, and sketched decision variables. Using these ideas, we show how to modify the standard Frank-Wolfe algorithm in order to scale

L. Ding · M. Udell (✉)
Operations Research and Information Engineering, Cornell University, Ithaca, NY, USA
e-mail: ld446@cornell.edu; udell@cornell.edu

© Springer Nature Switzerland AG 2018
P. Giselsson, A. Rantzer (eds.), *Large-Scale and Distributed Optimization*, Lecture Notes in Mathematics 2227,
https://doi.org/10.1007/978-3-319-97478-1_9

to enormous problems while preserving (up to constants) the optimal convergence rate.

To understand the challenges of huge scale optimization, let us start by recalling the original Frank-Wolfe algorithm. The Frank-Wolfe algorithm is designed to solve problems of the form

$$
\begin{aligned}
\text{minimize } & f(x) \\
\text{subject to } & x \in \Omega,
\end{aligned}
\tag{9.1}
$$

where f is a real valued convex differentiable function from $\mathbf{R}^n$ to $\mathbf{R}$, and the set Ω is a nonempty compact convex set in $\mathbf{R}^n$. Throughout the sequel, we let $x^\star \in \arg\min_{x \in \Omega} f(x)$ be an arbitrary solution to (9.1).

The Frank-Wolfe algorithm is presented as Algorithm 1 below. At each iteration, it computes the gradient of the objective $\nabla f(x)$ at the current iterate x, and finds a feasible point $v \in \Omega$ which maximizes $\nabla f(x)^T v$. The new iterate is taken to be a convex combination of the previous iterate and the point v.

The Frank-Wolfe algorithm can be used for optimization with matrix variables as well. With some abuse of notation, when x, $\nabla f(x)$, and v are matrices rather than vectors, we use the inner product $\nabla f(x)^T v$ to denote the matrix trace inner product $\mathbf{tr}(\nabla f(x)^T v)$.

Linear Optimization Subproblem The main bottleneck in implementing Frank-Wolfe is solving the linear optimization subproblem in Line 5 above:

$$
\begin{aligned}
\text{minimize } & \nabla f(x_{k-1})^T v \\
\text{subject to } & v \in \Omega.
\end{aligned}
\tag{9.2}
$$

Note that the objective of the subproblem (9.2) is linear even though the constraint set Ω may not be. Since Ω is compact, the solution to subproblem (9.2) always exists. Subproblem (9.2) can easily be solved when the feasible region has atomic structure [1]. We give three examples here.

Algorithm 1 Frank-Wolfe algorithm

1: **Input:** Objective function f and feasible region Ω
2: **Input:** A feasible starting point $x_{-1} \in \Omega$
3: **Input:** Stepsize sequence γ_k and tolerance level $\varepsilon > 0$
4: **for** $k = 0, 1, 2, \ldots$ **do**
5:     Compute $v_k = \arg\min_{v \in \Omega} \nabla f(x_{k-1})^T v$.
6:     **if** $(x_{k-1} - v_k)^T \nabla f(x_{k-1}) \le \varepsilon$ **then**
7:         break
8:     **end if**
9:     Update $x_k = (1 - \gamma_k)x_{k-1} + \gamma_k v_k$.
10: **end for**
11: **Output:** The last iteration result x

- *The feasible region is a one norm ball.* For some $\alpha > 0$,

$$\Omega = \{x \in \mathbf{R}^n \mid \|x\|_1 \leq \alpha\}.$$

Let $\{e_i\}_{i=1}^n$ to be the standard basis in $\mathbf{R}^n$, $S = \text{argmax}_i |\nabla f(x_{k-1})^T e_i|$ and $s_i = \text{sign}(\nabla f(x_{k-1})^T e_i)$. The solution v to subproblem (9.2) is any vector in the convex hull of $\{-\alpha s_i e_i \mid i \in S\}$:

$$v \in \mathbf{conv}(\{-\alpha s_i e_i \mid i \in S\}).$$

In practice, we generally choose $v = -\alpha s_i e_i$ for some $i \in S$.

- *The feasible region is a nuclear norm ball.* For some $\alpha > 0$,

$$\Omega = \{X \in \mathbf{R}^{m \times n} \mid \|X\|_* \leq \alpha\},$$

where $\|\cdot\|_*$ is the nuclear norm, i.e., the sum of the singular values. Here v, x_{k-1}, and $\nabla f(x_{k-1})$ are matrices in $\mathbf{R}^{m \times n}$, and we recall that the objective in problem (9.2), $v^T \nabla f(x_{k-1})$, should be understood as the matrix trace inner product $\mathbf{tr}(\nabla f(x_{k-1})^T v)$. Subproblem (9.2) in this case is

$$\begin{aligned} \text{minimize} \quad & \mathbf{tr}(\nabla f(x_{k-1})^T v) \\ \text{subject to} \quad & \|v\|_* \leq \alpha. \end{aligned} \tag{9.3}$$

Denote the singular values of $\nabla f(x_{k-1})$ as $\sigma_1 \geq \ldots, \geq \sigma_{\min(m,n)}$ and the corresponding singular vectors as $(u_1, v_1), \ldots, (u_{\min(m,n)}, v_{\min(m,n)})$. Let $S = \{i \mid \sigma_i = \sigma_1\}$ be the set of indices with maximal singular value. Then the solution to problem (9.2) is the convex hull of the singular vectors with maximal singular value, appropriately scaled:

$$\mathbf{conv}(\{-\alpha u_i v_i^T \mid i \in S\}).$$

In practice, we often take the solution $-\alpha u_1^T v_1$. This solution is easy to compute compared to the full singular value decomposition. Specifically, suppose $\nabla f(x_{k-1})$ is sparse, and let s be the number of non-zero entries in $\nabla f(x_{k-1})$. For any tolerance level $\epsilon > 0$, the number of arithmetic operations required to compute the top singular tuple (u_1, v_1) using the Lanczos algorithm such that $u_1^T \nabla f(x_{k-1}) v_1 \geq \sigma_1 - \epsilon$ is at most $\mathcal{O}\left(s \frac{\log(m+n)\sqrt{\sigma_1}}{\sqrt{\epsilon}}\right)$ with high probability [7].

- *The feasible region is a restriction of a nuclear norm ball.* For some $\alpha > 0$,

$$\Omega = \{X \in \mathbf{R}^{n \times n} \mid \|X\|_* \leq \alpha, \ X \succeq 0\},$$

where $X \succeq 0$ means X is symmetric and positive semidefinite, i.e., every eigenvalue of X is nonnegative. In this case the objective in problem (9.2) $v^T \nabla f(x_{k-1})$ should be understood as $\mathbf{tr}(\nabla f(x_{k-1})^T v)$, where v, x_{k-1}, and

$\nabla f(x_{k-1})$ are matrices in $\mathbf{S}^n$. The subproblem (9.2) in this case is just

$$
\begin{aligned}
\text{minimize } &\ \mathbf{tr}(\nabla f(x_{k-1})^T v) \\
\text{subject to } &\ \|v\|_* \leq \alpha \\
&\ v \succeq 0.
\end{aligned}
\tag{9.4}
$$

Denote the eigenvalues of $\nabla f(x_{k-1})$ as $\lambda_1 \geq \ldots, \geq \lambda_n$ and the corresponding eigenvectors as $v_1, \ldots, v_n$. Let $S = \{i \mid \lambda_i = \lambda_n\}$ be the set of indices with smallest eigenvalue. Then the solution to problem (9.2) is simply 0 if $\lambda_n \geq 0$, while if $\lambda_n \leq 0$, the solution set consists of the convex hull of the eigenvectors with smallest eigenvalue, appropriately scaled:

$$
\mathbf{conv}(\{\alpha v_i v_i^T \mid i \in S\}).
$$

In practice, we generally take $\alpha v_n v_n^T$ as a solution (if $\lambda_n \leq 0$). As in the previous case, this solution is easy to compute compared to the full eigenvalue decomposition. Specifically, suppose $\nabla f(x_{k-1})$ is sparse, and let s be the number of non-zero entries in $\nabla f(x_{k-1})$. For any tolerance level $\epsilon > 0$, the number of arithmetic operations required to compute the eigenvector v_n using the Lanczos algorithm such that $v_1^T \nabla f(x_{k-1}) v_1 \leq \lambda_n + \epsilon$ is at most $\mathcal{O}\left(s \frac{\log(2n)\sqrt{\max(|\lambda_1|,|\lambda_n|)}}{\sqrt{\epsilon}}\right)$ with high probability [3, Lemma 2].

Thus, after k iterations of the Frank-Wolfe algorithm, the sparsity or rank of the iterate x_k in the above three examples is bounded by k. This property has been noted and exploited by many authors [2, 3, 5].

The stopping criterion in Line 6 of Algorithm 1 bounds the suboptimality $f(x_{k-1}) - f(x^\star)$, where $x^\star \in \arg\min_{x \in \Omega} f(x)$. Indeed,

$$
\begin{aligned}
f(x_{k-1}) - f(x^\star) &\leq (x_{k-1} - x^\star)^T \nabla f(x_{k-1}) \\
&\leq (x_{k-1} - v_k)^T \nabla f(x_{k-1}),
\end{aligned}
$$

where the first inequality is due to convexity and the second line is due to optimality of v_k.

Matrix Completion To illustrate our previous points, let's consider the example of matrix completion. Keep this example in mind: we will return to this problem again in the coming sections to illustrate our methods.

We consider the optimization problem

$$
\begin{aligned}
\text{minimize } &\ f(\mathcal{A}X) \\
\text{subject to } &\ \|X\|_* \leq \alpha, \\
&\ X \in \mathcal{S}
\end{aligned}
\tag{9.5}
$$

with variable $X \in \mathbf{R}^{m \times n}$. Here $\mathcal{A} : \mathbf{R}^{m \times n} \to \mathbf{R}^d$ is a linear map and $\alpha > 0$ is a positive constant. The set $\mathcal{S}$ represents some additional information of the underlying problem. In this book chapter, the set $\mathcal{S}$ will be either $\mathbf{R}^{m \times n}$ or $\{X \in \mathbf{R}^{n \times n} \mid X \succeq 0\}$. In the first case, the feasible region of problem (9.5) is just the nuclear norm ball. In the second case, the feasible region is a restriction of the nuclear norm. In either case, the linear optimization subproblem can be solved efficiently as we just mentioned. The function $f : \mathbf{R}^d \to \mathbf{R}$ is a loss function that penalizes the misfit between the predictions $\mathcal{A}X$ of our model and our observations from the matrix.

For example, suppose we observe matrix entries c_{ij} with indices i, j in $\mathcal{O} \subset \{1, \ldots, m\} \times \{1, \ldots, n\}$ from a matrix $X^0 \in \mathbf{R}^{m \times n}$ corrupted by Gaussian noise E:

$$c_{ij} = (X^0)_{ij} + E_{ij}, \qquad E_{ij} \overset{iid}{\sim} N(0, \sigma^2)$$

for some $\sigma > 0$. A maximum likelihood formulation of problem (9.5) to recover X^0 would be

$$\begin{aligned} \text{minimize } & \sum_{(i,j) \in \mathcal{O}} (x_{ij} - c_{ij})^2 \\ \text{subject to } & \|X\|_* \leq \alpha. \end{aligned} \tag{9.6}$$

To rewrite this problem in the form of (9.5), we choose $\mathcal{A}$ so that $(\mathcal{A}X)_{ij} = x_{ij}$ for $(i, j) \in \mathcal{O}$, so the number of observations d is the cardinality of $\mathcal{O}$. Since there is no additional information of X^0, we set $\mathcal{S} = \mathbf{R}^{n \times m}$. The objective f is a sum of quadratic losses in this case.

Since the constraint region Ω is a nuclear norm ball when $\mathcal{S} = \mathbf{R}^{m \times n}$, we can apply Frank-Wolfe to this optimization problem. The resulting algorithm is shown as Algorithm 2. Here Line 5 computes the singular vectors with largest singular value, and Line 9 exploits the fact that at each iteration we can choose a rank one update.

Algorithm 2 Frank-Wolfe algorithm applied to matrix completion with nuclear ball constraint only

---

1: **Input:** Objective function f and $\alpha > 0$
2: **Input:** A feasible starting point $\|X_{-1}\| \leq \alpha$
3: **Input:** Stepsize sequence γ_k
4: **for** $k = 0, 1, 2, \ldots, K$ **do**
5:     Compute (u_k, v_k), the top singular vectors of $\nabla f(\mathcal{A}X_k)$.
6:     **if tr**$((X_{k-1} + \alpha u_k v_k^T)^T \nabla f(X_{k-1})) \leq \varepsilon$ **then**
7:         break the for loop.
8:     **end if**
9:     Update $X_k = (1 - \gamma_k)X_{k-1} - \gamma_k \alpha u_k v_k^T$.
10: **end for**
11: **Output:** The last iteration result X_K

However, there are three main challenges in the large scale setting that can pose difficulties in applying the Frank-Wolfe algorithm:

1. Solving the linear optimization subproblem (9.2) exactly,
2. computing the gradient ∇f, and
3. storing the decision variable x.

To understand why each of these steps might present a difficulty, consider again the matrix completion case with nuclear ball constraint only.

1. Due to Galois theory, it is not possible to exactly compute the top singular vector, even in exact arithmetic. Instead, we rely on iterative methods such as the QR algorithm with shifts, or the Lanczos method, which terminate with some approximation error. What error can we allow in an approximate solution of the linear optimization subproblem (9.2)? How will this error affect the performance of the Frank-Wolfe algorithm?
2. In many machine learning and statistics problems, the objective $f(X) = \sum_{i=1}^{d} f_i(X)$ is a sum over d observations, and each f_i measures the error in observation i. As we collect more data, computing ∇f exactly becomes more difficult, but approximating ∇f is generally easy. Can we use an approximate version of ∇f instead of the exact gradient?
3. Storing X, which requires $m \times n$ space in general, can be costly if n and m are large. One way to avoid using $\mathcal{O}(mn)$ memory is to store each updates (u_k, v_k). But this approach still uses $\mathcal{O}(mn)$ memory when the number of iterations $K \geq \min(m, n)$. Can we exploit structure in the solution $X^\star$ to reduce the memory requirements?

We provide a crucial missing piece to address the first challenge, and a gentle introduction to the ideas needed to tackle the second and third challenges. Specifically, we will show the following.

1. Frank-Wolfe type algorithms still converge when we use an approximate oracle to solve the linear optimization subproblem (9.2). In fact, the convergence rate is preserved up to a multiplicative user-specified constant.
2. Frank-Wolfe type algorithms still converge when the gradient is replaced by an approximate gradient, and the convergence rate is preserved in expectation.
3. Frank-Wolfe type algorithms are amenable to a matrix sketching procedure which can be used to reduce memory requirements, and the convergence rate is not affected.

Based on these ideas, we propose two new Frank-Wolfe Style algorithms which we call SVRF with approximate oracle (SVRF, pronounced as "tilde SVRF"), and Sketched SVRF (SSVRF). They can easily scale to extremely large problems.

The rest of this chapter describes how SSVRF addresses the three challenges listed above. To address the first challenge, we augment the Frank-Wolfe algorithm with an approximate oracle for the linear optimization subproblem, and prove a convergence rate in this setting. Numerical experiments confirm that using an approximate oracle reduces the time necessary to achieve a given error tolerance. To address the second challenge, we then present a Stochastic Variance Reduced Frank-Wolfe (SVRF) algorithm with approximate oracle, $\widetilde{\text{SVRF}}$. Finally, we show how to use the matrix sketching procedure of [10] to reduce the memory requirements of the algorithm. We call the resulting algorithm $\widetilde{\text{SSVRF}}$.

Notation We use $\| \cdot \|$ to denote the Euclidean norm when the norm is applied to a vector, and to denote the operator norm (maximum singular value) when applied to a matrix. We use $\| \cdot \|_F$ to denote the Frobenius norm and $\| \cdot \|_*$ to denote the nuclear norm (sum of singular values). The transpose of a matrix A and a vector v is denoted as A^T and v^T. The trace of a matrix $A \in \mathbf{R}^{n \times n}$ is the sum of all its diagonals, i.e., $\mathbf{tr}(A) = \sum_{i=1}^{n} A_{ii}$. The set of symmetric matrices in $\mathbf{R}^{n \times n}$ is denoted as $\mathbf{S}^n$. We use $X \succeq 0$ to mean that X is symmetric and positive semidefinite (psd). A convex function $f : \mathbf{R}^n \to \mathbf{R}$ is L-smooth if $\|\nabla f(x) - \nabla f(y)\| \leq L\|x - y\|$ for some finite $L \geq 0$. The diameter D of a set $\Omega \subset \mathbf{R}^n$ is defined as $D = \sup_{x,y \in \Omega} \|x - y\|$. For an arbitrary matrix Z, we define $[Z]_r$ to be the best rank r approximation of Z in Frobenius norm. For a linear operator $\mathcal{A} : \mathbf{R}^{m \times n} \to \mathbf{R}^l$, where $\mathbf{R}^{m \times n}$ and $\mathbf{R}^n$ are equipped with the trace inner product and the Euclidean inner product, the adjoint of $\mathcal{A}$ is denotes as $\mathcal{A}^* : \mathbf{R}^l \to \mathbf{R}^{m \times n}$.

9.2 Frank-Wolfe with Approximate Oracle

In this section, we address the first challenge: the linear optimization subproblem (9.2) can only be solved approximately. Most of the ideas in this section are drawn from [5]; we include this introduction for the sake of completeness.

We will show that the Frank-Wolfe algorithm with approximate subproblem oracle converges at the same rate as the one with exact subproblem oracle up to a user-specified multiplicative constant.

9.2.1 Algorithm and Convergence

As before, we seek to solve problem (9.1),

$$\text{minimize } f(x)$$
$$\text{subject to } x \in \Omega.$$

Let us introduce Algorithm 3, which we call Frank-Wolfe with approximate oracle. The only difference from the original Frank-Wolfe algorithm is the tolerance $\epsilon_k > 0$: in Line 5, we compute an approximate solution with tolerance ϵ_k rather than an exact solution.

Algorithm 3 Frank-Wolfe with approximate oracle

1: **Input:** Objective function f and feasible region Ω
2: **Input:** A feasible starting point $x_{-1} \in \Omega$
3: **Input:** Stepsize sequence γ_k and tolerance level ε and error sequence $\epsilon_k > 0$
4: **for** $k = 0, 1, 2, \ldots$ **do**
5:     Compute v_k such that $\nabla f(x_{k-1})^T v_k \leq \min_{v \in \Omega} \nabla f(x_{k-1})^T v + \epsilon_k$.
6:     **if** $(x_{k-1} - v_k)^T \nabla f(x_{k-1}) \leq \varepsilon$ **then**
7:         break
8:     **end if**
9:     Update $x_k = (1 - \gamma_k)x_{k-1} + \gamma_k v_k$.
10: **end for**
11: **Output:** The last iteration result x_k

There are a few variants on this algorithm that use different line search methods. The next iterate might be the point on the line determined by x_{k-1} and v_k with lowest objective value, or the point with best objective value on the polytope with vertices $x_{-1}, v_1, \ldots, v_k$. These variants may reduce the total number of iterations at the cost of an increased per-iteration complexity. When memory is plentiful and line search or polytope search is easy to implement, these techniques can be employed; otherwise, a predetermined stepsize rule, i.e., γ_k is determined as an input, e.g., $\gamma_k = \frac{2}{k+2}$ or γ_k is a constant, might be preferred. All these techniques enjoy the same complexity bounds as Algorithm 3 since within an iteration, starting from the same iterate x_k, the objective is guaranteed to decrease at least as much under each of these line search rules as using the predetermined stepsize rule in Algorithm 3.

The following theorem gives a guarantee on the primal convergence of the objective value when f is L-smooth.

Theorem 1 *Given an arbitrary $\delta > 0$, if f is L-smooth, Ω has diameter D, $\gamma_k = \frac{2}{k+2}$ and $\epsilon_k = \frac{LD^2}{2}\gamma_k\delta$, then the iterates x_k of Algorithm 3 satisfy*

$$f(x_k) - f(x^\star) \leq \frac{2LD^2}{k+2}(1 + \delta) \tag{9.7}$$

where $x^\star \in \arg\min_{x \in \Omega} f(x)$.

To start, recall an equivalent definition of L-smoothness [9, Theorem 2.1.5]. For completeness, we provide a short proof in the Appendix.

Proposition 1 *If the real valued differentiable convex function f with domain $\mathbf{R}^n$ is L-smooth, then for all $x, y \in \mathbf{R}^n$,*

$$f(x) \leq f(y) + \nabla f(y)^T(x - y) + \frac{L}{2}\|x - y\|^2.$$

Proof (Proof of Theorem 1)

Let $v_k^\star \in \arg\min_{v \in \Omega} f(x_{k-1})^T v$ in Line 5. Using the update equation $x_k = x_{k-1} + \gamma_k(v_k - x_{k-1})$, we have

$$
\begin{aligned}
f(x_k) - f(x^\star) &\leq f(x_{k-1}) - f(x^\star) + \nabla f(x_{k-1})^T (v_k - x_{k-1})\gamma_k \\
&\quad + \frac{L}{2}\gamma_k^2 \|v_k - x_{k-1}\|^2 \\
&\leq f(x_{k-1}) - f(x^\star) + \nabla f(x_{k-1})^T (v_k - x_{k-1})\gamma_k + \frac{LD^2}{2}\gamma_k^2 \\
&\leq f(x_{k-1}) - f(x^\star) + \nabla f(x_{k-1})^T (v_k^\star - x_{k-1})\gamma_k \\
&\quad + \frac{LD^2}{2}\gamma_k^2(1+\delta) \\
&\leq f(x_{k-1}) - f(x^\star) + \nabla f(x_{k-1})^T (x^\star - x_{k-1})\gamma_k \\
&\quad + \frac{LD^2}{2}\gamma_k^2(1+\delta) \\
&\leq (1-\gamma_k)(f(x_{k-1}) - f(x^\star)) + \frac{LD^2}{2}\gamma_k^2(1+\delta). \quad (9.8)
\end{aligned}
$$

The first inequality is due to Proposition 1. The second inequality uses the diameter D of Ω and the fact that x_k is feasible since $\gamma_k \in (0,1)$ and x_k is a convex combination of points in the convex set Ω. The third inequality uses the bound on the suboptimality of v_k in Line 5. The fourth inequality uses the optimality of $v_k^\star$ for $\min_{v \in \Omega} v^T \nabla f(x_{k-1})$ and fifth uses convexity of f. The conclusion of the above chain of inequalities is

$$
f(x_k) - f(x^\star) \leq (1-\gamma_k)(f(x_{k-1}) - f(x^\star)) + \frac{LD^2}{2}\gamma_k^2(1+\delta). \quad (9.9)
$$

Now we prove inequality (9.7) by induction. The base case $k = 0$ follows from (9.8) since $\gamma_0 = 1$. Now suppose inequality (9.7) is true for $k \leq s$. Then for $k = s+1$,

$$
\begin{aligned}
f(x_{s+1}) - f(x^\star) &\leq \left(1 - \frac{2}{s+2+1}\right)(f(x_s) - f(x^\star)) + \frac{LD^2}{2}\left(\frac{2}{s+2+1}\right)^2(1+\delta) \\
&= \frac{s+1}{s+2+1}(f(x_s) - f(x^\star)) + \frac{LD^2}{2}\left(\frac{2}{s+2+1}\right)^2(1+\delta) \\
&\leq \left(\frac{s+1}{s+2+1}\frac{2}{s+2} + \frac{2}{(s+2+1)^2}\right)LD^2(1+\delta) \\
&= \left(\frac{2s+2}{s+2} + \frac{2}{s+2+1}\right)\frac{LD^2}{s+2+1}(1+\delta) \\
&\leq \left(\frac{2s+2+2}{s+2}\right)\frac{LD^2}{s+2+1}(1+\delta) \\
&= \frac{2LD^2}{s+1+2}(1+\delta).
\end{aligned} \quad (9.10)
$$

We use (9.9) in the first inequality and the induction hypothesis in the second inequality to bound the term $f(x_s) - f(x^\star)$. The last line completes the induction. $\qquad\square$

9.2.2 Numerics

In this subsection, we demonstrate that Frank-Wolfe is robust to using an approximate oracle through numerical experiments.

The specific problem we will use as our case study is the following symmetric matrix completion problem which is a special case of problem (9.5). The symmetric matrix completion problem seeks to recover an underlying matrix $X^0 \succeq 0$ from a few noisy entries of X^0. Specifically, let $C = X^0 + E$ be a matrix of noisy observations of X^0, where E is a symmetric noise matrix. For each $i \geq j$, we observe C_{ij} independently with probability p. The quantity p is called the sample rate.

Let $\mathcal{O}$ be the set of observed entries and m be the number of entries observed. Note that if $(i, j) \in \mathcal{O}$, $(j, i) \in \mathcal{O}$ as well since our matrices are all symmetric.

The optimization problem we solve to recover X^0 is

$$
\begin{aligned}
\text{minimize } & f(X) := \tfrac{1}{2}\|P_{\mathcal{O}}(X) - P_{\mathcal{O}}(C)\|_F^2 \\
\text{subject to } & \|X\|_* \leq \alpha, \\
& X \succeq 0.
\end{aligned} \tag{9.11}
$$

Here the projection operator $P_{\mathcal{O}} : \mathbf{S}^n \to \mathbf{R}^m$ is

$$
[P_{\mathcal{O}}(Y)]_{ij} = \begin{cases} Y_{ij}, & \text{if } (i, j) \in \mathcal{O} \\ 0, & \text{if } (i, j) \notin \mathcal{O}. \end{cases}
$$

for any $Y \in \mathbf{S}^n$. By letting $\mathcal{A} = P_{\mathcal{O}}$, the set $\mathcal{S} = \{X \in \mathbf{R}^{n \times n} \mid X \succeq 0\}$ and $f(\cdot) = \|\cdot\|_F^2$, we see it is indeed a special case of problem (9.5).

The gradient at X_k is $\nabla f(X_k) = P_{\mathcal{O}}(X_k) - P_{\mathcal{O}}(C)$. As we discussed in the introduction, a solution to the linear optimization subproblem is

$$
V_k = \begin{cases} \alpha v_n v_n^T, & \text{if } \lambda_n(\nabla f(X_{k-1}) \leq 0 \\ 0, & \text{if } \lambda_n(\nabla f(X_{k-1})) > 0 \end{cases}
$$

where $\lambda_n(\nabla f(X_{k-1}))$ is the smallest eigenvalue of $\nabla f(X_{k-1})$.

When the sample rate $p < 1$ is fixed, i.e., independent of dimension n, the probability we observe all entries on the diagonal of C is very small. Hence the matrix $\nabla f(X_{k-1})$ is very unlikely to be positive definite, for any k. (Recall that a positive definite matrix has positive diagonal.) Let us suppose that at least one entry on the diagonal is not observed, so that $\lambda_n(\nabla f(X_{k-1})) \leq 0$ for every k. Thus Line 5 of Algorithm 3 reduces to finding an approximate eigenvector v such that

$$
\alpha v^T \nabla f(X_{k-1})v \leq \alpha \lambda_n(\nabla f(X_{k-1})) + \epsilon_k. \tag{9.12}
$$

However, the solver ARPACK [8], which is the default solver for iterative eigenvalue problems in a variety of languages (e.g., `eigs` in MATLAB), does not support specifying the approximation error in the form of (9.12). Instead, for a given tolerance ξ_k, it finds an approximate vector $v \in \mathbf{R}^n$ with unit two norm, i.e., $\|v\| = 1$, and an approximate eigenvalue $\lambda \in \mathbf{R}$, such that

$$\|\nabla f(X_{k-1})v - \lambda v\| \leq \xi_k \|\nabla f(X_{k-1})\|.$$

For simplicity, we assume that λ returned by our eigenvalue solver is the true smallest eigenvalue $\lambda_n(\nabla f(X_{k-1}))$, for any tolerance ξ_k. We will justify this assumption later through numerical experiments. In this case, the error ϵ_k is upper bounded by

$$\xi_k \alpha \|\nabla f(X_{k-1})\| \geq \epsilon_k. \tag{9.13}$$

This upper bound turns out to be very conservative for large ξ_k: $\xi_k \alpha \|\nabla f(X_{k-1})\|$ might be much larger than the actual error $\epsilon_k = \alpha v^T \nabla f(X_{k-1})v - \alpha \lambda_n(\nabla f(X_{k-1}))$, as we will see later.

In the experiments, we set the dimension $n = 1000$ and generated $X^0 = WW^T$, where $W \in \mathbf{R}^{n \times r}$ had independent standard normal distributed entries. We then added symmetric noise $E = \frac{1}{10} \times (L + L^T)$ to X^0 to get $C = X^0 + E$, where $L \in \mathbf{R}^{n \times n}$ had independent standard normal entries. We then sampled uniformly from the upper triangular part of C (including the diagonal) with probability $p = 0.8$.

In each experiment we solved problem (9.11) with $\alpha = \|X^0\|_*$. In real applications, one usually does not know $\|X^0\|_*$ in advance. In that case, one might solve problem (9.11) multiple times with different values of α and select the best α according to some criterion.

We ran nine experiments in total. In each experiment, we chose a rank r of X^0 in $\{10, 50, 100\}$ and ran Frank-Wolfe with approximate oracle with constant tolerance $\xi_k \in \{10^{-15}, 10^{-5}, 1\}$ using the step size rule $\gamma_k = \frac{2}{k+2}$, as required for Theorem 1, and terminated each experiment after 30 s; the qualitative performance of the algorithm is similar even after many more iterations. We emphasize that within an experiment, the tolerance ξ_k was the same for each iteration k. See the discussion above Fig. 9.4 for more details about the choice of ξ_k.

Figure 9.1 shows experimental results on the relationship between the relative objective $\frac{\|P_{\mathcal{O}}(X_k) - P_{\mathcal{O}}(C)\|_F^2}{\|P_{\mathcal{O}}(C)\|_F^2}$ (on a log scale) and the actual clock time under different combinations of rank r and tolerance ξ_k. For a fixed rank, the relative objective $\frac{\|P_{\mathcal{O}}(X_k) - P_{\mathcal{O}}(C)\|_F^2}{\|P_{\mathcal{O}}(C)\|_F^2}$ evolves similarly for any tolerance. When the underlying matrix has relatively high rank, using a lower tolerance allows faster convergence, at least for the moderate final relative objective achieved in these experiments. The per iteration cost is summarized in Table 9.1. In fact, these plots show no advantage to using a tighter tolerance in any setting.

One surprising feature of these graphs is the oscillation of relative error that occurs for the model with $r = 10$ once the relative error has reached 10^{-2} or so. This oscillation as the algorithm approaches the optimum is due to the stepsize rule

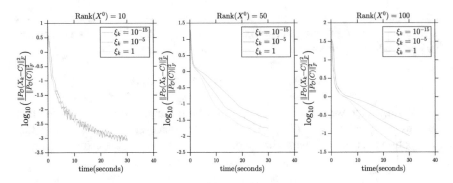

Fig. 9.1 The above plots demonstrate the relation between the relative objective value $\log\left(\frac{\|P_\mathcal{O}(X_k)-P_\mathcal{C}(C)\|_F^2}{\|P_\mathcal{O}(C)\|_F^2}\right)$ and the clock time for different combinations of rank $r = \text{Rank}(X^0)$ and tolerance parameters ξ_k

Table 9.1 Average per iteration time (seconds) of Algorithm 3 for problem (9.11)

	Rank$(X^0) = 10$	Rank$(X^0) = 50$	Rank$(X^0) = 100$
$\xi_k = 10^{-15}$	0.1136	0.1923	0.2400
$\xi_k = 10^{-5}$	0.0997	0.1376	0.1840
$\xi_k = 1$	0.1017	0.1099	0.1220

$\gamma_k = \frac{2}{k+2}$. To see how this stepsize leads to oscillation, suppose for simplicity that for some iterate k_0, $X_{k_0-1} = X^0$. We expect this iterate to have a very low objective value; indeed, in our experiments we found that the relative objective at X^0 is around 5×10^{-4} when $r = 10$. Then in the next iteration, we add V_{k_0} to X_{k_0-1} with step size $\frac{2}{k_0+2}$. Hence X_{k_0} is at least $\frac{2}{k_0+2}\alpha$ away from the true solution. This very likely will increase the relative objective since our p is 0.8. Suppose further that $V_{k_0+1} = -V_{k_0}$. Then we almost return to X^0 in the next iteration and again enjoy a small relative objective. For higher rank X^*, the oscillation begins at later iterations (not shown), as the algorithm approaches the solution.

Using line search eliminates the oscillation, but increases computation time for this problem. We do not consider linesearch further in this paper.

Our goal in this problem is not simply to find the solution of problem (9.11) but to produce a matrix X close to X^0. Hence we also study the numerical convergence of the relative error $\|X - X^0\|_F^2/\|X^0\|_F^2$. Figure 9.2 shows experimental results on the relationship between the relative error $\frac{\|X_k-X^0\|_F^2}{\|X^0\|_F^2}$ (on a log scale) and the actual clock time under different combinations of rank r and tolerance ξ_k. The evolution of $\frac{\|X_k-X^0\|_F^2}{\|X^0\|_F^2}$ is very similar to the evolution of $\frac{\|P_\mathcal{O}(X_k)-P_\mathcal{O}(C)\|_F^2}{\|P_\mathcal{O}(C)\|_F^2}$ in Fig. 9.1.

The assumption that the approximate eigenvalue λ returned by the eigenvalue solver is approximately equal to the true smallest eigenvalue λ_n (Eq. (9.12)) is

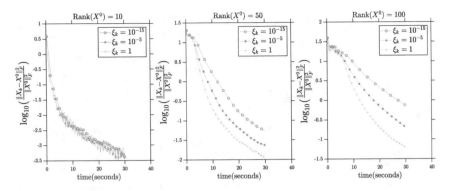

Fig. 9.2 The above plots demonstrate the relation between the relative distance to the solution $\log\left(\frac{\|X_k-X^0\|_F^2}{\|X^0\|_F^2}\right)$ and the clock time for different combinations of rank $r = \text{Rank}(X^0)$ and tolerance parameters ξ_k. We plot a marker on the line once every ten iterations (in this figure only)

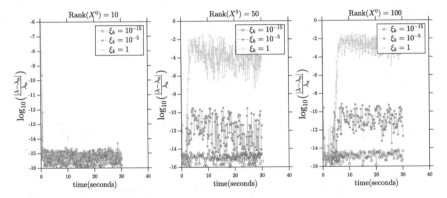

Fig. 9.3 The vertical axis is the relative difference between approximate eigenvalue λ with $\xi_k = \{10^{-15}, 10^{-5}, 1\}$ and the very accurate eigenvalue λ_n of $\lambda_n(\nabla f(X_{k-1}))$, computed with tolerance ξ_k equal to machine precision 10^{-16}

supported by Fig. 9.3. We computed the true eigenvalue λ_n by calling ARPACK with a very tight tolerance. It is interesting that for a low rank model, the estimate λ is very accurate even if ξ_k is large. The relative error in λ is about 10^{-2} on average when $\xi_k = 1$ for high rank models. However, this is not too large: the relative error in our iterate $\frac{\|X_k-X^0\|_F^2}{\|X^0\|_F^2}$ is also about 10^{-2}, hence these two errors are on the same scale.

Figure 9.4 shows the error $\epsilon_k = v^T \nabla f(X_{k-1})v - \alpha\lambda_n(\nabla f(X_{k-1}))$ achieved by our linear optimization subproblem solver. It can be seen that for a constant tolerance ξ_k, the error ϵ_k is also almost constant after some initial transient behavior. Hence controlling ξ_k indeed controls ϵ_k.

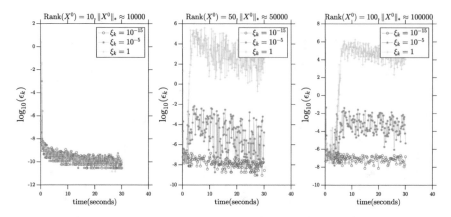

Fig. 9.4 The actual evolution of the error $\epsilon_k = \alpha v^T \nabla f(X_{k-1}) v - \alpha \lambda_n(\nabla f(X_{k-1}))$

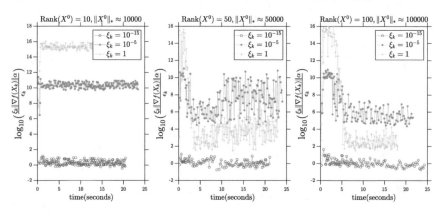

Fig. 9.5 Empirical evolution of the ratio $\frac{\xi_k \alpha \| \nabla f(X_{k-1}) \|}{\epsilon_k}$. Recall that $\xi_k \alpha \| \nabla f(X_{k-1}) \|$ is an upper bound of ϵ_k

Since our ϵ_k is approximately constant due to constant choice of ξ_k, rather than decreasing as required by the assumptions of Theorem 1, one might wonder whether the conclusion of Theorem 1 still holds. The answer is yes. In fact, we found that at each iteration throughout our numerical experiments, the inequality $\epsilon_k = \alpha v^T \nabla f(X_{k-1}) v - \alpha \lambda_n(\nabla f(X_{k-1})) \leq \gamma_k L D^2 \delta$ is satisfied, with $\delta = 1$ and $D = 2 \| X^0 \|_*$. Thus the conclusion of Theorem 1 is still satisfied in our numerical results, although we do not have explicit control over the error ϵ_k. We examine the accuracy of our bound $\xi_k \alpha \| \nabla f(X_{k-1}) \|$ on ϵ_k in Fig. 9.5. It shows that our bound is rather conservative for higher value of ξ_k.

9.3 Stochastic Variance Reduced Frank-Wolfe (SVRF) Algorithm with Approximate Oracle ($\widetilde{SVRF}$)

Having seen that Frank-Wolfe is robust to using an approximate oracle when solving linear optimization subproblem (9.2), we now turn to our second challenge: computing the gradient ∇f.

To formalize the challenge, we will consider the optimization problem

$$\begin{aligned} \text{minimize } & f(x) := \frac{1}{n} \sum_{i=1}^{n} f_i(x) \\ \text{subject to } & x \in \Omega, \end{aligned} \qquad (9.14)$$

where $x \in \mathbf{R}^m$. For each $i = 1, \ldots, n$, f_i is a convex continuously differentiable real valued function and Ω is a compact convex set in $\mathbf{R}^m$. This is a particular instance of problem (9.1).

Problem (9.14) is common in statistics and machine learning, where each f_i measures the error in observation i. Computing the gradient ∇f in this setting is a challenge, since the number of observations n can be enormous.

One way to address this challenge is to compute an approximation to the gradient rather than the exact gradient. We sample l elements $i_1, \ldots, i_l$ from the set $\{1, \ldots, n\}$ with replacement and compute the *stochastic gradient*

$$\tilde{\nabla} f(x) = \frac{1}{l} \sum_{j=1}^{l} \nabla f_{i_j}(x).$$

The parameter l is called the size of the minibatch $\{i_1, \ldots, i_l\}$. The computational benefit here is that we compute only $l \ll n$ derivatives. Intuitively, we expect this method to work since $\mathbf{E}[\tilde{\nabla} f(x)] = \nabla f(x)$.

Of course, the computational benefit does not come for free. This approach suffers one major drawback:

- the stochastic gradient $\tilde{\nabla} f(x)$ may have very large variance $\mathbf{var}(\|\tilde{\nabla} f(x)\|_2)$ even if x is near x^*. Large variance will destabilize any algorithm using $\tilde{\nabla} f(x)$, since even near the solution where $\|\nabla f(x)\|$ is small, $\|\tilde{\nabla} f(x)\|$ may be large.

One simple way to ensure that $\tilde{\nabla} f(x)$ concentrates near $\nabla f(x)$ is to increase the minibatch size l as $\mathbf{var}(\tilde{\nabla} f(x)) = \frac{1}{l} \mathbf{var}(\nabla f_i(x))$, where i is chosen uniformly from $\{1, \ldots, n\}$. But using a very large minibatch size l defeats the purpose of using a stochastic gradient.

Variance reduction techniques endeavor to avoid this tradeoff [6]. Instead of using a large minibatch at each iteration, they occasionally compute a full gradient and use it to reduce the variance of $\tilde{\nabla} f(x)$. The modified stochastic gradient is called the variance-reduced stochastic gradient. Johnson and Zhang [6] introduced one way to perform variance reduction. Specifically, they define a variance-reduced stochastic gradient at a point $x \in \Omega$ with respect to some *snapshot* $x_0 \in \Omega$ as

$$\tilde{\nabla} f(x; x_0) = \nabla f_i(x) - (\nabla f_i(x_0) - \nabla f(x_0)),$$

where i is sampled uniformly from $\{1, \ldots, n\}$. Notice we require the full gradient $\nabla f(x_0)$ at the snapshot, but only the gradient of the ith function $\nabla f_i(x)$ at the point x. In this case, we still have $\mathbf{E} \tilde{\nabla} f(x; x_0) = \nabla f(x)$, and the variance is

$$\mathbf{var}(\|\tilde{\nabla} f(x; x_0)\|_2) = \frac{1}{n} \sum_{i=1}^{n} \|\nabla f_i(x) - \nabla f_i(x_0) + (\nabla f(x_0) - \nabla f(x))\|_2^2.$$

If x and x_0 are near $x^\star$, the variance will be near zero and so indeed the variance is reduced. We can further reduce the variance using a minibatch by independently sampling l variance-reduced gradients $\tilde{\nabla} f(x; x_0)$ and taking their average.

Hazan and Luo [4, Theorem 1] introduced the stochastic variance reduced Frank-Wolfe (SVRF) algorithm, which augments the Frank-Wolfe algorithm with the variance reduction technique of Johnson and Zhang, and showed that it converges in expectation when an exact oracle is used for the linear optimization subproblem (9.2). As we will see in Theorem 2, the number of evaluation of full gradient and stochastic gradient is also considerably small.

As we saw in the previous section, Frank-Wolfe with an approximate oracle converges at the same rate as the one using an exact oracle for the linear optimization subproblem (9.2). One naturally wonders whether an approximate oracle is allowed when we use stochastic gradients. We will show below that the resulting algorithm, which we call SVRF with approximate oracle ($\widetilde{\text{SVRF}}$, pronounced as "tilde SVRF") and present as Algorithm 4, indeed works well. Note that when $\epsilon_k = 0$ for each k, Algorithm 4 reduces to SVRF.

Algorithm 4 SVRF with approximate oracle ($\widetilde{\text{SVRF}}$)

1: **Input:** Objective function $f = \frac{1}{n} \sum_{i=1}^{n} f_i$
2: **Input:** A feasible starting point $w_{-1} \in \Omega$
3: **Input:** Stepsize γ_k, minibatch size m_k, epoch length N_t and tolerance sequence ϵ_k
4: **Initialize:** Find x_0 s.t. $\nabla f(w_{-1})^T x_0 \leq \min_{x \in \Omega} \nabla f(w_{-1})^T x + \epsilon_0$.
5: **for** $t = 1, 2, \ldots, T$ **do**
6:     Take a snapshot $w_0 = x_{t-1}$ and compute gradient $\nabla f(w_0)$.
7:     **for** $k = 1$ to N_t **do**
8:         Compute g_k, the average of m_k iid samples of $\tilde{\nabla} f(w_{k-1}, w_0)$.
9:         Compute v_k s.t. $g_k^T v_k \leq \min_{v \in \Omega} g_k^T v + \epsilon_k$.
10:         Update $w_k := (1 - \gamma_k) w_{k-1} + \gamma_k v_k$.
11:     **end for**
12:     Set $x_t = w_{N_t}$.
13: **end for**
14: **Output:** The last iteration result x_T.

Algorithm 5 Stable $\widetilde{SVRF}$, k increasing in Line 7 of Algorithm 4

1: ...as Algorithm 4, except replacing the chunk from Line 7 to Line 12 with the following chunk
   and start k at $k = 1$ when $t = 1$.
2: **while** $k \leq N_t$ **do**
3:     Compute g_k, the average of m_k iid samples of $\tilde{\nabla} f(w_{k-1}, w_0)$.
4:     Compute v_k s.t. $g_k^T v_k \leq \min_{v \in \Omega} g_k^T v + \epsilon_k$.
5:     Update $w_k := (1 - \gamma_k)w_{k-1} + \gamma_k v_k$ and $k = k + 1$.
6: **end while**
7: Set $x_t = w_{N_t}$.

We give a quantitative description of the objective value convergence $f(x_k) - f(x^\star)$ for Algorithm 4 in Theorem 2. Moreover, we show that the convergence rate is the same as the one using the exact subproblem oracle up to a multiplicative user-specified constant.

In Algorithm 4, each time we take a snapshot, we let $k = 1$ again and the algorithm essentially restarts. Another option available is not to restart k. This modification is suggested and implemented in [4]; further, they observe this algorithmic variant is more stable. This modification ensures that the stepsize always decreases, and so intuitively should increase the stability.

We state this modification as Algorithm 5 below. We show it converges in expectation with the same rate as Algorithm 4, and that it converges almost surely. These results are new to the best of our knowledge, and theoretically justify why a diminishing stepsize makes the algorithm more stable: the optimality gap converges almost surely to 0 rather than just in expectation!

9.4 Theoretical Guarantees for $\widetilde{SVRF}$

We show below that $\widetilde{SVRF}$ has the same convergence rate as SVRF, up to constants depending on the error level δ. The proof is analogous to the one in Hazan and Luo [4, Theorem 1], with some additional care in handling the error term.

Theorem 2 *Suppose each f_i is L-smooth and Ω has diameter D. Then for any $\delta > 0$, Algorithms 4 and 5 with parameters*

$$\gamma_k = \frac{2}{k+1}, \quad m_k = 96(k+1), \quad N_t = 2^{t+3} - 2, \quad \epsilon_k = \frac{LD^2}{2}\gamma_k\delta$$

ensure that for any t,

$$\mathbf{E}[f(x_t) - f(x^\star)] \leq \frac{LD^2(1 + \delta)}{2^{t+1}}.$$

Moreover, for any k,

$$\mathbf{E}[f(w_k) - f(x^\star)] \le \frac{4LD^2(1+\delta)}{k+2}.$$

One might be concerned that $\widetilde{\text{SVRF}}$ is impractical, since the minibatch size required to compute the approximate gradient increases linearly with k. However, when the number of terms n in the objective is sufficiently large, in fact the complexity of $\widetilde{\text{SVRF}}$ is lower than that of Algorithm 3, Frank-Wolfe with approximate oracle. Under the parameter settings in Theorem 2, with a bit extra work, we see that $\widetilde{\text{SVRF}}$ requires $\mathcal{O}(\ln\left(\frac{LD^2(1+\delta)}{\epsilon}\right))$ full gradient evaluations, $\mathcal{O}\left(\frac{L^2D^4(1+\delta)^2}{\epsilon^2}\right)$ stochastic gradient evaluations, and the solution of $\mathcal{O}\left(\frac{LD^2(1+\delta)}{\epsilon}\right)$ linear optimization subproblems. As a comparison, Algorithm 3, Frank-Wolfe with approximate oracle, under the parameter settings in Theorem 1, requires $\mathcal{O}\left(\frac{LD^2(1+\delta)}{\epsilon}\right)$ full gradient evaluations and the solution of the same number of linear optimization subproblems. Suppose that the cost of computing the full gradient is n times the cost of computing one stochastic gradient. Then $\widetilde{\text{SVRF}}$ enjoys a smaller computational cost than Algorithm 3 if

$$\mathcal{O}\left(\ln\left(\frac{LD^2(1+\delta)}{\epsilon}\right)\right) + \frac{1}{n}\mathcal{O}\left(\frac{L^2D^4(1+\delta)^2}{\epsilon^2}\right) < \mathcal{O}\left(\frac{LD^2(1+\delta)}{\epsilon}\right),$$

which is satisfied for large n.

We begin the proof using the smoothness of f_i [9, Theorem 2.1.5].

Proposition 2 *Suppose a real valued function g is convex and L-smooth over its domain* $\mathbf{R}^n$. *Then g satisfies*

$$\|\nabla g(w) - \nabla g(v)\|^2 \le 2L(g(w) - g(v) - \nabla g(v)^T(w - v))$$

for all $w, v \in \mathbf{R}^n$.

Proof Consider $h(w) = g(w) - \nabla g(v)^T w$, which is also convex and L-smooth. The minimum of $h(w)$ occurs at $w = v$, since $\nabla h(v) = 0$. Hence

$$h(v) - h(w) \le h(w - \frac{1}{L}\nabla h(w)) - h(w)$$

$$\le -(\nabla g(w) - \nabla g(v))^T\left(\frac{1}{L}(\nabla g(w) - \nabla g(v))\right)$$

$$+ \frac{L}{2}\frac{1}{L^2}\|\nabla g(w) - \nabla g(v)\|^2$$

$$\le -\frac{1}{2L}\|\nabla g(w) - \nabla g(v)\|^2 \tag{9.15}$$

where the second inequality is due to the smoothness of h. Substitute $h(w) = g(w) - \nabla g(v)^T w$ back into the above inequality gives Proposition 2. $\qquad\square$

The second ingredient of the proof is bounding the variance of the reduced variance gradient $\tilde{\nabla} f(x_0, x)$ in terms of the difference between the current value and the optimal function value. Note that $\tilde{\nabla} f(x_0, x)$ is an unbiased estimator of $\nabla f(x)$. The proof relies on Proposition 2 and can found in Hazan and Luo [4, Lemma 1].

Lemma 1 *For any $x, x_0 \in \Omega$, we have*

$$\mathbf{E}[\|\tilde{\nabla} f(x; x_0) - \nabla f(x)\|^2] \le 6L(2\mathbf{E}[f(x) - f(x^\star)] + \mathbf{E}[f(x_0) - f(x^\star)]).$$

Proof

$$
\begin{aligned}
\mathbf{E}[\|\tilde{\nabla} f(x; x_0) - \nabla f(x)\|^2] &= \mathbf{E}[\|\nabla f_i(x) - \nabla f_i(x_0) + \nabla f(x_0) - \nabla f(x)\|^2]\\
&= \mathbf{E}[\|(\nabla f_i(x) - \nabla f_i(x^\star)) - (\nabla f_i(x_0) - \nabla f_i(x^\star))\\
&\quad + (\nabla f(x_0) - \nabla f(x^\star)) - (\nabla f(x) - \nabla f(x^\star))\|^2]\\
&\le 3\mathbf{E}[\|\nabla f_i(x) - \nabla f_i(x^\star)\|^2 + \|(\nabla f_i(x) - \nabla f_i(x^\star))\\
&\quad - (\nabla f(x_0) - \nabla f(x^\star))\|^2 + \|\nabla f(x) - \nabla f(x^\star)\|^2]\\
&\le 3\mathbf{E}[\|\nabla f_i(x) - \nabla f_i(x^\star)\|^2\\
&\quad + \|\nabla f_i(x_0) - \nabla f_i(x^\star)\|^2 + \|\nabla f(x) - \nabla f(x^\star)\|^2]
\end{aligned}
$$
$$(9.16)$$

where the first inequality is due to Cauchy-Schwarz and the fact that $2ab \le a^2 + b^2$ for any $a, b \in \mathbf{R}$. The second inequality is the variance $\mathbf{E}[\|(\nabla f_i(x) - \nabla f_i(x^\star)) - (\nabla f(x_0) - \nabla f(x^\star))\|^2]$ is less than its second moment $\mathbf{E}[\|(\nabla f_i(x) - \nabla f_i(x^\star))\|^2]$.

Now we apply Proposition 2 to the three terms above. For example, for the first term, we have

$$
\begin{aligned}
\mathbf{E}[\|\nabla f_i(x) - \nabla f_i(x^\star)\|^2] &\le 2L\mathbf{E}[f_i(x) - f_i(x^\star) - \nabla f_i(x^\star)^T(w - w^\star)]\\
&= 2L(f(x) - f(x^\star) - \nabla f(x^\star)^T(x - x^\star)) \quad (9.17)\\
&\le 2L(f(x) - f(x^\star))
\end{aligned}
$$

where the second inequality is due to the optimality of $x^\star$. Applying the proposition similarly to other two terms yields the lemma. $\qquad\square$

The key to the proof of Theorem 2 is the following lemma.

Lemma 2 *For any t and k in Algorithm 4 and 5, we have*

$$\mathbf{E}[f(w_k) - f(x^\star)] \le \frac{4LD^2(1 + \delta)}{k + 2}$$

if

$$\mathbf{E}[\|g_s - \nabla f(w_{s-1})\|^2] \leq \frac{L^2 D^2 (1+\delta)^2}{(s+1)^2}$$

for all $s \leq k$.

Proof The L-smoothness of f gives that for any $s \leq k$,

$$f(w_s) \leq f(w_{s-1}) + \nabla f(w_{s-1})^T (w_s - w_{s-1}) + \frac{L}{2} \|w_s - w_{s-1}\|^2.$$

Under Algorithm 4 or 5, we have $w_s = (1 - \gamma_s)w_{s-1} + \gamma_s v_s$. Plugging this in the above inequality gives

$$f(w_s) \leq f(w_{s-1}) + \gamma_s \nabla f(w_{s-1})^T (v_s - w_{s-1}) + \frac{L\gamma_s^2}{2} \|v_s - w_{s-1}\|^2.$$

Using the definition of the diameter of Ω, we can rearrange the previous inequality as

$$f(w_s) \leq f(w_{s-1}) + \gamma_s g_s^T (v_s - w_{s-1}) + \gamma_s (\nabla f(w_{s-1}) - g_s)^T (v_s - w_{s-1}) + \frac{LD^2 \gamma_s^2}{2}.$$

Since $g_s^T v_s \leq \min_{w \in \Omega} g_s^T w + \frac{\gamma_s \delta L D^2}{2} \leq g_s^T x^\star + \frac{\gamma_s \delta L D^2}{2}$, we arrive at

$$f(w_s) \leq f(w_{s-1}) + \gamma_s \nabla f(w_{s-1})^T (x^\star - w_{s-1}) \tag{9.18}$$
$$+ \gamma_s (\nabla f(w_{s-1}) - g_s)^T (v_s - x^\star) + \frac{LD^2 (1+\delta)\gamma_s^2}{2}.$$

By convexity, the term $\nabla f(w_{s-1})^T (x^\star - w_{s-1})$ is upper bounded by $f(x^\star) - f(w_{s-1})$, and Cauchy-Schwarz inequality yields that

$$|(\nabla f(w_{s-1}) - g_s)^T (v_s - x^\star)| \leq D\|g_s - \nabla f(w_{s-1})\|.$$

The assumption on $\|g_s - \nabla f(w_{s-1})\|^2$ gives $\mathbf{E}[\|g_s - \nabla f(w_{s-1})\|]$ is at most $\frac{LD(1+\delta)}{s+1}$ by Jensen's inequality. Recalling $\gamma_s = \frac{2}{s+1}$, we have

$$\mathbf{E}[f(w_s) - f(x^\star)]$$

$$\leq (1 - \gamma_s) \mathbf{E}[f(w_{s-1}) - f(x^\star)] + \frac{LD^2 \gamma_s^2 (1+\delta)}{2} + \frac{LD^2 \gamma_s^2 (1+\delta)}{2}$$

$$= (1 - \gamma_s) \mathbf{E}[f(w_{s-1}) - f(x^\star)] + LD^2 \gamma_s^2 (1+\delta).$$

We now prove $\mathbf{E}[f(w_k) - f(x^\star)] \le \frac{4LD^2(1+\delta)}{k+2}$ by induction. The base case $k = 1$ is simple by noting $\gamma_1 = 1$ and

$$\mathbf{E}[f(w_1) - f(x^\star)] \le (1 - \gamma_1)\,\mathbf{E}[f(w_0) - f(w^*)] + \gamma_1 LD^2(1+\delta) = LD^2(1+\delta).$$

Now suppose for $k = s-1$, $\mathbf{E}[f(w_{s-1}) - f(x^\star)] \le \frac{4LD^2(1+\delta)}{s+1}$. Then with $\gamma_s = \frac{2}{s+1}$, we have for $k = s$

$$\mathbf{E}[f(w_s) - f(x^\star)] \le \frac{4LD^2(1+\delta)}{s+1}\left(1 - \frac{2}{s+1} + \frac{1}{s+1}\right) \le \frac{4LD^2(1+\delta)}{s+2},$$

which completes the induction.   □

With this lemma, we are able to prove Theorem 2

Proof (Proof of Theorem 2) We proceed by induction. In the base case $t = 0$, we have

$$f(x_0) \le f(w_{-1}) + \nabla f(w_{-1})^T(x_0 - w_{-1}) + \frac{L}{2}\|w_{-1} - x_0\|^2$$

$$\le f(w_{-1}) + \nabla f(w_{-1})^T(x^\star - w_{-1}) + \frac{LD^2}{2} + \frac{LD^2\delta}{2}$$

$$\le f(x^\star) + \frac{LD^2(1+\delta)}{2},$$

where we use the L-smoothness in the first inequality, the near optimality of x_0 in the second inequality and convexity of f in the last inequality.

Now we assume that $\mathbf{E}[f(x_{t-1}) - f(x^\star)] \le \frac{LD^2(1+\delta)}{2^t}$ and we are in Algorithm 4. We consider iteration of the algorithm and use another induction to show $\mathbf{E}[f(w_k) - f(x^\star)] \le \frac{4LD^2(1+\delta)}{k+1}$ for any $k \le N_t$. The base case $w_0 = x_{t-1}$ is clearly satisfied because of the induction hypothesis $\mathbf{E}[f(x_{t-1}) - f(x^\star)] \le \frac{LD^2(1+\delta)}{2^t}$. Given the induction hypothesis $\mathbf{E}[f(w_{s-1} - f(x^\star))] \le \frac{4LD^2(1+\delta)}{s+1}$ for any $s \le k$, we have

$$\mathbf{E}[\|g_s - \nabla f(w_{s-1})\|^2]$$

$$\le \frac{6L}{m_s}\left(2\,\mathbf{E}[f(w_{s-1}) - f(x^\star)] + \mathbf{E}[f(w_0) - f(x^\star)]\right)$$

$$\le \frac{6L}{m_s}\left(\frac{8LD^2(1+\delta)}{s+1} + \frac{LD^2(1+\delta)}{2^t}\right)$$

$$\le \frac{6L}{m_s}\left(\frac{8LD^2(1+\delta)}{s+1} + \frac{8LD^2(1+\delta)}{s+1}\right)$$

$$= \frac{L^2D^2(1+\delta)}{(s+1)^2} \le \frac{L^2D^2(1+\delta)^2}{(s+1)^2}$$

where the first inequality use Lemma 1 and the fact that variance reduced by a factor m_s as g_s is the average of m_s iid samples of $\tilde{\nabla} f(w_{s-1}; w_0)$ and the second and third inequality are due to the two induction hypothesis and $s \leq N_t = 2^{t+3} - 2$. The last equality is due to the choice of m_s. Therefore, we see the condition of Lemma 2 is satisfied and the induction is completed.

Now suppose we are in the situation of Algorithm 5. The only difference here is that we don't restart k at 1. Assuming that $\mathbf{E}[f(x_{s-1}) - f(x^\star)] \leq \frac{LD^2(1+\delta)}{2^t}$ for all $s \leq t - 1$ and by inspecting previous argument, we only need to show $\mathbf{E}[f(w_k) - f(x^\star)] \leq \frac{4LD^2(1+\delta)}{k+1}$ for any $k \leq N_t$. Since our k is always increasing, we cannot directly employ our previous argument. By the structure of our algorithm, we can split the range of k into t cycles $\{1, \ldots, N_1\}, \{N_1 + 1, \ldots, N_2\}, \ldots, \{N_{t-1} + 1, \ldots, N_t\}$. Now within each cycle, we can apply the previous argument, and thus we indeed have $\mathbf{E}[f(w_k) - f(x^\star)] \leq \frac{4LD^2(1+\delta)}{k+2}$ for any $k \leq N_t$.

By the choice of N_t, we see

$$
\begin{aligned}
\mathbf{E}[f(w_{N_t}) - f(x^\star)] &= \mathbf{E}[f(x_t) - f(x^\star)] \\
&\leq \frac{4LD^2(1+\delta)}{N_t + 2} \\
&= \frac{LD^2(1+\delta)}{2^{t+1}}.
\end{aligned}
$$
                                                                                           $\square$

The authors of [4] mention that Algorithm 5 seems to be more stable than Algorithm 4. We give the following theoretical justification for this empirical observation.

Theorem 3 *Under the same assumption of Theorem 2, we have*

$$
\lim_{s \to \infty} f(w_s) = f(x^\star)
$$

with probability 1 for Algorithm 5.

The theorem asserts that the objective value will converge to the true optimum under almost any realization while Theorem 2 tells we have convergence in expectation.

The proof relies on the martingale convergence theorem, which we recall here.

Theorem 4 (Martingale Convergence Theorem) *Let $\{X_t\}_{t=1}^n$ be a sequence of real random variables and $\mathbf{E}_s$ to be the conditional expectation conditional on all $X_i, i \leq s - 1$, then if X_t is a supermartingale, i.e.,*

$$
\mathbf{E}_s(X_s) \leq X_{s-1}
$$

and for all t,

$$X_t \geq L$$

for some L. Then there is a random variable X that

$$X_s \to X \quad \text{almost surely.}$$

To make the presentation clear, we first prove a simple lemma in constructing a martingale.

Lemma 3 *Suppose a sequence of random variables* $\{X_s\}_{s=1}^{\infty}$ *and a deterministic sequence* $\{b_s\}_{s=1}^{\infty}$ *satisfy* $\mathbf{E}_s(X_s) \leq X_{s-1} + b_s$ *and* $X_s \geq L$ *for some* $L \in \mathbf{R}$ *for all s with probability 1. Furthermore, assume that* $\sum_{s=1}^{\infty} b_s = C < \infty$. *Then* $X_s + a_s$ *where* $a_s = C - \sum_{i=1}^{s} b_s$ *is a supermartingale.*

Proof The condition $X_s \geq 0$ is mainly used so that all our expectations make sense. We need to show $\mathbf{E}_s(X_s + a_s) \leq X_{s-1} + a_{s-1}$. Now by moving a_s to the RHS and use the definition of a_s, we see this inequality holds because of the assumption $\mathbf{E}_s(X_s) \leq X_{s-1} + b_s$. □

We now prove Theorem 3.

Proof (Proof of Theorem 3) Recall that $\mathbf{E}_s$ denotes the conditional expectation given all the past except the realization of s. Using inequality (9.18), we see that

$$\mathbf{E}_s f(w_s) \leq f(w_{s-1}) + \gamma_s \nabla f(w_{s-1})^T (x^\star - w_{s-1})$$
$$+ \gamma_s \mathbf{E}_s[(\nabla f(w_{s-1}) - g_s)^T (v_s - x^\star)] + \frac{LD^2(1+\delta)\gamma_s^2}{2}.$$

By convexity, the term $\nabla f(w_{s-1})^T (x^\star - w_{s-1})$ is upper bounded by $f(x^\star) - f(w_{s-1})$, and Cauchy-Schwarz inequality yields that $|(\nabla f(w_{s-1}) - g_s)^T (v_s - x^\star)| \leq D\|g_s - \nabla f(w_{s-1})\|$. Since $\mathbf{E}_s[\|g_s - \nabla f(w_{s-1})\|] \leq \sqrt{\mathbf{E}_s(\|g_s - \nabla f(w_{s-1})\|^2)}$ by Jensen's inequality. Using Lemma 1, we see that

$$\sqrt{\mathbf{E}_s[\|g_s - \nabla f(w_{s-1})\|^2]} \leq \sqrt{\frac{6L}{m_s}(2\mathbf{E}_s[f(w_{s-1}) - f(x^\star)] + \mathbf{E}_s[f(w_0) - f(x^\star)])}$$
$$\leq \sqrt{\frac{18LB}{m_s}}$$

where $B = \sup_{x \in \Omega} f(x) - f(x^\star)$ which is finite as Ω is compact.

Using all previous inequalities, we see that

$$\mathbf{E}_s[f(w_s) - f(x^\star)] \leq (1 - \gamma_s)(f(w_{s-1}) - f(x^\star)) + \frac{LD^2(1+\delta)\gamma_s^2}{2} + \gamma_s \sqrt{\frac{18LB}{m_s}}$$
$$\leq f(w_{s-1}) - f(x^\star) + \frac{LD^2(1+\delta)\gamma_s^2}{2} + \gamma_s \sqrt{\frac{18LB}{m_s}}$$

Since by our choice of γ_s and m_s, we know that by letting $b_s = \frac{LD^2\gamma_s^2}{2} + \gamma_s\sqrt{\frac{18LB}{m_s}}$, $X_s = f(w_s) - f(x^\star) \geq 0$, $a_s = \sum_{i=s+1}^{\infty} b_s$, the condition of Lemma 3 is satisfied and thus $X_s + a_s$ is indeed a super martingale.

Now using the martingale convergence theorem, we know that $X_s + a_s$ converges to a certain random variable X. Since $a_s \to 0$ as $s \to \infty$, $X_s \to X$ almost surely. $X_s \geq 0$ then implies $X \geq 0$. But $\mathbf{E}\, X \leq \mathbf{E}\, X_s + a_s$ for any s by the supermartingale property. Because $\mathbf{E}\, X_s \to 0$ by Theorem 2 and $a_s \to 0$, $\mathbf{E}\, X \leq \mathbf{E}\, X_s + a_s$ implies $\mathbf{E}\, X \leq 0$. Combine the fact $X \geq 0$ as we just argued, we see $X = 0$. This shows that $X_s \to 0$ almost surely which is what we need to prove. $\qquad\square$

The reason that the above argument does not work for Algorithm 4 is that once in a while we restart k and the sequence b_s we used above will be abandoned. More precisely, since the martingale convergence theorem does not tell when the sequence is about to converge, within tth cycle of $k \in \{N_t + 1, \ldots, N_{t+1}\}$, we don't know whether the sequence $f(w_s)$ has converged or not. When we enter a new cycle, we start fresh from $k = 1$ with a new b_s. By contrast, for Algorithm 5, we know that k is always increasing and we have only one sequence b_s. This observation explains why Algorithm 5 is likely to be more stable.

9.5 $\widetilde{SSVRF}$

In previous sections, we have seen how to augment the standard Frank Wolfe algorithm with

- an approximate oracle for linear optimization subproblem (9.2),
- stochastic variance reduced gradients.

Now we turn our attention to the third challenge we raised in the introduction, restricting our attention to the case where the decision variable $X \in \mathbf{R}^{m \times n}$ is a matrix: what if storing the decision variable X is also costly?

Of course, if the decision variable at the solution has no structure, there is no hope to store it more cheaply: in general, $m \times n$ space is required simply to output the solution to the problem. However, in many settings X at the solution may enjoy a low rank structure: X at the solution can be well approximated by a low rank matrix.

The idea introduced in [10] is designed to capture this low rank structure. It forms a linear sketch of the column and row spaces of the decision variable X, and then uses the sketched column and row spaces to recover the decision variable. The recovered decision variable approximates the original X well if a low rank structure is present.

The advantage of this procedure in the context of optimization is that the decision variable X may not be low rank at every iteration of the algorithm. However, so long as the *solution* is (approximately) low rank, we can use this procedure to sketch the decision variable and to recover the solution from this sketch, as introduced in [11].

Notably, we need not store the entire decision variable at each iteration, but only the sketch. Hence the memory requirements of the algorithm are substantially reduced.

Specifically, the sketch proposed in [10] is as follows. To sketch a matrix $X \in \mathbf{R}^{m \times n}$, draw two matrices with independent normal entries $\Psi \in \mathbf{R}^{n \times k}$ and $\Phi \in \mathbf{R}^{l \times m}$. We use Y^C and Y^R to capture the column space and the row space of X:

$$Y^C = X\Psi \in \mathbf{R}^{m \times k}, \qquad Y^R = \Phi X \in \mathbf{R}^{l \times n}. \tag{9.19}$$

In the optimization setting of matrix completion with Algorithm 2, we do not observe the matrix X directly. Rather, we observe a stream of rank one updates

$$X \leftarrow \beta_1 X + \beta_2 uv^T,$$

where β_1, β_2 are real scalars. In this setting, Y^C and Y^R can be updated as

$$Y^C \leftarrow \beta_1 Y^C + \beta_2 uv^T \Psi \in \mathbf{R}^{m \times k}, \quad Y^R \leftarrow \beta_1 Y^R + \beta_2 \Phi uv^T \in \mathbf{R}^{l \times n}. \tag{9.20}$$

This observation allows us to form the sketch Y^C and Y^R from the stream of updates.

We then reconstruct X and get the reconstructed matrix $\hat{X}$ by

$$Y^C = QR, \quad B = (\Phi Q)^\dagger Y^R, \quad \hat{X} = Q[B]_r, \tag{9.21}$$

where QR is the QR factorization of Y^C and $[\cdot]_r$ returns the best rank r approximation in Frobenius norm. Specifically, the best rank r approximation of a matrix Z is $U \Sigma V^T$, where U and V are right and left singular vectors corresponding to the r largest singular values of Z and Σ is a diagonal matrix with r largest singular values of Z. Note the matrix R is not used.

The following theorem [10, Theorem 5.1] guarantees that the resulting reconstruction approximates X well if X is approximately low rank.

Theorem 5 *Fix a target rank r. Let X be a matrix, and let (Y^C, Y^R) be a sketch as described in Eq. (9.19). The procedure (9.21) yields a rank-r matrix $\hat{X}$ with*

$$\mathbf{E} \|X - \hat{X}\|_F \le 3\sqrt{2}\|X - [X]_r\|_F.$$

In the paper [11], this matrix sketching procedure is combined with the original Frank-Wolfe (Algorithm 1). We show here that it also works well with SVRF, the stochastic version of Frank-Wolfe and an approximate subproblem oracle.

We use the following matrix completion problem, which is also a particular instance of problem (9.5), to illustrate this synthesis:

$$\begin{aligned} \text{minimize } & f(\mathcal{A}W) := \tfrac{1}{d} \sum_{i \in I} f_i(\mathcal{A}W) \\ \text{subject to } & \|W\|_* \le \alpha, \end{aligned} \tag{9.22}$$

where $d = |I|$ is the number of elements in I, $W \in \mathbf{R}^{m \times n}$, $\mathcal{A} : \mathbf{R}^{m \times n} \to \mathbf{R}^l$ is a linear map, and $\alpha > 0$ is a given constant. By setting $f = \sum_{i \in I} f_i$ and $\mathcal{S} = \mathbf{R}^{m \times n}$, we see it is indeed a special instance of problem (9.5). Since $\widetilde{\text{SVRF}}$ applied to problem (9.22) updates iterates W_k with a rank-one update at each inner loop iteration, the sketch matrices Y^C and Y^R can be updated using Eq. (9.20). In order to compute the gradient $\nabla(f \circ \mathcal{A})(W_k)$ at W_k, we can store the dual variable $z_k = \mathcal{A} W_k$ and compute the gradient from z_k as

$$\nabla(f \circ \mathcal{A})(W_k) = \mathcal{A}^*(\nabla f)(z_k).$$

Using linearity of $\mathcal{A}$, the dual variable can be updated as

$$z_k := (1 - \gamma_k)z_{k-1} + \gamma_k \mathcal{A}(-\alpha u_k v_k^*).$$

We can store the dual variable efficiently if $l = \mathcal{O}(n)$, and we can update it efficiently if the cost of applying $\mathcal{A}$ to a rank one matrix is $\mathcal{O}(l)$. In many settings we have $l = d$, the number of samples. This means that storing and updating the dual variable z_k could be as costly as computing the full gradient. However, in the oversampled setting, where $l = \mathcal{O}(n)$ while $d \gg \mathcal{O}(n)$, combining the techniques can be beneficial. In this setting, storing z_k is not too costly, and updating z_k is also efficient so long as applying $\mathcal{A}$ to a matrix costs $\mathcal{O}(l)$.

The combined algorithm, $\widetilde{\text{SSVRF}}$, is shown below as Algorithm 6.

Algorithm 6 $\widetilde{\text{SSVRF}}$

1: **Input:** Objective function $f \circ \mathcal{A} = \frac{1}{d} \sum_{i=1}^{d} f_i \circ \mathcal{A}$
2: **Input:** Stepsize γ_k, mini-batch size m_k, epoch length N_t and tolerance sequence ϵ_k
3: **Input:** Target rank r and maximum number of iteration T
4: **Initialize:** Set $x_{-1} = 0$, $Y^C = 0$, $Y^R = 0$ and draw $\Phi \in \mathbf{R}^{(4r+3) \times m}$, $\Psi \in \mathbf{R}^{n \times (2r+1)}$ with standard normal entries.
5: **for** $t = 1, 2, \ldots, T$ **do**
6:      Take a snapshot $z_0 = x_{t-1}$ and compute gradient $\nabla f(z_0)$
7:      **for** $k = 1$ to N_t **do**
8:          Compute $\tilde{\nabla}_k$, the average of m_k iid samples of $\tilde{\nabla} f(z_{k-1}, z_0)$
9:          Compute u, v such that
10:          $-\alpha \mathbf{tr}((\mathcal{A}^* \tilde{\nabla}_k)^T uv^T) \leq \min_{\|X\|_* \leq \alpha} \mathbf{tr}((\mathcal{A}^* \tilde{\nabla}_k)^T X) + \epsilon_k$
11:          Compute $h_k = \mathcal{A}(-\alpha uv^T)$
12:          Update $z_k := (1 - \gamma_k)z_{k-1} + \gamma_k h_k$
13:          Update $Y_k^C = (1 - \gamma_k)Y_{k-1}^C + \gamma_k(-\alpha uv^T)\Psi$
14:          Update $Y_k^R = (1 - \gamma_k)Y_{k-1}^R + \gamma_k \Phi(-\alpha uv^T)$
15:      **end for**
16:      Set $x_t = z_{N_t}$
17: **end for**
18: Compute QR factorization of the $Y_{N_T}^C = QR$ and compute $B = (\Phi Q)^\dagger Y_{N_T}^R$
19: Compute the top r many left and right singular vectors U, V of B and the diagonal matrix Σ with top r singular values.
20: **Output:** (U, Σ, V).

9.6 Theoretical Guarantees for $\widetilde{SSVRF}$

The following theorems are analogous to theorems in [11]. In this work, we introduce adaptations to cope with the approximate oracle and stochastic gradient.

Let us first instantiate some definitions. We assume for each i, $f_i \circ \mathcal{A}$ is L-smooth with respect to the Frobenius norm. Note that the diameter of the feasible region is bounded:

$$\sup_{\|X\|_*,\|Y\|_*\leq\alpha} \|X - Y\|_F \leq \sup_{\|X\|_*,\|Y\|_*\leq\alpha} \|X - Y\|_* \leq 2\alpha.$$

Hence the parameter D, the diameter of the feasible set in Theorem 2, can be replaced by 2α. For each t, we denote by $\hat{X}_t$ the matrix reconstructed using $Y^C_{N_t}$, $Y^R_{N_t}$:

$$Y^C_{N_t} = QR, \quad B = (\Phi Q)^\dagger Y^R_{N_t}, \quad \hat{X}_t = Q[B]_r.$$

The matrix $\hat{X}_t$ can be considered as the reconstruction of X_t (the snapshot, not the inner loop iterate) in $\widetilde{SSVRF}$. We use the same parameters as in Theorem 2 with D replaced by 2α to achieve the following theoretical guarantee:

Theorem 6 *Suppose we apply Algorithm 4 or 5 to the optimization problem (9.22) and that for a particular realization of the stochastic gradients, the iterates X_t converge to a matrix X_∞. Further suppose that in Algorithm 6, we use the same stochastic gradients.*
Then

$$\lim_{t\to\infty} \mathbf{E}_{\Psi,\Phi} \|\hat{X}_t - X_\infty\|_F \leq 3\sqrt{2}\|X_\infty - [X_\infty]_r\|_F.$$

Proof The proof exactly follows the proof of [10, Theorem 6]. □

When the solution set of optimization problem (9.22) contains only matrices with rank $\leq r$, we can prove a stronger guarantee for Algorithm 6:

Theorem 7 *Suppose that the solution set S_* of the optimization problem (9.22) contains only matrices with rank $\leq r$. Then Algorithm 6 attains*

$$\lim_{t\to\infty} \mathbf{E}\,\mathbf{dist}_F(\hat{X}_t, S_*) = 0,$$

where $\mathbf{dist}_F(X, S_*) = \inf_{Y\in S_*} \|X - Y\|_F$.

Proof The triangle inequality implies that

$$\mathbf{E}\,\mathbf{dist}_F(\hat{X}_t, S_*) \leq \mathbf{E}\,\|\hat{X}_t - X_t\|_F + \mathbf{E}\,\mathbf{dist}_F(X_t, S_*).$$

We claim that the second term, $\mathbf{E}\,\mathbf{dist}_F(X_t, S_*)$, converges to 0. If so, we may conclude that the first term converges to zero by the following inequality.

$$\mathbf{E}\,\|\hat{X}_t - X_t\|_F \le 3\sqrt{2}\,\mathbf{E}\,\|X_t - [X_t]_r\|_F$$

$$\le 3\sqrt{2}\,\mathbf{E}(\underset{F}{\mathbf{dist}}(X_t, S_*)) \to 0.$$

The first inequality is Theorem 5, and the second bound is due to the optimality of $[X_t]_r$.

It remains only to prove the claim $\mathbf{E}\,\mathbf{dist}_F(X_t, S_*) \to 0$. Let $g = f \circ \mathcal{A}$ and g_* to be the optimal value of g in program (9.22). Now fix a number $\epsilon > 0$. Define

$$E = \{X \in \mathbf{R}^{m \times n} : \|X\|_* \le \alpha \text{ and } \mathbf{dist}(X, S_*) \ge \epsilon\},$$

and $v = \inf\{g(X), X \in E\}$. If E is empty, then $v = +\infty$. Otherwise, the continuous function g attains the value v on the compact set E. In either case, $v > g_*$ because E contains no optimal point of (9.22). Thus

$$\mathbf{Prob}(X_t \in E) \le \mathbf{Prob}(g(X_t) - g_* > v - g^*) \le \frac{\mathbf{E}(g(X_t) - g^*)}{v - g^*},$$

where the first inequality is due to the optimality of v, and the second is just the Markov inequality. Notice

$$\mathbf{E}\,\underset{F}{\mathbf{dist}}(X_t, S_*) = \mathbf{E}\,\underset{F}{\mathbf{dist}}(X_t, S_*)\mathbf{1}_{\{X_t \in E\}} + \mathbf{E}\,\underset{F}{\mathbf{dist}}(X_t, S_*)\mathbf{1}_{\{X_t \notin E\}}$$

$$\le 2\alpha\,\mathbf{Prob}(X_t \in E) + \epsilon$$

$$\le 2\alpha\frac{\mathbf{E}(g(X_t) - g^*)}{v - g^*} + \epsilon,$$

where the inequality is due to the definition of E, and the feasible region is $\|X\|_* \le \alpha$. Since $\mathbf{E}(g(X_t)) \to g_*$ by Theorem 2, we know $\lim_{t \to \infty} \mathbf{E}\,\mathbf{dist}_F(X_t, S_*) \le \epsilon$ for any $\epsilon > 0$. Thus the claim is proved.   □

When the solution to the optimization problem (9.22) is unique and the function f has a strong curvature property, we can also bound the distance to the optimal solution in expectation.

Theorem 8 *Fix $\kappa > 0$ and $v \ge 1$. Suppose the unique solution $X^\star$ of (9.22) has rank less than or equal to r and*

$$f(\mathcal{A}X) - f(\mathcal{A}X^\star) \ge \kappa\|X - X^\star\|_F^v \tag{9.23}$$

for all $\|X\|_* \leq \alpha$. *Then we have the error bound*

$$\mathbf{E}\,\|\hat{X}_t - X^\star\|_F \leq 6\Big(\frac{4\kappa^{-1}L\alpha^2(1+\delta)}{2^{t+1}}\Big)^{\frac{1}{v}}$$

for all t.

Proof Let $g = f \circ \mathcal{A}$. The proof of Theorem 2 tells us that

$$\mathbf{E}(g(X_t) - g(X^\star)) \leq \frac{LD^2(1+\delta)}{2^{t+1}}.$$

Since the iterate X_t is feasible, the assumption in (9.23) gives us

$$\mathbf{E}(g(X_t) - g(X^\star)) \geq \kappa\,\mathbf{E}\,\|X_t - X^\star\|_F^v \tag{9.24}$$

$$\geq \kappa\,\mathbf{E}\,\|X_t - [X_t]_r\|_F^v$$

$$\geq \kappa[\mathbf{E}(\|X_t - [X_t]_r\|_F)]^v$$

$$\geq \frac{\kappa}{(3\sqrt{2})^v}(\mathbf{E}(\|X_t - \hat{X}_t\|_F))^v. \tag{9.25}$$

The second inequality is due to the optimality of $[X_t]_r$ and $X^\star$ has rank less then r. The third is because of Jensen's inequality and the last is from Theorem 5. We now conclude that

$$\mathbf{E}\,\|\hat{X}_t - X^\star\|_F \leq \mathbf{E}\,\|\hat{X}_t - X_t\| + \mathbf{E}\,\|X_t - X^\star\|$$

$$\leq 3\sqrt{2}\Big(\frac{\kappa^{-1}LD^2(1+\delta)}{2^{t+1}}\Big)^{1/v} + \Big(\frac{\kappa^{-1}LD^2(1+\delta)}{2^{t+1}}\Big)^{1/v}.$$

The last bound follows from inequality (9.24) and (9.25). To reach the final conclusion shown in the theorem, simplify the numerical constant, use the assumption that $v \geq 1$ and note that $D \leq 2\alpha$. $\qquad\square$

Acknowledgements This work was supported by DARPA Award FA8750-17-2-0101. The authors are grateful for helpful discussions with Joel Tropp, Volkan Cevher, and Alp Yurtsever.

Appendix

We prove the following simple proposition about L-smooth functions used in Sect. 9.2.

Proposition 3 *If f is a real valued differentiable convex function with domain $\mathbf{R}^n$ and satisfies $\|\nabla f(x) - \nabla f(y)\| \leq L\|x - y\|$, then for all $x, y \in \mathbf{R}^n$,*

$$f(x) \leq f(y) + \nabla f(y)^T(x - y) + \frac{L}{2}\|x - y\|^2.$$

Proof The inequality follows from the following computation:

$$
\begin{aligned}
f(x) - f(y) - \nabla f(y)^T(x - y) &= \int_0^1 (\nabla f(y + t(x - y)) \\
&\quad - \nabla f(y))^T (x - y)dt \\
&\leq \int_0^1 \|(\nabla f(y + t(x - y)) \\
&\quad - \nabla f(y))^T (x - y)\|dt \\
&\leq \int_0^1 \|(\nabla f(y + t(x - y)) \\
&\quad -\nabla f(y))\|\|(x - y)\|dt \\
&\leq \int_0^1 Lt\|x - y\|^2 dt \\
&= \frac{L}{2}\|x - y\|^2.
\end{aligned}
\tag{9.26}
$$

$\square$

References

1. V. Chandrasekaran, B. Recht, P.A. Parrilo, A.S. Willsky, The convex geometry of linear inverse problems. Found. Comput. Math. **12**(6), 805–849 (2012)
2. R.M. Freund, P. Grigas, R. Mazumder, An extended Frank-Wolfe method with "in- face" directions, and its application to low-rank matrix completion. SIAM J. Optim. **27**(1), 319–346 (2017)
3. E. Hazan, Sparse approximate solutions to semidefinite programs. Lect. Notes Comput. Sci. **4957**, 306–316 (2008)
4. E. Hazan, H. Luo, Variance-reduced and projection-free stochastic optimization, in *International Conference on Machine Learning* (2016), pp. 1263–1271
5. M. Jaggi, Revisiting Frank-Wolfe: projection-free sparse convex optimization, in *Proceedings of the 30th International Conference on Machine Learning ICML (1)* (2013), pp. 427–435
6. R. Johnson, T. Zhang, Accelerating stochastic gradient descent using predictive variance reduction, in *Advances in Neural Information Processing Systems* (2013), pp. 315–323
7. J. Kuczyóski, H. Wozóniakowski, Estimating the largest eigenvalue by the power and Lanczos algorithms with a random start. SIAM J. Matrix Anal. Appl. **13**(4), 1094–1122 (1992)

8. R.B. Lehoucq, D.C. Sorensen, C. Yang, *ARPACK Users' Guide: Solution of Large-Scale Eigenvalue Problems with Implicitly Restarted Arnoldi Methods* (SIAM, Philadelphia, 1998)
9. Y. Nesterov, *Introductory Lectures on Convex Optimization: A Basic Course*, vol. 87 (Springer, New York, 2013)
10. J.A. Tropp, A. Yurtsever, M. Udell, V. Cevher, Randomized single-view algorithms for low-rank matrix approximation (2016). arXiv preprint arXiv:1609.00048
11. A. Yurtsever, M. Udell, J.A. Tropp, V. Cevher, Sketchy decisions: convex low-rank matrix optimization with optimal storage (2017). arXiv preprint arXiv:1702.06838

Chapter 10
Decentralized Consensus Optimization and Resource Allocation

Angelia Nedić, Alexander Olshevsky, and Wei Shi

Abstract We consider the problems of consensus optimization and resource allocation, and we discuss decentralized algorithms for solving such problems. By "decentralized", we mean the algorithms are to be implemented in a set of networked agents, whereby each agent is able to communicate with its neighboring agents. For both problems, every agent in the network wants to collaboratively minimize a function that involves global information, while having access to only partial information. Specifically, we will first introduce the two problems in the context of distributed optimization, review the related literature, and discuss an interesting "mirror relation" between the problems. Afterwards, we will discuss some of the state-of-the-art algorithms for solving the decentralized consensus optimization problem and, based on the "mirror relationship", we then develop some algorithms for solving the decentralized resource allocation problem. We also provide some numerical experiments to demonstrate the efficacy of the algorithms and validate the methodology of using the "mirror relation".

Keywords Convex constrained problems · Consensus optimization · Resource allocation · Decentralized algorithms

AMS Subject Classifications 90C25, 90C30, 90C35

A. Nedić (✉) · W. Shi
School of Electrical, Computer, and Energy Engineering, Arizona State University, Tempe, AZ, USA
e-mail: Angelia.Nedich@asu.edu; Wilbur.Shi@asu.edu

A. Olshevsky
Department of Electrical Engineering, Boston University, Boston, MA, USA
e-mail: alexols@bu.edu

P. Giselsson, A. Rantzer (eds.), *Large-Scale and Distributed Optimization*, Lecture Notes in Mathematics 2227,
https://doi.org/10.1007/978-3-319-97478-1_10

10.1 Introduction

The decentralized consensus optimization problem is about seeking a solution to

$$\underset{x \in \mathbb{R}^p}{\text{minimize}} \ f(x) = \frac{1}{n} \sum_{i=1}^{n} f_i(x), \tag{10.1}$$

in a network of agents, labeled by $1, \ldots, n$. Each function $f_i : \mathbb{R}^p \to \mathbb{R}$ is assumed to be convex and held privately by agent i to encode its objective. This minimization problem has also appeared in centralized setting where a central entity has access to all the functions. In the decentralized setting, the problem is often reformulated by letting each agent have a "local copy" x_i of the argument x and then imposing the constraints $x_i = x_j$ for all agents i, j. The consensus is reflective of the fact that everyone's local "copy" is the same. The agents are connected through a communication network which can be time-varying. The agents want to collaboratively solve the problem, while each agent can only receive/send the information from/to its immediate neighbors (to be specified precisely soon).

On the other hand, the decentralized resource allocation problem is concerned with finding a solution to

$$\underset{\mathbf{x}=(x_1^{\top};\ldots;x_n^{\top}) \in \mathbb{R}^{n \times p}}{\text{minimize}} \ \mathbf{f}(\mathbf{x}) \triangleq \sum_{i=1}^{n} f_i(x_i) \tag{10.2a}$$

$$\text{subject to } \sum_{i=1}^{n} (x_i - r_i) = 0, \tag{10.2b}$$

$$x_j \in \Omega_j, \quad \Omega_j \subseteq \mathbb{R}^p, \quad \forall j = 1, \ldots, n. \tag{10.2c}$$

The problem is defined over a network of n agents. Note that we use f_i in both the consensual optimization problem (10.1) and the resource allocation problem (10.2), but these need not be the same. For each agent i, the vector $x_i \in \mathbb{R}^p$ is its local decision variable. The objective function $f_i : \mathbb{R}^p \to \mathbb{R}$ is convex, and the constraint set $\Omega_i \subseteq \mathbb{R}^p$ is a nonempty closed and convex set; both of which are privately known by agent i only. The equality constraints, $\sum_{i=1}^{n} (x_i - r_i) = 0$, are coupling the agents' decisions, where $r_i \in \mathbb{R}^p$ is a given resource demand vector for agent i. *We note that any partition of the total demand $r = \sum_{i=1}^{n} r_i$ would work, as the choice of the partition does not affect the problem in (10.2). We choose to use some partition $r_1, \ldots, r_n$ only to facilitate a more compact representation of the problem that we use later on.* Aside from this, we also note that, in a distributed multi-agent system, it is plausible to have each agent knowing its own resource demand r_i, in which case their total demand $r = \sum_{i=1}^{n} r_i$ does not have to be known by any entity. On contrary, when total network resource value r is given, then it introduces a coupling among the agents and it has to be somehow known by the agents in order to solve the problem in a distributed fashion.

To comply with the terminology used in the literature, in what follows, we will use "distributed" and "decentralized" interchangeably; unless otherwise stated, they both refer to "distributed without a center."[1]

10.1.1 Literature Review

Problems of the form (10.1) that require distributed computing have appeared in various domains including networked vehicle/UAV coordination/control, information processing and decision making in sensor networks, distributed estimation and learning. Some examples include distributed averaging [10, 58, 85], distributed spectrum sensing [3], formation control [57, 66], power system control [21, 62], statistical inference and learning [20, 53, 61]. In general, distributed optimization framework fits the scenarios where the data is collected and/or stored in a network of agents and having a fusion center is either inapplicable or unaffordable. In such scenarios, data processing and computing is to be performed in a distributed but collaborative manner by the agents within the network.

The research on distributed optimization dates back to 1980s [4, 77]. Due to the emergence of large-scale networks, the development of distributed algorithms for problem in (10.1) has received significant attention recently. Many efforts have been made to solve (10.1) in a master-slave structured isotropic network. The distributed algorithms designed over such special structure are usually fast in practice and mostly used in machine learning to handle big-data in a cluster of computers [9, 11]. Such scheme is "centralized" due to the use of a "master". Here, we focus on solving (10.1) in a decentralized fashion motivated by the applications mentioned above.

Some earlier methods include distributed incremental (sub)gradient methods [45, 46, 51, 63] and incremental proximal methods [5, 80], while a more recent work includes incremental aggregated gradient methods [24] and its proximal gradient variants [6]. All of the incremental methods require a special ring networks due to the nature of these methods. To handle a more general (possibly time-varying) networks, distributed subgradient algorithm was proposed in [50], while its stochastic variant was studied in [64] and its asynchronous variant in [44] with provable convergence rates. These algorithms are intuitive and simple but usually slow due to the fact that even if the objective functions are differentiable and strongly convex, these methods still need to use diminishing step-size to converge to a consensual solution. Other works on distributed algorithms that also require the use of diminishing step-sizes include [19, 48, 94]. With a fixed step-size,

---

[1] Even for special topologies such as a "star" shaped network, it is considered as decentralized since the agent at the "center" has a similar/same computational ability as the others in the network and is not responsible for any extra work of coordination.

these distributed methods can be fast, but they only converge to a neighborhood of the solution set. This phenomenon creates an exactness-speed dilemma. A different class of distributed approaches that bypasses this dilemma is based on introducing Lagrangian dual variables and working with the Lagrangian function. The resulting algorithms include distributed dual decomposition [75] and decentralized alternating direction method of multipliers (ADMM) [7, 40]. Specifically, the decentralized ADMM can employ a fixed step-size and it has nice provable rates [81]. Under the strong convexity assumption, the decentralized ADMM has been shown to have linear convergence[2] for time-invariant undirected graphs [71]. Building on (augmented) Lagrangian, a few improvements have been made via proximal-gradient [13], stochastic gradient [25], and second-order approximation [41, 42]. In particular, ADMM over a random undirected network has been shown to have $O(1/k)$ rate for convex functions [25]. However, de-synchronization and extensions of these methods to time-varying undirected graphs are more involved [25, 81], while their extensions to directed graphs are non-existent in the current literature.

Some distributed methods exist that do not (explicitly) use dual variables but can still converge to an exact consensual solution while using fixed step-sizes. In particular, work in [14] employs multi-consensus inner loop and Nesterov's acceleration method, which gives a proximal-gradient algorithm with a rate at least $O(1/k)$. By utilizing multi-consensus inner loop, *adapt-then-combine* (ATC) strategy, and Nesterov's acceleration, the algorithm proposed in [27] is shown to have $O\left(\ln(k)/k^2\right)$ rate under the assumption of bounded and Lipschitz gradients. For least squares, the general diffusion strategy (a generalization of ATC) can converge to the global minimizer [68]. Although it is unknown to the literature, the above algorithms that do not use dual variable but use fixed step-size are not likely to reach linear convergence even under the strong convexity assumption. References [72, 73] use a difference structure to cancel the steady state error in decentralized gradient descent [50, 89], thereby developing the algorithm EXTRA and its proximal-gradient variant. EXTRA converges at an $o(1/k)$ rate[3] when the objective function in (10.1) is convex, and it has a Q-linear rate when the objective function is strongly convex.

Another topic of interest is distributed optimization over *time-varying directed graphs*. The distributed algorithms over time-varying graphs require the use of doubly stochastic weight matrices, which are not easily constructed in a distributed

---

[2]Suppose that a sequence $\{x(k)\}$ converges to x^* in some norm $\|\cdot\|$. The convergence is: (i) Q-linear if there is $\lambda \in (0, 1)$ such that $\frac{\|x(k+1)-x^*\|}{\|x(k)-x^*\|} \leq \lambda$ for all k; (ii) R-linear if there is $\lambda \in (0, 1)$ and some positive constant C such that $\|x(k) - x^*\| \leq C\lambda^k$ for all k. Both of these rates are geometric. They are often referred to as global rates to distinguish them from the case when the given relations are valid for some sufficiently large k. The difference between these two types of geometric rate is in that Q-linear rate implies monotonic decrease of $\|x(k) - x^*\|$, while R-linear rate does not. We will use "geometric(ally)" and "linear(ly)" interchangeably when it does not cause confusion.

[3]A nonnegative sequence $\{a_k\}$ is said to be convergent to 0 at an $O(1/k)$ rate if $\limsup_{k\to\infty} ka_k < +\infty$. In contrast, it is said to have an $o(1/k)$ rate if $\limsup_{k\to\infty} ka_k = 0$.

fashion when the graphs are directed. To overcome this issue, Ref. [48] is the first to propose a different distributed approach, namely a subgradient-push algorithm that combines the distributed subgradient method [50] with the push-sum protocol [29]. While the subgradient-push eliminates the requirement of graph balancing [22], it suffers from a slow sublinear[4] convergence rate even for strongly convex smooth functions due to its employment of diminishing step-size [47]. On the other hand, noticing that EXTRA has satisfactory convergence rates for undirected graphs, Refs. [83, 91] combine EXTRA with the push-sum protocol [29] to produce DEXTRA (ExtraPush) algorithm in hope of making it work over directed graph. It turns out that for a time-invariant strongly connected directed graph, DEXTRA converges at an R-linear rate under strong convexity assumption but the step-size has to be carefully chosen in some interval (the feasible set of step-sizes for DEXTRA can be empty in some situations [83]). Recent paper [74] proposed an algorithm with diminishing step-size for nonconvex optimization over directed graphs based on the push-sum method [29] (showing convergence to a stationary point).

Another large class of algorithms studied in the literature, which has drawn a significant attention recently due to its (possible) wider applicability, crucially relies on tracking differences of gradients [36–39, 54, 55, 59, 79, 86, 87, 90, 93]. In the literature of multi-agent control, the structure of successive difference is first used in consensus algorithms to dynamically track the average of some time-varying quantities in a multi-agent system [90, 93]. Later, to the best of our knowledge, the idea of tracking the average of the gradient/Hessian quantity, which is time-varying in the process of optimization, is first introduced in Refs. [79, 90] for multi-agent optimization. In Refs. [79, 90], this "tracking technique" is specifically used to develop a second-order numerical method for solving (10.1) over undirected graphs. The idea of gradient tracking is then used in Refs. [86, 87] to allow uncoordinated step-sizes and in Refs. [36–38] to deal with nonconvex problems (using diminishing step-sizes). Very recently, Ref. [59], appearing simultaneously with Ref. [54], has shown that such class of algorithms can be geometrically fast under certain conditions/assumptions using constant step-sizes. Specifically, by using a version of the small-gain theorem, Ref. [54] is able to provide a unified scheme to obtain the geometric convergence rates for both time-varying undirected and directed graphs. Under the guidance of this small-gain theorem, in [39, 55] the geometric convergence rates are studied for uncoordinated step-sizes for both time-varying undirected and time-varying directed graphs.

A particular problem that falls under the preceding resource allocation formulation is the economic dispatch problem in which each f_i is a quadratic function and every constraint set Ω_i is a box, when the direct current power flow model is used [70]. Problems sharing similar forms have received extensive attention due to the emergence of smart city concepts. For example, Refs. [92] and [23] both consider

---

[4]When an algorithm has convergence rate of $O(\theta(k))$, we say that the rate is sublinear if $\lim_{k \to +\infty} \frac{\lambda^k}{\theta(k)} = 0$ for any constant $\lambda \in (0, 1)$. A typical sublinear rates include $O(1/k^p)$ with $p > 0$.

the economic dispatch in a smart grid with an extra consideration of a random wind power injection. Algorithms proposed in both references are accompanied with discussions of basic convergence properties. Some earlier theoretical papers which have focused on decentralized algorithm design for solving the "unconstrained version" ($\Omega_j = \mathbb{R}^p$, $\forall j$) of (10.2) are available in the literature [31, 43]. Reference [43] considers a class of algorithms that randomly pick pairs of neighbors to perform updates. Under convexity assumption, an $O(L/(k\lambda_2))$ rate on the objective optimality residual in expectation is derived over fixed graphs; under strong convexity assumption, an $O\left((1 - \kappa_{\mathbf{f}}^{-1}\lambda_2)^k\right)$ geometric rate is obtained also on the expectation of the objective optimality residual. Here, k is the number of iterations the concerned algorithm has performed, and L is the gradient Lipschitz constant for the objective function $\mathbf{f}$. The quantity $\kappa_{\mathbf{f}}$ is the condition number of the function $\mathbf{f}$ which is a scalar (no less than 1) defined as the ratio of the gradient Lipschitz constant L and the strong convexity constant μ of $\mathbf{f}$. The quantity λ_2 is the second smallest eigenvalue of a certain graph-dependent matrix. With a uniform assignment of probabilities, λ_2^{-1} scales at the order of $O(n^4)$ (though it is possible to considerably improve on this if the probabilities are chosen in a centralized way depending on the graph). Reference [31] gives an algorithm which is shown to have an $O(LBn^3/k)$ rate for the decay of the squared gradient consensus violation over time-varying graph sequences; here B is a constant which measures how long it takes for a time-varying graph sequence to be jointly connected. Reference [28] proposes a "consensus plus innovations" method for solving problem (10.2), and the convergence of the method is established for quadratic objectives f_i under a diminishing step size selection. Based on the alternating direction method of multipliers (ADMM), Ref. [13] provides a class of algorithms which can handle problem (10.2) with convergence guarantees. In particular, under the assumption that the objective functions are convex, the convergence properties are established; when the per-agent constraints (10.2c) are absent (i.e., $\Omega_j = \mathbb{R}^p, \forall j$), under the assumptions that the objective functions are strongly convex and have Lipschitz continuous gradients, a linear convergence (geometric) rate is shown. By using the ADMM, if a center (in a star-shaped network) is allowed to carry a part of computational tasks, a more general problem formulation beyond (10.2) can be handled. Such a formulation and its distributed algorithms have been found to be useful in Internet services over hybrid edge-cloud networks [26]. Reference [18] studies the special case when $\Omega_j = \mathbb{R}^p, \forall j$, and considers solving the problem over time-varying networks. Under the strong convexity and the gradient Lipschitz continuity of the objective function $\mathbf{f}$, the algorithm in Ref. [18] is proved to have a geometric convergence rate $O\left((1 - \kappa_{\mathbf{f}}^{-1}n^{-2})^k\right)$. In other words, for the algorithm in [18] to reach an ε-accuracy, the number of iterations needs to be of the order $O\left(\kappa_{\mathbf{f}}n^2 \ln(\varepsilon^{-1})\right)$. This translates to an $O(\kappa_{\mathbf{f}}n^2)$ scalability in the number n of agents, and it is the best scalability result (with the size n of the network) that currently exists in the literature. Reference [17] proposes a dual-based algorithm for (10.2) with a diminishing step size, for which an $O(1/\sqrt{k})$ convergence rate is shown. Recently, work in [1] has proposed a class of algorithms to handle the

Table 10.1 The convergence rates and scalability results for distributed resource allocation algorithms for problem (10.2), which is convex in all instances

Reference	Uncon. strongly convex	Unconstrained	Constrained	Scalability
[43]	Geometric	$O(1/k)$	–	$O(\kappa_f \lambda_2^{-1})$
[31]	–	$O(1/k)$	–	–
[13]	Geometric	–	–	–
[18]	Geometric	–	$O(1/k)$	$O(\kappa_f n^2)$
[17]	–	$O(1/\sqrt{k})$	$O(1/\sqrt{k})$	–
[1]	Geometric	$O(1/k)$	$O(1/k)$	–
[49]	Geometric	$o(1/k)$	$o(1/k)$	$O(n^2 + \sqrt{\kappa_f}n)$

The scalar L is the Lipschitz-gradient constant, while the condition number $\kappa_f = L/\mu$ where μ is the strong convexity constant for $\mathbf{f}$. The rates are given in terms of the number k of iterations, while the "scalability" column shows how the algorithm's geometric rate depends on the number of agents, n, and the condition number, κ_f. By saying "unconstrained" in the table, we mean that $\Omega_i = \mathbb{R}^p$, $\forall i$. The quantity λ_2 used in Ref. [43] is the second smallest eigenvalue of a certain graph-dependent matrix (see our literature review and Ref. [43] for more details)

resource sharing problem under conic constraints. The algorithms are built on a modified Lagrangian function and an ADMM-like scheme for seeking a saddle point of the Lagrangian function, which has been shown to have an $O(1/k)$ rate for convex agents' objective functions. A recent paper [88] proposes a distributed algorithm for solving problem (10.2) over time-varying directed networks and provides convergence guarantees. A continuous-time algorithm is proposed in [15], where convergence under general convexity assumption is ensured. Table 10.1 summarizes the most relevant references with the convergence rates and their scalability results for problem (10.2), and it compares them with the results discussed in this chapter.

A very recent work by Scaman et al. [69] proposes algorithms for decentralized consensus optimization for smooth and strongly convex objectives. By applying Nesterov's acceleration to the dual problem of the consensus optimization, the algorithms in [69] attain optimal geometric convergence rate of the first-order algorithms for decentralized consensus optimization. This implies that a decentralized resource allocation algorithm that scales in the order of $O(\sqrt{\kappa_f}n)$ with the number n of agents presumably exits. Nevertheless, to enjoy this rate/scalability improvement, one needs to know the strong convexity constant μ and the gradient Lipschitz constant L. The algorithms we discuss here only ask for knowing the parameter L. Reference [49] provides an algorithm for solving problem (10.2) with $o(1/k)$ convergence rate for general convex functions f_i (without requirements of strong convexity and smoothness), which is slightly better than the sub-linear convergence rates previously known in the literature. When the objective functions f_i are strongly convex and smooth, and $\Omega_i = \mathbb{R}^p$ for all i, a geometric convergence of the method is shown, with the scaling in the order of $O(n^2 + \sqrt{\kappa_f}n)$ in the number n of agents. This scaling in terms of the condition number κ_f is better than the best scaling previously known in the literature (see Ref. [18] in which the scaling is $O(\kappa_f n^2)$).

All of the aforementioned work deals with the first-order methods, which are also the focus of this chapter. So far we have discussed the upper bounds on the algorithms complexity. Regarding the lower bounds, Nesterov has shown in [56] that the convergence rate of any primal-only first-order method for unconstrained convex smooth problems can not exceed $\Omega(1/k^2)$; the scalability (condition number dependency) for smooth strongly convex functions can not be better than $\Omega(\sqrt{\kappa_f})$. Indeed, for decentralized consensus optimization, Ref. [27] gives an algorithm with $O(1/k^2)$ convergence rate under assumption of bounded gradients; Ref. [60] gives an algorithm that can achieve $O(1/k^{1.4})$ rate and $O(\kappa_f^{5/7})$ scalability. When agents do not have individual constraints and their objective functions are dual friendly,[5] the optimal rates/scalability established by Nesterov have been reached for decentralized consensus optimization in Refs. [69, 78]. A comprehensive comparison of the convergence rates achieved in consensus-based multi-agent optimization literature can also be found [78].

10.1.2 Notation and Basic Settings for Networks

Throughout the chapter, all vectors are considered in the column form. We let agent i hold a vector $x_i \in \mathbb{R}^p$. For problem (10.1), the variable x_i is a local copy of the variable x (here x is conceptual variable and it is never used in a decentralized system). The value of x_i at iteration/time k is denoted by $x_i(k)$. We also introduce a compact representation for the aggregation of local functions: $\mathbf{f}(\mathbf{x}) \triangleq \sum_{i=1}^n f_i(x_i)$, where its argument and gradient are defined as

$$
\mathbf{x} \triangleq \begin{pmatrix} - x_1^\top - \\ - x_2^\top - \\ \vdots \\ - x_n^\top - \end{pmatrix} \in \mathbb{R}^{n \times p} \quad \text{and} \quad \nabla \mathbf{f}(\mathbf{x}) \triangleq \begin{pmatrix} - (\nabla f_1(x_1))^\top - \\ - (\nabla f_2(x_2))^\top - \\ \vdots \\ - (\nabla f_n(x_n))^\top - \end{pmatrix} \in \mathbb{R}^{n \times p},
$$

respectively. Each row i of $\mathbf{x}$ and $\nabla \mathbf{f}(\mathbf{x})$ is associated with agent i. A matrix/vector with n rows is said to be *consensual* if all of its rows are identical. For instance, $\mathbf{x}$ is consensual if $x_1 = x_2 = \cdots = x_n$. We let $\mathbf{1}$ and $\mathbf{0}$ denote a column vector with all entries equal to one and zero, respectively.

For any $n \times p$ matrices A and B, their inner product is denoted as $\langle A, B \rangle = \text{Trace}(A^\top B)$. For a given matrix A, the Frobenius norm is given by $\|A\|_F$, the (entry-wise) max norm is given by $\|A\|_{\max}$, while the spectral norm is given by $\|A\|_2$ (the largest singular value $\sigma_{\max}\{A\}$). Given a positive (semi)definite square matrix $\mathbf{M}$, we define the $\mathbf{M}$-weighted (semi-)norm $\|A\|_\mathbf{M} = \sqrt{\langle A, \mathbf{M}A \rangle}$. For any matrix $V \in \mathbb{R}^{n \times p}$, we denote its average across the rows (a row corresponds to an agent) as

---

[5] See Ref. [78] for the definition of "dual friendly".

$\overline{V} = \frac{1}{n}\mathbf{1}^\top V \in \mathbb{R}^p$, and its consensus violation as $\check{V} = V - \mathbf{1}\overline{V}^\top = V - \frac{1}{n}\mathbf{1}\mathbf{1}^\top V = \left(I - \frac{1}{n}\mathbf{1}\mathbf{1}^\top\right) V \triangleq \mathbf{E}V$, where $\mathbf{E} = I - \frac{1}{n}\mathbf{1}\mathbf{1}^\top$ is a symmetric matrix. Note that since $\mathbf{E} = \mathbf{E}^\top\mathbf{E}$, we always have $\|A\|_{\mathbf{E}} = \|\mathbf{E}A\|_{\mathrm{F}}$.

For any $n \times p$ matrices A and B, in view of the definition of the consensus violation, we have for $C = A + B$,

$$\|\check{C}\|_{\mathrm{F}} = \|C\|_{\mathbf{E}} = \|A + B\|_{\mathbf{E}} \leq \|A\|_{\mathbf{E}} + \|B\|_{\mathbf{E}} = \|\check{A}\|_{\mathrm{F}} + \|\check{B}\|_{\mathrm{F}}.$$

We intend to use lower case symbols to represent (n-by-p dimensional) matrices that concatenate variables across agents, and use uppercase letters with subscripts such as W_{ij} to denote the mixing weights. For other quantities, we will try to follow the convention of using uppercase symbols for matrices while lowercase ones for vectors. For any matrix $A \in \mathbb{R}^{m \times n}$, $\mathrm{null}\{A\} \triangleq \{x \in \mathbb{R}^n \mid Ax = 0\}$ is the null space of A and $\mathrm{span}\{A\} \triangleq \{y \in \mathbb{R}^m \mid y = Ax, x \in \mathbb{R}^n\}$ is the linear span of all the columns of A. The largest eigenvalue of a symmetric positive semidefinite matrix A is denoted by $\lambda_{\max}\{A\}$, while its smallest non-zero eigenvalue is denoted by $\tilde{\lambda}_{\min}\{A\}$.

To model the underlying communication network among agents, we use the notion of graphs. All graphs are on the vertex set $\mathcal{V} = [n] \triangleq \{1, 2, \ldots, n\}$.

Time-Invariant Undirected Graphs The nodes can send some information (messages that contain variables we need in the numerical algorithms) to each other over (undirected) communication links. For simplicity, we assume that communication links are reliable, synchronized, and always in the status of ready to use (have no delays). For this case, a simple (no self-loop) undirected graph $\mathcal{G}^{\mathrm{un}} = \{\mathcal{V}, \mathcal{E}\}$ is used to describe the network connectivity, where $\mathcal{E}$ is the edge set. When no confusion will be caused, we will drop the superscript and simply use $\mathcal{G}$ to denote an undirected graph. We say that an $n \times n$ matrix A is compatible with the graph $\mathcal{G}$ when the following property holds: $\forall i, j \in [n]$, with $i \neq j$, the (i, j)-th entry of A is zero if $\{i, j\} \notin \mathcal{E}$. We let $\mathcal{N}_i$ denote the set of neighbors of agent i in the graph $\mathcal{G}$, i.e., $\mathcal{N}_i = \{j \in [n] \mid \{i, j\} \in \mathcal{E}\}$. The degree of agent i is defined as the cardinality of $\mathcal{N}_i$, denoted as $|\mathcal{N}_i|$. Note that $|\mathcal{E}| \leq 0.5n(n-1)$ for an undirected graph.

Time-Varying Undirected Graphs This class of graphs is used to model networks with duplex communicational links, with possible link losses or delays that might cause information exchange failures at certain time slots. In this case, each link may be considered as "on" or "off" at certain time slots, thus leading to a sequence of time-varying edge sets. Consider a time-varying undirected graph sequence $\{\mathcal{G}^{\mathrm{un}}(0), \mathcal{G}^{\mathrm{un}}(1), \ldots\}$. Every graph instance $\mathcal{G}^{\mathrm{un}}(k)$ is over the vertex set $\mathcal{V}$ and a set of time-varying edges $\mathcal{E}(k)$. The unordered pair of vertices $\{i, j\} \in \mathcal{E}(k)$ if and only if agents j and i exchange information at time (iteration) k. The set of neighbors of agent i at time k is defined as $\mathcal{N}_i(k) = \{j \mid \{i, j\} \in \mathcal{E}(k)\}$.

Time-Varying Directed Graphs A directed graph is used to model networks that contains at least one unidirectional information flow. A unidirectional information flow means that there exists certain pairs of agents (i, j) between which effective information transmission from i to j is guaranteed while the reverse way from j to i is not necessarily provided. In this model, if an arc (ordered pair of agents) (j, i) exists at time k, then agent j can send information reliably sent to agent i at time k. Consider a time-varying graph sequence $\{\mathcal{G}^{\text{dir}}(0), \mathcal{G}^{\text{dir}}(1), \ldots\}$. Every graph instance $\mathcal{G}^{\text{dir}}(k)$ consists of a static set of agents $\mathcal{V}$ and a set $\mathcal{A}(k)$ of time-varying links. The set of in- and out-neighbors of agent i at time k are defined as $\mathcal{N}_i^{\text{in}}(k) = \{j | (j, i) \in \mathcal{A}(k)\}$ and $\mathcal{N}_i^{\text{out}}(k) = \{j | (i, j) \in \mathcal{A}(k)\}$, respectively. At time k, the in- and out- degree of agent i is defined as $|\mathcal{N}_i^{\text{in}}(k)|$ and $|\mathcal{N}_i^{\text{out}}(k)|$, respectively. For a directed graph with an edge set $\mathcal{A}$, we always have $|\mathcal{A}| \leq n(n-1)$.

10.2 Decentralized Consensus Optimization

In this section, we discuss some algorithms for solving the consensus optimization problem (10.1) under different networking conditions: time-invariant undirected graphs, time-varying undirected graphs, and time-varying directed graphs.

10.2.1 Over Time-Invariant Undirected Graphs

There have been many algorithms proposed for time-invariant undirected graphs. Convergence rates that are typical in classical centralized deterministic optimization (without using Nesterov's acceleration) have been achieved for decentralized optimization over a time-invariant undirected graph.

For the algorithms discussed in this subsection, we will make use of the following assumption. One of the most popular and fast algorithms for solving (10.1) over time-invariant undirected graphs is based on the alternating direction method of multipliers (ADMM) [7]. This algorithm is summarized as follows.

Algorithm 1 Decentralized ADMM (time-invariant undirected graphs)

1: Select a parameter $c > 0$ and initialize variables $x_i(0) \in \mathbb{R}^p$, $y_i(0) = \mathbf{0} \in \mathbb{R}^p$;
2: **for** $k = 0, 1, \cdots$ **do**, every agent i in parallel

3: $\quad x_i(k+1) = \underset{x_i \in \mathbb{R}^p}{\operatorname{argmin}} f_i(x_i) + c|\mathcal{N}_i| \left(x_i + \frac{1}{2c|\mathcal{N}_i|} y_i(k) - \frac{1}{2} x_i(k) - \frac{1}{2|\mathcal{N}_i|} \sum_{j \in \mathcal{N}_i} x_j(k) \right)^2$;

4: $\quad y_i(k+1) = y_i(k) + c \left(|\mathcal{N}_i| x_i(k+1) - \sum_{j \in \mathcal{N}_i} x_j(k+1) \right)$;

5: **end for**

This algorithm achieves geometric convergence under certain functional conditions (each local objective function is strongly convex and has Lipschitz continuous gradient). If we further know the parameters of strong convexity and Lipschitz gradient, optimal algorithmic parameter can be determined based on optimizing an upper error bound on the convergence rate (see [71] for more details). When the parameters are properly selected, this algorithm is efficient in terms of the number of iterations needed to reach a certain accuracy. However, a drawback of this algorithm is in the update that requires to solve a convex optimization problem (a system of nonlinear equations) at each iteration. In some situations when the computational resources are limited, a gradient based update may be preferable. An algorithm, known as EXTRA, performing a gradient-based update at each iteration has been developed in [73]. The algorithm is given as follows.

Algorithm 2 EXTRA (time-invariant undirected graphs)

1: Select a parameter $\alpha > 0$ and mixing weights $W_{ij} \in \mathbb{R}$ and $\widetilde{W}_{ij} \in \mathbb{R}$ for all $\{i, j\} \in \mathcal{E}$;
2: For all i, initialize variables $x_i(0) \in \mathbb{R}^p$ and set $x_i(1) = \sum\limits_{j \in \mathcal{N}_i \bigcup\{i\}} W_{ij} x_j(0) - \alpha \nabla f_i(x_i(0))$;
3: **for** $k = 0, 1, \cdots$ **do**, every agent i in parallel
4:     $x_i(k+2) = x_i(k+1) + \sum\limits_{j \in \mathcal{N}_i \bigcup\{i\}} W_{ij} x_j(k+1)$
        $\qquad - \sum\limits_{j \in \mathcal{N}_i \bigcup\{i\}} \widetilde{W}_{ij} x_j(k) - \alpha \left(\nabla f_i(x_i(k+1)) - \nabla f_i(x_i(k)) \right)$;
5: **end for**

The algorithm uses two mixing matrices $\mathbf{W} = [W_{ij}]$ and $\widetilde{\mathbf{W}} = [\widetilde{W}_{ij}]$. Some approaches to construct $\mathbf{W}$ can be found in Section 2.4 of [73]. Usually $\widetilde{\mathbf{W}}$ can be chosen as $(I + \widetilde{\mathbf{W}})/2$. More results on selecting $\widetilde{\mathbf{W}}$ to accelerate convergence can be found in [73]. Under the condition that the original objective function f (not each individual local objective f_i) in (10.1) is strongly convex and its gradients are Lipschitz continuous, EXTRA algorithm has a geometric convergence rate. More details including an intuition used in the development of this algorithm, the convergence guarantees under basic convexity assumption, and some numerical examples can be found in [73]. A follow-on work in [72, 82] builds on the EXTRA algorithm by considering a proximal-gradient variant, tolerance to delays and asynchronous updates.

Another algorithm that received considerable attention recently is the Augmented Distributed Gradient Method (Aug-DGM) algorithm [87], given as follows.

Algorithm 3 Aug-DGM (time-invariant undirected graphs)

1: Select parameters $\alpha_i > 0$ for all i and mixing weights $W_{ij} \in \mathbb{R}$ for all $\{i, j\} \in \mathcal{E}$;
2: For all i, initialize variables $x_i(0) \in \mathbb{R}^p$ and $y_i(0) = \nabla f_i(x_i(0))$;
3: **for** $k = 0, 1, \cdots$ **do**, every agent i in parallel
4:     $x_i(k+1) = \sum\limits_{j \in \mathcal{N}_i \bigcup\{i\}} W_{ij} \left(x_j(k) - \alpha_j y_j(k) \right)$;
5:     $y_i(k+1) = \sum\limits_{j \in \mathcal{N}_i \bigcup\{i\}} W_{ij} \left(y_j(k) + \nabla f_j(x_j(k+1)) - \nabla f_j(x_j(k)) \right)$;
6: **end for**

This algorithm employs a "gradient tracking" technique and allows the agents to use different step sizes. Its convergence has been studied in [87] for smooth convex objective functions. Although uncoordinated step sizes are not difficult to implement in augmented Lagrangian-based methods [33, 35] over undirected graphs, such algorithms have not been considered for time-varying directed graphs.

The algorithms employing gradient tracking turn out to be extendable to both time-varying and directed graphs [39, 54, 55]. The Aug-DGM shares some similarity with the "Adapt-Then-Combine (ATC)" diffusion strategy in optimization (see [33, 55, 68]).

10.2.2 Over Time-Varying Undirected Graphs

Here, we focus on a **D**istributed **I**nexact method with **G**radient track**ing** (DIG-ing) [54]. This algorithm is the first one known to have a geometric convergence rate for time-varying graphs. To introduce the algorithm and provide some insights into its iterations, we give a result stating the optimality conditions for problem (10.1).

Proposition 1 ([73]) *Let* $\mathbf{W} \in \mathbb{R}^{n \times n}$ *be such that* $\mathrm{null}\{I - \mathbf{W}\} = \mathrm{span}\{\mathbf{1}\}$. *Let* $\mathbf{x}^*$ *satisfy the following conditions:* $\mathbf{x}^* = \mathbf{W}\mathbf{x}^*$ *(consensus) and* $\mathbf{1}^\top \nabla \mathbf{f}(\mathbf{x}^*) = 0$ *(optimality). Then, all the rows of* $\mathbf{x}^*$ *are equal to a vector* $(x^*)^\top$ *for some* x^* *that is an optimal solution of the problem* (10.1).

To provide the idea behind the DIGing algorithm, let us focus on the case of a static graph for the moment. Consider the distributed gradient descent (DGD) method, given as follows:

$$\mathbf{x}(k+1) = \mathbf{W}\mathbf{x}(k) - \alpha \nabla \mathbf{f}(\mathbf{x}(k)),$$

where $\mathbf{W}$ is a doubly stochastic mixing matrix and $\alpha > 0$ is a fixed step-size. The mixing part "$\mathbf{W}\mathbf{x}(k)$" is necessary for reaching consensus, while DGD exhibits undesirable behavior due to its use of the gradient direction, "$-\alpha \nabla \mathbf{f}(\mathbf{x}(k))$". To see this, let us break the update into steps per agent: for every agent i, we have $x_i(k+1) = W_{ii}x_i(k) + \sum_{j \in \mathcal{N}_i} W_{ij}x_j(k) - \alpha \nabla f_i(x_i(k))$, where $\mathcal{N}_i$ is the set of the neighbors of agent i in the given graph. Thus, each agent is updating using only the gradient of its local objective function f_i. Suppose now that the values $x_i(k)$ have reached consensus and that $x_i(k) = x^*$ for all i and some solution x^* of the problem (10.1). Then, the mixing part gives $W_{ii}x_i(k) + \sum_{j \in \mathcal{N}_i} W_{ij}x_j(k) = x^*$ for all i. However, the gradient-based term gives $-\alpha \nabla f_i(x_i(k))$ for all i, which need not be zero in general, thus resulting in $x_i(k+1)$ that will move away from the solution x^* (recall that a solution to the problem (10.1) is at a point x where $\sum_{j=1}^n \nabla f_j(x) = \mathbf{0}$ $\forall i$ and not necessarily a point where $\nabla f_i(x) = \mathbf{0}$ $\forall i$—which may not exist).

Conceptually, one (non-distributed) scheme that bypasses this limitation is the update

$$\mathbf{x}(k+1) = \mathbf{W}\mathbf{x}(k) - \alpha \frac{1}{n} \mathbf{1}\mathbf{1}^\top \nabla \mathbf{f}(\mathbf{x}(k)), \tag{10.3}$$

which can be implemented if every agent has access to the average of all the gradients $\nabla f_j(x_j(k))$, $j = 1, \ldots, n$ (evaluated at each agent's local copy). One can verify that if (10.3) converges, its limit point $\mathbf{x}(\infty)$ satisfies the optimality conditions as given in Proposition 1. However, the update in (10.3) is not distributed among the agents as it requires a central entity to provide the average of the gradients.

Nevertheless, one may approximate the update in (10.3) through a surrogate direction that tracks the gradient average. To track the average of the gradients, namely, $\frac{1}{n}\mathbf{1}\mathbf{1}^\top \nabla \mathbf{f}(\mathbf{x}(k))$, we introduce a variable $\mathbf{y}(k)$ that is updated as follows:

$$\mathbf{y}(k+1) = \mathbf{W}\mathbf{y}(k) + \nabla \mathbf{f}(\mathbf{x}(k+1)) - \nabla \mathbf{f}(\mathbf{x}(k)), \tag{10.4}$$

with initialization $\mathbf{y}(0) = \nabla \mathbf{f}(\mathbf{x}(0))$ and where each row i of $\mathbf{y}(k) \in \mathbb{R}^{n \times p}$ is associated with agent i. A similar technique has been used in [30, 65] for tracking some network-wide aggregate quantities and in [93] for dynamically tracking the average state of a multi-agent system. If $\mathbf{x}(k+1)$ converges to some point $\mathbf{x}(\infty)$ and the underlying graph is connected, then it can be seen that the sequence $\mathbf{y}(k)$ generated by the gradient tracking procedure (10.4) will converge to the point $\mathbf{y}(\infty)$ given by $\mathbf{y}(\infty) = \mathbf{1}\mathbf{1}^\top \nabla \mathbf{f}(\mathbf{x}(\infty))/n$, which is exactly what we need in view of (10.3). Replacing $\frac{1}{n}\mathbf{1}\mathbf{1}^\top \nabla \mathbf{f}(\mathbf{x}(k))$ in (10.3) by its dynamic approximation $\mathbf{y}(k)$ is exactly what we use to construct the DIGing algorithm. Furthermore, to accommodate time-varying graphs, the static weight matrix $\mathbf{W}$ is replaced by a time varying matrix $\mathbf{W}(k)$, thus resulting in the DIGing algorithm, as given below.

Algorithm 4 DIGing (time-varying undirected graphs)

1: Select a parameter $\alpha > 0$ for all i;
2: For all i, initialize variables $x_i(0) \in \mathbb{R}^p$ and $y_i(0) = \nabla f_i(x_i^0)$;
3: **for** $k = 0, 1, \cdots$ **do**, every agent i in parallel select mixing weights $W_{ij}(k) \in \mathbb{R}$ for all $j \in \mathcal{N}_i(k) \bigcup \{i\}$
4: $\quad x_i(k+1) = \sum\limits_{j \in \mathcal{N}_i(k) \bigcup \{i\}} W_{ij}(k)x_j(k) - \alpha y_i(k)$;
5: $\quad y_i(k+1) = \sum\limits_{j \in \mathcal{N}_i(k) \bigcup \{i\}} W_{ij}y_j(k) + \nabla f_i(x_i(k+1)) - \nabla f_i(x_i(k))$;
6: **end for**

Each matrix $\mathbf{W}(k)$ is compatible with the graph $\mathcal{G}^{un}(k)$. The initialization of DIGing uses an arbitrary $\mathbf{x}(0) \in \mathbb{R}^{n \times p}$ and $\mathbf{y}(0) = \nabla \mathbf{f}(\mathbf{x}(0))$. At each iteration k, the algorithm maintains two variables $\mathbf{x}(k)$ and $\mathbf{y}(k) \in \mathbb{R}^p$, which are updated as follows:

$$\mathbf{x}(k+1) = \mathbf{W}(k)\mathbf{x}(k) - \alpha \mathbf{y}(k),$$
$$\mathbf{y}(k+1) = \mathbf{W}(k)\mathbf{y}(k) + \nabla \mathbf{f}(\mathbf{x}(k+1)) - \nabla \mathbf{f}(\mathbf{x}(k)).$$

When the graph is static, the DIGing algorithm has relationships with some of the existing algorithms such as EXTRA and basic primal-dual approach. However, it appears to be difficult to obtain geometric convergence analysis for DIGing by simply adapting the classical primal-dual analysis (more detailed comments can be found in Subsection 2.2 of [54]). Compared to the Aug-DGM algorithm, DIGing only needs one round of communication per iteration. However, it is found that the ATC structure in Aug-DGM can accelerate the convergence (see [55]). To get the benefit from the ATC structure but avoid two rounds of communication per iteration, one can replace either x-update or y-update by its ATC analogue. One can also correspondingly find "Combine-Then-Adapt (CTA)" variants of DIGing, which should have presumably geometric convergence.

10.2.2.1 Geometric Convergence

We state the linear convergence rate of DIGing over time-varying undirected graphs. Consider a sequence of time-varying undirected graphs $\{\mathcal{G}^{un}(k)\}$ with $\mathcal{G}^{un}(k) = \{\mathcal{V}, \mathcal{E}(k)\}$, and a mixing matrix sequence $\{\mathbf{W}(k)\}$, where the matrix $\mathbf{W}(k)$ is compatible with the graph $\mathcal{G}^{un}(k)$ for each k. To formally describe the assumptions we make on the graphs and on the mixing matrices, we define the graphs

$$\mathcal{G}_b^{un}(k) \triangleq \left\{ \mathcal{V}, \mathcal{E}(k) \bigcup \mathcal{E}(k+1) \bigcup \cdots \bigcup \mathcal{E}(k+b-1) \right\},$$

for any $k = 0, 1, \ldots$ and any $b = 1, 2, \ldots$. We also define the matrices

$$\mathbf{W}_b(k) \triangleq \mathbf{W}(k)\mathbf{W}(k-1) \cdots \mathbf{W}(k-b+1) \text{ for any } k = 0, 1, \ldots \text{ and any } b = 0, 1, \ldots,$$

where $\mathbf{W}_b(k) = I$ for any needed $k < 0$ and $\mathbf{W}_0(k) = I$ for any k.

The basic assumption that we impose on the weight matrices is as follows.

Assumption 1 (Mixing Matrix Sequence $\{\mathbf{W}(k)\}$) *For any $k = 0, 1, \ldots,$ the mixing matrix $\mathbf{W}(k) = [W_{ij}(k)] \in \mathbb{R}^{n \times n}$ satisfies the following relations:*

  i. *(Decentralized property) If $i \neq j$ and the edge $\{i, j\} \notin \mathcal{E}(k)$, then $W_{ij}(k) = 0$;*
 ii. *(Double stochasticity) $\mathbf{W}(k)\mathbf{1} = \mathbf{1}$ and $\mathbf{1}^\top \mathbf{W}(k) = \mathbf{1}^\top$;*
iii. *(Joint spectrum property) There exists a positive integer B such that*

$$\sup_{k \geq B-1} \delta(k) < 1 \text{ where } \delta(k) = \sigma_{\max} \left\{ \mathbf{W}_B(k) - \frac{1}{n}\mathbf{1}\mathbf{1}^\top \right\} \text{ for all } k = 0, 1, \ldots.$$

Several different mixing rules exist that yield the matrix sequences which have property (iii) (see Subsection 2.4 of [73]). In particular, the following two assumptions taken together imply Assumption 1 [52].

Assumption 2 ($\tilde{B}$-Connected Graph Sequence) *The time-varying undirected graph sequence $\{\mathcal{G}^{un}(k)\}$ is $\tilde{B}$-connected. Specifically, there exists some positive integer $\tilde{B}$ such that the undirected graph $\mathcal{G}_{\tilde{B}}^{un}(t\tilde{B})$ is connected for all $t = 0, 1, \ldots.$*

Assumption 2 is weaker than the assumption that each $\mathcal{G}^{un}(k)$ is connected. It has been often used in multi-agent coordination and distributed optimization [48].

Assumption 3 (Mixing Matrix Sequence $\{\mathbf{W}(k)\}$) *For any $k = 0, 1, \ldots,$ the mixing matrix $\mathbf{W}(k) = [W_{ij}(k)] \in \mathbb{R}^{n \times n}$ satisfies*

i. *(Double stochasticity)* $\mathbf{W}(k)\mathbf{1} = \mathbf{1}$ *and* $\mathbf{1}^\top \mathbf{W}(k) = \mathbf{1}^\top$;

ii. *(Positive diagonal)* *For all i,* $\mathbf{W}_{ii}(k) > 0$;

iii. *(Edge utilization)* *If $\{i, j\} \in \mathcal{E}(k)$, then $W_{ij}(k) > 0$; otherwise $W_{ij}(k) = 0$;*

iv. *(Non-vanishing weights)* *There is a $\tau > 0$ such that if $W_{ij}(k) > 0$, then $W_{ij}(k) \geq \tau$.*

Assumption 3 is strong but typical for multi-agent coordination and optimization. For undirected graph it can be fulfilled, for example, by using Metropolis weights:

$$
W_{ij}(k) = \begin{cases} 1/\left(1 + \max\{d_i(k), d_j(k)\}\right), & \text{if } (j, i) \in \mathcal{E}(k), \\ 0, & \text{if } (j, i) \notin \mathcal{E}(k)) \text{ and } j \neq i, \\ 1 - \sum_{l \in \mathcal{N}_i(k)} W_{il}(k), & \text{if } j = i, \end{cases}
$$

where $d_i(k) = |\mathcal{N}_i(k)|$ is the degree of agent i at time k. In this case, Assumption 3 is satisfied with $\tau = 1/n$.

Now let us consider the graph sequence $\{\mathcal{G}^{\mathrm{un}}(k)\}$ under Assumption 2. If the constant B of Assumption 1 is chosen as $B \geq 2\tilde{B} - 1$, then for any $k = 0, 1, \ldots,$ the union of a B-length consecutive clip of the edge set sequence from time index k, $\bigcup_{b=k}^{k+B-1} \mathcal{E}(b)$, is always a super set of $\bigcup_{b=\lceil k/\tilde{B}\rceil \tilde{B}}^{\lceil k/\tilde{B}\rceil \tilde{B}+\tilde{B}-1} \mathcal{E}(b)$ and the graph $\mathcal{G}_{\tilde{B}}^{\mathrm{un}}(\lceil k/\tilde{B}\rceil \tilde{B})$ is connected by assumption, where $\lceil \cdot \rceil$ denotes the ceiling function (rounding a real number to the smallest succeeding integer). Thus, it can be seen that the graph $\mathcal{G}_B^{\mathrm{un}}(k)$ is connected. With such a choice of B, Assumptions 2 and 3 together imply Assumption 1. Thus, we assume $B = 2\tilde{B} - 1$ whenever we invoke Assumption 2.

We use the following assumption on the objective functions, which is standard for deriving geometric rate of gradient algorithms for convex minimization problems.

Assumption 4 (Smoothness and Strong Convexity) *Assume that*

(a) *For every agent i, its objective $f_i : \mathbb{R}^p \to \mathbb{R}$ is differentiable and has Lipschitz continuous gradients, i.e., there exists a Lipschitz constant $L_i \in (0, +\infty)$ such that*

$$
\|\nabla f_i(x) - \nabla f_i(y)\|_{\mathrm{F}} \leq L_i \|x - y\|_{\mathrm{F}} \text{ for any } x, y \in \mathbb{R}^p.
$$

(b) *For every agent i, its objective $f_i : \mathbb{R}^p \to \mathbb{R}$ satisfies*

$$
f_i(x) \geq f_i(y) + \langle \nabla f_i(y), x - y \rangle + \frac{\mu_i}{2}\|x - y\|_{\mathrm{F}}^2 \text{ for any } x, y \in \mathbb{R}^p,
$$

where $\mu_i \geq 0$ and at least one μ_i is nonzero.

When Assumption 4(a) holds, we say that each ∇f_i is L_i-Lipschitz (continuous). When $\mu_i > 0$ in Assumption 4(b) holds, we will say that f_i is μ_i-strongly convex. We define $L \triangleq \max_i\{L_i\}$, which is the Lipschitz constant of $\nabla \mathbf{f}(\mathbf{x})$, and $\bar{L} \triangleq (1/n) \sum_{i=1}^{n} L_i$ which is the Lipschitz constant of $\nabla f(x)$. Also, we introduce $\hat{\mu} \triangleq \max_i\{\mu_i\}$, $\bar{\mu} \triangleq (1/n) \sum_{i=1}^{n} \mu_i$, and $\bar{\kappa} \triangleq L/\bar{\mu}$. Assumption 4(b) implies the $\bar{\mu}$-strong convexity of $f(x)$. Under this assumption, there is a unique optimal solution to problem (10.1).

We now state main results on the convergence rates of DIGing. The first theorem gives an explicit convergence rate for DIGing in terms of the network parameters (B, n, and δ), objective parameters ($\bar{\mu}$ and $\bar{\kappa} = \frac{L}{\bar{\mu}}$), and the algorithmic step-size (α).

Theorem 1 (DIGing: Explicit Geometric Rate Over Time-Varying Graphs [54]) *Let Assumptions 1 and 4 hold, and let*

$$\delta = \sup_{k \geq B-1} \left\{ \sigma_{\max} \left\{ \mathbf{W}_B(k) - \frac{1}{n} \mathbf{1}\mathbf{1}^{\top} \right\} \right\} \quad and \quad J_1 = 3\bar{\kappa} B^2 \left(1 + 4\sqrt{n}\sqrt{\bar{\kappa}} \right).$$

Then, for any step-size $\alpha \in \left(0, \frac{1.5(1-\delta)^2}{\bar{\mu} J_1} \right]$, the sequence $\{\mathbf{x}(k)\}$ generated by DIGing algorithm converges to the matrix $\mathbf{x}^* = \mathbf{1}(x^*)^{\top}$ at a global R-linear (geometric) rate of $O(\lambda^k)$, where x^* is the unique optimal solution of problem (10.1) and the parameter λ is given by

$$\lambda = \begin{cases} \sqrt[2B]{1 - \frac{\alpha\bar{\mu}}{1.5}}, & if \, \alpha \in \left(0, \frac{1.5\left(\sqrt{J_1^2+(1-\delta^2)J_1}-\delta J_1\right)^2}{\bar{\mu} J_1 (J_1+1)^2} \right], \\[4mm] \sqrt[B]{\sqrt{\frac{\alpha\bar{\mu} J_1}{1.5}} + \delta}, & if \, \alpha \in \left(\frac{1.5\left(\sqrt{J_1^2+(1-\delta^2)J_1}-\delta J_1\right)^2}{\bar{\mu} J_1 (J_1+1)^2}, \frac{1.5(1-\delta)^2}{\bar{\mu} J_1} \right]. \end{cases}$$

Theorem 1 explicitly characterizes the rate of DIGing in terms of the parameter δ which measures the convergence speed of consensus. Apparently, λ is a piecewise function of α: it is monotonically decreasing with α in the first interval, and monotonically increasing in the second interval. Hence, λ reaches the optimal value at the joint of the two intervals. One of the reason that we would like to depict the step-size region in terms of δ is that, in practice, the consensus speed in networks may be faster than the worst-case analytical results provided here. Thus, there are chances that δ is experimentally measurable instead of having it determined analytically. An explicit expression of δ in terms of n can be found in [52] if Assumption 3 is used to ensure that Assumption 1 holds. Other possible choices of β, η, α, and λ exist and may give a tighter bound. The following two corollaries provide some results showing how the geometric rate parameter λ scales with the number n of agents.

Corollary 1 (DIGing: Polynomial Network Scalability [54])   *Let Assumptions 2–4 hold. Also, let the step-size be given by*

$$\alpha(\tau) = \frac{3\tau^2}{128B^2n^{4.5}L\sqrt{\bar{\kappa}}} - \frac{1.5}{\bar{\mu}}\left(\frac{\tau^2}{128B^2n^{4.5}\bar{\kappa}^{1.5}}\right)^2,\qquad(10.5)$$

where τ is the smallest nonzero positive element of the nonnegative matrices $\mathbf{W}(k)$ for all k (cf. Assumption 3). Then, the sequence $\{x(k)\}$ generated by DIGing converges at a global R-linear rate of $O\left((\lambda(\tau))^k\right)$ with

$$\lambda(\tau) = \sqrt[B]{1 - \frac{\tau^2}{128B^2n^{4.5}\bar{\kappa}^{1.5}}}.$$

Corollary 1 shows explicitly how the linear convergence rate of the DIGing algorithm depends on the condition number $\bar{\kappa}$, time-varying graph connectivity constant B, and the network size n. To reach ε-accuracy, the iteration complexity is of the order $O\left(\tau^{-2}B^3n^{4.5}\bar{\kappa}^{1.5}\ln\frac{1}{\varepsilon}\right)$, which is polynomial in the number n of agents. Corollary 1 is obtained from Theorem 1 and the bound $\delta \leq 1 - \tau/(2n^2)$ from [52], which may be very conservative since it applies to a rather general class of graphs. Moreover, any further advances in "consensus theory" deriving improved convergence bounds on the rate of consensus algorithm would immediately translate into improvements through the use of Corollary 1.

Corollary 2 (DIGing: Iteration Complexity Under Lazy Metropolis Mixing [54])   *Let Assumptions 2 and 4 hold. Assume that the graphs are time-invariant, undirected and connected (i.e., $B = 1$). Let each $\mathbf{W}(k)$ be a lazy Metropolis matrix, i.e.,*

$$W_{ij}(k) = \begin{cases} 1/\left(2\max\{d_i(k), d_j(k)\}\right), & \text{if } \{i, j\} \in \mathcal{E}, \\ 0, & \text{if } \{i, j\} \notin \mathcal{E} \text{ and } j \neq i, \\ 1 - \sum_{l \in \mathcal{N}_i(k)} W_{il}(k), & \text{if } j = i. \end{cases}$$

Then, with the step-size $\alpha(\frac{2}{71})$ (see (10.5) with $\tau = \frac{2}{71}$ and $B = 1$), the sequence $\{x(k)\}$ generated by DIGing converges at a global R-linear rate of $O(\lambda^k)$, where $\lambda = 1 - \frac{1}{161312n^{4.5}\bar{\kappa}^{1.5}}$.

Corollary 2 indicates that the number of iterations needed to reach ε-accuracy is of the order $O\left(n^{4.5}\bar{\kappa}^{1.5}\ln\frac{1}{\varepsilon}\right)$.

The bound $\delta \leq 1 - \tau/(2n^2)$ can be conservative. For example, for the complete graph this bound tells us that $1 - \delta$ is bounded below by $1/n^3$, whereas one can see that for the complete graph $1 - \delta$ is actually bounded away from zero by a constant. However, in general, this bound cannot be improved, in the sense that there are graphs for which it is essentially tight. For example, on the (fixed) line or ring graph, the bound implies that $1 - \delta$ is lower bounded by $1/n^2$, which is the correct scaling up to a constant. This can be seen by observing that it may take two random walks at least $\Omega(n^2)$ steps to intersect on these graphs and, thus, the spectral gap has to be at least that much (for more details on making such arguments rigorous see the introductory chapter of [34]). For instances when the graph is *fixed*, the quantity δ can be bounded accurately using a connection between δ and the average hitting time (Theorem 11.11 of [32]), which in turn can be bounded in terms of graph resistance [12]. Using this connection, the following scalings may be obtained:

- On the complete graph, $\delta \leq 1 - \Omega(1)$.
- On the line and ring graphs, we have $\delta \leq 1 - \Omega(1/n^2)$.
- On the 2D grid and on the complete binary tree, $\delta \leq 1 - \Omega(1/(n \log n))$.
- On any regular graph, $\delta \leq 1 - \Omega(1/n^2)$—a consequence of the hitting time bounds of [16].
- On the star graph and on the two-star graph (defined to be two stars on $n/2$ nodes with a connection between the centers), $\delta \leq 1 - \Omega(1/n^2)$.

10.2.3 Over Time-Varying Directed Graphs

For directed graphs, distributed algorithms are desirable to work with mixing matrices that need not be doubly stochastic. This can be accomplished through the use of push-sum protocol, which relaxes the requirement of doubly stochastic mixing matrices to column stochastic matrices to achieve average consensus. We discuss an algorithm, termed Push-DIGing (Algorithm 5), that uses push-sum protocol for tracking the gradient average in time [54].

Suppose the agents communicate over a static strongly connected directed graph. Each agent i initially holds row i of $\mathbf{x}(0) \in \mathbb{R}^{n \times p}$ and would like to compute the average $\frac{1}{n} \mathbf{1}^\top \mathbf{x}(0)$. A possible decentralized approach is to construct a doubly stochastic matrix $\mathbf{W}$ and perform updates $\mathbf{x}(k + 1) = \mathbf{W}\mathbf{x}(k)$ starting from $\mathbf{x}(0)$. However, a construction of a doubly stochastic matrix needs a weight balancing procedure and it can be costly [22]. This becomes even less realistic when the graph is time-varying, as the maintenance of a doubly stochastic matrix sequence needs a real-time weight balancing process.

Alternatively, if every agent knows its out-degree, it is possible for the agents to construct a column stochastic matrix $\mathbf{C}$ and perform the following steps, initialized with $\mathbf{u}(0) = \mathbf{x}(0)$ and $\mathbf{v}(0) = \mathbf{1}$, to achieve the average (push-sum protocol [29]):

$$\mathbf{u}\text{-update: } \mathbf{u}(k + 1) = \mathbf{C}\mathbf{u}(k);$$
$$\mathbf{v}\text{-update: } \mathbf{v}(k + 1) = \mathbf{C}\mathbf{v}(k); \; \mathbf{V}(k + 1) = \text{diag}\{\mathbf{v}(k + 1)\};$$
$$\mathbf{x}\text{-update: } \mathbf{x}(k + 1) = (\mathbf{V}(k + 1))^{-1}\mathbf{u}(k + 1).$$

Noticing that $\bar{u}(k + 1) = \bar{u}(k)$, rows of $\mathbf{u}(k)$ are moving towards scaled averages with uneven scaling ratios across the vertices caused by the non-double stochasticity of $\mathbf{C}$ (the ratios are actually the elements of a right eigenvector of $\mathbf{C}$ corresponding to the eigenvalue 1). At the same time, $\mathbf{V}(k)$ is recording these ratios. By applying the recorded ratio inverse $(\mathbf{V}(k))^{-1}$ on $\mathbf{u}(k)$, the algorithm recovers the unscaled average of the rows in $\mathbf{x}(k)$. We formally state the Push-DIGing algorithm as follows.

Algorithm 5 Push-DIGing (time-varying directed graphs)

1: Set algorithmic parameter $\alpha > 0$ for all i and
2: For all i, initialize variables $x_i(0) \in \mathbb{R}^p$ and $y_i(0) = \nabla f_i(x_i(0))$;
3: **for** $k = 0, 1, \cdots$ **do**, every agent i in parallel decide mixing weights $W_{ij}(k) \in \mathbb{R}$ for all $j \in \mathcal{N}_i(k) \bigcup \{i\}$
4: $\quad u_i(k+1) = \displaystyle\sum_{j \in \mathcal{N}_i^{\text{in}}(k) \bigcup \{i\}} C_{ij}(k) \left(u_j(k) - \alpha y_j(k) \right)$;
5: $\quad v_i(k+1) = \displaystyle\sum_{j \in \mathcal{N}_i^{\text{in}}(k) \bigcup \{i\}} C_{ij}(k) v_j(k)$;
6: $\quad x_i(k+1) = u_i(k+1)/v_i(k+1)$;
7: $\quad y_i(k+1) = \displaystyle\sum_{j \in \mathcal{N}_i^{\text{in}}(k) \bigcup \{i\}} C_{ij}(k) y_j(k) + \nabla f_i(x_i(k+1)) - \nabla f_i(x_i(k))$;
8: **end for**

At every iteration k, each agent i sends its $u_i(k) - \alpha y_i(k)$, $y_i(k)$, and $v_i(k)$ all scaled by $C_{ij}(k)$ to each of its out-neighbors $j \in \mathcal{N}_i^{\text{out}}(k)$, and receives the corresponding messages from its in-neighbors $j \in \mathcal{N}_i^{\text{in}}(k)$. Then, each agent i updates its own $u_i(k+1)$ by summing its own $C_{ii}(k)(u_i(k) - \alpha y_i(k))$ and the received $C_{ij}(k)(u_j(k) - \alpha y_j(k))$ from its in-neighbors $\mathcal{N}_i^{\text{in}}(k)$; a similar strategy applies to the update of $v_i(k+1)$; then $x_i(k+1)$ is given by scaling $u_i(k+1)$ with $(v_i(k+1))^{-1}$; finally each agent i updates its own $y_i(k+1)$ by summing its own $C_{ii} y_i(k)$ and the received $C_{ij}(k) y_j(k)$ from its in-neighbors, and accumulating its current local gradient $\nabla f_i(x_i(k+1))$ and subtracting its previous local gradient $\nabla f_i(x_i(k))$ (in order to filter in only the new information contained in the most recent gradient). Unlike DIGing for undirected graphs in which each agent scales the received variables ($x_i(k)$ and $y_i(k)$) and then sums them up, in Push-DIGing, the variables ($u_i(k) - \alpha y_i(k)$, $y_i(k)$, and $v_i(k)$) are scaled before being sent out. This is due to the fact that, over directed graphs, usually a scaling weight $\mathbf{C}_{ij}$ can only be conveniently determined by the out-degree information of agent j which is not available to agent i.

10.2.3.1 Geometric Convergence

Consider a time-varying graph sequence $\{\mathcal{G}^{\text{dir}}(k)\}$, for which we assume that the following assumption hold.

Assumption 5 ($\tilde{B}_\Theta$-Strongly Connected Graph Sequence) *There exists an integer $\tilde{B}_\Theta > 0$ such that for any $t = 0, 1, \ldots$, the directed graph*

$$\mathcal{G}_{\tilde{B}_\Theta}^{\mathrm{dir}}(t\tilde{B}_\Theta) \triangleq \left\{ \mathcal{V}, \bigcup_{\ell = t\tilde{B}_\Theta}^{(t+1)\tilde{B}_\Theta - 1} \mathcal{A}(\ell) \right\}$$

is strongly connected.

Note that Assumption 5 implies that $\mathcal{G}_{B_\Theta}^{\mathrm{dir}}(k)$, with $B_\Theta = 2\tilde{B}_\Theta - 1$ is strongly connected for all $k = 0, 1, \ldots$.

Assumption 6 (Mixing Matrix Sequence $\{\mathbf{C}(k)\}$) *For any $k = 0, 1, \ldots$, the mixing matrix $\mathbf{C}(k) = [C_{ij}(k)] \in \mathbb{R}^{n \times n}$ is given by*

$$C_{ij}(k) = \frac{1}{d_j^{\mathrm{out}}(k) + 1} \text{ if } (j, i) \in \mathcal{A}(k), \text{ and otherwise } C_{ij}(k) = 0,$$

where $d_j^{\mathrm{out}}(k) = |\mathcal{N}_j^{\mathrm{out}}(k)|$ is the out-degree of agent j at time k.

We may have other options for the weights $C_{ij}(k)$. In the existing literature on push-sum consensus protocol, the best understood are the matrices relying on the out-degree information (as in Assumption 6).

Next, we provide a convergence rate estimate for the Push-DIGing algorithm.

Theorem 2 (Push-DIGing: Explicit Geometric Rate over Time-Varying Directed Graphs [54]) *Let Assumptions 4–6 hold. Let B be a large enough integer constant such that*

$$\delta \triangleq Q_1 \left(1 - \frac{1}{n^{(2+nB_\Theta)nB_\Theta}} \right)^{\frac{B-1}{nB_\Theta}} < 1.$$

Also, let the step-size α be chosen in the interval $\left(0, \frac{1.5(1-\delta)^2}{\bar{\mu} J_2} \right]$, where the constant J_2 is given by

$$J_2 = 3Q_1 \|\mathbf{V}^{-1}\|_{\max}^1 \bar{\kappa} B (\delta + Q_1(B-1))(1 + \sqrt{n}) \left(1 + 4\sqrt{n}\sqrt{\bar{\kappa}} \right).$$

Then, the sequence $\{\mathbf{x}(k)\}$ generated by Push-DIGing converges to the matrix with all rows equal to $(x^*)^\top$ at a global R-linear rate $O(\lambda^k)$, where x^* is the solution to the problem and λ is given by

$$\lambda = \begin{cases} \sqrt[2B]{1 - \frac{\alpha\bar{\mu}}{1.5}}, & \text{if } \alpha \in \left(0, \frac{1.5 \left(\sqrt{J_2^2 + (1-\delta^2)J_2} - \delta J_2 \right)^2}{\bar{\mu} J_2 (J_2+1)^2} \right], \\[4ex] \sqrt[B]{\sqrt{\frac{\alpha\bar{\mu} J_2}{1.5}} + \delta}, & \text{if } \alpha \in \left(\frac{1.5 \left(\sqrt{J_2^2 + (1-\delta^2)J_2} - \delta J_2 \right)^2}{\bar{\mu} J_2 (J_2+1)^2}, \frac{1.5(1-\delta)^2}{\bar{\mu} J_2} \right]. \end{cases}$$

We note that the bound is pessimistic as it applies to an arbitrary sequence of directed graphs subject only to the string connectivity condition imposed by Assumption 5. In this case, the parameter λ does not have a polynomial growth with n.

10.3 Resource Allocation and Its Connection to Consensus Optimization

We now turn our attention to the resource allocation problem (10.2). We let agent i hold a local variable x_i, a function f_i, and a constraint set Ω_i of the problem (10.2).

Our basic assumption is that problem (10.2) is convex, given as follows.

Assumption 7 (Functional Properties) *For any $i \in [n]$, the function $f_i : \mathbb{R}^p \to \mathbb{R}$ is convex while the set $\Omega_i \subseteq \mathbb{R}^p$ is nonempty, closed and convex.*

We define g_i as the indicator function of the set Ω_i, namely,

$$g_i(x_i) = \begin{cases} 0, & \text{if } x_i \in \Omega_i, \\ +\infty, & \text{if } x_i \notin \Omega_i. \end{cases}$$

We also define a composite function h_i for agent i, as follows:

$$h_i \triangleq f_i + g_i : \mathbb{R}^p \to \mathbb{R} \bigcup \{+\infty\}, \quad \forall i = 1, \ldots, n.$$

By Assumption 7, the functions $g_i : \mathbb{R}^p \to \mathbb{R} \bigcup \{+\infty\}$ are proper, closed, and convex. Since the domain of f_i is $\mathbb{R}^p$, the functions $h_i = f_i + g_i$ are also proper, closed, and convex. By Assumption 7, the subdifferential sets $\partial h_i(x_i)$ satisfy

$$\partial h_i(x_i) = \partial f_i(x_i) + \partial g_i(x_i) \qquad \text{for all } x \in \mathbb{R}^p \qquad (10.6)$$

(see Theorem 23.8 of [67]). Note that $\partial g_i(x_i)$ coincides with the normal cone of Ω_i at x_i, so that $\partial h_i(x_i) \neq \emptyset$ for all $x_i \in \Omega_i$.

We introduce

$$\mathbf{g}(\mathbf{x}) \triangleq \sum_{i=1}^{n} g_i(x_i), \qquad \text{and} \qquad \mathbf{h}(\mathbf{x}) \triangleq \sum_{i=1}^{n} h_i(x_i), \qquad (10.7)$$

where $[x_1^\top; x_2^\top; \ldots; x_n^\top]$. Similarly, we define a matrix $\mathbf{r}$ by using the vectors r_i, $i = 1, \ldots, n$. Using $\widetilde{\nabla} f_i(x_i)$ to denote a subgradient of f_i at x_i, we construct a matrix $\widetilde{\nabla} \mathbf{f}(\mathbf{x})$ of subgradients $\widetilde{\nabla} f_i(x_i)$, as follows:

$$\widetilde{\nabla} \mathbf{f}(\mathbf{x}) \triangleq \begin{pmatrix} - (\widetilde{\nabla} f_1(x_1))^\top - \\ - (\widetilde{\nabla} f_2(x_2))^\top - \\ \vdots \\ - (\widetilde{\nabla} f_n(x_n))^\top - \end{pmatrix} \in \mathbb{R}^{n \times p}. \qquad (10.8)$$

Similarly, the matrices $\widetilde{\nabla}\mathbf{g}(\mathbf{x})$ and $\widetilde{\nabla}\mathbf{h}(\mathbf{x})$ are constructed from subgradients of g_i and $h_i = f_i + g_i$ at x_i, respectively. Each row i of $\mathbf{x}$, $\mathbf{r}$, $\widetilde{\nabla}\mathbf{f}(\mathbf{x})$, $\widetilde{\nabla}\mathbf{g}(\mathbf{x})$, and $\widetilde{\nabla}\mathbf{h}(\mathbf{x})$ corresponds to the information pertinent only to agent i.

Let us consider for the moment that we are dealing with the resource allocation problem (10.2) over a time-invariant undirected graph $\mathcal{G}$. Let $\mathbf{L}_{\mathcal{G}}$ denote the (standard) Laplacian matrix associated with the graph $\mathcal{G}$, i.e., $\mathbf{L}_{\mathcal{G}} = D - J$, where D is the diagonal matrix with diagonal entries $D_{ii} = d_i$ and d_i being the number of edges incident to node i, while J is the graph adjacency matrix (with $J_{ij} = 1$ when $\{i, j\} \in \mathcal{E}$ and $J_{ij} = 0$ otherwise). Note that $\mathbf{L}_{\mathcal{G}}$ is compatible with $\mathcal{G}$, symmetric and positive semidefinite.

In the algorithms that will be discussed, we use the matrix $\mathbf{L} = [Ł_{ij}]$ whose behavior is "close to or the same as" that of $\mathbf{L}_{\mathcal{G}}$, in the sense of the following assumption.

Assumption 8 (Graph Connectivity and Null/Span Property) *The graph $\mathcal{G}$ is connected and a matrix $\mathbf{L}$ is compatible with the graph $\mathcal{G}$. Furthermore, $\mathbf{L} = U^\top U$ for some full row-rank matrix $U \in \mathbb{R}^{(n-1) \times n}$ and $\mathrm{null}\{U\} = \{a\mathbf{1} \in \mathbb{R}^n \mid a \in \mathbb{R}\}$.*

Assumption 8 states that $\mathrm{null}\{U\} = \mathrm{null}\{\mathbf{L}\}$ implying that $x_1 = x_2 = \cdots = x_n \Leftrightarrow \mathbf{L}\mathbf{x} = 0 \Leftrightarrow U\mathbf{x} = 0$. The matrix $\mathbf{L}$ may be chosen in several different ways:

- Observing that the graph Laplacian $\mathbf{L}_{\mathcal{G}}$ satisfies Assumption 8, we can set $\mathbf{L} = \mathbf{L}_{\mathcal{G}}$. In this case, if each agent knows the number of its neighbors (its degree), the matrix $\mathbf{L}$ can be constructed without any communication among the agents.
- We can set $\mathbf{L} = \mathbf{L}_{\mathcal{G}}/\lambda_{\max}\{\mathbf{L}_{\mathcal{G}}\}$. The network needs $\lambda_{\max}\{\mathbf{L}_{\mathcal{G}}\}$ to configure this matrix, which can be done by preprocessing to retrieve $\lambda_{\max}\{\mathbf{L}_{\mathcal{G}}\}$ (see [76]).
- We can choose $\mathbf{L} = 0.5(I - W)$ where W is a symmetric doubly stochastic matrix that is compatible with the graph $\mathcal{G}$ and $\lambda_{\max}\{W - \mathbf{1}\mathbf{1}^\top/n\} < 1$. This matrix can be constructed in the network through a few rounds of local interactions between the agents by using local strategies for determining W, such as the Metropolis-Hasting rule (can be done in one round of local interactions [2, 84]).

10.3.1 The Resource Allocation and Consensus Optimization Problems

Here, we compare the first-order optimality conditions for problem (10.2) and for consensus optimization. For this, using the notation introduced in the preceding section we rewrite the resource allocation problem (10.2), as follows:

$$\min_{\mathbf{x} \in \mathbb{R}^{n \times p}} \ \mathbf{h}(\mathbf{x}) = \sum_{i=1}^{n} h_i(x_i),$$
$$\text{s.t.} \quad \mathbf{1}^\top(\mathbf{x} - \mathbf{r}) = 0. \tag{10.9}$$

By using the Lagrangian function, we provide the optimality conditions for problem (10.9) in a special form, as given in the following lemma.

Lemma 1 (First-Order Optimality Condition for (10.9) **[49])** *Let Assumptions 7 and 8 hold, and let $c \neq 0$ be a given scalar. Then, $\mathbf{x}^*$ is an optimal solution of* (10.9) *if and only if there exists a matrix $\mathbf{q}^* \in \mathbb{R}^{(n-1) \times p}$ such that the pair $(\mathbf{x}^*, \mathbf{q}^*)$ satisfies the following relations:*

$$\mathbf{x}^* - \mathbf{r} + cU^\top \mathbf{q}^* = \mathbf{0}, \tag{10.10a}$$

$$U \widetilde{\nabla} \mathbf{h}(\mathbf{x}^*) = \mathbf{0}, \tag{10.10b}$$

where U is the matrix defined in Assumption 8.

The optimality conditions for the resource optimization problem have an interesting connection with those for the consensus optimization problem. To expose this relation, we re-write the consensus optimization problem in the following form:

$$\min_{\mathbf{x} \in \mathbb{R}^{n \times p}} \ \mathbf{h}(\mathbf{x}) = \sum_{i=1}^{n} h_i(x_i), \tag{10.11}$$
$$\text{s.t.} \quad x_1 = x_2 = \cdots = x_n.$$

The local objective of each agent in (10.11) is the same as that in (10.9). Unlike the resource allocation problem, instead of having $\sum_{i=1}^{n}(x_i - r_i) = 0$ as constraints, here we have the consensus constraints, i.e., $x_1 = x_2 = \cdots = x_n$.

The first-order optimality condition of (10.11) is stated in the following lemma.

Lemma 2 (First-Order Optimality Condition for (10.11) **[49])** *Let Assumptions 7 and 8 hold, and let $c \neq 0$ be a given scalar. Then, $\mathbf{x}^*$ is an optimal solution of* (10.11) *if and only if there exists a matrix $\mathbf{q}^* \in \mathbb{R}^{(n-1) \times p}$ such that the pair $(\mathbf{x}^*, \mathbf{q}^*)$ satisfies the following relations:*

$$\widetilde{\nabla} \mathbf{h}(\mathbf{x}^*) + cU^\top \mathbf{q}^* = \mathbf{0}, \tag{10.12a}$$

$$U\mathbf{x}^* = \mathbf{0}, \tag{10.12b}$$

where U is the matrix defined in Assumption 8.

10.3.2   The Mirror Relationship

The Lagrangian dual problem of the resource allocation problem is a consensus optimization problem (see Section 7.3 of [9] or the discussion around equations (4)–(6) in [13]). As pointed out in [13], a distributed optimization method that can solve the consensus optimization problem may also be used for the resource allocation problem through solving the dual of the resource allocation problem.

Table 10.2 Summary of optimality conditions

Opt. cond. of resource allocation, (10.10)	Opt. cond. of consensus optimization, (10.12)
$\mathbf{x}^* - \mathbf{r} = -cU^\top \mathbf{q}^*$	$\widetilde{\nabla}\mathbf{h}(\mathbf{x}^*) = -cU^\top \mathbf{q}^*$
$U\widetilde{\nabla}\mathbf{h}(\mathbf{x}^*) = \mathbf{0}$	$U\mathbf{x}^* = \mathbf{0}$

Here, however, we will provide more special relations for the two problems leading to a class of resource allocation algorithms following the design of a class of decentralized consensus optimization algorithms. The relations are based on the optimality conditions for these two problems, which are summarized in Table 10.2. These conditions share the same form, in the sense that if in (10.12), we replace $\widetilde{\nabla}\mathbf{h}(\mathbf{x}^*)$ by $\mathbf{x}^* - \mathbf{r}$ and $\mathbf{x}^*$ by $\widetilde{\nabla}\mathbf{h}(\mathbf{x}^*)$, it will result in (10.10). Formally, both conditions can be written as $\mathcal{A}(\mathbf{x}^*) = -cU^\top \mathbf{q}^*$, $U\mathcal{B}(\mathbf{x}^*) = \mathbf{0}$ for some $\mathcal{A} : \mathbb{R}^{n \times p} \to \mathbb{R}^{n \times p}$ and $\mathcal{B} : \mathbb{R}^{n \times p} \to \mathbb{R}^{n \times p}$. The only difference between relations (10.10) and (10.12) is in the exchange of the roles of the span space of $U^\top$ and the null space of U. In the resource allocation problem, the vector $\widetilde{\nabla}\mathbf{h}(\mathbf{x})$ needs to be consensual while the rows of $\mathbf{x} - \mathbf{r}$ summing to 0. In the consensus optimization problem, the vector $\mathbf{x}$ needs to be consensual while the rows of $\widetilde{\nabla}\mathbf{h}(\mathbf{x})$ are summing to $\mathbf{0}$. Thus, in an iterative consensus optimization algorithm with iteration index k, we may substitute the image of the "subgradient map" $\widetilde{\nabla}\mathbf{h}(\mathbf{x}(k))$ by that of some other map $\mathcal{A}(\mathbf{x}(k))$ and substitute the image of the identity map $\mathbf{x}(k)$ by that of some other map $\mathcal{B}(\mathbf{x}(k))$, with hope to obtain an algorithm for solving the resource allocation problem. For that to happen, we would need $\mathcal{A}(\mathbf{x}(k))$ to approach the span space of $U^\top$ and $\mathcal{B}(\mathbf{x}(k))$ to approach the null space of U, as $k \to \infty$.

10.4 Decentralized Resource Allocation Algorithms

Let us introduce the solution set of the resource allocation problem (10.2), denoted by $\mathcal{X}^*$. We make the following assumption for problem (10.2).

Assumption 9 *The solution set $\mathcal{X}^*$ of the resource allocation problem* (10.2) *is nonempty. Furthermore, a Slater condition is satisfied, i.e., there is a point $\tilde{\mathbf{x}}$ which satisfies the linear constraints in* (10.2) *and lies in the relative interior of the set constraint* $\Omega = \Omega_1 \times \cdots \times \Omega_n$.

The set $\mathcal{X}^*$ is nonempty, for example, when the constraint set Ω of the resource allocation problem (10.2) is compact, or when the objective function satisfies some growth condition. Under Assumptions 7 and 9, the strong duality holds for problem (10.2) and its Lagrangian dual problem, and the dual optimal set is nonempty (see Proposition 6.4.2 of [8]). Thus, under Assumptions 7 and 9, the primal-dual optimal pairs for problem (10.2) exist and they satisfy the optimality conditions (10.10).

10.4.1   Over Time-Invariant Undirected Graphs

The basic algorithm we will consider is Mirror-P-EXTRA of [49], given as follows.

Algorithm 6 Mirror-P-EXTRA (time-invariant undirected graphs)

1: Each agent i chooses its own parameter $\beta_i > 0$ and the same parameter $c > 0$;
2: Each agent i initializes with $x_i(0) \in \Omega_i$, $s_i(0) = \widetilde{\nabla} f_i(x_i(0))$, and $y_i(-1) = \mathbf{0}$;
3: **for** $k = 0, 1, \dots$ **do**, each agent i
4: $\quad y_i(k) = y_i(k-1) + \sum_{j \in \mathcal{N}_i \bigcup\{i\}} \mathbf{L}_{ij} s_j(k)$;
5: $\quad x_i(k+1) = \underset{x_i \in \Omega_i}{\operatorname{argmin}} \left\{ f_i(x_i) - \langle s_i(k), x_i \rangle + \frac{1}{2\beta_i} \|x_i - r_i + 2cy_i(k) - cy_i(k-1)\|_2^2 \right\}$;
6: $\quad s_i(k+1) = s_i(k) - \frac{1}{\beta_i}(x_i(k+1) - r_i + 2cy_i(k) - cy_i(k-1))$;
7: **end for**

The algorithm is motivated by P-EXTRA algorithm of [72] for a consensus optimization problem. Its convergence can be guaranteed under Assumptions 7–9 and a condition on the parameters β_i and $c > 0$, as provided in the following theorem.

Theorem 3 (Convergence of Mirror-P-EXTRA [49])   *Let Assumptions 7–9 hold. Let the parameters β_i and $c > 0$ be such that $B - c\mathbf{L} \succcurlyeq \mathbf{0}$, where B is the diagonal matrix with entries $\beta_1, \dots, \beta_n$ on its main diagonal. Then, the sequence $\{\mathbf{x}^k\}$ converges to a point in the optimal solution set $\mathcal{X}^*$.*

With additional assumptions on the objective function, we can provide stronger convergence rate results. In particular, a geometric convergence rate can be shown for the case when $\boldsymbol{\Omega}$ is the full space, i.e., in which case $\mathbf{h} = \mathbf{f}$. The geometric rate is obtained under strong convexity and the Lipschitz gradient property of the objective function $\mathbf{f}$, which are formally imposed by the following assumptions.

Assumption 10   *The gradient map ∇f_i of each function $f_i : \mathbb{R}^p \to \mathbb{R}$ is L_i-Lipschitz.*

Assumption 11   *Each function f_i is μ_i-strongly convex.*

Under these assumptions, the following convergence rate result is valid for the iterates of Mirror-P-EXTRA.

Theorem 4 (Linear Rate of Mirror-P-EXTRA [49])   *Let $\Omega_i = \mathbb{R}^p$ for all i, and let Assumption 8, 10 and 11 hold. Also, assume that the parameters $\beta_i, i = 1, \dots, n$ and $c > 0$ are such that $B - c\mathbf{L} \succcurlyeq \mathbf{0}$, where B is the diagonal matrix with entries $\beta_1, \dots, \beta_n$ on its main diagonal. Then, the sequence $\{\mathbf{x}(k)\}$ converges to the optimal solution $\mathbf{x}^*$ of problem (10.2) at an R-linear rate $O\left(\left(\frac{1}{1+\delta}\right)^k\right)$ for some small[6] enough $\delta > 0$.*

---
[6]A description of the range for δ can be found in [49].

When $B = \beta I$, further refinements of Theorem 4 can be provided. These results make use of the condition number $\kappa_{\mathbf{f}}$ of the objective function $\mathbf{f}$ given by

$$\kappa_{\mathbf{f}} = \frac{L}{\mu} \quad \text{where } \mu = \min_i \mu_i \ \text{ and } \ L = \max_i L_i.$$

The following corollary provides more details about the scalability of the algorithm in terms of the condition number $\kappa_{\mathbf{f}}$ of the function $\mathbf{f}$ and the condition number $\kappa_{\mathbf{L}} = \frac{1}{\tilde{\lambda}_{\min}\{\mathbf{L}\}}$ of the matrix $\mathbf{L}$.

Corollary 3 (The Scalability of Mirror-P-EXTRA [49]) *Let assumptions of Theorem 4 hold, and let $B = \beta I$. Then, for Mirror-P-EXTRA to reach an ε-accuracy, the number of iterations needed is of the order $O\left(\left(\kappa_{\mathbf{L}} + \sqrt{\kappa_{\mathbf{L}}\kappa_{\mathbf{f}}}\right)\ln\left(\frac{1}{\varepsilon}\right)\right)$.*

A more detailed result that captures the scalability in terms of the number n of agents can be obtained assuming a specific structure on the matrix $\mathbf{L}$. In particular, when using lazy Metropolis matrix,

$$W_{ij} = \begin{cases} 1/\left(2\max\{|\mathcal{N}_i|, |\mathcal{N}_j|\}\right), & \text{if } \{i, j\} \in \mathcal{E}, \\ 0, & \text{if } \{i, j\} \notin \mathcal{E} \text{ and } j \neq i, \\ 1 - \sum_{l \in \mathcal{N}_i} W_{il}, & \text{if } j = i, \end{cases}$$

we have the following result.

Corollary 4 (Agent Number Dependency [49]) *Under the assumptions of Corollary 3, by letting $\mathbf{L} = 0.5(I - W)$ where W is a lazy Metropolis matrix, and by setting $c = \Theta\left(n\mu^{-0.5}L^{-0.5}\right)$ and $\beta = c$ in Mirror-P-EXTRA, to reach an ε-accuracy, the number of iterations needed is of the order of $O\left((n^2 + n\kappa_{\mathbf{f}}^{0.5})\ln\left(\varepsilon^{-1}\right)\right)$.*

Corollary 4 can obtained from Corollary 3 and the fact that $\kappa_{\mathbf{L}} = O(n^2)$ under the choice $\mathbf{L} = 0.5(I - W)$. Other bounds on $\kappa_{\mathbf{L}}$, which may lead to different scalability result for other classes of graphs, can be found in Remark 2 of [54].

Although Mirror-P-EXTRA scales favorably with the condition number of the graph and the objective function, the computational cost per-iteration is high. To address this issue, we consider a gradient-based scheme which has a low per-iteration cost. This scheme is the most efficient for the unconstrained case $(\Omega_i = \mathbb{R}^p)$.

10.4.1.1 Unconstrained Case: Mirror-EXTRA

We consider the resource allocation without local constraints, i.e.,

$$\begin{aligned} \min_{\mathbf{x}} \ \mathbf{f}(\mathbf{x}) &= \sum_{i=1}^n f_i(x_i), \\ \text{s.t.} \ \mathbf{1}^\top(\mathbf{x} - \mathbf{r}) &= \mathbf{0}. \end{aligned} \tag{10.13}$$

In this case, the Mirror-EXTRA algorithm of [49] can be applied, which is given below.

Algorithm 7 Mirror-EXTRA (time-invariant undirected graphs)

1: A parameter c is selected such that $c \in \left(0, \frac{1}{2L\lambda_{\max}\{L\}}\right)$;
2: Each agent i initializes with $x_i(0) \in \mathbb{R}^p$ and $y_i(-1) = \mathbf{0}$;
3: **for** $k = 0, 1, \ldots$ **do**, each agent i
4:     $y_i(k) = y_i(k-1) + \sum_{j \in \mathcal{N}_i \cup \{i\}} L_{ij} \nabla f_j \nabla f_j \nabla f_j(x_j(k))$;
5:     $x_i(k+1) = r_i - 2cy_i(k) + cy_i(k-1)$;
6: **end for**

This algorithm has a lower per-iteration cost in the update of $x_i(k)$ than that of Mirror-P-EXTRA, as it requires a gradient evaluation instead of a "prox"-type update. We next provide a basic convergence result for the iterates of Mirror-EXTRA.

Theorem 5 (Convergence of Mirror-EXTRA [49]) *Let $\Omega_i = \mathbb{R}^p$ for all i, and let Assumptions 7, 8, and 10 hold. Also, assume that problem (10.13) has a solution, i.e., $\mathcal{X}^* \neq \emptyset$. If the parameter c is chosen such that $0 < c < \frac{1}{2L\lambda_{\max}\{L\}}$, then the sequence $\{\mathbf{x}^k\}$ converges to a point in $\mathcal{X}^*$.*

In Theorem 5, the condition $c \leq \frac{1}{2L\lambda_{\max}\{L\}}$ seemingly requires knowledge of the graph $\mathcal{G}$ structure. However, such a requirement can be avoided by using, for example, $L = 0.5(I - W)$, where W is a symmetric stochastic matrix, in which case one may use $c \leq \frac{1}{2L}$.

The convergence rate of Mirror-EXTRA is R-linear, as given in the following theorem.

Theorem 6 (Linear Rate of Mirror-EXTRA [49]) *Let $\Omega_i = \mathbb{R}^p$ for all i, and let Assumptions 8, 10 and 11 hold. Furthermore, suppose that L is such that $\lambda_{\max}\{L\} \leq 1$ and the parameter c satisfies $c \in (0, \frac{1}{2L})$. Then, the iterate sequence $\{\mathbf{x}(k)\}$ generated by Algorithm 7 converges to the optimal solution $\mathbf{x}^*$ of problem (10.2) at an R-linear rate, i.e., $\|\mathbf{x}^{k+1} - \mathbf{x}^*\|_F^2 = O\left(\frac{1}{(1+\delta)^k}\right)$ for some small[7] enough $\delta > 0$.*

Theorem 6 implies the following scalability result for Mirror-EXTRA.

Corollary 5 (Scalability of Mirror-EXTRA [49]) *Under the conditions of Theorem 6, for Algorithm 7 to reach an ε-accuracy, the number of iterations needed is of the order $O\left(\kappa_L \kappa_f \ln\left(\frac{1}{\varepsilon}\right)\right)$.*

Note that $O\left(\kappa_L \kappa_f \ln\left(\frac{1}{\varepsilon}\right)\right)$ scalability of Mirror-EXTRA also coincides with that of the algorithm proposed in [18]. The complexity of the Mirror-EXTRA is slightly worse than the complexity of Mirror-P-EXTRA, which is

---

[7]A more specific description of the range for δ can be found in [49].

$O\left((\kappa_{\mathbf{L}} + \sqrt{\kappa_{\mathbf{L}}\kappa_{\mathbf{f}}})\ln\left(\frac{1}{\varepsilon}\right)\right)$. Thus, the Mirror-EXTRA has a lower per-iteration cost at the expense of less favorable scalability.

10.4.1.2 Projection Gradient-Based Algorithm: Mirror-PG-EXTRA

Since Algorithm 7 cannot be applied to constrained problems, we consider another algorithm that uses a gradient-projection update and, thus, has a lower computational cost per iteration as compared to Algorithm 6. The algorithm is given below (Algorithm 8).

Algorithm 8 Mirror-PG-EXTRA (time-invariant undirected graphs)

1: Each agent uses a common parameter $c \in (0, \frac{1}{2L\lambda_{\max}\{\mathbf{L}\}})$;
2: Each agent chooses its own parameter β_i such that $B \succcurlyeq c\mathbf{L}$;
3: Each agent i initializes with $x_i^0 \in \Omega_i$, $s_i^0 = 0$, and $y_i^{-1} = 0$;
4: **for** $k = 0, 1, \ldots$ **do**, each agent i
5: $\quad y_i^k = y_i^{k-1} + \sum_{j \in \mathcal{N}_i \bigcup\{i\}} \mathbf{L}_{ij}(\nabla \mathbf{f}_j(x_j^k) + s_j^k)$;
6: $\quad x_i^{k+1} = \mathcal{P}_{\Omega_i}\left\{r_i - 2cy_i^{k+1} + cy_i^k + \beta_i s_i^k\right\}$;
7: $\quad s_i^{k+1} = s_i^k - \frac{1}{\beta_i}(x_i^{k+1} - r_i + 2cy_i^k - cy_i^{k-1})$;
8: **end for**

In Algorithm 8, the matrix B is the diagonal matrix with entries $B_{ii} = \beta_i$ on its diagonal, while $\mathcal{P}_{\Omega_i}[\cdot]$ is the projection operator on the set Ω_i with respect to the Euclidean norm. In the absence of the per-agent-constraints, Algorithm 8 reduces to Algorithm 7. Algorithm 8 is proposed merely to serve as an alternative for Mirror-P-EXTRA when the prox-type update is costly. We have not analyzed its convergence properties. We will evaluate it numerically in our simulations, which are reported later in Sect. 10.5. We believe that Algorithm 8 has an $O(1/k)$ convergence rate.

10.4.2 Over Time-Varying Directed Graphs

In our simulations, we will also use Mirror-Push-DIGing of [49]. The details of its construction are omitted here and they can be found in [49]. The implementation of Mirror-Push-DIGing is provided below.

Algorithm 9 Mirror-Push-DIGing for time-varying directed graphs

---

1: Each agent i chooses the same parameter $\alpha > 0$ and picks any $x_i(0) \in \Omega_i$;
2: Each agent i initializes with $v_i(0) = 1$, $s_i(0) = \frac{1}{\alpha}\tilde{\nabla}f_i(x_i(0)) + x_i(0)$, and $y_i(0) = x_i(0) - r_i$;
3: **for** $k = 0, 1, \ldots$ **do**, each agent i
4: $\quad v_i(k+1) = \sum_{j \in \mathcal{N}_i^{\text{in}}(k)\bigcup\{i\}} C_{ij}(k)v_j(k)$;
5: $\quad s_i(k+1) = \sum_{j \in \mathcal{N}_i^{\text{in}}(k)\bigcup\{i\}} C_{ij}(k)\left(s_j(k) - x_j(k) - y_j(k)\right) + x_i(k)$;
6: $\quad x_i(k+1) = \arg\min_{x_i \in \Omega_i}\left\{f_i(x_i) + \frac{\alpha}{2v_i(k+1)}\|x_i - s_i(k+1)\|_2^2\right\}$;
7: $\quad y_i(k+1) = \sum_{j \in \mathcal{N}_i^{\text{in}}(k)\bigcup\{i\}} C_{ij}(k)y_j(k) + x_i(k+1) - x_i(k)$;
8: **end for**

Here, $\mathcal{N}_i^{in}(k)$ denotes the set of in-neighbors of agent i at iteration k. Each agent j sends the quantities $C_{ij}(k)v_j(k)$, $C_{ij}(k)\left(s_j(k) - x_j(k) - y_j(k)\right)$ and $C_{ij}(k)y_j(k)$ to its out-neighbors. Upon receiving these quantities from its in-neighbors $j \in \mathcal{N}_i^{in}(k)$, each agent i sums these quantities over $j \in \mathcal{N}_i^{in}(k)$ and combines them with its own information, at different steps in different ways. We have not performed convergence analysis of Algorithm 9. Based on what we have obtained in Ref. [54], a geometric convergence can be obtained when each $\Omega_i = \mathbb{R}^p$. We have some numerical tests of this algorithm later in Sect. 10.5.

10.4.3 Decentralized Resource Allocation with Local Couplings

We consider a generalization of the resource allocation problem with local linear coupling of multiple resources, given as follows:

$$
\begin{aligned}
\min_{\mathbf{x}\in\mathbb{R}^p} \ & \mathbf{h}(\mathbf{x}) = \sum_{i=1}^{n} h_i(x_i), \\
\text{s.t.} \ & \sum_{i=1}^{n} (A_i x_i - r_i) = \mathbf{0},
\end{aligned}
\tag{10.14}
$$

where $x_i \in \mathbb{R}^{p_i}$, $A_i \in \mathbb{R}^{m \times p_i}$, $r_i \in \mathbb{R}^m$, and $p = \sum_{i=1}^n p_i$. Here, we have $h_i \triangleq f_i + g_i$ with g_i being the indicator function of the local constraint $x_i \in \Omega_i$.

It can be seen that the optimality conditions for (10.14) are given by

$$
\begin{cases}
\tilde{\nabla} h_i(x_i^*) = A_i^\top \xi_i^*, & \text{for all } i = 1, 2, \ldots, n, & (10.15a) \\[2mm]
\xi_1^* = \xi_2^* = \ldots = \xi_n^*, & & (10.15b) \\[2mm]
\displaystyle\sum_{i=1}^{n}(A_i x_i^* - r_i) = \mathbf{0}, & & (10.15c)
\end{cases}
$$

where $\xi_i \in \mathbb{R}^m$ are auxiliary variables. Again, here we can spot the consensus relation (10.15b) and the "summing to zero" relation (10.15c), similar forms of which have appeared in the optimality condition of the original resource allocation problem. In particular, the original resource allocation problem corresponds to the case $A_i = I$ for all i, and the above conditions reduce to the optimality conditions in (10.10).

To simplify notation, define

$$
\mathbf{x} \triangleq \begin{pmatrix} x_1 \\ x_2 \\ \vdots \\ x_n \end{pmatrix} \in \mathbb{R}^p; \quad
\mathbf{A} \triangleq \begin{pmatrix} A_1 & 0 & \cdots & 0 \\ 0 & A_2 & \cdots & 0 \\ \vdots & \vdots & \ddots & \vdots \\ 0 & 0 & \cdots & A_n \end{pmatrix} \in \mathbb{R}^{(mn)\times p}; \quad
\mathbf{r} \triangleq \begin{pmatrix} r_1 \\ r_2 \\ \vdots \\ r_n \end{pmatrix} \in \mathbb{R}^{mn}.
\tag{10.16}
$$

Unlike the notation in the preceding sections, here we cannot use the compact notation in a matrix form since the vectors x_i's have different dimensions. Also, define

$$\tilde{\nabla}\mathbf{h}(\mathbf{x}) \triangleq \left(\tilde{\nabla}h_1(x_1); \tilde{\nabla}h_2(x_2); \cdots ; \tilde{\nabla}h_n(x_n)\right) \in \mathbb{R}^p,$$

$$\boldsymbol{\xi} \triangleq (\xi_1; \xi_2; \cdots ; \xi_n) \in \mathbb{R}^{mn},$$

$$\mathbf{L} \triangleq \mathbf{L} \otimes I \in \mathbb{R}^{(mn)\times(mn)},$$

where "$\otimes$" represents the Kronecker product. In this notation, the optimality conditions in (10.15) can be represented compactly in the following form:

$$\begin{cases} \tilde{\nabla}\mathbf{h}(\mathbf{x}^*) = \mathbf{A}^\top \boldsymbol{\xi}^*, & \text{(10.17a)} \\ \mathbf{L}\boldsymbol{\xi}^* = \mathbf{0}, & \text{(10.17b)} \\ (\mathbf{1} \otimes I_m)^\top (\mathbf{A}\mathbf{x}^* - \mathbf{r}) = \mathbf{0}, & \text{(10.17c)} \end{cases}$$

Next we want to construct sequences $\{\mathbf{x}(k)\}$ and $\{\boldsymbol{\xi}(k)\}$ that converge, respectively, to points $\mathbf{x}^*$ and $\boldsymbol{\xi}^*$ that satisfy the conditions in (10.17). One such a construction is as follows:

$$\mathbf{A}\mathbf{x}(k+1) - \mathbf{r} + c\mathbf{y}(k+1) + (\mathbf{B} - c\mathbf{L})(\boldsymbol{\xi}(k+1) - \boldsymbol{\xi}(k)) = \mathbf{0}, \quad \text{(10.18a)}$$

$$\tilde{\nabla}\mathbf{h}(\mathbf{x}(k+1)) = \mathbf{A}^\top \boldsymbol{\xi}(k+1), \quad \text{(10.18b)}$$

$$\tilde{\nabla}\mathbf{h}(\mathbf{x}(k)) = \mathbf{A}^\top \boldsymbol{\xi}(k), \quad \text{(10.18c)}$$

$$\mathbf{y}(k+1) = \mathbf{y}(k) + \mathbf{L}\boldsymbol{\xi}(k+1), \quad \text{(10.18d)}$$

where $\mathbf{B} = \text{blkdiag}\{B_1, B_2, \ldots, B_n\} \in \mathbb{R}^{(mn)\times(mn)}$ is the block diagonal matrix with each B_i a matrix selected by agent i. Each matrix step size B_i is maintained by agent i and $\mathbf{B} - c\mathbf{L} \succ 0$ should be fulfilled. One can verify that if the sequences produced by the updates in (10.18) converge, then the sequence $\{(\mathbf{x}(k), \boldsymbol{\xi}(k)\}$ converge to $\{(\mathbf{x}^*, \boldsymbol{\xi}^*)\}$ that satisfies the optimality conditions in (10.17).

The relations in (10.18) can be equivalently written in the following form:

$$\mathbf{x}(k+1) = \arg\min_{\mathbf{x}\in\Omega} \phi^k(\mathbf{x}), \quad \text{(10.19a)}$$

$$\boldsymbol{\xi}(k+1) = \boldsymbol{\xi}(k) - c\mathbf{B}^{-1}(2\mathbf{y}(k) - \mathbf{y}(k-1)) - \mathbf{B}^{-1}(\mathbf{A}\mathbf{x}(k+1) - \mathbf{r}), \quad \text{(10.19b)}$$

$$\mathbf{y}(k+1) = \mathbf{y}(k) + \mathbf{L}\boldsymbol{\xi}(k+1), \quad \text{(10.19c)}$$

where $\boldsymbol{\Omega} = \Omega_1 \times \cdots \times \Omega_n$ and

$$\phi^k(\mathbf{x}) = \mathbf{f}(\mathbf{x}) + \|\mathbf{x}\|^2_{\mathbf{A}^\top \mathbf{B}^{-1}\mathbf{A}} + \langle 2c\mathbf{A}^\top\mathbf{B}^{-1}(2\mathbf{y}(k) - \mathbf{y}(k-1)) - \mathbf{A}^\top\mathbf{B}^{-1}\mathbf{r} - \mathbf{A}^\top\boldsymbol{\xi}(k), \mathbf{x}\rangle.$$

The method is initialized with $\boldsymbol{\xi}(0) \in \mathbb{R}^p$, $\mathbf{x}(0) = \arg\min_{\mathbf{x}\in\boldsymbol{\Omega}}\{\mathbf{f}(\mathbf{x}) + \langle \mathbf{A}^\top\boldsymbol{\xi}(0), \mathbf{x}\rangle\}$, $\mathbf{y}(-1) = \mathbf{0}$, and $\mathbf{y}(0) = \mathbf{L}\boldsymbol{\xi}(0)$. The corresponding per-agent implementation is given below.

Algorithm 10 Mirror-P-EXTRA handling local couplings

1: Each agent i chooses its own matrix $B_i \in \mathbb{R}^{m\times m}$ and the same parameter $c > 0$;
2: Each agent i initializes with arbitrary $\xi_i(0) \in \mathbb{R}^{p_i}$ and with $y_i(-1) = \mathbf{0}$;
3: Each agent sets $x_i(0) = \arg\min_{x_i\in\Omega_i}\{f_i(x_i) + \langle A_i^\top\xi_i(0), x_i\rangle\}$;
4: **for** $k = 0, 1, \ldots$ **do**, each agent i
5:     $y_i(k) = y_i(k-1) + \sum_{j\in\mathcal{N}_i\bigcup\{i\}}\mathbf{L}_{ij}\xi_j(k)$;
6:     $x_i(k+1) = \arg\min_{x_i\in\Omega_i}\phi_i^k(x_i)$;
7:     $\xi_i(k+1) = \xi_i(k) - cB_i^{-1}(2y_i(k) - y_i(k-1)) - B_i^{-1}(A_ix_i(k+1) - r_i)$;
8: **end for**

The function $\phi_i^k(x_i)$ in the update of $x_i(k+1)$ is given by

$$\phi_i^k(x_i) = \Big\{ f_i(x_i) + \|x_i\|^2_{A_i^\top B_i^{-1}A_i}$$
$$+ \langle -A_i^\top B_i^{-1}r_i - A_i^\top\xi_i(k) + 2cA_i^\top B_i^{-1}(2y_i(k) - y_i(k-1)), x_i\rangle\Big\}.$$

It can be seen that this algorithm reduces to Mirror-P-EXTRA (Algorithm 6) when $A_i = I$ for all i. We have not analyzed the convergence properties of this algorithm. We believe that its convergence behavior is similar to that of Mirror-P-EXTRA, since it is a generalization of Mirror-P-EXTRA (Algorithm 6 here). Our intuition is based on the results we have obtained in [49] for Mirror-P-EXTRA.

10.5 Numerical Experiments

In this section, we report some numerical experiments for the consensus optimization and resource allocation problems. In Sect. 10.5.1, we discuss some experimental results that illustrate the behavior of the DIGing and Push-DIGing algorithms. In Sect. 10.5.2, we report our experimental results for the behavior of MIrror-P-EXTRA, MIrror-PG-EXTRA, and Mirror-Push-DIGing algorithms.

10.5.1 Consensus Optimization

We present some simulation results obtained for a decentralized estimation problem where the agents collaboratively estimate a vector x, which have been reported in [49]. No agent can directly observe the vector x, but every agent can obtain some noisy linear observations of the vector: each agent $i \in \{1, \ldots, n\}$ has an observation y_i of the form $y_i = M_i x + e_i$, with $y_i \in \mathbb{R}^{m_i}$ and $M_i \in \mathbb{R}^{m_i \times p}$ being known data, while $x \in \mathbb{R}^p$ is unknown, and $e_i \in \mathbb{R}^{m_i}$ is (unknown) measurement noise. As an objective function, the Huber loss function is used, which is given by:

$$
H_\xi(a) = \begin{cases} \frac{1}{2}a^2, & \text{if } |a| \le \xi, \quad (\ell_2^2\text{-zone}), \\ \xi(|a| - \frac{1}{2}\xi), & \text{otherwise}, \quad (\ell_1\text{-zone}), \end{cases}
$$

which will allow us to observe both sublinear and linear convergence rates for the simulated algorithms. The resulting agent optimization problem is given by

$$
\min_{x \in \mathbb{R}^p} f(x) = \frac{1}{n} \sum_{i=1}^{n} \left\{ \sum_{j=1}^{m_i} H_\xi(M_{i,j} x - y_{i,j}) \right\},
$$

with $M_{i,j}$ and $y_{i,j}$ being, respectively, the j-th row of the matrix M_i and the vector y_i.

In all reported results, the parameter ξ in the Huber loss function is $\xi = 2$. The number of agents is $n = 12$ and their measurements are scalars (i.e., $m_i = 1$ for all i). The dimension of the unknown vector x is $p = 3$. For every agent i, the observation matrix M_i and the observation noise e_i are generated according to the standard normal distribution, which are subsequently re-normalized in order to have the Lipschitz constant $L_i = 1$ for all i. All algorithms are initiated with $x_i(0) = \mathbf{0}$ for all i, which is located in the ℓ_1-zone of the Huber loss function for all i. The optimal solution x^* is artificially constructed to ensure that it belongs to the ℓ_2^2-zone of the Huber loss function and satisfies $\|x^* - x_i(0)\| = 300$ for all i.

Three graph cases have been used in the experiments, namely time-invariant directed graphs, time-varying undirected graphs, and time-varying directed graphs. Figure 10.1 illustrates the graphs, and the details about the graph constructions can be found in [49]. The push-gradient method [47], DIGing, DIGing-ATC (Aug-DGM with step-size matrix αI), and Push-DIGing are shown in Figs. 10.1 and 10.2, which appeared in [49] and are reproduced here with a permission.[8] We can see that DIGing and its variants have R-linear rates, while push-gradient method has a sublinear rate, see [49] for more detailed comparison of the algorithms.

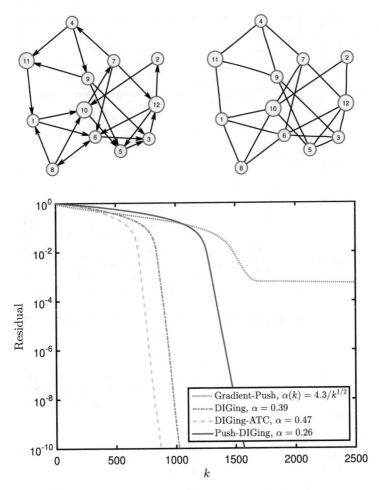

Fig. 10.1 The plots on the top are showing the underlying directed and undirected graphs for experiments. The plot at the bottom shows the residuals $\frac{\|\mathbf{x}(k)-\mathbf{x}^*\|_F}{\|\mathbf{x}(0)-\mathbf{x}^*\|_F}$ for the time-invariant directed graph illustrated on the top-left. Step-sizes have been *hand-optimized* to give faster convergence and more accurate solution for all algorithms

10.5.2 Resource Allocation

The experiments are conducted over a fixed undirected connected graph with $n = 100$ vertices and $|\mathcal{E}| = 198$ edges. The "connectivity ratio" of the undirected graph is defined by $r_c = |\mathcal{E}|/(0.5n(n-1)) = 0.04$ (the average degree is 3.96), where

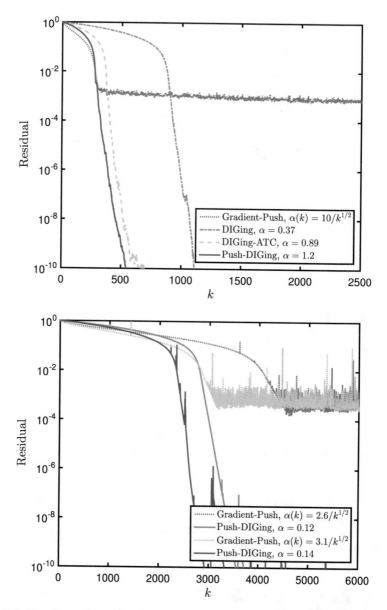

Fig. 10.2 The plots at the top and the bottom are showing the residuals $\frac{\|\mathbf{x}(k)-\mathbf{x}^*\|_F}{\|\mathbf{x}(0)-\mathbf{x}^*\|_F}$ for a time-varying undirected graph sequence and a time-varying directed graph sequence, respectively. Step-sizes have been *hand-optimized* to give faster convergence and more accurate solution for all algorithms

$0.5n(n-1)$ is the maximum number of edges an undirected graph can have. We generate the graphs randomly.[9] The objective function f_i of agent i is given by

$$f_i(x_i) = 0.5 x_i^\top H_i^\top H_i x_i + b_i x_i,$$

where $H_i \in \mathbb{R}^{2 \times 2}$ and $b_i \in \mathbb{R}^{2 \times 2}$ have the entries generated by the normal distribution with zero mean and unit variance. For each agent i, the local constraint set is a box,

$$\Omega_i = \{(\omega_{i,1}, \omega_{i,2}) \in \mathbb{R}^2 | 0 \leq \omega_{i,1} \leq \overline{\omega}_{i,1}, \ 0 \leq \omega_{i,2} \leq \overline{\omega}_{i,2}\},$$

where the interval boundaries $\overline{\omega}_{i,j}$ are randomly generated following the uniform distribution over the interval $[1, 2]$ for all i and $j = 1, 2$.

For each agent $i \in [n]$, the resource vector r_i is the mean value of the interval constraints, i.e., $r_i = [\overline{\omega}_{i,1}/2; \overline{\omega}_{i,2}/2] \in \mathbb{R}^2$. If the optimal solution of the problem does not hit the boundary of the set Ω, multiple trials are made to obtain the problem for which the constraint set Ω is active at the optimal solution. Mirror-EXTRA cannot be applied to solve such problems, so we implement Mirror-PG-EXTRA that uses a gradient-projection update.

In the experiments, we use the matrix $\mathbf{L} = 0.5(I - W)$ for both Mirror-P-EXTRA and Mirror-PG-EXTRA, where W is generated according to the Metropolis-Hasting rule. For Mirror-P-EXTRA we use $c = 0.01/(\mu L \tilde{\lambda}_{min}\{\mathbf{L}\})^{0.5}$, which is based on Corollary 3. The constant factor 0.01 is hand-tuned and found to be effective for most of our randomly generated graphs and randomly generated $\mathbf{f}$, $\mathbf{r}$, and Ω. To verify the viability of using different parameters β_i across agents, we choose $\beta_i = \phi_i c \lambda_{max}\{\mathbf{L}\}$ for each i, where ϕ_i is a random variable following the uniform distribution over the interval $[1, 1.5]$. For Mirror-PG-EXTRA we set $c = 0.5/L$ and $\beta_i = \phi_i c$ for each agent i, where ϕ_i is a random variable following the uniform distribution over the interval $[1, 1.5]$.

To compare the competitiveness of the algorithms of Sect. 10.4 with those in the existing literature, we implement the DPDA-S algorithm of [1]. DPDA-S has one system-level parameter γ and two per-agent parameters, τ_i and κ_i. Based on the recommended parametric structure as given in Remark II.1 of [1], which sets τ_i and κ_i automatically to produce another parameter c_i, we tuned the parameter c_i and γ to obtain one plot for this algorithm. In another plot for this algorithm, we hand-optimized all the parameters to achieve a better performance. In Mirror-P-EXTRA, there is a system-level parameter c which can be set based on the per-agent parameters β_i. Compared to Mirror-EXTRA, DPDA-S requires a finer tuning of its parameters to achieve a competitive performance. The convergence curves are shown in Fig. 10.3.

---

[9]To guarantee connectedness of the graph, we first grow a random tree; then uniformly randomly add edges into the graph to reach the specified connectivity ratio.

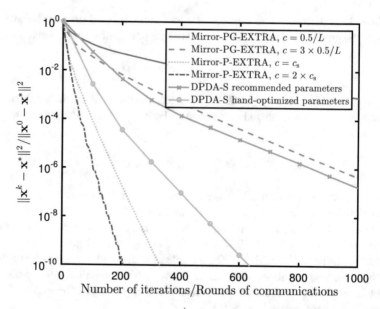

Fig. 10.3 Plots of the normalized residual $\frac{\|\mathbf{x}^k - \mathbf{x}^*\|_F}{\|\mathbf{x}^0 - \mathbf{x}^*\|_F}$. The step size for Mirror-P-EXTRA, $c_s = 0.01/(\mu L \tilde{\lambda}_{\min}\{\mathbf{L}\})^{0.5}$, is based on Corollary 3. The constant 0.01 in the numerator is hand-tuned and found to be effective for most of our randomly generated graphs ($r_c = 4/n = 4/100$) and randomly generated $\mathbf{f}$, $\mathbf{r}$, and Ω. For Mirror-PG-EXTRA, a step size larger than $3 \times 0.5/L$ will lead to divergence in the current trial. For Mirror-P-EXTRA, the parameter $2 \times c_s$ gives the fastest convergence speed in the current trial

To show the viability of the Mirror-Push-DIGing algorithm for time-varying directed graphs, we randomly randomly generate a directed graph with $n = 100$ vertices. Such a graph can have at most $n(n - 1) = 9900$ directed links. We define the connectivity ratio of a directed graph on n vertices by $r_c = |\mathcal{A}|/n(n - 1)$ where $\mathcal{A}$ is the arc set. For a time-varying graph sequence $\mathcal{G}_{TV}^{\dir}(k) = \{\mathcal{V}, \mathcal{A}(k)\}$, the activation ratio at time k is defined as $r_a(k) = |\mathcal{A}(k)|/|\mathcal{A}|$, where $\mathcal{A}$ is the set of links of the underlying(fixed) digraph $\mathcal{A} = \bigcup_k \mathcal{A}(k)$. Thus, $r_a(k) \times r_c$ is equal to the connectivity ratio of $\mathcal{G}_{TV}^{\dir}(k)$ at time k. When $r_a(k)$ is a constant for all k, we drop the time index and simply use r_a. In the experiment, we run Mirror-Push-DIGing over a few different time-varying directed graph sequences. The graph sequences are generated by uniformly randomly activating the arcs of the above mentioned underlying digraph. The results are plotted in Fig. 10.4. More details of settings for r_c, r_a, and the step sizes α are given in the legend of Fig. 10.4. The step sizes for Mirror-Push-DIGing are roughly hand tuned to obtain relatively fast convergence.

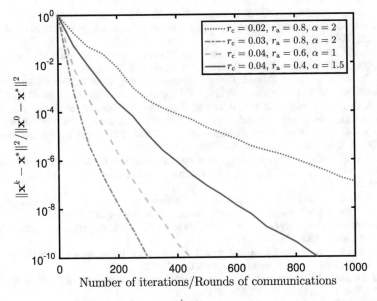

Fig. 10.4 Plots of the normalized residual $\frac{\|x^k-x^*\|_F}{\|x^0-x^*\|_F}$. The step sizes for Mirror-Push-DIGing are roughly hand tuned. r_c is the connectivity ratio and r_a is the activation ratio

Acknowledgements The work of A.N. and A.O. was supported by the Office of Naval Research grant N000014-16-1-2245. The work of A.O. was also supported by the NSF award CMMI-1463262 and AFOSR award FA-95501510394.

References

1. N.S. Aybat, E.Y. Hamedani, A distributed ADMM-like method for resource sharing under conic constraints over time-varying networks (2016). arXiv preprint arXiv:1611.07393
2. T. Başar, S.R. Etesami, A. Olshevsky, Convergence time of quantized metropolis consensus over time-varying networks. IEEE Trans. Autom. Control **61**(12), 4048–4054 (2016)
3. J. Bazerque, G. Giannakis, Distributed spectrum sensing for cognitive radio networks by exploiting sparsity. IEEE Trans. Signal Process. **58**, 1847–1862 (2010)
4. D. Bertsekas, Distributed asynchronous computation of fixed points. Math. Program. **27**(1), 107–120 (1983)
5. D.P. Bertsekas, Incremental proximal methods for large scale convex optimization. Math. Program. **129**, 163–195 (2011)
6. D.P. Bertsekas, Incremental aggregated proximal and augmented Lagrangian algorithms. Technical report, Laboratory for Information and Decision Systems Report LIDS-P-3176, MIT (2015)
7. D.P. Bertsekas, J. Tsitsiklis, *Parallel and Distributed Computation: Numerical Methods*, 2nd edn. (Athena Scientific, Nashua, 1997)
8. D. Bertsekas, A. Nedić, A. Ozdaglar, *Convex Analysis and Optimization* (Athena Scientific, Belmont, 2003)

9. S. Boyd, N. Parikh, E. Chu, B. Peleato, J. Eckstein, Distributed optimization and statistical learning via the alternating direction method of multipliers. Found. Trends Mach. Learn. 3(1), 1–122 (2011)
10. K. Cai, H. Ishii, Average consensus on arbitrary strongly connected digraphs with time-varying topologies. IEEE Trans. Autom. Control 59(4), 1066–1071 (2014)
11. V. Cevher, S. Becker, M. Schmidt, Convex optimization for big data: scalable, randomized, and parallel algorithms for big data analytics. IEEE Signal Process. Mag. 31(5), 32–43 (2014)
12. A.K. Chandra, P. Raghavan, W.L. Ruzzo, R. Smolensky, P. Tiwari, The electrical resistance of a graph captures its commute and cover times. Comput. Complex. 6(4), 312–340 (1996)
13. T.-H. Chang, M. Hong, X. Wang, Multi-agent distributed optimization via inexact consensus ADMM. IEEE Trans. Signal Process. 63(2), 482–497 (2015)
14. A. Chen, A. Ozdaglar, A fast distributed Proximal-Gradient method, in *The 50th Annual Allerton Conference on Communication, Control, and Computing (Allerton)* (2012), pp. 601–608
15. A. Cherukuri, J. Cortés, Initialization-free distributed coordination for economic dispatch under varying loads and generator commitment. Automatica 74, 183–193 (2016)
16. D. Coppersmith, U. Feige, J. Shearer, Random walks on regular and irregular graphs. SIAM J. Discret. Math. 9(2), 301–308 (1996)
17. T.T. Doan, C.L. Beck, Distributed primal dual methods for economic dispatch in power networks (2016). arXiv preprint arXiv:1609.06287
18. T.T. Doan, A. Olshevsky, Distributed resource allocation on dynamic networks in quadratic time. Syst. Control Lett. 99, 57–63 (2017)
19. J. Duchi, A. Agarwal, M. Wainwright, Dual averaging for distributed optimization: convergence analysis and network scaling. IEEE Trans. Autom. Control 57(3), 592–606 (2012)
20. P. Forero, A. Cano, G. Giannakis, Consensus-based distributed support vector machines. J. Mach. Learn. Res. 59, 1663–1707 (2010)
21. L. Gan, U. Topcu, S. Low, Optimal decentralized protocol for electric vehicle charging. IEEE Trans. Power Syst. 28(2), 940–951 (2013)
22. B. Gharesifard, J. Cortes, Distributed strategies for generating weight-balanced and doubly stochastic digraphs. Eur. J. Control 18(6), 539–557 (2012)
23. F. Guo, C. Wen, J. Mao, Y. Song, Distributed economic dispatch for smart grids with random wind power. IEEE Trans. Smart Grid 7(3), 1572–1583 (2016)
24. M. Gurbuzbalaban, A. Ozdaglar, P. Parrilo, On the convergence rate of incremental aggregated gradient algorithms (2015). Available on arxiv at http://arxiv.org/abs/1506.02081
25. M. Hong, T. Chang, Stochastic proximal gradient consensus over random networks (2015). arXiv preprint arXiv:1511.08905
26. H. Huang, Q. Ling, W. Shi, J. Wang, Collaborative resource allocation over a hybrid cloud center and edge server network. Int. J. Comput. Math. 35(4), 421–436 (2017)
27. D. Jakovetic, J. Xavier, J. Moura, Fast distributed gradient methods. IEEE Trans. Autom. Control 59, 1131–1146 (2014)
28. S. Kar, G. Hug, Distributed robust economic dispatch in power systems: a consensus + innovations approach, in *IEEE Power and Energy Society General Meeting* (2012), pp. 1–8
29. D. Kempe, A. Dobra, J. Gehrke, Gossip-based computation of aggregate information, in *Proceedings of the 44th Annual IEEE Symposium on Foundations of Computer Science* (2003), pp. 482–491
30. J. Koshal, A. Nedić, U.V. Shanbhag, Distributed algorithms for aggregative games on graphs. Oper. Res. 64(3), 680–704 (2016)
31. H. Lakshmanan, D.P. De Farias, Decentralized resource allocation in dynamic networks of agents. SIAM J. Optim. 19(2), 911–940 (2008)
32. D.A. Levin, Y. Peres, E.L. Wilmer, *Markov Chains and Mixing Times* (American Mathematical Society, Providence, 2009)
33. Z. Li, W. Shi, M. Yan, A decentralized proximal-gradient method with network independent step-sizes and separated convergence rates (2017). arXiv preprint arXiv:1704.07807
34. T. Lindvall, *Lectures on the Coupling Method* (Dover Publications, Newburyport, 2002)

35. Q. Ling, W. Shi, G. Wu, A. Ribeiro, Dlm: decentralized linearized alternating direction method of multipliers. IEEE Trans. Signal Process. **63**(15), 4051–4064 (2015)
36. P. Lorenzo, G. Scutari, Distributed nonconvex optimization over networks, in *IEEE International Workshop on Computational Advances in Multi-Sensor Adaptive Processing (CAMSAP)* (2015), pp. 229–232
37. P. Lorenzo, G. Scutari, Distributed nonconvex optimization over time-varying networks, in *IEEE International Conference on Acoustics, Speech and Signal Processing (ICASSP)* (2016), pp. 4124–4128
38. P. Lorenzo, G. Scutari, NEXT: in-network nonconvex optimization, in *IEEE Transactions on Signal and Information Processing over Networks* (2016)
39. Q. Lü, H. Li, Geometrical convergence rate for distributed optimization with time-varying directed graphs and uncoordinated step-sizes (2016). arXiv preprint arXiv:1611.00990
40. G. Mateos, J. Bazerque, G. Giannakis, Distributed sparse linear regression. IEEE Trans. Signal Process. **58**, 5262–5276 (2010)
41. A. Mokhtari, W. Shi, Q. Ling, A. Ribeiro, DQM: decentralized quadratically approximated alternating direction method of multipliers (2015). arXiv preprint arXiv:1508.02073
42. A. Mokhtari, W. Shi, Q. Ling, A. Ribeiro, A decentralized second-order method with exact linear convergence rate for consensus optimization (2016). arXiv preprint arXiv:1602.00596
43. I. Necoara, Random coordinate descent algorithms for multi-agent convex optimization over networks. IEEE Trans. Autom. Control **58**(8), 2001–2012 (2013)
44. A. Nedić, Asynchronous broadcast-based convex optimization over a network. IEEE Trans. Autom. Control **56**(6), 1337–1351 (2011)
45. A. Nedić, D.P. Bertsekas, Convergence rate of incremental subgradient algorithms, in *Stochastic Optimization: Algorithms and Applications* (Kluwer Academic Publishers, Boston, 2000), pp. 263–304
46. A. Nedić, D.P. Bertsekas, Incremental subgradient methods for nondifferentiable optimization. SIAM J. Optim. **12**(1), 109–138 (2001)
47. A. Nedić, A. Olshevsky, Stochastic gradient-push for strongly convex functions on time-varying directed graphs (2014). arXiv preprint arXiv:1406.2075
48. A. Nedić, A. Olshevsky, Distributed optimization over time-varying directed graphs. IEEE Trans. Autom. Control **60**(3), 601–615 (2015)
49. A. Nedić, A. Olshevsky, W. Shi, Improved convergence rates for distributed resource allocation (2017). arXiv preprint arXiv:
50. A. Nedić, A. Ozdaglar, Distributed subgradient methods for multi-agent optimization. IEEE Trans. Autom. Control **54**, 48–61 (2009)
51. A. Nedić, D.P. Bertsekas, V. Borkar, Distributed asynchronous incremental subgradient methods, in *Proceedings of the March 2000 Haifa Workshop "Inherently Parallel Algorithms in Feasibility and Optimization and Their Applications"* (Elsevier, Amsterdam, 2001)
52. A. Nedić, A. Olshevsky, A. Ozdaglar, J. Tsitsiklis, On distributed averaging algorithms and quantization effects. IEEE Trans. Autom. Control **54**(11), 2506–2517 (2009)
53. A. Nedić, A. Olshevsky, C. Uribe, Fast convergence rates for distributed non-Bayesian learning (2015). arXiv preprint arXiv:1508.05161
54. A. Nedić, A. Olshevsky, W. Shi, Achieving geometric convergence for distributed optimization over time-varying graphs (2016). arXiv preprint arXiv:1607.03218
55. A. Nedić, A. Olshevsky, W. Shi, C.A. Uribe, Geometrically convergent distributed optimization with uncoordinated step-sizes (2016). arXiv preprint arXiv:1609.05877
56. Y. Nesterov, *Introductory Lectures on Convex Optimization: A Basic Course*, vol. 87 (Springer Science and Business Media, Berlin, 2013)
57. A. Olshevsky, Efficient information aggregation strategies for distributed control and signal processing, Ph.D. thesis, Massachusetts Institute of Technology, 2010
58. A. Olshevsky, Linear time average consensus on fixed graphs and implications for decentralized optimization and multi-agent control (2016). arXiv preprint arXiv:1411.4186v6
59. G. Qu, N. Li, Harnessing smoothness to accelerate distributed optimization (2016). arXiv preprint arXiv:1605.07112

60. G. Qu, N. Li, Accelerated distributed Nesterov gradient descent (2017). arXiv preprint arXiv:1705.07176
61. M. Rabbat, R. Nowak, Distributed optimization in sensor networks, in *Proceedings of the 3rd International Symposium on Information Processing in Sensor Networks* (ACM, New York, 2004), pp. 20– 27
62. S. Ram, V. Veeravalli, A. Nedić, Distributed non-autonomous power control through distributed convex optimization, in *INFOCOM* (2009), pp. 3001–3005
63. S. Ram, A. Nedić, V. Veeravalli, Incremental stochastic subgradient algorithms for convex optimization. SIAM J. Optim. **20**(2), 691–717 (2009)
64. S.S. Ram, A. Nedić, V. Veeravalli, Distributed stochastic subgradient projection algorithms for convex optimization. J. Optim. Theory Appl. **147**(3), 516–545 (2010)
65. S.S. Ram, A. Nedić, V.V. Veeravalli, A new class of distributed optimization algorithms: application to regression of distributed data. Optim. Methods Softw. **27**(1), 71–88 (2012)
66. W. Ren, Consensus based formation control strategies for multi-vehicle systems, in *Proceedings of the American Control Conference* (2006), pp. 4237–4242
67. R. Rockafellar, *Convex Analysis* (Princeton University Press, Princeton, 1970)
68. A. Sayed, Diffusion adaptation over networks. Acad. Press Libr. Signal Process. **3**, 323–454 (2013)
69. K. Scaman, F. Bach, S. Bubeck, Y. Lee, L. Massoulié, Optimal algorithms for smooth and strongly convex distributed optimization in networks (2017). arXiv preprint arXiv:1702.08704
70. H. Seifi, M. Sepasian, Electric Power System Planning: Issues, Algorithms and Solutions (Springer Science and Business Media, Berlin, 2011)
71. W. Shi, Q. Ling, K. Yuan, G. Wu, W. Yin, On the linear convergence of the ADMM in decentralized consensus optimization. IEEE Trans. Signal Process. **62**(7), 1750–1761 (2014)
72. W. Shi, Q. Ling, G. Wu, W. Yin, A proximal gradient algorithm for decentralized composite optimization. IEEE Trans. Signal Process. **63**(22), 6013–6023 (2015)
73. W. Shi, Q. Ling, G. Wu, W. Yin, EXTRA: an exact first-order algorithm for decentralized consensus optimization. SIAM J. Optim. **25**(2), 944–966 (2015)
74. Y. Sun, G. Scutari, D. Palomar, Distributed nonconvex multiagent optimization over time-varying networks (2016). arXiv preprint arXiv:1607.00249
75. H. Terelius, U. Topcu, R. Murray, Decentralized multi-agent optimization via dual decomposition, in *18th IFAC World Congress* (2011)
76. T.M.D. Tran, A.Y. Kibangou, Distributed estimation of graph Laplacian Eigen-values by the alternating direction of multipliers method. IFAC Proc. Vol. **47**(3), 5526–5531 (2014)
77. J. Tsitsiklis, D. Bertsekas, M. Athans, Distributed asynchronous deterministic and stochastic gradient optimization algorithms. IEEE Trans. Autom. Control **31**(9), 803–812 (1986)
78. C.A. Uribe, S. Lee, A. Gasnikov, A. Nedić, Optimal algorithms for distributed optimization (2017). arXiv preprint arXiv:1712.00232
79. D. Varagnolo, F. Zanella, A. Cenedese, G. Pillonetto, L. Schenato, Newton-Raphson consensus for distributed convex optimization. IEEE Trans. Autom. Control **61**(4), 994–1009 (2016)
80. M. Wang, D.P. Bertsekas, Incremental constraint projection-proximal methods for non-smooth convex optimization. Lab for Information and Decision Systems Report LIDS-P- 2907, MIT (2013).
81. E. Wei, A. Ozdaglar, On the $O(1/k)$ convergence of asynchronous distributed alternating direction method of multipliers (2013). arXiv preprint arXiv:1307.8254
82. T. Wu, K. Yuan, Q. Ling, W. Yin, A.H. Sayed, Decentralized consensus optimization with asynchrony and delays, in *50th Asilomar Conference on Signals, Systems and Computers* (2016), pp. 992–996
83. C. Xi, U. Khan, DEXTRA: a fast algorithm for optimization over directed graphs. IEEE Trans. Autom. Control **PP**(99), 1–1 (2017)
84. L. Xiao, S. Boyd, S. Lall, Distributed average consensus with time-varying metropolis weights (2006), available: http://www.stanford.edu/boyd/papers/avg_metropolis
85. L. Xiao, S. Boyd, S. Kim, Distributed average consensus with least-mean-square deviation. J. Parallel Distrib. Comput. **67**(1), 33–46 (2007)

86. J. Xu, Augmented distributed optimization for networked systems. Ph.D. thesis, Nanyang Technological University, 2016
87. J. Xu, S. Zhu, Y. Soh, L. Xie, Augmented distributed gradient methods for multi-agent optimization under uncoordinated constant stepsizes, in *Proceedings of the 54th IEEE Conference on Decision and Control (CDC)* (2015), pp. 2055–2060
88. T. Yang, J. Lu, D. Wu, J. Wu, G. Shi, Z. Meng, K.H. Johansson, A distributed algorithm for economic dispatch over time-varying directed networks with delays. IEEE Trans. Ind. Electron. **64**(6), 5095–5106 (2017)
89. K. Yuan, Q. Ling, W. Yin, On the convergence of decentralized gradient descent (2013). arXiv preprint arXiv:1310.7063
90. F. Zanella, D. Varagnolo, A. Cenedese, G. Pillonetto, L. Schenato, Multidimensional Newton-Raphson consensus for distributed convex optimization, in *American Control Conference (ACC), 2012* (IEEE, Piscataway, 2012), pp. 1079–1084
91. J. Zeng, W. Yin, ExtraPush for convex smooth decentralized optimization over directed networks. J. Comput. Math. **35**(4), 381–394 (2017)
92. Y. Zhang, G. Giannakis, Efficient decentralized economic dispatch for microgrids with wind power integration, in *6th Annual IEEE Green Technologies Conference (GreenTech)* (2014), pp. 7–12
93. M. Zhu, S. Martinez, Discrete-time dynamic average consensus. Automatica **46**(2), 322–329 (2010)
94. M. Zhu, S. Martinez, On distributed convex optimization under inequality and equality constraints. IEEE Trans. Autom. Control **57**(1), 151–164 (2012)

Chapter 11
Communication-Efficient Distributed Optimization of Self-concordant Empirical Loss

Yuchen Zhang and Lin Xiao

Abstract We consider distributed convex optimization problems originating from sample average approximation of stochastic optimization, or empirical risk minimization in machine learning. We assume that each machine in the distributed computing system has access to a local empirical loss function, constructed with i.i.d. data sampled from a common distribution. We propose a communication-efficient distributed algorithm to minimize the overall empirical loss, which is the average of the local empirical losses. The algorithm is based on an inexact damped Newton method, where the inexact Newton steps are computed by a distributed preconditioned conjugate gradient method. We analyze its iteration complexity and communication efficiency for minimizing self-concordant empirical loss functions, and discuss the results for ridge regression, logistic regression and binary classification with a smoothed hinge loss. In a standard setting for supervised learning where the condition number of the problem grows with square root of the sample size, the required number of communication rounds of the algorithm does not increase with the sample size, and only grows slowly with the number of machines.

Keywords Empirical risk minimization · Distributed optimization · Inexact Newton methods · Self-concordant functions

AMS Subject Classifications 65F08, 68W15, 68W20, 90C25

Y. Zhang
Stanford University, Stanford, CA, USA
e-mail: zhangyuc@cs.stanford.edu

L. Xiao (✉)
Microsoft Research, Redmond, WA, USA
e-mail: lin.xiao@microsoft.com

© Springer Nature Switzerland AG 2018
P. Giselsson, A. Rantzer (eds.), *Large-Scale and Distributed Optimization*, Lecture Notes in Mathematics 2227,
https://doi.org/10.1007/978-3-319-97478-1_11

11.1   Introduction

Many optimization problems in data science (including statistics, machine learning, data mining, etc.) are formulated with a large amount of data as input. They are typically solved by iterative algorithms which need to access the whole dataset or at least part of it during each iteration. With the amount of data we collect and process growing at a fast pace, it happens more often that the dataset involved in an optimization problem cannot fit into the memory or storage of a single computer (machine). To solve such "big data" optimization problems, we need to use distributed algorithms that rely on inter-machine communication.

We focus on distributed optimization problems arising from *sample average approximation* (SAA) of stochastic optimization. Consider the problem

$$\underset{w \in \mathbb{R}^d}{\text{minimize}} \quad \mathbb{E}_z[\phi(w, z)], \tag{11.1}$$

where z is a random vector whose probability distribution is supported on a set $\mathcal{Z} \subset \mathbb{R}^p$, and the cost function $\phi : \mathbb{R}^d \times \mathcal{Z} \to \mathbb{R}$ is convex in w for every $z \in \mathcal{Z}$. In general, evaluating the expected objective function with respect to z is intractable, even if the distribution is given. The idea of SAA is to approximate the solution to (11.1) by solving a deterministic problem defined over a large number of i.i.d. (independent and identically distributed) samples generated from the distribution of z (see, e.g., [51, Chapter 5]). Suppose our distributed computing system consists of m machines, and each has access to n samples $z_{i,1}, \ldots, z_{i,n}$, for $i = 1, \ldots, m$. Then each machine can evaluate a local empirical loss function

$$f_i(w) = \frac{1}{n} \sum_{j=1}^{n} \phi(w, z_{i,j}), \qquad i = 1, \ldots, m.$$

Our goal is to minimize the overall empirical loss defined with all mn samples:

$$f(w) = \frac{1}{m} \sum_{i=1}^{m} f_i(w) = \frac{1}{mn} \sum_{i=1}^{m} \sum_{j=1}^{n} \phi(w, z_{i,j}). \tag{11.2}$$

In machine learning applications, the probability distribution of z is usually unknown, and the SAA approach is referred to as *empirical risk minimization* (ERM). As a concrete example, we consider ERM of linear predictors for supervised learning. In this case, each sample has the form $z_{i,j} = (x_{i,j}, y_{i,j}) \in \mathbb{R}^{d+1}$, where $x_{i,j} \in \mathbb{R}^d$ is a feature vector and $y_{i,j}$ can be a target response in $\mathbb{R}$ (for regression) or a discrete label (for classification). We omit the subscripts i, j in the following examples:

- linear regression: $\phi(w, (x, y)) = (y - w^T x)^2$ where $y \in \mathbb{R}$;
- logistic regression: $\phi(w, (x, y)) = \log(1 + \exp(-y(w^T x)))$ where $y \in \{\pm 1\}$;
- hinge loss: $\phi(w, (x, y)) = \max\{0, \ 1 - y(w^T x)\}$, where $y \in \{\pm 1\}$.

For numerical stability and statistical generalization purposes, we often add a regularization term $(\lambda/2)\|w\|_2^2$ (where $\lambda > 0$) to make the empirical loss function strongly convex. In other words, we modify the definition of $f_i(w)$ as

$$f_i(w) = \frac{1}{n} \sum_{j=1}^{n} \phi(w, z_{i,j}) + \frac{\lambda}{2}\|w\|_2^2, \qquad i = 1, \ldots, m. \tag{11.3}$$

For example, when ϕ is the hinge loss, this yields the *support-vector machine* [13].

In the distributed setting, each function f_i can be accessed only at machine i. We consider distributed algorithms that alternate between a local computation procedure at each machine and a communication round involving simple Map-Reduce type of operations [14, 35]. Compared with local computation at each machine, the cost of inter-machine communication is much higher in terms of both delay and energy consumption (see, e.g., [5, 48]); thus it is often considered as the bottleneck for distributed computing. Our goal is to develop *communication-efficient* distributed algorithms, which try to use a minimal number of communication rounds to reach certain precision in minimizing f.

11.1.1 Communication Efficiency of Distributed Algorithms

We assume that each communication round requires only simple Map-Reduce type of operations, such as broadcasting a vector in $\mathbb{R}^d$ to the m machines and computing the sum or average of m vectors in $\mathbb{R}^d$. Typically, if a distributed iterative algorithm takes T iterations to converge, then it communicates at least T rounds (usually one or two communication rounds per iteration). Therefore, we can measure the communication efficiency of a distributed algorithm by its iteration complexity $T(\epsilon)$, which is the number of iterations required by the algorithm to find a solution w_T such that $f(w_T) - f(w_\star) \leq \epsilon$.

For a concrete discussion, we make the following assumption:

Assumption 1 *The function $f : \mathbb{R}^d \to \mathbb{R}$ is twice continuously differentiable, and there exist constants $L \geq \lambda > 0$ such that*

$$\lambda I \preceq f''(w) \preceq LI, \qquad \forall\, w \in \mathbb{R}^d,$$

where $f''(w)$ denotes the Hessian of f at w, and I is the $d \times d$ identity matrix.

Functions that satisfy Assumption 1 are often called L-smooth and λ-strongly convex. The value $\kappa = L/\lambda \geq 1$ is called the *condition number* of f, which is a key quantity in characterizing the complexity of iterative algorithms. We focus on ill-conditioned cases where $\kappa \gg 1$.

A straightforward approach for minimizing f is using a distributed implementation of the classical gradient descent method. More specifically, at each iteration k, each machine computes the local gradient $f_i'(w_k) \in \mathbb{R}^d$ and sends it to a master node to compute $f'(w_k) = (1/m) \sum_{i=1}^m f_i'(w_k)$. The master node takes a gradient step to compute w_{k+1}, and broadcasts it to each machine for the next iteration. The iteration complexity of this method is the same as the classical gradient method: $O(\kappa \log(1/\epsilon))$, which is linear in the condition number κ (see, e.g., [36]). If we use Nesterov's accelerated gradient methods [36, 37], then the iteration complexity can be improved to $O(\sqrt{\kappa} \log(1/\epsilon))$.

Another popular approach for distributed optimization is to use the alternating direction method of multipliers (ADMM); see, e.g., [11, Section 8]. Under the assumption that each local function f_i is L-smooth and λ-strongly convex, ADMM can also achieve linear convergence, and the best known complexity is $O(\sqrt{\kappa} \log(1/\epsilon))$ [18]. This turns out to be on the same order as for accelerated gradient methods.

The polynomial dependence of the iteration complexity on the condition number can be unsatisfactory. For machine learning applications, both the precision ϵ and the regularization parameter λ should decrease as the overall sample size mn increases, typically on the order of $\Theta(1/\sqrt{mn})$ (see, e.g., [9, 47]). This translates into the condition number κ being $\Theta(\sqrt{mn})$. In this case, the iteration complexity, and thus the number of communication rounds, scales as $(mn)^{1/4}$ for both accelerated gradient methods and ADMM. This suggests that the number of communication rounds grows with the total sample size.

Despite the rich literature on distributed optimization (see, e.g., [1, 6, 11, 16, 20, 41, 42, 49, 56]), most algorithms involve high communication cost. In particular, their iteration complexity have similar or worse dependency on the condition number than the methods discussed above. In fact, it has been shown recently [2, 43] that $O(\sqrt{\kappa} \log(1/\epsilon))$ is the lower bound for the iteration complexity of distributed optimization using only first-order oracles. Thus in order to obtain better communication efficiency, we need to look into additional problem structure and alternative optimization methods.

First, we note that the above discussion on iteration complexity does not exploit the fact that each function f_i comes from SAA of a stochastic optimization problem. Since the data $z_{i,j}$ are i.i.d. samples from a common distribution, the local empirical loss functions $f_i(w) = (1/n) \sum_{j=1}^n \phi(w, z_{i,j})$ will be similar to each other if the local sample size n is large. Under this assumption, Zhang et al. [56] studied a one-shot averaging scheme that approximates the minimizer of f by simply averaging the minimizers of f_i. For a fixed condition number, the one-shot approach is communication efficient because it achieves optimal dependence on the overall sample size mn (in the sense of statistical lower bounds). But their conclusion does not hold if the regularization parameter λ decreases to zero as n goes to infinity [50].

Exploiting the stochastic nature alone seems not enough to overcome the ill-conditioning in the regime of first-order methods. This motivates the development of distributed second-order methods. Shamir et al. [50] proposed a distributed approximate Newton-type (DANE) method. Their method takes advantage of the fact that, under the stochastic assumptions of SAA, the Hessians $f_1'', f_2'', \ldots, f_m''$

are similar to each other. For quadratic loss functions, DANE is shown to converge in $\widetilde{O}\big((L/\lambda)^2 n^{-1} \log(1/\epsilon)\big)$ iterations with high probability, where the notation $\widetilde{O}(\cdot)$ hides additional logarithmic factors involving m and d. If $\lambda \sim 1/\sqrt{mn}$, then the iteration complexity becomes $\widetilde{O}(m \log(1/\epsilon))$, which scales linearly with the number of machines m, not the total sample size mn. The dependence on m is worse than that of AFG and ADMM, but m is usually much smaller than n (say, tens versus millions). However, the analysis in [50] does not extend to non-quadratic functions.

11.1.2 Outline of Our Approach

We propose a communication-efficient distributed second-order method for minimizing the overall empirical loss f defined in (11.2). Our method is based on an *inexact* damped Newton method. Assume f is strongly convex and has continuous second derivatives. In the *exact* damped Newton method (see, e.g., [36, Section 4.1.5]), we first choose an initial point $w_0 \in \mathbb{R}^d$, and then repeat

$$w_{k+1} = w_k - \frac{1}{1 + \delta(w_k)} \Delta w_k, \qquad k = 0, 1, 2, \ldots, \tag{11.4}$$

where $\Delta w_k = [f''(w_k)]^{-1} f'(w_k)$ is the Newton step, and $\delta(w_k)$ is the Newton decrement, defined as

$$\delta(w_k) = \sqrt{f'(w_k)^T [f''(w_k)]^{-1} f'(w_k)} = \sqrt{(\Delta w_k)^T f''(w_k) \Delta w_k} \,. \tag{11.5}$$

Since f is the average of $f_1, \ldots, f_m$, its gradient and Hessian can be written as

$$f'(w_k) = \frac{1}{m} \sum_{i=1}^{m} f_i'(w_k), \qquad f''(w_k) = \frac{1}{m} \sum_{i=1}^{m} f_i''(w_k). \tag{11.6}$$

In order to compute Δw_k in a distributed setting, the naive approach would require all the machines to send their gradients and Hessians to a master node. However, the task of transmitting the Hessians (which are $d \times d$ matrices) can be prohibitive for large dimension d. A better alternative is to use the conjugate gradient (CG) method to compute Δw_k as the solution to a linear system $f''(w_k) \Delta w_k = f'(w_k)$. Each iteration of the CG method requires a matrix-vector product of the form

$$f''(w_k) v = \frac{1}{m} \sum_{i=1}^{m} f_i''(w_k) v,$$

where v is some vector in $\mathbb{R}^d$. More specifically, the master node can broadcast the vector v to each machine, and each machine computes the product $f_i''(w_k)v$, which is a vector in $\mathbb{R}^d$, and sends it back to the master node. The master node then forms the average vector $f''(w_k)v$ and performs the CG update. Due to the iterative nature of the CG method, we can only compute the Newton step approximately with limited number of communication rounds.

The overall method has two levels of loops: the outer-loop of the damped Newton method, and the inner loop of the CG method for computing the inexact Newton steps. A similar approach (using a distributed truncated Newton method) was proposed in [29, 57] for ERM of linear predictors, and it was reported to perform very well in practice. However, the total number of CG iterations (each takes a round of communication) may still be high.

We first consider the outer loop complexity. It is well-known that Newton-type methods have asymptotic superlinear convergence. However, in classical analysis of Newton's method (see, e.g., [10, Section 9.5.3]), the number of steps needed to reach the superlinear convergence zone still depends on the condition number; more specifically, it scales quadratically in κ. In order to obtain better global iteration complexity, we resort to the machinery of self-concordant functions [36, 38]. For self-concordant empirical losses, we show that the iteration complexity of the inexact damped Newton method has a much weaker dependence on the condition number.

Second, we consider the inner loop complexity. The convergence rate of the CG method also depends on the condition number κ: it takes $O(\sqrt{\kappa}\log(1/\varepsilon))$ CG iterations to compute an ε-precise Newton step. Thus we arrive at the dilemma that the overall complexity of the CG-powered inexact Newton method is no better than accelerated gradient methods or ADMM. To overcome this difficulty, we exploit the stochastic nature of the problem and propose to use a preconditioned CG (PCG) method for solving the Newton system. Roughly speaking, if the local Hessians $f_1''(w_k), \ldots, f_m''(w_k)$ are "similar" to each other, then we can use any local Hessian $f_i''(w_k)$ as a preconditioner. Without loss of generality, let $P = f_1''(w_k) + \mu I$, where μ is an estimate of the spectral norm $\|f_1''(w_k) - f''(w_k)\|_2$. Then we use CG to solve the pre-conditioned linear system

$$P^{-1}f''(w_k)\Delta w_k = P^{-1}f'(w_k),$$

where the preconditioning (multiplication by P^{-1}) can be computed locally at machine 1 (the master node). The convergence rate of PCG depends on the condition number of the matrix $P^{-1}f''(w_k)$, which is close to 1 if the spectral norm $\|f_1''(w_k) - f''(w_k)\|_2$ is small.

To exactly characterize the similarity between $f_1''(w_k)$ and $f''(w_k)$, we rely on a stochastic analysis in the framework of SAA or ERM. We show that with high probability, $\|f_1''(w_k) - f''(w_k)\|_2$ decreases as $\widetilde{O}(\sqrt{d/n})$ in general, and $\widetilde{O}(\sqrt{1/n})$ for quadratic loss. Therefore, when n is large, the preconditioning is very effective and the PCG method converges to a sufficient precision within a small number of

Table 11.1 Communication efficiency of several distributed algorithms for ERM of linear predictors, when the regularization parameter λ in (11.3) is on the order of $1/\sqrt{mn}$

Algorithm	Number of communication rounds $\widetilde{O}(\cdot)$	
	Ridge regression (quadratic loss)	Binary classification (logistic loss, smoothed hinge loss)
Accelerated gradient	$(mn)^{1/4}\log(1/\epsilon)$	$(mn)^{1/4}\log(1/\epsilon)$
ADMM	$(mn)^{1/4}\log(1/\epsilon)$	$(mn)^{1/4}\log(1/\epsilon)$
DANE [50]	$m\log(1/\epsilon)$	$(mn)^{1/2}\log(1/\epsilon)$
DiSCO (this paper)	$m^{1/4}\log(1/\epsilon)$	$m^{3/4}d^{1/4}+m^{1/4}d^{1/4}\log(1/\epsilon)$

All results are deterministic or high probability upper bounds, except that the last one, DiSCO for binary classification, is a bound in expectation (with respect to the randomness in generating the i.i.d. samples)

iterations. The stochastic assumption is also critical for obtaining an initial point w_0 which further brings down the overall iteration complexity.

Combining the above ideas, we propose an algorithm for Distributed Self-Concordant Optimization (DiSCO, which also stands for Distributed SeCond-Order method, or Distributed Stochastic Convex Optimization). We show that several popular empirical loss functions in machine learning, including ridge regression, regularized logistic regression and a (new) smoothed hinge loss, are actually self-concordant. For ERM with these loss functions, Table 11.1 lists the number of communication rounds required by DiSCO and several other algorithms to find an ϵ-optimal solution. As the table shows, the communication cost of DiSCO weakly depends on the number of machines m and on the feature dimension d, and is independent of the local sample size n (excluding logarithmic factors). Comparing to DANE [50], DiSCO not only improves the communication efficiency on quadratic loss, but also has better guarantees for non-quadratic loss functions. Moreover, the dependence of DiSCO's complexity on ϵ can be further improved to $\log\log(1/\epsilon)$, as a result of superlinear convergence when the Newton steps are computed more accurately.

The rest of this chapter is organized as follows. In Sect. 11.2, we review the definition of self-concordant functions, and show that several popular empirical loss functions used in machine learning are either self-concordant or can be well approximated by self-concordant functions. In Sect. 11.3, we analyze the iteration complexity of an inexact damped Newton method for minimizing self-concordant functions. In Sect. 11.4, we show how to compute the inexact Newton step using a distributed PCG method, describe the overall DiSCO algorithm, and discuss its communication complexity. In Sect. 11.5, we present our main theoretical results based on a stochastic analysis, and apply them to linear regression and classification. In Sect. 11.6, we report experiment results to illustrate the advantage of DiSCO in communication efficiency, compared with several other distributed algorithms. Finally, we discuss the extension of DiSCO to distributed minimization of composite loss functions in Sect. 11.7, and conclude in Sect. 11.8.

11.2 Self-concordant Empirical Loss

The theory of self-concordant functions was developed by Nesterov and Nemirovski for the analysis of interior-point methods [38]. Roughly speaking, a function is called self-concordant if its third derivative can be controlled, in a specific way, by its second derivative. Suppose the function $f : \mathbb{R}^d \to \mathbb{R}$ has continuous third derivatives. We use $f''(w) \in \mathbb{R}^{d \times d}$ to denote its Hessian at w, and for any $u \in \mathbb{R}^d$ define $f'''(w)[u] \in \mathbb{R}^{d \times d}$ as

$$f'''(w)[u] = \lim_{t \to 0} \frac{1}{t} \left(f''(w + tu) - f''(w) \right).$$

Definition 1 A function $f : \mathbb{R}^d \to \mathbb{R}$ is self-concordant if there exists a constant $M_f \geq 0$ such that the inequality

$$\left| u^T (f'''(w)[u])u \right| \leq M_f \left(u^T f''(w)u \right)^{3/2}$$

holds for any $w \in \text{dom}(f)$ and $u \in \mathbb{R}^d$. In particular, a self-concordant function with parameter $M_f = 2$ is called *standard* self-concordant.

The reader may refer to the books [36, 38] for detailed treatment of self-concordance. In particular, the following lemma [36, Corollary 4.1.2] states that any self-concordant function can be rescaled to become standard self-concordant.

Lemma 1 *If f is self-concordant with parameter M_f, then $(M_f^2/4)f$ is standard self-concordant (with parameter 2).*

In the rest of this section, we show that several popular regularized empirical loss functions for linear regression and binary classification are either self-concordant or can be well approximated by self-concordant functions.

First, we consider regularized linear regression (ridge regression) with

$$f(w) = \frac{1}{N} \sum_{i=1}^{N} (y_i - w^T x_i)^2 + \frac{\lambda}{2} \|w\|_2^2.$$

To simplify notation, here we use a single subscript i from 1 to $N = mn$, instead of the double subscripts $\{i, j\}$ used in the introduction. Since f is a quadratic function, its third derivatives are all zero. Therefore, it is self-concordant with parameter 0, and by definition is also standard self-concordant.

For binary classification, we consider the regularized empirical loss function

$$\ell(w) = \frac{1}{N} \sum_{i=1}^{N} \varphi(y_i w^T x_i) + \frac{\gamma}{2} \|w\|_2^2, \tag{11.7}$$

where $x_i \in \mathcal{X} \subset \mathbb{R}^d$, $y_i \in \{-1, 1\}$, and $\varphi : \mathbb{R} \to \mathbb{R}$ is a convex surrogate function for the binary loss function which returns 0 if $y_i = \text{sign}(w^T x_i)$ and 1 otherwise. We further assume that the elements of $\mathcal{X}$ are bounded, i.e., $\sup_{x \in \mathcal{X}} \|x\|_2 \leq B$ for some finite B. Under this assumption, the following lemma shows that the regularized loss function ℓ is self-concordant.

Lemma 2 *Assume that $\gamma > 0$ and there exist $Q > 0$ and $\alpha \in [0, 1)$ such that $|\varphi'''(t)| \leq Q(\varphi''(t))^{1-\alpha}$ for every $t \in \mathbb{R}$. Then:*

(a) The function $\ell(w)$ defined in (11.7) is self-concordant with parameter $\frac{B^{1+2\alpha} Q}{\gamma^{1/2+\alpha}}$.

(b) The scaled function $f(w) = \frac{B^{2+4\alpha} Q^2}{4\gamma^{1+2\alpha}} \ell(w)$ is standard self-concordant.

Proof We need to bound the third derivative of ℓ appropriately. Using Eq. (11.7), the fact $y_i \in \{-1, +1\}$, and the assumption on φ, we have

$$\left| u^T (\ell'''(w)[u])u \right| = \frac{1}{N} \sum_{i=1}^{N} \left| \varphi'''(y_i w^T x_i)(y_i u^T x_i)^3 \right|$$

$$\leq \frac{Q}{N} \sum_{i=1}^{N} \left(\varphi''(y_i w^T x_i) \right)^{1-\alpha} \left| u^T x_i \right|^3$$

$$= \frac{Q}{N} \sum_{i=1}^{N} \left((u^T x_i)^2 \varphi''(y_i w^T x_i) \right)^{1-\alpha} \left| u^T x_i \right|^{1+2\alpha}.$$

Using concavity of $(\cdot)^{1-\alpha}$ when $\alpha \in [0, 1)$ and Jensen's inequality, we have

$$\left| u^T (\ell'''(w)[u])u \right| \leq Q \left(\frac{1}{N} \sum_{i=1}^{N} (u^T x_i)^2 \varphi''(y_i w^T x_i) \right)^{1-\alpha} \left| u^T x_i \right|^{1+2\alpha}$$

$$= Q \left(u^T \ell''(w)u - \gamma \|u\|_2 \right)^{1-\alpha} \left| u^T x_i \right|^{1+2\alpha}$$

$$\leq Q \left(u^T \ell''(w)u \right)^{1-\alpha} \left| u^T x_i \right|^{1+2\alpha}.$$

Next using the Cauchy-Schwarz inequality $|u^T x_i| \leq \|u\|_2 \|x_i\|_2 \leq B\|u\|_2$, we obtain

$$\left| u^T (\ell'''(w)[u])u \right| \leq B^{1+2\alpha} Q \left(u^T \ell''(w)u \right)^{1-\alpha} (\|u\|_2)^{1+2\alpha}.$$

Since ℓ is γ-strongly convex, we have $u^T \ell''(w)u \geq \gamma \|u\|_2^2$. Thus, we can upper bound $\|u\|_2$ by $\|u\|_2 \leq \gamma^{-1/2}(u^T \ell''(w)u)^{1/2}$. Substituting this inequality into the above upper bound completes the proof of part (a). Part (b) follows from Lemma 1. $\qquad \square$

It is important to note that the self-concordance of ℓ essentially relies on the regularization parameter γ being positive. If $\gamma = 0$, then the function will no longer be self-concordant, as pointed out by Bach [4] on logistic regression. The condition in Lemma 2 applies to a broad class of empirical loss functions. Next, we take the logistic loss and a smoothed hinge loss as two concrete examples.

Logistic Regression For logistic regression, we minimize the objective function (11.7) where φ is the logistic loss defined as $\varphi(t) = \log(1 + e^{-t})$. We can calculate the second and the third derivatives of $\varphi(t)$:

$$\varphi''(t) = \frac{e^t}{(e^t + 1)^2}, \qquad \varphi'''(t) = \frac{e^t(1 - e^t)}{(e^t + 1)^3} = \frac{1 - e^t}{1 + e^t}\varphi''(t).$$

Since $\left|\frac{1 - e^t}{1 + e^t}\right| \le 1$ for all $t \in \mathbb{R}$, we conclude that $|\varphi'''(t)| \le \varphi''(t)$ for all $t \in \mathbb{R}$. This implies that the condition in Lemma 2 holds with $Q = 1$ and $\alpha = 0$. Therefore, the regularized empirical loss ℓ is self-concordant with parameter $B/\sqrt{\gamma}$, and the scaled loss function $f(w) = (B^2/(4\gamma))\ell(w)$ is standard self-concordant.

Smoothed Hinge Loss In classification tasks, it is sometimes more favorable to use the hinge loss $\varphi(t) = \max\{0, 1 - t\}$ than using the logistic loss. We consider a family of smoothed hinge loss functions φ_p parametrized by a positive number $p \ge 3$. The function is defined by

$$\varphi_p(t) = \begin{cases} \frac{3}{2} - \frac{p-2}{p-1} - t & \text{for } t < -\frac{p-3}{p-1}, \\ \frac{3}{2} - \frac{p-2}{p-1} - t + \frac{(t+(p-3)/(p-1))^p}{p(p-1)} & \text{for } -\frac{p-3}{p-1} \le t < 1 - \frac{p-3}{p-1}, \\ \frac{p+1}{p(p-1)} - \frac{t}{p-1} + \frac{1}{2}(1-t)^2 & \text{for } 1 - \frac{p-3}{p-1} \le t < 1, \\ \frac{(2-t)^p}{p(p-1)} & \text{for } 1 \le t < 2, \\ 0 & \text{for } t \ge 2. \end{cases}$$

(11.8)

We plot the function φ_p for $p = 3, 5, 10, 20$ in Fig. 11.1. As the plot shows, $\varphi_p(t)$ is zero for $t > 2$, and is linear with unit slope for $t < -\frac{p-3}{p-1}$. These two linear zones are connected by three smooth non-linear segments on the interval $[-\frac{p-3}{p-1}, 2]$.

The third derivative of $\varphi_p(t)$ is nonzero only when $t \in [-\frac{p-3}{p-1}, 1 - \frac{p-3}{p-1}]$ and when $t \in [1, 2]$. On the first interval, we have

$$\varphi_p''(t) = \left(t + \frac{p-3}{p-1}\right)^{p-2}, \qquad \varphi_p'''(t) = (p-2)\left(t + \frac{p-3}{p-1}\right)^{p-3},$$

and on the second interval, we have

$$\varphi_p''(t) = (2 - t)^{p-2}, \qquad \varphi_p'''(t) = -(p-2)(2-t)^{p-3}.$$

Fig. 11.1 The standard hinge loss $\varphi(t) = \max\{0, 1 - t\}$ and the smoothed hinge loss φ_p defined in (11.8) with different parameters $p = 3, 5, 10, 20$

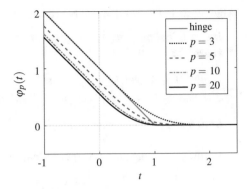

For both cases we have the inequality

$$|\varphi_p'''(t)| \le (p - 2)(\varphi_p''(t))^{1 - \frac{1}{p-2}},$$

which satisfies the condition of Lemma 2 with $Q = p - 2$ and $\alpha = \frac{1}{p-2}$. Therefore, according to Lemma 2, the regularized empirical loss ℓ is self-concordant with parameter

$$M_p = \frac{(p - 2)B^{1+\frac{2}{p-2}}}{\gamma^{\frac{1}{2}+\frac{1}{p-2}}}, \tag{11.9}$$

and the scaled loss function $f(w) = (M_p^2/4)\ell(w)$ is standard self-concordant.

11.3   Inexact Damped Newton Method

In this section, we propose and analyze an inexact damped Newton method (see Algorithm 1) for minimizing self-concordant functions. Without loss of generality, we assume that the objective function $f : \mathbb{R}^d \to \mathbb{R}$ is *standard* self-concordant.

In Algorithm 1, if we let $\epsilon_k = 0$ for all $k \ge 0$, then $v_k = [f''(w_k)]^{-1} f'(w_k)$ is the exact Newton step and δ_k is the Newton decrement defined in (11.5), so the algorithm reduces to the exact damped Newton method given in (11.4). But here we allow the computation of the Newton step (hence also the Newton decrement) to be inexact and contain approximation errors. The explicit account of approximation errors is essential for distributed optimization, in particular, if the Newton system is solved by iterative algorithms such as the conjugate gradient method.

Algorithm 1 Inexact damped Newton method

input: initial point w_0 and specification of a nonnegative sequence $\{\epsilon_k\}$.
repeat for $k = 0, 1, 2, \ldots$:

1. *find $v_k \in \mathbb{R}^d$ such that $\|f''(w_k)v_k - f'(w_k)\|_2 \le \epsilon_k$.*
2. *compute $\delta_k = \sqrt{v_k^T f''(w_k)v_k}$ and update $w_{k+1} = w_k - \frac{1}{1+\delta_k}v_k$.*

until a stopping criterion is satisfied.

Before presenting the convergence analysis, we introduce two auxiliary functions

$$\omega(t) = t - \log(1 + t), \qquad t \ge 0,$$
$$\omega_*(t) = -t - \log(1 - t), \qquad 0 \le t < 1.$$

These two functions are conjugate function of each other, and they are very useful for characterizing the properties of self-concordant functions; see [36, Section 4.1.4] for a detailed account. Here, we simply note that $\omega(0) = \omega_*(0) = 0$, both of them are strictly increasing for $t \ge 0$, and $\omega_*(t) \to \infty$ as $t \to 1$.

We also need to define two auxiliary vectors

$$\widetilde{u}_k = [f''(w_k)]^{-1/2} f'(w_k),$$
$$\widetilde{v}_k = [f''(w_k)]^{1/2} v_k.$$

The norm of the first vector, $\|\widetilde{u}_k\|_2 = \sqrt{f'(w_k)^T [f''(w_k)]^{-1} f'(w_k)}$, is the exact Newton decrement. The norm of the second one is $\|\widetilde{v}_k\|_2 = \delta_k$, which is the approximate Newton decrement computed during each iteration of Algorithm 1. Note that we do *not* compute $\widetilde{u}_k$ or $\widetilde{v}_k$ in Algorithm 1. They are introduced here solely for the purpose of convergence analysis. The following Theorem is proven in Appendix 1.

Theorem 1 *Suppose that $f : \mathbb{R}^d \to \mathbb{R}$ is a standard self-concordant function and Assumption 1 holds. If we choose the sequence $\{\epsilon_k\}_{k \ge 0}$ in Algorithm 1 as*

$$\epsilon_k = \beta(\lambda/L)^{1/2}\|f'(w_k)\|_2, \quad \text{with } \beta = 1/20, \tag{11.10}$$

then:

(a) *For any $k \ge 0$, we have $f(w_{k+1}) \le f(w_k) - \frac{1}{2}\omega(\|\widetilde{u}_k\|_2)$.*
(b) *If $\|\widetilde{u}_k\|_2 \le 1/6$, then we have $\omega(\|\widetilde{u}_{k+1}\|_2) \le \frac{1}{2}\omega(\|\widetilde{u}_k\|_2)$.*

As mentioned before, when $\epsilon_k = 0$, the vector $v_k = [f''(w_k)]^{-1} f'(w_k)$ becomes the exact Newton step. In this case, we have $\widetilde{v}_k = \widetilde{u}_k$, and it can be shown that $f(w_{k+1}) \le f(w_k) - \omega(\|\widetilde{u}_k\|_2)$ for all $k \ge 0$ and the exact damped Newton method has quadratic convergence when $\|\widetilde{u}_k\|_2$ is small (see [36, Section 4.1.5]). With the approximation error ϵ_k specified in (11.10), we have

$$\|\widetilde{v}_k - \widetilde{u}_k\|_2 \leq \|(f''(w_k))^{-1/2}\|_2 \|f''(w_k)v_k - f'(w_k)\|_2$$
$$\leq \lambda^{-1/2}\epsilon_k$$
$$= \beta L^{-1/2}\|f'(w_k)\|_2$$
$$\leq \beta\|\widetilde{u}_k\|_2,$$

which implies

$$(1 - \beta)\|\widetilde{u}_k\|_2 \leq \|\widetilde{v}_k\|_2 \leq (1 + \beta)\|\widetilde{u}_k\|_2. \tag{11.11}$$

Appendix 1 shows that when β is sufficiently small, the above inequality leads to the conclusion in part (a). Compared with the exact damped Newton method, the guaranteed reduction of the objective value per iteration is cut by half.

Part (b) of Theorem 1 suggests a linear rate of convergence when $\|\widetilde{u}_k\|_2$ is small. The overall iteration complexity of Algorithm 1 is given in the following corollary.

Corollary 1 *Suppose that $f : \mathbb{R}^d \to \mathbb{R}$ is a standard self-concordant function and Assumption 1 holds. If we choose the sequence $\{\epsilon_k\}$ in Algorithm 1 as in (11.10), then for any $\epsilon > 0$, we have $f(w_k) - f(w_\star) \leq \epsilon$ whenever $k \geq K$ where*

$$K = \left\lceil \frac{2(f(w_0) - f(w_\star))}{\omega(1/6)} \right\rceil + \left\lceil \log_2\left(\frac{2\omega(1/6)}{\epsilon}\right) \right\rceil. \tag{11.12}$$

Here $\lceil t \rceil$ denotes the smallest nonnegative integer that is larger than or equal to t.

Proof Since $\omega(t)$ is strictly increasing for $t \geq 0$, part (a) of Theorem 1 implies that if $\|\widetilde{u}_k\|_2 > 1/6$, one step of Algorithm 1 decreases the value of f by at least a constant $\frac{1}{2}\omega(1/6)$. So within at most $K_1 = \left\lceil \frac{2(f(w_0)-f(w_\star))}{\omega(1/6)} \right\rceil$ iterations, we are guaranteed that $\|\widetilde{u}_k\|_2 \leq 1/6$.

According to [36, Theorem 4.1.13], if $\|\widetilde{u}_k\|_2 < 1$, then we have

$$\omega(\|\widetilde{u}_k\|_2) \leq f(w_k) - f(w_\star) \leq \omega_*(\|\widetilde{u}_k\|_2). \tag{11.13}$$

Moreover, it is easy to check that $\omega_*(t) \leq 2\,\omega(t)$ for $0 \leq t \leq 1/6$. Therefore, using part (b) of Theorem 1, we conclude that when $k \geq K_1$,

$$f(w_k) - f(w_\star) \leq 2\omega(\|\widetilde{u}_k\|_2) \leq 2(1/2)^{k-K_1}\omega(\|\widetilde{u}_{K_1}\|_2) \leq 2(1/2)^{k-K_1}\omega(1/6).$$

Bounding the right-hand side of above inequality by ϵ, we have $f(w_k) - f(w_\star) \leq \epsilon$ whenever $k \geq K_1 + \left\lceil \log_2\left(\frac{2\omega(1/6)}{\epsilon}\right) \right\rceil = K$, which is the desired result.

Notice that for $\epsilon > 2\,\omega(1/6)$, it suffices to have $k \geq K_1$. □

If we set the tolerances ϵ_k in Algorithm 1 to be small enough, then super-linear convergence can be established. The following theorem and corollary are proven in Appendix 2.

Theorem 2 *Suppose that $f : \mathbb{R}^d \to \mathbb{R}$ is a standard self-concordant function and Assumption 1 holds. If we choose the sequence $\{\epsilon_k\}_{k \geq 0}$ in Algorithm 1 as*

$$\epsilon_k = \frac{\lambda^{1/2}}{2} \min\left\{ \frac{\omega(r_k)}{2}, \frac{\omega^{3/2}(r_k)}{10} \right\}, \quad \text{where} \quad r_k = L^{-1/2}\|f'(w_k)\|_2, \quad (11.14)$$

then:

(a) *For any $k \geq 0$, we have $f(w_{k+1}) \leq f(w_k) - \frac{1}{2}\omega(\|\tilde{u}_k\|_2)$.*
(b) *If $\|\tilde{u}_k\|_2 \leq 1/8$, then we have $\omega(\|\tilde{u}_{k+1}\|_2) \leq \sqrt{6}\,\omega^{3/2}(\|\tilde{u}_k\|_2)$.*

Part (b) of Theorem 2 suggests superlinear convergence when $\|\tilde{u}_k\|_2$ is small. In classical analysis of inexact Newton methods [17, 19], asymptotic superlinear convergence occurs with $\epsilon_k \sim \|f'(w_k)\|_2^{3/2}$ (in fact with $\epsilon \sim \|f'(w_k)\|_2^s$ for any $s > 1$). This agrees with our analysis since $\omega(t) = O(t)$ when t is not too small. As $t \to 0$, we have $\omega(t) \sim t^2$, which means that the tolerance ϵ_k in (11.14) needs to be proportional to $\|f'(w_k)\|_2^3$. This result is very conservative asymptotically. However, using the properties of ω and the associated self-concordance analysis, we are able to derive a much better *global* complexity result.

Corollary 2 *Suppose that $f : \mathbb{R}^d \to \mathbb{R}$ is a standard self-concordant function and Assumption 1 holds. If we choose the sequence $\{\epsilon_k\}$ in Algorithm 1 as in (11.14), then for any $\epsilon \leq 1/(3e)$, we have $f(w_k) - f(w_\star) \leq \epsilon$ whenever*

$$k \geq \left\lceil \frac{2(f(w_0) - f(w_\star))}{\omega(1/8)} \right\rceil + \left\lceil \frac{\log\log(1/(3\epsilon))}{\log(3/2)} \right\rceil. \quad (11.15)$$

Recently Lu [31] improved the complexity of the inexact damped Newton method by showing that its first stage (where $\|\tilde{u}_k\|_2 > 1/6$) also enjoy a linear rate of convergence. As a result, we can replace the first part of K in (11.12), denoted as $K_1 = \left\lceil \frac{2\Delta_0}{\omega(1/6)} \right\rceil$ where $\Delta_0 = f(w_0) - f(w_\star)$, by

$$K_1 = \min\left\{ \left\lceil \frac{2\left(1 + \omega_*^{-1}(\Delta_0)\right)}{1 - \omega_*^{-1}(\Delta_0)} \log\left(\frac{2\Delta_0}{\omega(1/6)} \right) \right\rceil, \left\lceil \frac{2\Delta_0}{\omega(1/6)} \right\rceil \right\}.$$

Here ω_*^{-1} denote the inverse function of ω_*, not its reciprocal $1/\omega_*$. The same improvement can apply to the bound in (11.15) by replacing $\omega(1/6)$ with $\omega(1/8)$. In addition, Tran-Dinh et al. [53] developed a proximal Newton method to minimize the sum of two functions, where one is convex and self-concordant and the other is convex but can be nonsmooth. Lu [31] showed that an inexact variant of the proximal Newton method also enjoys similar complexity and extended the results to a randomized block coordinate proximal Newton method.

11.3.1 Stopping Criteria

We discuss two stopping criteria for Algorithm 1. The first one is based on strong convexity of f, which leads to the inequality (see, e.g., [36, Theorem 2.1.10])

$$f(w_k) - f(w_\star) \leq \frac{1}{2\lambda} \|f'(w_k)\|_2^2.$$

Therefore, we can use the stopping criterion $\|f'(w_k)\|_2 \leq \sqrt{2\lambda\epsilon}$, which implies $f(w_k) - f(w_\star) \leq \epsilon$. However, this choice can be too conservative in practice (see discussions in [10, Section 9.1.2]).

Another choice for the stopping criterion is based on self-concordance. Using the fact that $\omega_*(t) \leq t^2$ for $0 \leq t \leq 0.68$ (see [10, Section 9.6.3]), we have

$$f(w_k) - f(w_\star) \leq \omega_*(\|\widetilde{u}_k\|_2) \leq \|\widetilde{u}_k\|_2^2, \tag{11.16}$$

provided that $\|\widetilde{u}_k\|_2 \leq 0.68$. Since we do not compute $\|\widetilde{u}_k\|_2$ (the exact Newton decrement) directly in Algorithm 1, we can use δ_k as an approximation. Using the inequality (11.11), and noticing that $\|\widetilde{v}_k\|_2 = \delta_k$, we conclude that

$$\delta_k \leq (1 - \beta)\sqrt{\epsilon}$$

implies $f(w_k) - f(w_\star) \leq \epsilon$ whenever $\epsilon \leq 0.68^2 \approx 0.46$. Since δ_k is computed at each iteration of Algorithm 1, this can serve as a good stopping criterion.

11.3.2 Scaling for Non-standard Self-concordant Functions

In many applications, we need to deal with empirical loss functions that are not standard self-concordant; see the examples in Sect. 11.2. Suppose a regularized loss function ℓ is self-concordant with parameter $M_\ell > 2$. By Lemma 1, the scaled function $f = \eta\ell$ with $\eta = M_\ell^2/4$ is standard self-concordant. We can apply Algorithm 1 to minimize the scaled function f, and examine the updates in terms of ℓ and the scaling constant η. In particular, using the sequence $\{\epsilon_k\}$ defined in (11.10), the condition for computing v_k in Algorithm 1 is

$$\|f''(w_k)v_k - f'(w_k)\|_2 \leq \beta(\lambda/L)^{1/2}\|f'(w_k)\|_2.$$

Let λ_ℓ and L_ℓ be the strong convexity and smoothness parameters of ℓ. With the scaling, we have $\lambda = \eta\lambda_\ell$ and $L = \eta L_\ell$, thus their ratio (the condition number) does not change. Therefore the above condition is equivalent to

$$\|\ell''(w_k)v_k - \ell'(w_k)\|_2 \leq \beta(\lambda_\ell/L_\ell)^{1/2}\|\ell'(w_k)\|_2. \tag{11.17}$$

In other words, the precision requirement for v_k in Algorithm 1 is *scaling invariant*.

The damped Newton step of Algorithm 1 can be rewritten in terms of ℓ as

$$w_{k+1} = w_k - \frac{v_k}{1 + \sqrt{\eta} \cdot \sqrt{v_k^T \ell''(w_k) v_k}}. \tag{11.18}$$

Here, the factor η explicitly appears in the update. By using a larger scaling factor η, the algorithm chooses a smaller step size. This adjustment is intuitive because the convergence of Newton-type methods rely on local smoothness conditions. By multiplying a large constant to ℓ (in order to make it standard self-concordant), the function's Hessian becomes less smooth, so that the step size should shrink.

In terms of complexity analysis, if we target to obtain $\ell(w_k) - \ell(w_\star) \le \epsilon$, then the iteration bound in (11.12) becomes

$$\left\lceil \frac{2\eta \big(\ell(w_0) - \ell(w_\star) \big)}{\omega(1/6)} \right\rceil + \left\lceil \log_2 \left(\frac{2\,\omega(1/6)}{\eta \epsilon} \right) \right\rceil. \tag{11.19}$$

For ERM problems in supervised learning, the self-concordant parameter M_ℓ, and hence the scaling factor $\eta = M_\ell^2/4$, can grow with the number of samples. For example, the regularization parameter γ in (11.7) often scales as $1/\sqrt{N}$ where $N = mn$ is the total number of samples. Lemma 2 suggests that η grows on the order of $\sqrt{mn}$. A larger η will render the second term in (11.19) less relevant, but the first term grows with the sample size mn. In order to counter the effect of the growing scaling factor, we need to choose the initial point w_0 judiciously to guarantee a small initial gap. This will be explained further in the following sections.

11.4 The DiSCO Algorithm

In this section, we adapt the inexact damped Newton method (Algorithm 1) to a distributed system, in order to minimize

$$f(w) = \frac{1}{m} \sum_{i=1}^{m} f_i(w), \tag{11.20}$$

where each function f_i can only be evaluated locally at machine i (see background in Sect. 11.1). This involves two questions: (1) how to set the initial point w_0 and (2) how to compute the inexact Newton step v_k in a distributed manner.

In accordance with the averaging structure in (11.20), we choose the initial point based on averaging. More specifically, we let $w_0 = (1/m) \sum_{i=1}^{m} \widehat{w}_i$, where

$$\widehat{w}_i = \arg \min_{w \in \mathbb{R}^d} \left\{ f_i(w) + \frac{\rho}{2} \|w\|_2^2 \right\}, \qquad i = 1, \dots, m. \tag{11.21}$$

Here $\rho \geq 0$ is a regularization parameter, which we will discuss in detail in the context of the stochastic analysis in Sect. 11.5. Roughly speaking, if each local function f_i is constructed with n i.i.d. samples as in (11.3), then we can choose $\rho \sim 1/\sqrt{n}$ to make $\mathbb{E}[f(w_0) - f(w_\star)]$ decreasing as $O(1/\sqrt{n})$. In this section, we simply regard it as an input parameter.

Here we comment on the computational cost of solving (11.21) locally at each machine. Suppose each f_i has the form in (11.3), then the local optimization problems in (11.21) become

$$\widehat{w}_i = \arg\min_{w \in \mathbb{R}^d} \left\{ \frac{1}{n} \sum_{j=1}^{n} \phi(w, z_{i,j}) + \frac{\lambda + \rho}{2} \|w\|_2^2 \right\}, \qquad i = 1, \ldots, m. \qquad (11.22)$$

The finite average structure of the above objective function can be effectively exploited by the stochastic average gradient (SAG) method [25, 44] or its newer variant SAGA [15]. Each step of these methods processes only one component function $\phi(w, z_{i,j})$, picked at random. Suppose f_i is L-smooth, then SAG(A) returns an ϵ-optimal solution with $O\left((n + \frac{L+\rho}{\lambda+\rho}) \log(1/\epsilon)\right)$ steps of stochastic updates. For ERM of linear predictors, we can also use the stochastic dual coordinate ascent (SDCA) method [45], which has the same complexity. Using accelerated stochastic coordinate methods [30, 46, 55] may further reduce the complexity.

After all local solutions $\widehat{w}_i$ are computed, we need one round of inter-machine communication (an all-reduce operation) to form their average w_0 as the starting point in the inexact damped Newton method.

11.4.1 Distributed Computing of the Inexact Newton Step

In each iteration of Algorithm 1, we need to compute an inexact Newton step v_k such that $\|f''(w_k)v_k - f'(w_k)\|_2 \leq \epsilon_k$. This boils down to solving the Newton system $f''(w_k)v_k = f'(w_k)$ approximately. When the objective f has the averaging form (11.20), its Hessian and gradient are given in (11.6). In the setting of distributed optimization, we propose to use a preconditioned conjugate gradient (PCG) method to solve the Newton system.

To simplify notation, we use H to represent $f''(w_k)$ and use H_i to represent $f_i''(w_k)$. Without loss of generality, we define a preconditioning matrix using the local Hessian at the first machine (the master node):

$$P = H_1 + \mu I,$$

Algorithm 2 Distributed PCG algorithm for computing v_k, r_k and δ_k

input: $w_k \in \mathbb{R}^d$ and $\mu \geq 0$. Let $H = f''(w_k)$ and $P = f_1''(w_k) + \mu I$.

communication: The master machine broadcasts w_k to all other machines to compute $f_i'(w_k)$, for $i = 1, \ldots, m$; then it forms $f'(w_k) = (1/m) \sum_{i=1}^m f_i'(w_k)$.

initialization: Compute ϵ_k given in (11.10) and set

$$v^{(0)} = 0, \qquad r^{(0)} = f'(w_k), \qquad s^{(0)} = P^{-1} r^{(0)}, \qquad u^{(0)} = s^{(0)}.$$

repeat for $t = 0, 1, 2 \ldots$,

1. **communication:** The master machine broadcasts $u^{(t)}$ to other machines to compute $f_i''(w_k) u^{(t)}$, and then form a vector $H u^{(t)} = (1/m) \sum_{i=1}^m f_i''(w_k) u^{(t)}$.

2. Compute $\alpha_t = \frac{\langle r^{(t)}, s^{(t)} \rangle}{\langle u^{(t)}, H u^{(t)} \rangle}$ and update:

$$v^{(t+1)} = v^{(t)} + \alpha_t u^{(t)}, \qquad r^{(t+1)} = r^{(t)} - \alpha_t H u^{(t)}, \qquad s^{(t+1)} = P^{-1} r^{(t+1)},$$

and also $H v^{(t+1)} = H v^{(t)} + \alpha_t H u^{(t)}$.

3. Compute $\beta_t = \frac{\langle s^{(t+1)}, r^{(t+1)} \rangle}{\langle s^{(t)}, r^{(t)} \rangle}$ and update:

$$u^{(t+1)} = s^{(t+1)} + \beta_t u^{(t)}.$$

until $\|r^{(t+1)}\|_2 \leq \epsilon_k$

return: $v_k = v^{(t+1)}$, $r_k = r^{(t+1)}$, and $\delta_k = \sqrt{(v^{(t+1)})^T H v^{(t+1)}}$.

where $\mu > 0$ is a small regularization parameter which we will discuss later. Instead of solving the equation $H v_k = f'(w_k)$ directly, the PCG method effectively solves the preconditioned linear system

$$P^{-1} H v_k = P^{-1} f'(w_k).$$

The distributed implementation of the PCG method is given in Algorithm 2.

In Algorithm 2, the master machine carries out the main steps of the classical PCG algorithm (see, e.g., [21, Section 10.3]), and all machines (including the master) compute the local gradients and Hessians and compute the matrix-vector products $f_i''(w_k) u^{(t)}$. Communication between the master and other machines are used to form the gradient $f'(w_k) = (1/m) \sum_{i=1}^m f_i'(w_k)$ and $H u^{(t)} = (1/m) \sum_{i=1}^m f_i''(w_k) u^{(t)}$. We note that the overall Hessian $H = f''(w_k)$ is never formed and the master machine only stores and updates the vectors $H u^{(t)}$ and $H v^{(t)}$. The vector $H v^{(t)}$ is used to compute the approximate Newton decrement δ_k as in the last line of Algorithm 2.

As explained in Sect. 11.1.2, the motivation for preconditioning is that when H_1 is sufficiently close to H, the condition number of $P^{-1} H$ might be close to 1, which is much smaller than that of H itself. As a result, the PCG method may converge much faster than CG without preconditioning. The following lemma characterizes the extreme eigenvalues of $P^{-1} H$ based on the closeness between H_1 and H.

Lemma 3 *Suppose Assumption 1 holds. If $\|H_1 - H\|_2 \le \mu$, then we have*

$$\sigma_{\max}(P^{-1}H) \le 1, \tag{11.23}$$

$$\sigma_{\min}(P^{-1}H) \ge \frac{\lambda}{\lambda + 2\mu}. \tag{11.24}$$

Here $\| \cdot \|_2$ denotes the spectral norm of a matrix, and $\sigma_{\max}(\cdot)$ and $\sigma_{\min}(\cdot)$ denote the largest and smallest eigenvalues, respectively, of a diagonalizable matrix.

Proof Since both P and H are symmetric and positive definite, all eigenvalues of $P^{-1}H$ are positive real numbers (see, e.g., [22, Section 7.6]). The eigenvalues of $P^{-1}H$ are identical to that of $P^{-1/2}HP^{-1/2}$. Thus, it suffices to prove inequalities (11.23) and (11.24) for the matrix $P^{-1/2}HP^{-1/2}$. To prove inequality (11.23), we need to show that $H \preceq P = H_1 + \mu I$. This is equivalent to $H - H_1 \preceq \mu I$, which is a direct consequence of the assumption $\|H_1 - H\|_2 \le \mu I$.

Similarly, the second inequality (11.24) is equivalent to $H \succeq \frac{\lambda}{\lambda+2\mu}(H_1 + \mu I)$, which is the same as $\frac{2\mu}{\lambda}H - \mu I \succeq H_1 - H$. Since $H \succeq \lambda I$ (by Assumption 1), we have $\frac{2\mu}{\lambda}H - \mu I \succeq \mu I$. The additional assumption $\|H_1 - H\|_2 \le \mu I$ implies $\mu I \succeq H_1 - H$, which complete the proof. $\qquad\square$

Recall that the condition number of a symmetric positive definite matrix is the ratio between its largest and smallest eigenvalues. By Assumption 1, the condition number of the Hessian matrix H can be bounded by $\kappa(H) = L/\lambda$. Lemma 3 establishes that the condition number of the preconditioned linear system is

$$\kappa(P^{-1}H) = \frac{\sigma_{\max}(P^{-1}H)}{\sigma_{\min}(P^{-1}H)} \le 1 + \frac{2\mu}{\lambda}, \tag{11.25}$$

provided that $\|H_1 - H\|_2 \le \mu$. When μ is small (i.e., comparable with λ), the condition number $\kappa(P^{-1}H)$ is close to one and can be much smaller than $\kappa(H)$. Based on classical convergence analysis of the CG method (see, e.g., [3, 32]), the following lemma shows that Algorithm 2 terminates in $\widetilde{O}(\sqrt{1 + \mu/\lambda})$ iterations.

Lemma 4 *Suppose Assumption 1 holds and assume that $\|H_1 - H\|_2 \le \mu$. Let*

$$T_\mu = \left\lceil \sqrt{1 + \frac{2\mu}{\lambda}} \log\left(\frac{2\sqrt{L/\lambda}\|f'(w_k)\|_2}{\epsilon_k} \right) \right\rceil.$$

Then Algorithm 2 terminates in T_μ iterations and v_k satisfies $\|Hv_k - f'(w_k)\|_2 \le \epsilon_k$.

The proof of Lemma 4 is given in Appendix 3. When the tolerance ϵ_k is chosen as in (11.10), the iteration bound T_μ is independent of $f'(w_k)$, i.e.,

$$T_\mu = \left\lceil \sqrt{1 + \frac{2\mu}{\lambda}} \log\left(\frac{2L}{\beta\lambda}\right) \right\rceil. \tag{11.26}$$

Under Assumption 1, we always have $\|H_1 - H\|_2 \le L$. If we choose $\mu = L$, then Lemma 4 implies that Algorithm 2 terminates in $\widetilde{O}(\sqrt{L/\lambda})$ iterations, where the notation $\widetilde{O}(\cdot)$ hides logarithmic factors. In practice, however, the matrix norm $\|H_1 - H\|_2$ is usually much smaller than L due to the stochastic nature of f_i. Thus, we can choose μ to be a tight upper bound on $\|H_1 - H\|_2$, and expect the algorithm terminating in $\widetilde{O}(\sqrt{1 + \mu/\lambda})$ iterations. In Sect. 11.5, we show that if the local empirical losses f_i are constructed with n i.i.d. samples from the same distribution, then $\|H_1 - H\|_2 \sim 1/\sqrt{n}$ with high probability. As a consequence, the iteration complexity of Algorithm 2 is upper bounded by $\widetilde{O}(1 + \lambda^{-1/2} n^{-1/4})$.

We wrap up this section by discussing the computation and communication complexities of Algorithm 2. The bulk of computation is at the master machine, especially computing the vector $s^{(t)} = P^{-1} r^{(t)}$ in Step 3, which is equivalent to minimize the quadratic function $(1/2)s^T P s - s^T r^{(t)}$. Using $P = f_1''(w_k) + \mu I$ and the form of f_1 in (11.3), this is equivalent to

$$s^{(t)} = \arg\min_{s \in \mathbb{R}^d} \left\{ \frac{1}{n} \sum_{j=1}^n \frac{s^T \phi''(w_k, z_{i,j}) s}{2} - \langle r^{(t)}, s \rangle + \frac{\lambda + \mu}{2} \|s\|_2^2 \right\}. \tag{11.27}$$

This problem has the same structure as (11.22), and an ϵ-optimal solution can be obtained with $O\left((n + \frac{L+\mu}{\lambda+\mu}) \log(1/\epsilon)\right)$ stochastic-gradient type of steps or less; see the discussions after Eq. (11.22).

In practice, we found that it is not necessary to solve the minimization problem in (11.27) to high precision, and usually running SAGA [15] or SDCA [45] for a few epochs would be sufficient. On the other hand, the convergence of the PCG method can be sped up by replacing the Fletcher-Reeves formula $\beta_t = \frac{\langle s^{(t+1)}, r^{(t+1)} \rangle}{\langle s^{(t)}, r^{(t)} \rangle}$ in Algorithm 2 with the Polak-Ribière formula

$$\beta_t = \frac{\langle s^{(t+1)}, r^{(t+1)} - r^{(t)} \rangle}{\langle s^{(t)}, r^{(t)} \rangle}.$$

The reason is that solving (11.27) approximately implies that we actually use slightly different preconditioners between the PCG iterations, and the Polak-Ribière formula is well-known for having better performance with variable preconditioning.

Algorithm 3 DiSCO

> **input:** parameters $\rho, \mu \geq 0$ and precision $\epsilon > 0$.
> **initialize:** compute $\widehat{w}_i$ for $i = 1, \ldots, m$ as in (11.21) and let $w_0 = (1/m) \sum_{i=1}^{m} \widehat{w}_i$.
> **repeat** for $k = 0, 1, 2, \ldots$
> 1. Run Algorithm 2: given w_k and ϵ_k, compute v_k and δ_k.
> 2. Update $w_{k+1} = w_k - \frac{1}{1+\delta_k} v_k$.
> **until** $\delta_k \leq (1-\beta)\sqrt{\epsilon}$.
> **output:** $\widehat{w} = w_{k+1}$.

11.4.2 DiSCO and Its Communication Efficiency

Combining the initialization scheme and distributed PCG method for computing the inexact Newton steps, we are ready to present the DiSCO algorithm.

Next we study the communication efficiency of DiSCO. Recall that by one round of communication, the master machine broadcasts a message of $O(d)$ bits to all machines, and every machine processes the message and sends another message of $O(d)$ bits back to the master. The following theorem gives an upper bound on the number of communication rounds taken by the DiSCO algorithm.

Theorem 3 *Assume that f is a standard self-concordant function and it satisfies Assumption 1. Suppose the input parameter μ in Algorithm 3 is an upper bound on $\|f_1''(w_k) - f''(w_k)\|_2$ for all $k \geq 0$. Then for any $\epsilon > 0$, in order to find a solution $\widehat{w}$ satisfying $f(\widehat{w}) - f(w_\star) < \epsilon$, the total number of communication rounds T satisfies*

$$ T \leq 1 + \left(\left\lceil \frac{2(f(w_0) - f(w_\star))}{\omega(1/6)} \right\rceil + \left\lceil \log_2 \left(\frac{2\omega(1/6)}{\epsilon} \right) \right\rceil \right) \left(2 + \sqrt{1 + \frac{2\mu}{\lambda}} \log \left(\frac{2L}{\beta\lambda} \right) \right). $$

Ignoring logarithmic terms and universal constants, T is bounded by

$$ \widetilde{O} \left((f(w_0) - f(w_\star) + \log(1/\epsilon)) \sqrt{1 + 2\mu/\lambda} \right). $$

Proof First we notice that the number of communication rounds in each call of Algorithm 2 is no more than $1 + T_\mu$, where T_μ is given in (11.26), and the extra 1 accounts for the communication round to form $f'(w_k)$ before the PCG iterations. Corollary 1 states that in order to guarantee $f(w_k) - f(w_\star) \leq \epsilon$, the total number of calls of Algorithm 2 in DiSCO is bounded by K given in (11.12). Thus the total number of communication rounds is bounded by $1 + K(1 + T_\mu)$, where the extra one count is for computing the initial point w_0 by averaging the local solutions $\widehat{w}_i$. □

Algorithm 4 Adaptive DiSCO

> **input:** *parameters $\rho \geq 0$ and $\mu_0 > 0$, and precision $\epsilon > 0$.*
> **initialize:** *compute $\widehat{w}_i$ for $i = 1, \ldots, m$ as in (11.21) and let $w_0 = (1/m) \sum_{i=1}^{m} \widehat{w}_i$.*
> **repeat** *for $k = 0, 1, 2, \ldots$*
>
> 1. *Run Algorithm 2 up to T_{μ_k} PCG iterations, with output v_k, δ_k and r_k.*
> 2. **if** $\|r_k\|_2 > \epsilon_k$ **then**
>   > *set $\mu_k := 2\mu_k$ and go to Step 1;*
>   **else**
>   > *set $\mu_{k+1} := \mu_k/2$ and go to Step 3.*
> 3. *Update $w_{k+1} = w_k - \frac{1}{1+\delta_k} v_k$.*
>
> **until** $\delta_k \leq (1 - \beta)\sqrt{\epsilon}$.
> **output:** $\widehat{w} = w_{k+1}$.

In practice, it can be hard to give a good a priori estimate of μ that satisfies the condition in Theorem 3. Instead, we want to adjust the value of μ adaptively while running the algorithm. Inspired by a line search procedure studied in [37], we propose an adaptive DiSCO method, described in Algorithm 4.

Next we bound the number of communication rounds required by Algorithm 4.

Theorem 4 *Assume that f is a standard self-concordant function and it satisfies Assumption 1. Let $\mu_{\max}$ be the largest value of μ_k generated by Algorithm 4, i.e., $\mu_{\max} = \max\{\mu_0, \mu_1, \ldots, \mu_K\}$ where K is the number of outer iterations. Then for any $\epsilon > 0$, in order to find a solution $\widehat{w}$ satisfying $f(\widehat{w}) - f(w_\star) < \epsilon$, the total number of communication rounds T is bounded by*

$$\tilde{O}\left(\left(f(w_0) - f(w_\star) + \log_2(1/\epsilon)\right)\sqrt{1 + 2\mu_{\max}/\lambda}\right).$$

Proof Let n_k be the number of calls to Algorithm 2 during the kth iteration of Algorithm 4. We have $\mu_{k+1} = (1/2)\mu_k 2^{n_k-1} = \mu_k 2^{n_k-2}$, which implies $n_k = 2 + \log_2(\mu_{k+1}/\mu_k)$. The total number of calls to Algorithm 2 is

$$N_K = \sum_{k=0}^{K-1} n_k = \sum_{k=0}^{K-1}\left(1 + \log_2 \frac{\mu_{k+1}}{\mu_k}\right) = 2K + \log_2 \frac{\mu_K}{\mu_0} \leq 2K + \log_2 \frac{\mu_{\max}}{\mu_0}.$$

Since Algorithm 2 involves no more than $T_{\mu_{\max}} + 1$ communication rounds, we have

$$T \leq 1 + N_K(T_{\mu_{\max}} + 1) \leq 1 + \left(2K + \log_2 \frac{\mu_{\max}}{\mu_0}\right)(T_{\mu_{\max}} + 1).$$

We can always pick a relatively large μ_0 so that $\log_2(\mu_{\max}/\mu_0)$ is small. Therefore, the above bound is very similar to the result of Theorem 3, except for replacing μ by

$\mu_{\max}$. Plugging in the expression of K in (11.12) and $T_{\mu_{\max}}$ in (11.26), and ignoring logarithmic terms and universal constants, we obtain the desired result. □

From the above proof, we see that the average number of calls to Algorithm 2 during each iteration is $N_K/K = 2 + (1/K)\log_2(\mu_K/\mu_0)$. Suppose μ_K is not too large, then this is roughly twice as many as the non-adaptive Algorithm 3. In general, we can update μ_k in Algorithm 4 as

$$\mu_k := \begin{cases} \theta_{\text{inc}}\mu_k & \text{if } \|r_k\|_2 > \epsilon_k, \\ \mu_k/\theta_{\text{dec}} & \text{if } \|r_k\|_2 \leq \epsilon_k, \end{cases}$$

with any $\theta_{\text{inc}} > 1$ and $\theta_{\text{dec}} \geq 1$. We used $\theta_{\text{inc}} = \theta_{\text{dec}} = 2$ to simplify presentation.

11.4.3 A Simple Variant Without PCG Iterations

We consider a simple variant of DiSCO where the approximate Newton step v_k is computed without using the PCG method. Instead, we simply set

$$v_k = P^{-1}f'(w_k) = \left(f_1''(w_k) + \mu I\right)^{-1}f'(w_k). \tag{11.28}$$

This is equivalent to setting $v_k = s^{(0)}$ in the initialization phase of Algorithm 2, without executing any PCG step.

A similar approach has been studied by Byrd et al. [7], where sub-sampled Hessians are used in a Newton-CG method, but not as a preconditioner. Instead, they proposed to use the sub-sampled Hessian to replace the full Hessian in solving the search direction, i.e., finding Δw_k by solving $f_1''(w_k)\Delta w_k = f'(w_k)$ using the CG method. Other related work include iterative Hessian sketch [40] and using sub-sampled Hessians for local function approximations in distributed optimization [34]. The algorithms in [7] and [34] have linear rate of convergence, but their iteration complexity still depends on the condition number.

Here we examine the theoretical conditions under which this variant of DiSCO enjoys a low iteration complexity that is independent of the condition number. Recall the two auxiliary vectors defined in Sect. 11.3:

$$\tilde{u}_k = H^{-1/2}f'(w_k), \qquad \tilde{v}_k = H^{1/2}v_k.$$

The norm of their difference can be bounded as

$$\begin{aligned}
\|\tilde{v}_k - \tilde{u}_k\|_2 &= \left\|H^{1/2}P^{-1}f'(w_k) - \tilde{u}_k\right\|_2 = \left\|H^{1/2}P^{-1}H^{1/2}\tilde{u}_k - \tilde{u}_k\right\|_2 \\
&\leq \left\|I - H^{1/2}P^{-1}H^{1/2}\right\|_2 \cdot \|\tilde{u}_k\|_2 = \left\|I - P^{-1}H\right\|_2 \cdot \|\tilde{u}_k\|_2.
\end{aligned}$$

From Lemma 3, we know that when $\|H_1 - H\|_2 \le \mu$, the eigenvalues of $P^{-1}H$ are located within the interval $[\frac{\lambda}{\lambda+2\mu}, 1]$. Therefore, we have

$$\|\tilde{v}_k - \tilde{u}_k\|_2 \le \left(1 - \frac{\lambda}{\lambda+2\mu}\right) \|\tilde{u}_k\|_2 = \frac{2\mu}{\lambda+2\mu} \|\tilde{u}_k\|_2.$$

The above inequality implies

$$\left(1 - \frac{2\mu}{\lambda+2\mu}\right) \|\tilde{u}_k\|_2 \le \|\tilde{v}_k\|_2 \le \left(1 + \frac{2\mu}{\lambda+2\mu}\right) \|\tilde{u}_k\|_2.$$

This inequality has the same form as (11.11), which is responsible to obtain the desired low complexity result if $\frac{2\mu}{\lambda+\mu}$ is sufficiently small. Indeed, if $\frac{2\mu}{\lambda+2\mu} \le \beta = \frac{1}{20}$ as specified in (11.10), the same convergence rate and complexity result stated in Theorem 1 and Corollary 1 apply. Since each iteration of this method does not use PCG, it requires only two communication rounds: one to compute $f'(w_k)$ and the other to form the matrix-vector product $f''(w)k)v_k$ in order to compute $\delta_k = (v_k^T f''(w_k)v_k)^{1/2}$. The result can be summarized by the following corollary.

Corollary 3 *Assume that f is a standard self-concordant function and it satisfies Assumption 1. In the DiSCO algorithm, we compute the inexact Newton step using (11.28). Suppose $\frac{2\mu}{\lambda+2\mu} \le \frac{1}{20}$ and $\|f_1''(w_k) - f''(w_k)\|_2 \le \mu$ for all $k \ge 0$. Then for any $\epsilon > 0$, in order to find a solution $\hat{w}$ satisfying $f(\hat{w}) - f(w_\star) < \epsilon$, the total number of communication rounds T satisfies*

$$T \le 1 + 2\left(\left\lceil \frac{2(f(w_0) - f(w_\star))}{\omega(1/6)} \right\rceil + \left\lceil \log_2\left(\frac{2\omega(1/6)}{\epsilon}\right) \right\rceil\right). \qquad (11.29)$$

In Corollary 3, the requirement on μ, which upper bounds $\|f_1''(w_k) - f''(w_k)\|_2$ for all $k \ge 0$, is quite strong. In particular, it requires μ to be a small fraction of λ in order to satisfy $\frac{2\mu}{\lambda+2\mu} \le \frac{1}{20}$. As we will see from the stochastic analysis in the next section, the spectral bound μ decreases on the order of $1/\sqrt{n}$. Therefore, in the standard setting where the regularization parameter $\lambda \sim 1/\sqrt{mn}$, the condition in Corollary 3 cannot be satisfied, and the convergence of this simple variant may be slow. In contrast, DiSCO with PCG iterations is much more tolerant of a relatively large μ, and can achieve superlinear convergence with a smaller ϵ_k.

11.5 Stochastic Analysis

From Theorems 3 and 4 in the previous section, we see that the communication complexity of the DiSCO algorithm mainly depends on two quantities: the initial objective gap $f(w_0) - f(w_\star)$ and the upper bound μ on the spectral norms

$\|f_1''(w_k) - f''(w_k)\|_2$ for all $k \geq 0$. As we discussed in Sect. 11.3.2, the initial gap $f(w_0) - f(w_\star)$ may grow with the number of samples due to the scaling required to make the objective function standard self-concordant. On the other hand, the upper bound μ may decrease as the number of samples increases based on the intuition that the local Hessians and the global Hessian become similar to each other. In this section, we show how to exploit the stochastic origin of the problem (SAA or ERM, as explained in Sect. 11.1) to mitigate the effect of objective scaling and quantify the notion of similarity between the local and global Hessians. These lead to improved complexity results.

We focus on the setting of distributed optimization of *regularized* empirical loss. That is, our goal is to minimize $f(w) = (1/m) \sum_{i=1}^{m} f_i(w)$, where

$$f_i(w) = \frac{1}{n} \sum_{j=1}^{n} \phi(w, z_{i,j}) + \frac{\lambda}{2} \|w\|_2^2, \qquad i = 1, \ldots, m. \qquad (11.30)$$

We assume that $z_{i,j}$ are i.i.d. samples from a common distribution. Our theoretical analysis relies on refined assumptions on the smoothness of the loss function ϕ. In particular, we assume that for any z in the sampling space $\mathcal{Z}$, the function $\phi(\cdot, z)$ has bounded first derivative in a compact set, and its second derivatives are bounded and Lipschitz continuous. We formalize these statements in the following assumption.

Assumption 2 *There are finite constants (V_0, G, L, M), such that for any $z \in \mathcal{Z}$:*

(a) $\phi(w, z) \geq 0$ *for all* $w \in \mathbb{R}^d$, *and* $\phi(0, z) \leq V_0$;
(b) $\|\phi'(w, z)\|_2 \leq G$ *for any* $\|w\|_2 \leq \sqrt{2V_0/\lambda}$;
(c) $\|\phi''(w, z)\|_2 \leq L - \lambda$ *for any* $w \in \mathbb{R}^d$;
(d) $\|\phi''(u, z) - \phi''(w, z)\|_2 \leq M\|u - w\|_2$ *for any* $u, w \in \mathbb{R}^d$.

For the functions f_i defined in (11.30), Assumption 2(c) implies $\lambda I \preceq f_i''(w) \preceq LI$ for $i = 1, \ldots, m$, which in turn implies Assumption 1.

Recall that the initial point for DiSCO is obtained as $w_0 = (1/m) \sum_{i=1}^{m} \widehat{w}_i$, where each $\widehat{w}_i$ is the solution to a regularized local optimization problem given in (11.21). The following lemma shows that the expected value of the initial gap $f(w_0) - f(w_\star)$ decreases with order $1/\sqrt{n}$ as the local sample size n increases. The proof uses the notion and techniques of *uniform stability* for analyzing the generalization performance of ERM [9]. See Appendix 4 for the proof.

Lemma 5 *Suppose that Assumption 2 holds and the optimal solution $w_\star$ is bounded, i.e., $\mathbb{E}[\|w_\star\|_2^2] \leq D^2$ for some constant $D > 0$. If we choose $\rho = \sqrt{6}G/(\sqrt{n}D)$ in (11.21) to compute $\widehat{w}_i$, then the initial point $w_0 = (1/m) \sum_{i=1}^{m} \widehat{w}_i$ satisfies*

$$\max\{\|w_\star\|_2, \|w_0\|_2\} \leq \sqrt{2V_0/\lambda}, \qquad (11.31)$$

and

$$\mathbb{E}[f(w_0) - f(w_\star)] \le \frac{\sqrt{6}GD}{\sqrt{n}}. \tag{11.32}$$

Here the expectation is with respect to the randomness in generating the i.i.d. data.

Next, we show that with high probability, $\|f_i''(w) - f''(w)\|_2 \sim \sqrt{d/n}$ for any $i \in \{1, \ldots, m\}$ and for any vector w in an ℓ_2-ball. Thus, if the number of samples n is large, the Hessian matrix of f can be approximated well by that of f_i. The proof uses random matrix concentration theories [33], which is given in Appendix 5.

Lemma 6 *Suppose Assumption 2 holds. For any $r > 0$ and any $i \in \{1, \ldots, m\}$, we have with probability at least $1 - \delta$,*

$$\sup_{\|w\|_2 \le r} \left\| f_i''(w) - f''(w) \right\|_2 \le \mu_{r,\delta},$$

where

$$\mu_{r,\delta} = \min \left\{ L, \sqrt{\frac{32L^2d}{n} \left(\log\left(1 + \frac{rM\sqrt{2n}}{L}\right) + \frac{\log(md/\delta)}{d} \right)} \right\}. \tag{11.33}$$

If $\phi(w, z_{i,j})$ are quadratic functions in w, then we have $M = 0$ in Assumption 2. In this case, Lemma 6 implies $\|f_i''(w) - f''(w)\|_2 \sim \sqrt{1/n}$. For general non-quadratic loss, Lemma 6 implies $\|f_i''(w) - f''(w)\|_2 \sim \sqrt{d/n}$. We use this lemma to obtain an upper bound on the spectral norm of the Hessian distances $\|f_1''(w_k) - f''(w_k)\|_2$, where the vectors w_k are generated by Algorithm 1 or 3.

Corollary 4 *Suppose Assumption 2 holds and the sequence $\{w_k\}_{k \ge 0}$ is generated by Algorithm 1. Let $r = \left(\frac{2V_0}{\lambda} + \frac{2G}{\lambda}\sqrt{\frac{2V_0}{\lambda}}\right)^{1/2}$. Then with probability at least $1 - \delta$, we have for all $k \ge 0$,*

$$\left\| f_1''(w_k) - f''(w_k) \right\|_2 \le \min \left\{ L, \sqrt{\frac{32L^2d}{n} \left(\log\left(1 + \frac{rM\sqrt{2n}}{L}\right) + \frac{\log(md/\delta)}{d} \right)} \right\}.$$

Proof We begin by upper bounding the ℓ_2-norm of w_k, for $k = 0, 1, 2 \ldots$, generated by Algorithm 1. By Theorem 1, we have $f(w_k) \le f(w_0)$ for all $k \ge 0$. By Assumption 2(a), we have $\phi(w, z) \ge 0$ for all $w \in \mathbb{R}^d$ and $z \in \mathcal{Z}$. As a consequence,

$$(\lambda/2)\|w_k\|_2^2 \le f(w_k) \le f(w_0) \le f(0) + G\|w_0\|_2 \le V_0 + G\|w_0\|_2.$$

Substituting $\|w_0\|_2 \le \sqrt{2V_0/\lambda}$ (see Lemma 5) into the above inequality yields

$$\|w_k\|_2 \le \left(\frac{2V_0}{\lambda} + \frac{2G}{\lambda}\sqrt{\frac{2V_0}{\lambda}}\right)^{1/2} = r.$$

Thus, we have $\|w_k\|_2 \le r$ for all $k \ge 0$. Applying Lemma 6 finishes the proof. $\quad\square$

Here we remark that the dependence on d of the spectral upper bound in Corollary 4 comes from $\mu_{r,\delta}$ in Lemma 6, where the bound needs to hold for all points in a d-dimensional ball with radius r. However, for the analysis of DiSCO, we only need the matrix concentration bound to hold for a finite number of vectors $w_0, w_1, \ldots, w_K$, instead of for all vectors satisfying $\|w\|_2 \le r$. Thus we conjecture that the spectral bound in Corollary 4, especially its dependence on the dimension d, is too conservative and very likely can be tightened.

We are now ready to present the main results of our stochastic analysis.

Theorem 5 *Suppose Assumption 2 hold. Assume that the regularized empirical loss function f is standard self-concordant, and its minimizer $w_\star = \arg\min_w f(w)$ satisfies $\mathbb{E}[\|w_\star\|_2^2] \le D^2$ for some constant $D > 0$. Let the input parameters to Algorithm 3 be $\rho = \sqrt{6}G/(\sqrt{n}D)$ and $\mu = \mu_{r,\delta}$ in (11.33) with*

$$r = \left(\frac{2V_0}{\lambda} + \frac{2G}{\lambda}\sqrt{\frac{2V_0}{\lambda}}\right)^{1/2}, \qquad \delta = \frac{GD}{\sqrt{n}} \cdot \frac{\sqrt{\lambda/(4L)}}{4V_0 + 2G^2/\lambda}. \tag{11.34}$$

Then for any $\epsilon > 0$, the expected number of communication rounds T required to reach $f(\widehat{w}) - f(w_\star) \le \epsilon$ satisfies

$$\mathbb{E}[T] \le 1 + \left(C_1 + \frac{6}{\omega(1/6)} \cdot \frac{GD}{\sqrt{n}}\right)\left(2 + C_2\left(1 + 2\sqrt{\frac{32L^2d\,C_3}{\lambda^2 n}}\right)^{1/2}\right),$$

where C_1, C_2, C_3 are given as

$$C_1 = \left(1 + \left\lceil \log_2\left(\frac{2\omega(1/6)}{\epsilon}\right)\right\rceil\right)\left(1 + \frac{1}{\sqrt{n}} \cdot \frac{GD}{4V_0 + 2G^2/\lambda}\right),$$

$$C_2 = \log\left(\frac{2L}{\beta\lambda}\right),$$

$$C_3 = \log\left(1 + \frac{rM\sqrt{2n}}{L}\right) + \frac{\log(dm/\delta)}{d}.$$

The expectation is with respect to the randomness in generating the i.i.d. data sets.

The proof of Theorem 5 is given in Appendix 6. Ignoring numerical constants and logarithmic terms, we have

$$\mathbb{E}[T] \le \tilde{O}\left(\left(\log(1/\epsilon) + \frac{GD}{n^{1/2}}\right)\left(1 + \frac{L^{1/2}d^{1/4}}{\lambda^{1/2}n^{1/4}}\right)\right).$$

According to Theorem 5, we need to set the two parameters ρ and μ in Algorithm 3 appropriately to obtain the desired communication efficiency. Using the adaptive DiSCO method given in Algorithm 4, we can avoid the explicit specification of $\mu = \mu_{r,\delta}$ defined in (11.33) and (11.34), and obtain the same complexity.

The expectation bound on the rounds of communication given in Theorem 5 is obtained by combining two consequences of averaging over a large number of i.i.d. local samples. One is the expected reduction of the initial gap $f(w_0) - f(w_\star)$ (Lemma 5), which helps to mitigate the effect of objective scaling required to make f standard self-concordant. The other is a high-probability bound that characterizes the similarity between the local and global Hessians (Corollary 4). If the empirical loss f is standard self-concordant without scaling, then we can regard $f(w_0) - f(w_\star)$ as a constant, and only need to use Corollary 4 to obtain a high-probability bound. This is demonstrated for linear regression in Sect. 11.5.1.

For applications where the loss function needs to be scaled to be standard self-concordant, the convexity parameter λ as well as the constants (V_0, G, L, M) in Assumption 2 also need to be scaled. If the scaling factor grows with n, then we need to rely on Lemma 5 to balance the effects of scaling. As a result, we only obtain bounds on the expected number of communication rounds (not with high probability). These are demonstrated in Sect. 11.5.2 for binary classification.

11.5.1   Application to Linear Regression

We consider linear regression with quadratic regularization (ridge regression). More specifically, we minimize the overall empirical loss function

$$f(w) = \frac{1}{mn}\sum_{i=1}^{m}\sum_{j=1}^{n}(y_{i,j} - w^T x_{i,j})^2 + \frac{\lambda}{2}\|w\|_2^2, \tag{11.35}$$

where the i.i.d. instances $(x_{i,j}, y_{i,j})$ are sampled from $\mathcal{X} \times \mathcal{Y}$. We assume that $\mathcal{X} \subset \mathbb{R}^d$ and $\mathcal{Y} \subset \mathbb{R}$ are bounded: there exist constants B_x and B_y such that $\|x\|_2 \le B_x$ and $|y| \le B_y$ for any $(x, y) \in \mathcal{X} \times \mathcal{Y}$. It can be shown that the least-squares loss $\phi(w, (x, y)) = (y - w^T x)^2$ satisfies Assumption 2 with

$$V_0 = B_y^2, \qquad G = 2B_x\left(B_y + B_x B_y \sqrt{2/\lambda}\right), \qquad L = \lambda + 2B_x^2, \qquad M = 0.$$

Theorems 5 gives an expectation bound on the number of communication rounds required by DiSCO. For linear regression, however, we can obtain a stronger result.

Since f is a quadratic function, it is self-concordant with parameter 0, and by definition also standard self-concordant (with parameter 2). In this case, we do not need to rescale the objective function, and can regard the initial gap $f(w_0) - f(w_\star)$ as a constant. As a consequence, we can directly apply Theorem 3 and Corollary 4 to obtain a high probability bound on the communication complexity. In particular, Theorem 3 states that if

$$\left\| f_1''(w_k) - f''(w_k) \right\|_2 \le \mu, \quad \text{for all} \quad k = 0, 1, 2, \ldots, \tag{11.36}$$

then the number of communication rounds T is bounded as

$$T \le 1 + \left(\left\lceil \frac{2(f(w_0) - f(w_\star))}{\omega(1/6)} \right\rceil + \left\lceil \log_2 \left(\frac{2\omega(1/6)}{\epsilon} \right) \right\rceil \right) \left(2 + \sqrt{1 + \frac{2\mu}{\lambda}} \log \left(\frac{2L}{\beta\lambda} \right) \right).$$

Since there is no scaling, the initial gap $f(w_0) - f(w_\star)$ can be considered as a constant. For example, we can simply pick $w_0 = 0$ and have

$$f(0) - f(w_\star) \le f(0) = \frac{1}{N} \sum_{i=1}^{N} y_i^2 \le B_y^2.$$

By Corollary 4 and the fact that $M = 0$ for quadratic functions, the condition (11.36) holds with probability at least $1 - \delta$ if we choose

$$\mu = \sqrt{\frac{32L^2 d}{n}} \sqrt{\frac{\log(md/\delta)}{d}} = \frac{8L}{\sqrt{n}} \sqrt{2\log(md/\delta)}. \tag{11.37}$$

Further using $L \le \lambda + 2B_x^2$, we obtain the following corollary.

Corollary 5 *Suppose we apply DiSCO (Algorithm 3) to minimize f defined in (11.35) with the input parameter μ given in (11.37), and let T be the total number of communication rounds required to find an ϵ-optimal solution. With probability at least $1 - \delta$, we have*

$$T = \tilde{O} \left(\left(1 + \frac{B_x}{\lambda^{1/2} n^{1/4}} \right) \log \left(\frac{1}{\epsilon} \right) \log \left(\frac{md}{\delta} \right) \right). \tag{11.38}$$

We note that the same conclusion also holds for the adaptive DiSCO algorithm (Algorithm 4), where we do not need to specify the input parameter μ based on (11.37).

The communication complexity guaranteed by Corollary 5 is strictly better than that of distributed implementation of the accelerated gradient method and ADMM (cf. Table 11.1). If we choose $\lambda = \Theta(1/\sqrt{mn})$, then Corollary 5 implies

$$T = \widetilde{O}\left(m^{1/4}\log(1/\epsilon)\right)$$

with high probability. The DANE algorithm [50], under the same setting, converges in $\widetilde{O}(m\log(1/\epsilon))$ iterations with high probability (and each iteration requires two rounds of communication). Thus DiSCO enjoys a better communication efficiency.

11.5.2 Application to Binary Classification

For binary classification, we consider the following regularized empirical loss

$$\ell(w) = \frac{1}{mn}\sum_{i=1}^{m}\sum_{j=1}^{n}\varphi(y_{i,j}w^T x_{i,j}) + \frac{\gamma}{2}\|w\|_2^2, \tag{11.39}$$

where $x_{i,j} \in \mathcal{X} \subset \mathbb{R}^d$, $y_{i,j} \in \{-1, 1\}$, and $\varphi : \mathbb{R} \to \mathbb{R}$ is a convex surrogate function for the binary loss. We further assume that the elements of $\mathcal{X}$ are bounded, i.e., $\sup_{x \in \mathcal{X}} \|x\|_2 \leq B$ for some finite B. Under these assumptions, Lemma 2 gives conditions on φ for ℓ to be self-concordant. As we have seen in Sect. 11.2, the function ℓ usually needs to be scaled by a large factor to become standard self-concordant. Next we discuss the theoretical implications for logistic regression and the smoothed hinge loss constructed in Sect. 11.2.

11.5.2.1 Logistic Regression

For logistic regression, we have $\varphi(t) = \log(1 + e^{-t})$. In Sect. 11.2, we have shown that the logistic loss satisfies the condition of Lemma 2 with $Q = 1$ and $\alpha = 0$. Consequently, with the factor $\eta = B^2/(4\gamma)$, the rescaled function $f = \eta\ell$ is standard self-concordant. If we express f in the standard form

$$f(w) = \frac{1}{mn}\sum_{i=1}^{m}\sum_{j=1}^{n}\phi(y_{i,j}w^T x_{i,j}) + \frac{\lambda}{2}\|w\|_2^2, \tag{11.40}$$

then we have $\phi(w, (x, y)) = \eta\varphi(yw^T x)$ and $\lambda = \eta\gamma$. Then Assumption 2 holds with

$$V_0 = \eta\log(2), \qquad G = \eta B, \qquad L = \eta(B^2/4 + \gamma), \qquad M = \eta B^3/10,$$

which all contain the scaling factor η. Plugging these scaled constants into Theorems 5, we have the following corollary.

Corollary 6 *For logistic regression, the number of communication rounds required by DiSCO to find an ϵ-optimal solution satisfies*

$$\mathbb{E}[T] = \tilde{O}\left(\left(\log(1/\epsilon) + \frac{B^3 D}{\gamma n^{1/2}}\right)\left(1 + \frac{B d^{1/4}}{\gamma^{1/2} n^{1/4}}\right)\right).$$

In the specific case when $\gamma = \Theta(1/\sqrt{mn})$, Corollary 6 implies

$$\mathbb{E}[T] = \tilde{O}\left(m^{3/4} d^{1/4} + m^{1/4} d^{1/4} \log(1/\epsilon)\right).$$

This result appeared in Table 11.1. If we ignore logarithmic terms, then the expected number of communication rounds is independent of the sample size n, and only grows slowly with the number of machines m.

11.5.2.2 Smoothed Hinge Loss

We consider minimizing ℓ in (11.39) where the loss function φ is the smoothed hinge loss defined in (11.8), which depends on a parameter $p \geq 3$. Using Lemma 2, we have shown in Sect. 11.2 that ℓ is self-concordant with parameter M_p given in (11.9). As a consequence, by choosing

$$\eta = \frac{M_p^2}{4} = \frac{(p-2)^2 B^{2+\frac{4}{p-2}}}{4\gamma^{1+\frac{2}{p-2}}},$$

the function $f = \eta \ell$ is standard self-concordant. If we express f in the form of (11.40), then $\phi(w, (x, y)) = \eta \varphi_p(y w^T x)$ and $\lambda = \eta \gamma$. It is easy to verify that Assumption 2 holds with

$$V_0 = \eta, \qquad G = \eta B, \qquad L = \eta(B^2 + \lambda), \qquad M = \eta(p-2)B^3.$$

If we choose $p = 2 + \log(1/\gamma)$, then Theorems 5 implies the following result.

Corollary 7 *For the smoothed hinge loss φ_p defined in (11.8) with $p = 2 + \log(1/\gamma)$, the total number of communication rounds required by DiSCO to find an ϵ-optimal solution satisfies*

$$\mathbb{E}[T] = \tilde{O}\left(\left(\log(1/\epsilon) + \frac{B^3 D}{\gamma n^{1/2}}\right)\left(1 + \frac{B d^{1/4}}{\gamma^{1/2} n^{1/4}}\right)\right).$$

Thus, the smoothed hinge loss enjoys the same communication efficiency as the logistic loss.

11.6 Numerical Experiments

In this section, we conduct numerical experiments to compare the DiSCO algorithm with several state-of-the-art distributed optimization algorithms: the ADMM algorithm (see, e.g., [11]), the accelerated full gradient method (AFG) [36, Section 2.2], the L-BFGS quasi-Newton method (see, e.g., [39, Section 7.2]), and the more recent DANE algorithm [50].

The algorithms ADMM, AFG and L-BFGS are well known and each has a rich literature. In particular, using ADMM for empirical risk minimization in a distributed setting is straightforward; see [11, Section 8]. For AFG and L-BFGS, we use the simple distributed implementation discussed in Sect. 11.1.1: at each iteration k, each machine computes the local gradients $f_i'(w_k)$ and sends it to the master machine to form $f'(w_k) = (1/m) \sum_{i=1}^{m} f_i'(w_k)$, and the master machine executes the main steps of the algorithm to compute w_{k+1}. The iteration complexities of these algorithms stay the same as their classical analysis for a centralized implementation, and each iteration usually involves one or two rounds of communication.

Here we briefly describe the DANE (Distributed Approximate NEwton) algorithm proposed by Shamir et al. [50]. Each iteration of DANE takes two rounds of communication to compute w_{k+1} from w_k. The first round of communication is used to compute the gradient $f'(w_k) = (1/m) \sum_{i=1}^{m} f_i'(w_k)$. Then each machine solves the local minimization problem

$$v_{k+1,i} = \arg \min_{w \in \mathbb{R}^d} \left\{ f_i(w) - \langle f_i'(w_k) - f'(w_k), w \rangle + \frac{\mu}{2} \|w - w_k\|_2^2 \right\},$$

and take a second round of communication to compute $w_{k+1} = (1/m) \sum_{i=1}^{m} v_{k+1,i}$. Here $\mu \geq 0$ is a regularization parameter with a similar role as in DiSCO.

11.6.1 Experiment Setup

For comparison, we solve three binary classification tasks using logistic regression. They are obtained from the LIBSVM website [12] and summarized in Table 11.2. These datasets are selected to cover different relations between the sample size $N = mn$ and the feature dimensionality d: including cases with $N \gg d$ (Covtype [8]), $N \approx d$ (RCV1 [27]) and $N \ll d$ (News20 [23, 24]). For each dataset, our goal is to minimize the regularized empirical loss function:

$$\ell(w) = \frac{1}{N} \sum_{i=1}^{N} \log(1 + \exp(-y_i(w^T x_i))) + \frac{\gamma}{2} \|w\|_2^2,$$

Table 11.2 Summary of three binary classification datasets from the LIBSVM website [12]

Dataset name	Number of samples	Number of features	Sparsity
Covtype	581,012	54	22%
RCV1	20,242	47,236	0.16%
News20	19,996	1,355,191	0.04%

where $x_i \in \mathbb{R}^d$ and $y_i \in \{-1, 1\}$. The data have been normalized so that $\|x_i\| = 1$ for all $i = 1, \ldots, N$. The regularization parameter is set to be $\gamma = 10^{-5}$.

We describe some implementation details. In Sect. 11.5.2, the theoretical analysis suggests that we scale the function ℓ by a factor $\eta = B^2/(4\gamma)$. Here we have $B = 1$ due to the normalization of the data. In practice, we find that DiSCO converges faster without rescaling. Thus, we use $\eta = 1$ for all experiments. For Algorithm 3, we choose the input parameters $\mu = m^{1/2}\mu_0$, where μ_0 is chosen by trial and error. In particular, we used $\mu_0 = 0$ for Covtype, $\mu_0 = 4 \times 10^{-4}$ for RCV1, and $\mu_0 = 2 \times 10^{-4}$ for News20. For the distributed PCG method (Algorithm 2), we choose the stopping precision to be $\epsilon_k = \|f'(w_k)\|_2/10$.

Among other methods in comparison, we manually tune the penalty parameter of ADMM and the regularization parameter μ for DANE in order to obtain their best performance. For AFG, we used an adaptive line search scheme [28, 37] to speed up its convergence. For L-BFGS, we adopted the memory size 30 (number of most recent iterates and gradients stored) as a general rule of thumb suggested in [39].

11.6.2 Performance Evaluation

It is important to note that different algorithms take different number of communication rounds per iteration. ADMM requires one round of communication per iteration. For AFG and L-BFGS, each iteration consists of at least two rounds of communications: one for finding the descent direction, and another one or more for searching the stepsize. For DANE, there are also two rounds of communications per iteration, for computing the gradient and for aggregating the local solutions (as explained at the beginning of this section). For DiSCO, each iteration in the inner loop takes one round of communication, and there is an additional round of communication at the beginning of each inner loop. Since we are interested in the communication efficiency of the algorithms, we plot their progress in reducing the objective value with respect to the number of communication rounds taken.

We want to evaluate DiSCO not only on w_k, but also during each inner iteration of the PCG for calculating v_k. To this end, we follow Eq. (11.18) to define an intermediate solution $\widehat{w}_k^t$ for each iteration t in Algorithm 2:

$$\widehat{w}_k^t = w_k - \frac{v^{(t)}}{1 + \sqrt{\eta}\sqrt{(v^{(t)})^T \ell''(w_k)v^{(t)}}},$$

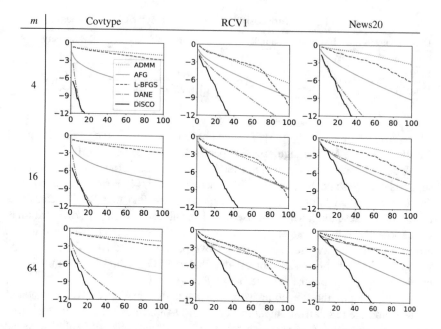

Fig. 11.2 Comparing DiSCO with other distributed optimization algorithms. We splits each dataset evenly to m machines, with $m \in \{4, 16, 64\}$. Each plot above shows the reduction of the logarithmic gap $\log_{10}(\ell(\widehat{w}) - \ell(w_\star))$ (the vertical axis) versus the number of communication rounds (the horizontal axis) taken by each algorithm

and evaluate the associated objective function $\ell(\widehat{w}_k^t)$. This function value is treated as a measure of progress after each round of communication.

We plot the results of ADMM, AFG, L-BFGS, DANE and DiSCO in Fig. 11.2. According to the plots, DiSCO converges substantially faster than ADMM and AFG. It is also notably faster than L-BFGS and DANE. In particular, the convergence speed (and the communication efficiency) of DiSCO is more robust to the number of machines in the distributed system. For $m = 4$, the performance of DiSCO is somewhat comparable to that of DANE. As m grows to 16 and 64, the convergence of DANE becomes significantly slower, while the performance of DiSCO only degrades slightly. This coincides with the theoretical analysis: the iteration complexity of DANE is proportional to m, but the complexity of DiSCO is proportional to $m^{1/4}$.

Since both DANE and DiSCO take a regularization parameter μ, we study their sensitivity to the choices of this parameter. Figure 11.3 shows the performance of DANE and DiSCO with the value of μ varying from 10^{-5} to 128×10^{-5}. We observe that the curves of DiSCO are relatively smooth and stable. In contrast, the curves of DANE exhibit sharp valley at particular values of μ. This suggests that DiSCO is more robust to non-optimal choices of parameters.

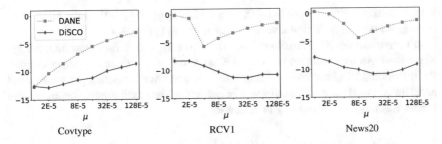

Fig. 11.3 Comparing the sensitivity of DiSCO and DANE with respect to the regularization parameter μ, when the datasets are split on $m = 16$ machines. We varied μ from 10^{-5} to 128×10^{-5}. The vertical axis is the logarithmic gap $\log_{10}(\ell(\widehat{w}) - \ell(w_\star))$ after 40 rounds of communications

11.7 Extension to Distributed Composite Minimization

Thus far, we have studied the problem of minimizing empirical loss functions that are standard self-concordant. In this section, we sketch how to extend DiSCO to solve distributed composite minimization problems. In composite minimization, we consider the minimization of

$$F(w) = f(w) + \Psi(w), \tag{11.41}$$

where f is a standard self-concordant function taking the form of (11.2), and Ψ is a closed convex function with a simple structure (see discussions in [37]). For solving the Lasso [52], for example, the ℓ_1-penalty $\Psi(w) = \sigma\|w\|_1$ with $\sigma > 0$ is nonsmooth but admits a simple proximal mapping.

We modify Algorithms 1 and 2 to minimize the composite function F. To modify Algorithm 1, we update w_{k+1} using an inexact version of the proximal-Newton method (see, e.g., [26, 53]). More specifically, the two steps in each iteration of Algorithm 1 are replaced with:

1. Find a vector v_k that is an approximate solution of

$$\underset{v \in \mathbb{R}^d}{\text{minimize}} \quad \left\{ \frac{1}{2} v^T f''(w_k) v - v^T f'(w_k) + \Psi(w_k - v) \right\}. \tag{11.42}$$

2. Update $w_{k+1} = w_k - \dfrac{1}{1 + \sqrt{v^T f''(w_k) v_k}} v_k$.

Note that for $\Psi \equiv 0$, the above proximal-Newton method reduces to Algorithm 1. Since v_k only needs to be an inexact solution to problem (11.42), we need a measure to quantify the approximation error. For this purpose, we define the following gradient mapping

$$g(v_k) = \arg \min_{g \in \mathbb{R}^d} \left\{ \frac{L}{2}\|g\|_2^2 + \langle f''(w_k)v_k - f'(w_k), g \rangle + \Psi(w_k - v_k + g) \right\}.$$

If v_k is an exact minimizer of (11.42), then we have $\|g(v_k)\|_2 = 0$. In the distributed setting, we only need to find a vector v_k such that $\|g(v_k)\|_2 \leq \epsilon_k$.

It remains to devise a distributed algorithm to compute an inexact minimizer v_k. Since the objective function in (11.42) is not quadratic, we can no longer employ the distributed PCG method in Algorithm 2. Instead, we propose a preconditioned accelerated proximal gradient method. In particular, we modify the algorithm on the master machine in Algorithm 2 as follows:

$$
\begin{aligned}
v^{(t+1)} = \arg\min_{v \in \mathbb{R}^d} \Big\{ & \frac{1}{2}(v - s^{(t)})^T [f_1''(w_k) + \mu I](v - s^{(t)}) \\
& + \langle f''(w_k)s^{(t)} - f'(w_k), v - s^{(t)} \rangle + \Psi(w_k + v) \Big\},
\end{aligned} \tag{11.43}
$$

$$
s^{(t+1)} = v^{(t+1)} + \frac{\sqrt{1 + 2\mu/\lambda} - 1}{\sqrt{1 + 2\mu/\lambda} + 1}(v^{(t+1)} - v^{(t)}),
$$

where $s^{(t+1)}$ is an auxiliary vector. We output $v_k = v^{(t+1)}$ once the condition $\|g(v^{(t+1)})\|_2 \leq \epsilon_k$ is satisfied. Each update takes one round of communication to compute the vector $f''(w_k)s^{(t)}$. Then, the sub-problem (11.43) is locally solved by the master machine. This problem has similar structure as problems (11.21) and (11.27), and can be solved in $O\big((n + \frac{L+\mu}{\lambda+\mu}) \log(1/\epsilon)\big)$ time using the randomized algorithms proposed in [15, 45, 54].

If we replace the first term on the right-hand side of (11.43) by $(L/2)\|v - v^{(t)}\|_2^2$ and set $\mu = L$, then the above algorithm is exactly the accelerated proximal gradient algorithm [28, 37], which converges in $\widetilde{O}(\sqrt{L/\lambda})$ iterations. By utilizing the similarity between $f_1''(w_k)$ and $f''(w_k)$, and assuming $\|f_1''(w_k) - f''(w_k)\|_2 \leq \mu$ for all $k \geq 0$, it can be shown that the algorithm in (11.43) converges in $\widetilde{O}(1 + \sqrt{\mu/\lambda})$ iterations, which is of the same order as the PCG algorithm for minimizing quadratic functions. Under the same assumptions on f, we can obtain similar guarantees on the overall communication efficiency as stated in Theorems 5.

11.8 Conclusions

We considered distributed convex optimization problems originating from SAA or ERM, which involve a large amount of i.i.d. data stored at different machines in a distributed computing system. For algorithms based on first-order methods, including accelerated gradient methods and ADMM, the required number of communication rounds grows with the condition number of the objective function. In the context of machine learning, the condition number itself often grows with the number of samples due to weaker regularization required. This causes the total number of communication rounds to grow with the overall sample size.

In this chapter, we proposed and analyzed DiSCO, a communication-efficient distributed algorithm for minimizing self-concordant empirical loss functions, and discussed its application to linear regression and classification. DiSCO is based on an inexact damped Newton method, where the inexact Newton steps are computed by a distributed preconditioned conjugate gradient method. In a standard setting for supervised learning, its required number of communication rounds does not increase with the sample size, but only grows slowly with the number of machines in the distributed system. We summarized the three main thrusts in our approach:

- *Self-concordant analysis.* We showed that several popular empirical loss functions used in machine learning are either self-concordant or can be well approximated by self-concordant functions. We gave complexity analysis of the inexact damped Newton method, and characterized the conditions for both linear and superlinear convergence.
- *Preconditioned conjugate gradient (PCG) method.* We proposed a distributed implementation of the PCG method for computing the inexact Newton step. In particular, the statistical preconditioner based on similarity between local and global Hessians is very effective in reducing the number of communication rounds, both in theory and practice.
- *Stochastic analysis of communication efficiency.* Our main theoretical results combine two consequences of averaging over a large number of i.i.d. samples. One is the expected reduction of the initial objective value, which counters the effect of objective scaling required to make the objective function standard self-concordant. The other is a matrix concentration bound that characterizes the similarity between the local and global Hessians.

Our numerical experiments on real datasets confirmed the superior communication efficiency of the DiSCO algorithm. In addition, we also proposed an extension for solving distributed optimization problems with composite empirical loss functions.

Appendix 1: Proof of Theorem 1

First, we notice that Step 2 of Algorithm 1 is equivalent to

$$w_{k+1} - w_k = \frac{v_k}{1 + \delta_k} = \frac{v_k}{1 + \|\tilde{v}_k\|_2},$$

which implies

$$\|[f''(w_k)]^{1/2}(w_{k+1} - w_k)\|_2 = \frac{\|\tilde{v}_k\|_2}{1 + \|\tilde{v}_k\|_2} < 1. \tag{11.44}$$

When inequality (11.44) holds, Nesterov [36, Theorem 4.1.8] has shown that

$$f(w_{k+1}) \le f(w_k) + \langle f'(w_k), w_{k+1} - w_k \rangle + \omega_*(\|[f''(w_k)]^{1/2}(w_{k+1} - w_k)\|_2).$$

Here we recall the definitions of the pair of conjugate functions

$$\omega(t) = t - \log(1+t), \qquad t \geq 0,$$
$$\omega_*(t) = -t - \log(1-t), \qquad 0 \leq t < 1.$$

Using the definition of ω and ω_*, and with some algebraic operations, we obtain

$$
\begin{aligned}
f(w_{k+1}) &\leq f(w_k) - \frac{\langle \tilde{u}_k, \tilde{v}_k \rangle}{1 + \|\tilde{v}_k\|_2} - \frac{\|\tilde{v}_k\|_2}{1 + \|\tilde{v}_k\|_2} + \log(1 + \|\tilde{v}_k\|_2) \\
&= f(w_k) - \omega(\|\tilde{u}_k\|_2) + \left(\omega(\|\tilde{u}_k\|_2) - \omega(\|\tilde{v}_k\|_2) \right) + \frac{\langle \tilde{v}_k - \tilde{u}_k, \tilde{v}_k \rangle}{1 + \|\tilde{v}_k\|_2}.
\end{aligned}
$$
$$(11.45)$$

By the second-order mean-value theorem, we have

$$\omega(\|\tilde{u}_k\|_2) - \omega(\|\tilde{v}_k\|_2) = \omega'(\|\tilde{v}_k\|_2)(\|\tilde{u}_k\|_2 - \|\tilde{v}_k\|_2) + \frac{1}{2}\omega''(t)\,(\|\tilde{u}_k\|_2 - \|\tilde{v}_k\|_2)^2$$

for some t satisfying

$$\min\{\|\tilde{u}_k\|_2, \|\tilde{v}_k\|_2\} \leq t \leq \max\{\|\tilde{u}_k\|_2, \|\tilde{v}_k\|_2\}.$$

Using the inequality (11.11), we can upper bound the second derivative $\omega''(t)$ as

$$\omega''(t) = \frac{1}{(1+t)^2} \leq \frac{1}{1+t} \leq \frac{1}{1 + \min\{\|\tilde{u}_k\|_2, \|\tilde{v}_k\|_2\}} \leq \frac{1}{1 + (1-\beta)\|\tilde{u}_k\|_2}.$$

Therefore,

$$
\begin{aligned}
\omega(\|\tilde{u}_k\|_2) - \omega(\|\tilde{v}_k\|_2) &= \frac{(\|\tilde{u}_k\|_2 - \|\tilde{v}_k\|_2)\|\tilde{v}_k\|_2}{1 + \|\tilde{v}_k\|_2} + \frac{1}{2}\omega''(t)\,(\|\tilde{u}_k\|_2 - \|\tilde{v}_k\|_2)^2 \\
&\leq \frac{\|\tilde{u}_k - \tilde{v}_k\|_2\|\tilde{v}_k\|_2}{1 + (1-\beta)\|\tilde{u}_k\|_2} + \frac{(1/2)\|\tilde{u}_k - \tilde{v}_k\|_2^2}{1 + (1-\beta)\|\tilde{u}_k\|_2} \\
&\leq \frac{\beta(1+\beta)\|\tilde{u}_k\|_2^2 + (1/2)\beta^2\|\tilde{u}_k\|_2^2}{1 + (1-\beta)\|\tilde{u}_k\|_2}.
\end{aligned}
$$

In addition, we have

$$\frac{\langle \tilde{v}_k - \tilde{u}_k, \tilde{v}_k \rangle}{1 + \|\tilde{v}_k\|_2} \leq \frac{\|\tilde{u}_k - \tilde{v}_k\|_2\|\tilde{v}_k\|_2}{1 + \|\tilde{v}_k\|_2} \leq \frac{\beta(1+\beta)\|\tilde{u}_k\|_2^2}{1 + (1-\beta)\|\tilde{u}_k\|_2}.$$

Combining the two inequalities above, and using the relation $t^2/(1+t) \le 2\omega(t)$ for all $t \ge 0$, we obtain

$$\omega(\|\tilde{u}_k\|_2) - \omega(\|\tilde{v}_k\|_2) + \frac{\langle \tilde{v}_k - \tilde{u}_k, \tilde{v}_k \rangle}{1 + \|\tilde{v}_k\|_2} \le \left(2\beta(1+\beta) + (1/2)\beta^2\right)$$

$$\frac{\|\tilde{u}_k\|_2^2}{1 + (1-\beta)\|\tilde{u}_k\|_2}$$

$$= \left(\frac{2\beta + (5/2)\beta^2}{(1-\beta)^2}\right) \frac{(1-\beta)^2 \|\tilde{u}_k\|_2^2}{1 + (1-\beta)\|\tilde{u}_k\|_2}$$

$$\le \left(\frac{2\beta + (5/2)\beta^2}{(1-\beta)^2}\right) 2\omega\big((1-\beta)\|\tilde{u}_k\|_2\big)$$

$$\le \left(\frac{4\beta + 5\beta^2}{1-\beta}\right) \omega(\|\tilde{u}_k\|_2).$$

In the last inequality above, we used the fact that for any $t \ge 0$ we have

$$\omega((1-\beta)t) \le (1-\beta)\omega(t),$$

which is the result of convexity of $\omega(t)$ and $\omega(0) = 0$. Substituting the above upper bound into inequality (11.45) yields

$$f(w_{k+1}) \le f(w_k) - \left(1 - \frac{4\beta + 5\beta^2}{1-\beta}\right) \omega(\|\tilde{u}_k\|_2). \tag{11.46}$$

With inequality (11.46), we are ready to prove the conclusions of Theorem 1. In particular, Part (a) of Theorem 1 holds for any $0 \le \beta \le 1/10$.

For part (b), we assume that $\|\tilde{u}_k\|_2 \le 1/6$. According to [36, Theorem 4.1.13], when $\|\tilde{u}_k\|_2 < 1$, it holds that for every $k \ge 0$,

$$\omega(\|\tilde{u}_k\|_2) \le f(w_k) - f(w_\star) \le \omega_*(\|\tilde{u}_k\|_2). \tag{11.47}$$

Combining this sandwich inequality with inequality (11.46), we have

$$\omega(\|\tilde{u}_{k+1}\|_2) \le f(w_{k+1}) - f(w_\star)$$

$$\le f(w_k) - f(w_\star) - \omega(\|\tilde{u}_k\|_2) + \frac{4\beta + 5\beta^2}{1-\beta}\omega(\|\tilde{u}_k\|_2)$$

$$\le \omega_*(\|\tilde{u}_k\|_2) - \omega(\|\tilde{u}_k\|_2) + \frac{4\beta + 5\beta^2}{1-\beta}\omega(\|\tilde{u}_k\|_2). \tag{11.48}$$

It is easy to verify that $\omega_*(t) - \omega(t) \leq 0.26\,\omega(t)$ for all $t \leq 1/6$, and

$$(4\beta + 5\beta^2)/(1 - \beta) \leq 0.23, \qquad \text{if} \quad \beta \leq 1/20.$$

Applying these two inequalities to inequality (11.48) completes the proof.

It should be clear that other combinations of the value of β and bound on $\|\widetilde{u}_k\|_2$ are also possible. For example, for $\beta = 1/10$ and $\|\widetilde{u}_k\|_2 \leq 1/10$, we have $\omega(\|\widetilde{u}_{k+1}\|_2) \leq 0.65\,\omega(\|\widetilde{u}_k\|_2)$.

Appendix 2: Proof of Theorem 2 and Corollary 2

First we prove Theorem 2. We start with the inequality (11.45), and upper bound the last two terms on its right-hand side. Since $\omega'(t) = t/(1+t) < 1$, we have

$$\omega(\|\widetilde{u}_k\|_2) - \omega(\|\widetilde{v}_k\|_2) \leq \big|\|\widetilde{u}_k\|_2 - \|\widetilde{v}_k\|_2\big| \leq \|\widetilde{u}_k - \widetilde{v}_k\|_2.$$

In addition, we have

$$\frac{\langle \widetilde{v}_k - \widetilde{u}_k, \widetilde{v}_k \rangle}{1 + \|\widetilde{v}_k\|_2} \leq \frac{\|\widetilde{v}_k\|_2}{1 + \|\widetilde{v}_k\|_2}\|\widetilde{u}_k - \widetilde{v}_k\|_2 \leq \|\widetilde{u}_k - \widetilde{v}_k\|_2.$$

Applying these two bounds to (11.45), we obtain

$$f(w_{k+1}) \leq f(w_k) - \omega(\|\widetilde{u}_k\|_2) + 2\|\widetilde{u}_k - \widetilde{v}_k\|_2. \tag{11.49}$$

Next we bound $\|\widetilde{u}_k - \widetilde{v}_k\|_2$ using the approximation tolerance ϵ_k specified in (11.14),

$$
\begin{aligned}
\|\widetilde{u}_k - \widetilde{v}_k\|_2 &= \left\| [f''(w_k)]^{-1/2} f'(w_k) - [f''(w_k)]^{1/2} v_k \right\|_2 \\
&= \left\| [f''(w_k)]^{-1/2}\big(f''(w_k)v_k - f'(w_k)\big) \right\|_2 \\
&\leq \lambda^{-1/2} \left\| f''(w_k)v_k - f'(w_k) \right\|_2 \\
&\leq \lambda^{-1/2} \epsilon_k \\
&= \frac{1}{2}\min\left\{ \frac{\omega(r_k)}{2}, \frac{\omega^{3/2}(r_k)}{10} \right\}.
\end{aligned}
$$

Combining the above inequality with (11.49), and using $r_k = L^{-1/2}\|f'(w_k)\|_2 \leq \|\widetilde{u}_k\|_2$ with the monotonicity of ω, we arrive at

$$f(w_{k+1}) \leq f(w_k) - \omega(\|\widetilde{u}_k\|_2) + \min\left\{ \frac{\omega(\widetilde{u}_k)}{2}, \frac{\omega^{3/2}(\widetilde{u}_k)}{10} \right\}. \tag{11.50}$$

Part (a) of the theorem follows immediately from inequality (11.50).

For part (b), we assume that $\|\tilde{u}_k\|_2 \leq 1/8$. Combining (11.47) with (11.50), we have

$$\omega(\|\tilde{u}_{k+1}\|_2) \leq f(w_{k+1}) - f(w_\star) \leq f(w_k) - f(w_\star) - \omega(\|\tilde{u}_k\|_2) + \frac{\omega^{3/2}(\|\tilde{u}_k\|_2)}{10}$$

$$\leq \omega_*(\|\tilde{u}_k\|_2) - \omega(\|\tilde{u}_k\|_2) + \frac{\omega^{3/2}(\|\tilde{u}_k\|_2)}{10}. \tag{11.51}$$

Let $h(t) = \omega_*(t) - \omega(t)$ and consider only $t \geq 0$. Notice that $h(0) = 0$ and $h'(t) = \frac{2t^2}{1-t^2} < \frac{128}{63}t^2$ for $t \leq 1/8$. Thus, we conclude that $h(t) \leq \frac{128}{189}t^3$ for $t \leq 1/8$. We also notice that $\omega(0) = 0$ and $\omega'(t) = \frac{t}{1+t} \geq \frac{8}{9}t$ for $t \leq 1/8$. Thus, we have $\omega(t) \geq \frac{4}{9}t^2$ for $t \leq 1/8$. Combining these results, we obtain

$$\omega_*(t) - \omega(t) \leq \frac{128}{189}t^3 = \frac{128}{189}(t^2)^{3/2} \leq \frac{128}{189}\left(\frac{9}{4}\omega(t)\right)^{3/2} \leq \left(\sqrt{6} - \frac{1}{10}\right)\omega^{3/2}(t).$$

Applying this inequality to the right-hand side of (11.51) completes the proof.

Next we prove Corollary 2. By part (a) of Theorem 2, if $\omega(\|\tilde{u}_k\|_2) \geq 1/8$, then each iteration of Algorithm 1 decreases the function value at least by the constant $\frac{1}{2}\omega(1/8)$. So within at most $K_1 = \left\lceil \frac{2(f(w_0)-f(w_\star))}{\omega(1/8)} \right\rceil$ iterations, we are guaranteed to have $\|\tilde{u}_k\|_2 \leq 1/8$. Part (b) of Theorem 2 implies $6\,\omega(\|\tilde{u}_{k+1}\|_2) \leq (6\,\omega(\|\tilde{u}_k\|_2))^{3/2}$ when $\|\tilde{u}_k\|_2 \leq 1/8$, and hence

$$\log\bigl(6\,\omega(\|\tilde{u}_k\|_2)\bigr) \leq \left(\frac{3}{2}\right)^{k-K_1} \log\left(6\,\omega(1/8)\right), \qquad k \geq K_1.$$

Note that both sides of the above inequality are negative. Therefore, after $k \geq K_1 + \frac{\log\log(1/(3\epsilon))}{\log(3/2)}$ iterations (assuming $\epsilon \leq 1/(3e)$), we have

$$\log\bigl(6\,\omega(\|\tilde{u}_k\|_2)\bigr) \leq \log(1/(3\epsilon)) \log(6\,\omega(1/8)) \leq -\log(1/(3\epsilon)),$$

which implies $\omega(\|\tilde{u}_k\|_2) \leq \epsilon/2$. Finally using (11.47) and the fact that $\omega_*(t) \leq 2\,\omega(t)$ for $t \leq 1/8$, we obtain

$$f(w_k) - f(w_\star) \leq \omega_*(\|\tilde{u}_k\|_2) \leq 2\,\omega(\|\tilde{u}_k\|_2) \leq \epsilon.$$

This completes the proof of Corollary 2.

Appendix 3: Proof of Lemma 4

It suffices to show that the algorithm terminates at iteration $t \leq T_\mu - 1$, because when the algorithm terminates, it outputs a vector v_k which satisfies

$$\|Hv_k - f'(w_k)\|_2 = \|r^{(t+1)}\|_2 \leq \epsilon_k.$$

Denote by $v^* = H^{-1}f'(w_k)$ the solution of the linear system $Hv_k = f'(w_k)$. By the classical analysis on the preconditioned conjugate gradient method (see, e.g., [3, 32]), Algorithm 2 has the following convergence property:

$$(v^{(t)} - v^*)^T H(v^{(t)} - v^*) \leq 4 \left(\frac{\sqrt{\kappa} - 1}{\sqrt{\kappa} + 1} \right)^{2t} (v^*)^T Hv^*, \tag{11.52}$$

where $\kappa = 1 + 2\mu/\lambda$ is the condition number of $P^{-1}H$ given in (11.25). For the left-hand side of inequality (11.52), we have

$$(v^{(t)} - v^*)^T H(v^{(t)} - v^*) = (r^{(t)})^T H^{-1} r^{(t)} \geq \frac{\|r^{(t)}\|_2^2}{L}.$$

For the right-hand side of inequality (11.52), we have

$$(v^*)^T Hv^* = (f'(w_k))^T H^{-1} f'(w_k) \leq \frac{\|f'(w_k)\|_2^2}{\lambda}.$$

Combining the above two inequalities with inequality (11.52), we obtain

$$\|r^{(t)}\|_2 \leq 2\sqrt{\frac{L}{\lambda}} \left(\frac{\sqrt{\kappa} - 1}{\sqrt{\kappa} + 1} \right)^t \|f'(w_k)\|_2 \leq 2\sqrt{\frac{L}{\lambda}} \left(1 - \sqrt{\frac{\lambda}{\lambda + 2\mu}} \right)^t \|f'(w_k)\|_2.$$

To guarantee that $\|r^{(t)}\|_2 \leq \epsilon_k$, it suffices to have

$$t \geq \frac{\log \left(\frac{2\sqrt{L/\lambda}\|f'(w_k)\|_2}{\epsilon_k} \right)}{-\log \left(1 - \sqrt{\frac{\lambda}{\lambda+2\mu}} \right)} \geq \sqrt{1 + \frac{2\mu}{\lambda}} \log \left(\frac{2\sqrt{L/\lambda}\|f'(w_k)\|_2}{\epsilon_k} \right),$$

where in the last inequality we used $-\log(1 - x) \geq x$ for $0 < x < 1$. Comparing with the definition of T_μ in (11.26), this is the desired result.

Appendix 4: Proof of Lemma 5

First, we prove Inequality (11.31). Recall that $w_\star$ and $\widehat{w}_i$ minimizes $f(w)$ and $f_i(w) + (\rho/2)\|w\|_2^2$ respectively. Since both functions are λ-strongly convex, we have

$$\frac{\lambda}{2}\|w_\star\|_2^2 \le f(w_\star) \le f(0) \le V_0,$$

$$\frac{\lambda}{2}\|\widehat{w}_i\|_2^2 \le f_i(\widehat{w}_i) + \frac{\rho}{2}\|\widehat{w}_i\|_2^2 \le f_i(0) \le V_0,$$

where we also used Assumption 2(a) in the first inequality on both lines. These two inequalities imply $\|w_\star\|_2 \le \sqrt{2V_0/\lambda}$ and $\|\widehat{w}_i\|_2 \le \sqrt{2V_0/\lambda}$. Then the inequality (11.31) follows since w_0 is the average over $\widehat{w}_i$ for $i = 1, \ldots, m$.

Next we prove inequality (11.32). Let z be a random variable in $\mathcal{Z} \subset \mathbb{R}^p$ with an unknown probability distribution. We define a regularized population risk:

$$R(w) = \mathbb{E}_z[\phi(w, z)] + \frac{\lambda + \rho}{2}\|w\|_2^2.$$

Let S be a set of n i.i.d. samples in $\mathcal{Z}$ from the same distribution. We define a regularized empirical risk

$$r_S(w) = \frac{1}{n}\sum_{z \in S}\phi(w, z) + \frac{\lambda + \rho}{2}\|w\|_2^2,$$

and its minimizer

$$\widehat{w}_S = \arg\min_w r_S(w).$$

The following lemma states that the population risk of $\widehat{w}_S$ is very close to its empirical risk. The proof is based on the notion of *stability* of regularized ERM [9].

Lemma 7 *Suppose Assumption 2 holds and S is a set of n i.i.d. samples in $\mathcal{Z}$. Then*

$$\mathbb{E}_S\big[R(\widehat{w}_S) - r_S(\widehat{w}_S)\big] \le \frac{2G^2}{\rho n}.$$

Proof Let $S = \{z_1, \ldots, z_n\}$. For any $k \in \{1, \ldots, n\}$, we define a modified training set $S^{(k)}$ by replacing z_k with another sample $\widetilde{z}_k$, which is drawn from the same distribution and is independent of S. The empirical risk on $S^{(k)}$ is defined as

$$r_S^{(k)}(w) = \frac{1}{n}\sum_{z \in S^{(k)}}\phi(w, z) + \frac{\lambda + \rho}{2}\|w\|_2^2.$$

Let $\widehat{w}_S^{(k)} = \arg\min_w r_S^{(k)}(w)$. Since both r_S and $r_S^{(k)}$ are ρ-strongly convex, we have

$$r_S(\widehat{w}_S^{(k)}) - r_S(\widehat{w}_S) \geq \frac{\rho}{2}\|\widehat{w}_S^{(k)} - \widehat{w}_S\|_2^2,$$

$$r_S^{(k)}(\widehat{w}_S) - r_S^{(k)}(\widehat{w}_S^{(k)}) \geq \frac{\rho}{2}\|\widehat{w}_S^{(k)} - \widehat{w}_S\|_2^2.$$

Summing the above two inequalities, and noticing that

$$r_S(w) - r_S^{(k)}(w) = \frac{1}{n}(\phi(w, z_k) - \phi(w, \widetilde{z}_k)),$$

we have

$$\|\widehat{w}_S^{(k)} - \widehat{w}_S\|_2^2 \leq \frac{1}{\rho n}\left(\phi(\widehat{w}_S^{(k)}, z_k) - \phi(\widehat{w}_S^{(k)}, \widetilde{z}_k) - \phi(\widehat{w}_S, z_k) + \phi(\widehat{w}_S, \widetilde{z}_k)\right).$$

$$(11.53)$$

By Assumption 2(b) and the facts $\|\widehat{w}_S\|_2 \leq \sqrt{2V_0/\lambda}$ and $\|\widehat{w}_S^{(k)}\|_2 \leq \sqrt{2V_0/\lambda}$, we have

$$\left|\phi(\widehat{w}_S^{(k)}, z) - \phi(\widehat{w}_S, z)\right| \leq G\|\widehat{w}_S^{(k)} - \widehat{w}_S\|_2, \qquad \forall z \in \mathcal{Z}.$$

Combining the above Lipschitz condition with (11.53), we obtain

$$\|\widehat{w}_S^{(k)} - \widehat{w}_S\|_2^2 \leq \frac{2G}{\rho n}\|\widehat{w}_S^{(k)} - \widehat{w}_S\|_2.$$

As a consequence, we have $\|\widehat{w}_S^{(k)} - \widehat{w}_S\|_2 \leq \frac{2G}{\rho n}$, and therefore

$$\left|\phi(\widehat{w}_S^{(k)}, z) - \phi(\widehat{w}_S, z)\right| \leq \frac{2G^2}{\rho n}, \qquad \forall z \in \mathcal{Z}. \qquad (11.54)$$

In the terminology of learning theory, this means that empirical minimization over the regularized loss r_S has *uniform stability* $2G^2/(\rho n)$ with respect to the loss function ϕ; see [9].

For any fixed $k \in \{1, \ldots, n\}$, since $\widetilde{z}_k$ is independent of S, we have

$$\mathbb{E}_S\left[R(\widehat{w}_S) - r_S(\widehat{w}_S)\right] = \mathbb{E}_S\left[\mathbb{E}_{\widetilde{z}_k}[\phi(\widehat{w}_S, \widetilde{z}_k)] - \frac{1}{n}\sum_{j=1}^n \phi(\widehat{w}_S, z_j)\right]$$

$$= \mathbb{E}_{S,\widetilde{z}_k}[\phi(\widehat{w}_S, \widetilde{z}_k) - \phi(\widehat{w}_S, z_k)]$$

$$= \mathbb{E}_{S,\widetilde{z}_k}[\phi(\widehat{w}_S, \widetilde{z}_k) - \phi(\widehat{w}_S^{(k)}, \widetilde{z}_k)],$$

where the second equality used the fact that $\mathbb{E}_S[\phi(\widehat{w}_S, z_j)]$ has the same value for all $j = 1, \ldots, n$, and the third equality used the symmetry between the pairs (S, z_k) and $(S^{(k)}, \widetilde{z}_k)$ (also known as the *renaming* trick; see [9, Lemma 7]). Combining the above equality with (11.54) yields the desired result.   □

Next, we consider a distributed system with m machines, where each machine has a local dataset S_i of size n, for $i = 1, \ldots, m$. To simplify notation, we denote the local regularized empirical loss function and its minimizer by r_i and $\widehat{w}_i$, respectively. We would like to bound the excessive error when applying $\widehat{w}_i$ to a different dataset S_j. Notice that

$$\mathbb{E}_{S_i,S_j}\left[r_j(\widehat{w}_i) - r_j(\widehat{w}_j)\right] = \underbrace{\mathbb{E}_{S_i,S_j}\left[r_j(\widehat{w}_i) - r_i(\widehat{w}_i)\right]}_{v_1} + \underbrace{\mathbb{E}_{S_i,S_j}\left[r_i(\widehat{w}_i) - r_j(\widehat{w}_R)\right]}_{v_2}$$

$$+ \underbrace{\mathbb{E}_{S_j}\left[r_j(\widehat{w}_R) - r_j(\widehat{w}_j)\right]}_{v_3}, \tag{11.55}$$

where $\widehat{w}_R$ is the minimizer of $R(w)$. Since S_i and S_j are independent, we have

$$v_1 = \mathbb{E}_{S_i}\left[\mathbb{E}_{S_j}[r_j(\widehat{w}_i)] - r_i(\widehat{w}_i)\right] = \mathbb{E}_{S_i}\left[R(\widehat{w}_i) - r_i(\widehat{w}_i)\right] \leq \frac{2G^2}{\rho n},$$

where the inequality is due to Lemma 7. For the second term in (11.55), we have

$$v_2 = \mathbb{E}_{S_i}\left[r_i(\widehat{w}_i) - \mathbb{E}_{S_j}[r_j(\widehat{w}_R)]\right] = \mathbb{E}_{S_i}\left[r_i(\widehat{w}_i) - r_i(\widehat{w}_R)\right] \leq 0.$$

It remains to bound the third term v_3. We first use the strong convexity of r_j to obtain (see, e.g., [36, Theorem 2.1.10])

$$r_j(\widehat{w}_R) - r_j(\widehat{w}_j) \leq \frac{\|r_j'(\widehat{w}_R)\|_2^2}{2\rho}, \tag{11.56}$$

where $r_j'(\widehat{w}_R)$ denotes the gradient of r_j at $\widehat{w}_R$. If we index the elements of S_j by $z_1, \ldots, z_n$, then

$$r_j'(\widehat{w}_R) = \frac{1}{n}\sum_{k=1}^{n}\left(\phi'(\widehat{w}_R, z_k) + (\lambda + \rho)\widehat{w}_R\right). \tag{11.57}$$

By the optimality condition of $\widehat{w}_R = \arg\min_w R(w)$, we have for any $k \in \{1, \ldots, n\}$,

$$\mathbb{E}_{z_k}\left[\phi'(\widehat{w}_R, z_k) + (\lambda + \rho)\widehat{w}_R\right] = 0.$$

Therefore, according to (11.57), the gradient $r_j(\widehat{w}_R)$ is the average of n independent and zero-mean random vectors. Combining (11.56) and (11.57) with the definition of v_3 in (11.55), we have

$$
\begin{aligned}
v_3 &\leq \frac{\mathbb{E}_{S_j}\left[\sum_{k=1}^{n} \|\phi'(\widehat{w}_R, z_k) + (\lambda + \rho)\widehat{w}_R\|_2^2\right]}{2\rho n^2} \\
&= \frac{\sum_{k=1}^{n} \mathbb{E}_{S_j}\left[\|\phi'(\widehat{w}_R, z_k) + (\lambda + \rho)\widehat{w}_R\|_2^2\right]}{2\rho n^2} \\
&\leq \frac{\sum_{k=1}^{n} \mathbb{E}[\|\phi'(\widehat{w}_R, z_k)\|_2^2]}{2\rho n^2} \\
&\leq \frac{G^2}{2\rho n}.
\end{aligned}
$$

In the equality above, we used the fact that $\phi'(\widehat{w}_R, z_k) + (\lambda + \rho)\widehat{w}_R$ are i.i.d. zero-mean random variables, so the variance of their sum equals the sum of their variances. The last inequality above is due to Assumption 2(b) and the fact that $\|\widehat{w}_R\|_2 \leq \sqrt{2V_0/(\lambda + \rho)} \leq \sqrt{2V_0/\lambda}$. Combining the upper bounds for v_1, v_2 and v_3, we have

$$
\mathbb{E}_{S_i, S_j}\left[r_j(\widehat{w}_i) - r_j(\widehat{w}_j)\right] \leq \frac{3G^2}{\rho n}. \tag{11.58}
$$

Recall the definition of f as

$$
f(w) = \frac{1}{mn} \sum_{i=1}^{m} \sum_{k=1}^{n} \phi(w, z_{i,k}) + \frac{\lambda}{2}\|w\|_2^2,
$$

where $z_{i,k}$ denotes the kth sample at machine i. Let $r(w) = (1/m)\sum_{j=1}^{m} r_j(w)$, then

$$
r(w) = f(w) + \frac{\rho}{2}\|w\|_2^2. \tag{11.59}
$$

We compare the value $r(\widehat{w}_i)$, for any $i \in \{1, \ldots, m\}$, with the minimum of $r(w)$:

$$
\begin{aligned}
r(\widehat{w}_i) - \min_w r(w) &= \frac{1}{m}\sum_{j=1}^{m} r_j(\widehat{w}_i) - \min_w \frac{1}{m}\sum_{j=1}^{m} r_j(w) \\
&\leq \frac{1}{m}\sum_{j=1}^{m} r_j(\widehat{w}_i) - \frac{1}{m}\sum_{j=1}^{m} \min_w r_j(w) \\
&= \frac{1}{m}\sum_{j=1}^{m} \left(r_j(\widehat{w}_i) - r_j(\widehat{w}_j)\right).
\end{aligned}
$$

Taking expectation with respect to all the random data sets $S_1, \ldots, S_m$, we obtain

$$\mathbb{E}[r(\widehat{w}_i) - \min_w r(w)] \leq \frac{1}{m} \sum_{j=1}^n \mathbb{E}[r_j(\widehat{w}_i) - r_j(\widehat{w}_j)] \leq \frac{3G^2}{\rho n}, \qquad (11.60)$$

where the last inequality is due to (11.58). Finally, we bound the expected value of $f(\widehat{w}_i)$:

$$\mathbb{E}[f(\widehat{w}_i)] \leq \mathbb{E}[r(\widehat{w}_i)] \leq \mathbb{E}\left[\min_w r(w)\right] + \frac{3G^2}{\rho n}$$

$$\leq \mathbb{E}\left[f(w_\star) + \frac{\rho}{2}\|w_\star\|_2^2\right] + \frac{3G^2}{\rho n}$$

$$\leq \mathbb{E}[f(w_\star)] + \frac{\rho D^2}{2} + \frac{3G^2}{\rho n},$$

where the first inequality holds because of (11.59), the second inequality is due to (11.60), and the last inequality follows from the assumption that $\mathbb{E}[\|w_\star\|_2] \leq D^2$. Choosing $\rho = \sqrt{6G^2/(nD^2)}$ results in

$$\mathbb{E}[f(\widehat{w}_i) - f(w_\star)] \leq \frac{\sqrt{6}GD}{\sqrt{n}}, \qquad i = 1, \ldots, m.$$

Since $w_0 = (1/m) \sum_{i=1}^m \widehat{w}_i$, we can use the convexity of the function f to conclude that $\mathbb{E}[f(w_0) - f(w_\star)] \leq \sqrt{6}GD/\sqrt{n}$, which is the desired result.

Appendix 5: Proof of Lemma 6

We consider the regularized empirical loss functions f_i defined in (11.30). For any two vectors $u, w \in \mathbb{R}^d$ satisfying $\|u - w\|_2 \leq \varepsilon$, Assumption 2(d) implies

$$\|f_i''(u) - f_i''(w)\|_2 \leq M\varepsilon.$$

Let $B(0, r)$ be the ball in $\mathbb{R}^d$ with radius r, centered at the origin. Let $N_\varepsilon^{\text{cov}}(B(0, r))$ be the *covering number* of $B(0, r)$ by balls of radius ε, i.e., the minimum number of balls of radius ε required to cover $B(0, r)$. We also define $N_\varepsilon^{\text{pac}}(B(0, r))$ as the *packing number* of $B(0, r)$, i.e., the maximum number of disjoint balls whose centers belong to $B(0, r)$. It is easy to verify that

$$N_\varepsilon^{\text{cov}}(B(0, r)) \leq N_{\varepsilon/2}^{\text{pac}}(B(0, r)) \leq (1 + 2r/\varepsilon)^d.$$

Therefore, there exist a set of points $U \subseteq \mathbb{R}^d$ with cardinality at most $(1 + 2r/\varepsilon)^d$, such that for any vector $w \in B(0, r)$, we have

$$\min_{u \in U} \| f_i''(w) - f_i''(u) \|_2 \leq M\varepsilon. \tag{11.61}$$

We consider an arbitrary point $u \in U$ and the associated Hessian matrices for the functions f_i defined in (11.30). We have

$$f_i''(u) = \frac{1}{n} \sum_{j=1}^n \left(\phi''(u, z_{i,j}) + \lambda I \right), \qquad i = 1, \ldots, m.$$

The components of the above sum are i.i.d. matrices that are upper bounded by LI. By the matrix Hoeffding's inequality [33, Corollary 4.2], we have

$$\mathbb{P}\left[\| f_i''(u) - \mathbb{E}[f_i''(u)] \|_2 > t \right] \leq d \cdot e^{-\frac{nt^2}{2L^2}}.$$

Note that $\mathbb{E}[f_1''(w)] = \mathbb{E}[f''(w)]$ for any $w \in B(0, r)$. Using the triangular inequality and inequality (11.61), we obtain

$$\| f_1''(w) - f''(w) \|_2 \leq \| f_1''(w) - \mathbb{E}[f_1''(w)] \|_2 + \| f''(w) - \mathbb{E}[f''(w)] \|_2$$

$$\leq 2 \max_{i \in \{1,\ldots,m\}} \| f_i''(w) - \mathbb{E}[f_i''(w)] \|_2$$

$$\leq 2 \max_{i \in \{1,\ldots,m\}} \left(\max_{u \in U} \| f_i''(u) - \mathbb{E}[f_i''(u)] \|_2 + M\varepsilon \right). \tag{11.62}$$

Applying the union bound, we have with probability at least

$$1 - md(1 + 2r/\varepsilon)^d \cdot e^{-\frac{nt^2}{2L^2}},$$

the inequality $\| f_i''(u) - \mathbb{E}[f_i''(u)] \|_2 \leq t$ holds for every $i \in \{1, \ldots, m\}$ and every $u \in U$. Combining this probability bound with inequality (11.62), we have

$$\mathbb{P}\left[\sup_{w \in B(0,r)} \| f_1''(w) - f''(w) \|_2 > 2t + 2M\varepsilon \right] \leq md(1 + 2r/\varepsilon)^d \cdot e^{-\frac{nt^2}{2L^2}}. \tag{11.63}$$

As the final step, we choose $\varepsilon = \frac{\sqrt{2}L}{\sqrt{n}M}$ and then choose t to make the right-hand side of inequality (11.63) equal to δ. This yields the desired result.

Appendix 6: Proof of Theorem 5

Suppose Algorithm 3 terminates in K iterations. Let t_k be the number of conjugate gradient steps in each call of Algorithm 2, for $k = 0, 1, \ldots, K - 1$. For any given $\mu > 0$, we define T_μ as in (11.26). Let $\mathcal{A}$ denotes the event that $t_k \leq T_\mu$ for all $k \in \{0, \ldots, K - 1\}$. Let $\mathcal{A}_c$ be the complement of $\mathcal{A}$, i.e., the event that $t_k > T_\mu$ for some $k \in \{0, \ldots, K - 1\}$. In addition, let the probabilities of the events $\mathcal{A}$ and $\mathcal{A}_c$ be $1 - \delta$ and δ respectively. By the law of total expectation, we have

$$\mathbb{E}[T] = \mathbb{E}[T|\mathcal{A}]\mathbb{P}(\mathcal{A}) + \mathbb{E}[T|\mathcal{A}_c]\mathbb{P}(\mathcal{A}_c) = (1 - \delta)\mathbb{E}[T|\mathcal{A}] + \delta\,\mathbb{E}[T|\mathcal{A}_c].$$

When the event $\mathcal{A}$ happens, we have $T \leq 1 + K(T_\mu + 1)$ where T_μ is given in (11.26); otherwise we have $T \leq 1 + K(T_L + 1)$, where

$$T_L = \sqrt{2 + \frac{2L}{\lambda} \log\left(\frac{2L}{\beta\lambda}\right)} \tag{11.64}$$

bounds the number of PCG iterations in Algorithm 2 when the event $\mathcal{A}_c$ happens. Since Algorithm 2 always ensures $\|f''(w_k)v_k - f'(w_k)\|_2 \leq \epsilon_k$, the outer iteration count K shares the same bound in (11.12), which depends on $f(w_0) - f(w_\star)$. Notice that $f(w_0) - f(w_\star)$ is a random variable depending on the random generation of the datasets. However, T_μ and T_L are deterministic constants. So we have

$$\mathbb{E}[T] \leq 1 + (1 - \delta)\mathbb{E}[K(T_\mu + 1)|\mathcal{A}] + \delta\,\mathbb{E}[K(T_L + 1)|\mathcal{A}_c]$$
$$= 1 + (1 - \delta)(T_\mu + 1)\mathbb{E}[K|\mathcal{A}] + \delta(T_L + 1)\mathbb{E}[K|\mathcal{A}_c]. \tag{11.65}$$

Next we bound $\mathbb{E}[K|\mathcal{A}]$ and $\mathbb{E}[K|\mathcal{A}_c]$ separately. To bound $\mathbb{E}[K|\mathcal{A}]$, we use

$$\mathbb{E}[K] = (1 - \delta)\mathbb{E}[K|\mathcal{A}] + \delta\,\mathbb{E}[K|\mathcal{A}_c] \geq (1 - \delta)\mathbb{E}[K|\mathcal{A}]$$

to obtain

$$\mathbb{E}[K|\mathcal{A}] \leq \mathbb{E}[K]/(1 - \delta). \tag{11.66}$$

In order to bound $\mathbb{E}[K|\mathcal{A}_c]$, we derive a deterministic bound on $f(w_0) - f(w_\star)$. By Lemma 5, we have $\|w_0\|_2 \leq \sqrt{2V_0/\lambda}$, which together with Assumption 2(b) yields

$$\|f'(w)\|_2 \leq G + \lambda\|w\|_2 \leq G + \sqrt{2\lambda V_0}.$$

Combining with the strong convexity of f, we obtain

$$f(w_0) - f(w_\star) \leq \frac{1}{2\lambda}\|f'(w_0)\|_2^2 \leq \frac{1}{2\lambda}\left(G + \sqrt{2\lambda V_0}\right)^2 \leq 2V + \frac{G^2}{\lambda}.$$

Therefore by Corollary 1,

$$K \leq K_{\max} = 1 + \frac{4V_0 + 2G^2/\lambda}{\omega(1/6)} + \left\lceil \log_2 \left(\frac{2\omega(1/6)}{\epsilon} \right) \right\rceil, \tag{11.67}$$

where the additional 1 counts compensate for removing one $\lceil \cdot \rceil$ operator in (11.12).

Using inequality (11.65), the bound on $\mathbb{E}[K|\mathcal{A}]$ in (11.66) and the bound on $\mathbb{E}[K|\mathcal{A}_c]$ in (11.67), we obtain

$$\mathbb{E}[T] \leq 1 + (T_\mu + 1)\mathbb{E}[K] + \delta(T_L + 1)K_{\max}.$$

Now we can bound $\mathbb{E}[K]$ by Corollary 1 and Lemma 5. More specifically,

$$\mathbb{E}[K] \leq \frac{\mathbb{E}[2(f(w_0) - f(w_\star))]}{\omega(1/6)} + \left\lceil \log_2 \left(\frac{2\omega(1/6)}{\epsilon} \right) \right\rceil + 1 = C_0 + \frac{2\sqrt{6}}{\omega(1/6)} \cdot \frac{GD}{\sqrt{n}},$$

where $C_0 = 1 + \lceil \log_2(2\omega(1/6)/\epsilon) \rceil$. With the choice of δ in (11.34) and the definition of T_L in (11.64), we have

$$\delta(T_L + 1)K_{\max} = \frac{GD}{\sqrt{n}} \cdot \frac{\sqrt{\lambda/(4L)}}{4V_0 + 2G^2/\lambda} \left(2 + \sqrt{2 + \frac{2L}{\lambda} \log \left(\frac{2L}{\beta\lambda} \right)} \right) \left(C_0 + \frac{4V_0 + 2G^2/\lambda}{\omega(1/6)} \right)$$

$$= \left(\frac{C_0}{\sqrt{n}} \cdot \frac{GD}{4V_0 + 2G^2/\lambda} + \frac{1}{\omega(1/6)} \cdot \frac{GD}{\sqrt{n}} \right) \left(\sqrt{\frac{\lambda}{L}} + C_2\sqrt{\frac{\lambda}{2L} + \frac{1}{2}} \right)$$

$$\leq \left(\frac{C_0}{\sqrt{n}} \cdot \frac{GD}{4V_0 + 2G^2/\lambda} + \frac{1}{\omega(1/6)} \cdot \frac{GD}{\sqrt{n}} \right) \left(2 + C_2\sqrt{1 + \frac{2\mu}{\lambda}} \right)$$

$$= \left(\frac{C_0}{\sqrt{n}} \cdot \frac{GD}{4V_0 + 2G^2/\lambda} + \frac{1}{\omega(1/6)} \cdot \frac{GD}{\sqrt{n}} \right) (T_\mu + 1),$$

where $C_2 = \log(2L/(\beta\lambda))$. Putting everything together, we have

$$\mathbb{E}[T] \leq 1 + \left(C_0 + \frac{C_0}{\sqrt{n}} \cdot \frac{GD}{4V_0 + 2G^2/\lambda} + \frac{2\sqrt{6}+1}{\omega(1/6)} \cdot \frac{GD}{\sqrt{n}} \right) (T_\mu + 1)$$

$$\leq 1 + \left(C_1 + \frac{6}{\omega(1/6)} \cdot \frac{GD}{\sqrt{n}} \right) (T_\mu + 1).$$

Replacing T_μ by its expression in (11.26) and applying Corollary 4, we obtain the desired result.

References

1. A. Agarwal, J.C. Duchi, Distributed delayed stochastic optimization, in *Advances in Neural Information Processing Systems (NIPS) 24* (2011), pp. 873–881
2. Y. Arjevani, O. Shamir, Communication complexity of distributed convex learning and optimization, in *Advances in Neural Information Processing Systems (NIPS) 28* (2015), pp. 1756–1764
3. M. Avriel, *Nonlinear Programming: Analysis and Methods* (Prentice-Hall, Upper Saddle River, 1976)
4. F. Bach, Self-concordant analysis for logistic regression. Electron. J. Stat. **4**, 384–414 (2010)
5. R. Bekkerman, M. Bilenko, J. Langford, *Scaling Up Machine Learning: Parallel and Distributed Approaches* (Cambridge University Press, Cambridge, 2011)
6. D.P. Bertsekas, J.N. Tsitsiklis, *Parallel and Distributed Computation: Numerical Methods* (Prentice-Hall, Upper Saddle River, 1989)
7. R.H. Bird, G.M. Chin, W. Neveitt, J. Nocedal, On the use of stochastic Hessian information in optimization methods for machine learning. SIAM J. Optim. **21**(3), 977–955 (2011)
8. J.A. Blackard, D.J. Dean, C.W. Anderson, Covertype data set, in *UCI Machine Learning Repository*, ed. by K. Bache, M. Lichman (School of Information and Computer Sciences, University of California, Irvine, 2013). http://archive.ics.uci.edu/ml
9. O. Bousquet, A. Elisseeff, Stability and generalization. J. Mach. Learn. Res. **2**, 499–526 (2002)
10. S. Boyd, L. Vandenberghe, *Convex Optimization* (Cambridge University Press, Cambridge, 2004)
11. S. Boyd, N. Parikh, E. Chu, B. Peleato, J. Eckstein, Distributed optimization and statistical learning via the alternating direction method of multipliers. Found. Trends Mach. Learn. **3**(1), 1–122 (2010)
12. C.-C. Chang, C.-J. Lin, Libsvm: a library for support vector machines. ACM Trans. Intell. Syst. Technol. **2**(3), 27 (2011)
13. C. Cortes, V. Vapnik, Support-vector networks. Mach. Learn. **20**(3), 273–297 (1995)
14. J. Dean, S. Ghemawat, MapReduce: simplified data processing on large clusters. Commun. ACM **51**(1), 107–113 (2008)
15. A. Defazio, F. Bach, S. Lacoste-Julien, SAGA: a fast incremental gradient method with support for non-strongly convex composite objectives, in *Advances in Neural Information Processing Systems (NIPS) 27* (2014), pp. 1646–1654
16. O. Dekel, R. Gilad-Bachrach, O. Shamir, L. Xiao, Optimal distributed online prediction using mini-batches. J. Mach. Learn. Res. **13**(1), 165–202 (2012)
17. R.S. Dembo, S.C. Eisenstat, T. Steihaug, Inexact Newton methods. SIAM J. Numer. Anal. **19**(2), 400–408 (1982)
18. W. Deng, W. Yin, On the global and linear convergence of the generalized alternating direction method of multipliers. J. Sci. Comput. **66**(3), 889–916 (2016)
19. J.E. Dennis, J.J. Moreó, A characterization of superlinear convergence and its application to quasi-Newton methods. Math. Comput. **28**(126), 549–560 (1974)
20. J.C. Duchi, A. Agarwal, M.J. Wainwright, Dual averaging for distributed optimization: convergence analysis and network scaling. IEEE Trans. Autom. Control **57**(3), 592–606 (2012)
21. G.H. Golub, C.F. Van Loan, *Matrix Computations*, 3rd edn. (The John Hopkins University Press, Baltimore, 1996)
22. R.A. Horn, C.R. Johnson, *Matrix Analysis* (Cambridge University Press, Cambridge, 1985)
23. S.S. Keerthi, D. DeCoste, A modified finite Newton method for fast solution of large scale linear svms. J. Mach. Learn. Res. **6**, 341–361 (2005)
24. K. Lang, Newsweeder: learning to filter netnews, in *Proceedings of the Twelfth International Conference on Machine Learning (ICML)* (1995), pp. 331–339
25. N. Le Roux, M. Schmidt, F. Bach, A stochastic gradient method with an exponential convergence rate for finite training sets, in *Advances in Neural Information Processing Systems (NIPS) 25* (2012), pp. 2672–2680

26. J.D. Lee, Y. Sun, M. Saunders, Proximal Newton-type methods for minimizing composite functions. SIAM J. Optim. **24**(3), 1420–1443 (2014)
27. D.D. Lewis, Y. Yang, T. Rose, F. Li, RCV1: a new benchmark collection for text categorization research. J. Mach. Learn. Res. **5**, 361–397 (2004)
28. Q. Lin, L. Xiao, An adaptive accelerated proximal gradient method and its homotopy continuation for sparse optimization. Comput. Optim. Appl. **60**(3), 633–674 (2015)
29. C.-Y. Lin, C.-H. Tsai, C.-P. Lee, C.-J. Lin, Large-scale logistic regression and linear support vector machines using Spark, in *Proceedings of the IEEE Conference on Big Data*, Washington, 2014
30. Q. Lin, Z. Lu, L. Xiao, An accelerated randomized proximal coordinate gradient method and its application to regularized empirical risk minimization. SIAM J. Optim. **25**(4), 2244–2273 (2015)
31. Z. Lu, Randomized block proximal dampled Newton method for composite self-concordant minimization. SIAM J. Optim. **27**(3), 1910–1942 (2017)
32. D.G. Luenberger, *Introduction to Linear and Nonlinear Programming* (Addison-Wesley, New York, 1973)
33. L. Mackey, M.I. Jordan, R.Y. Chen, B. Farrell, J.A. Tropp et al., Matrix concentration inequalities via the method of exchangeable pairs. Ann. Probab. **42**(3), 906–945 (2014)
34. D. Mahajan, N. Agrawal, S.S. Keerthi, S. Sundararajan, L. Bottou, An efficient distributed learning algorithm based on effective local functional approximation. arXiv:1310.8418
35. MPI Forum, MPI: a message-passing interface standard, version 3.0 (2012), http://www.mpi-forum.org
36. Y. Nesterov, *Introductory Lectures on Convex Optimization: A Basic Course* (Kluwer, Boston, 2004)
37. Y. Nesterov, Gradient methods for minimizing composite functions. Math. Program. Ser. B **140**, 125–161 (2013)
38. Y. Nesterov, A. Nemirovski, *Interior Point Polynomial Time Methods in Convex Programming* (SIAM, Philadelphia, 1994)
39. J. Nocedal, S.J. Wright, *Numerical Optimization*, 2nd edn. (Springer, New York, 2006)
40. M. Pilanci, M.J. Wainwright, Iterative Hessian sketch: fast and accurate solution approximation for constrained least-squares. J. Mach. Learn. Res. **17**(53), 1–38 (2016)
41. S.S. Ram, A. Nedić, V.V. Veeravalli, Distributed stochastic subgradient projection algorithms for convex optimization. J. Optim. Theory Appl. **147**(3), 516–545 (2010)
42. B. Recht, C. Re, S. Wright, F. Niu, Hogwild: a lock-free approach to parallelizing stochastic gradient descent, in *Advances in Neural Information Processing Systems* (2011), pp. 693–701
43. K. Scaman, F. Bach, S. Bubeck, Y.T. Lee, L. Massoulie6, Optimal algorithms for smooth and strongly convex distributed optimization in networks, in *Proceedings of the 34th International Conference on Machine Learning (ICML)*, Sydney (2017), pp. 3027–3036
44. M. Schmidt, N. Le Roux, F. Bach, Minimizing finite sums with the stochastic average gradient. Math. Program. **162**, 83–112 (2017)
45. S. Shalev-Shwartz, T. Zhang, Stochastic dual coordinate ascent methods for regularized loss minimization. J. Mach. Learn. Res. **14**, 567–599 (2013)
46. S. Shalev-Shwartz, T. Zhang, Accelerated proximal stochastic dual coordinate ascent for regularized loss minimization. Math. Program. **155**(1), 105–145 (2015)
47. S. Shalev-Shwartz, O. Shamir, N. Srebro, K. Sridharan, Stochastic convex optimization, in *Proceedings of the 22nd Annual Conference on Learning Theory (COLT)* (2009)
48. J. Shalf, S. Dosanjh, J. Morrison, Exascale computing technology challenges, in *Proceedings of the 9th International Conference on High Performance Computing for Computational Science, VECPAR'10*, Berkeley (Springer, Berlin, 2011), pp. 1–25
49. O. Shamir, N. Srebro, On distributed stochastic optimization and learning, in *Proceedings of the 52nd Annual Allerton Conference on Communication, Control, and Computing* (2014)

50. O. Shamir, N. Srebro, T. Zhang, Communication efficient distributed optimization using an approximate Newton-type method, in *Proceedings of the 31st International Conference on Machine Learning (ICML), JMLR: W&CP*, vol. 32 (2014)
51. A. Shapiro, D. Dentcheva, A. Ruszczyński, *Lectures on Stochastic Programming: Modeling and Theory.* MPS-SIAM Series on Optimization (SIAM-MPS, Philadelphia, 2009)
52. R. Tibshirani, Regression shrinkage and selection via the lasso. J. R. Stat. Soc. Ser. B (Methodol.) **58**, 267–288 (1996)
53. Q. Tran-Dinh, A. Kyrillidis, V. Cevher, Composite self-concordant minimization. J. Mach. Learn. Res. **16**, 371–416 (2015)
54. L. Xiao, T. Zhang, A proximal stochastic gradient method with progressive variance reduction. SIAM J. Optim. **24**(4), 2057–2075 (2014)
55. Y. Zhang, L. Xiao, Stochastic primal-dual coordinate method for regularized empirical risk minimization. J. Mach. Learn. Res. **18**(84), 1–42 (2017)
56. Y. Zhang, J.C. Duchi, M.J. Wainwright, Communication-efficient algorithms for statistical optimization. J. Mach. Learn. Res. **14**, 3321–3363 (2013)
57. Y. Zhuang, W.-S. Chin, Y.-C. Juan, C.-J. Lin, Distributed Newton method for regularized logistic regression. Technical Report, Department of Computer Science, National Taiwan University (2014)

Chapter 12
Numerical Construction of Nonsmooth Control Lyapunov Functions

Robert Baier, Philipp Braun, Lars Grüne, and Christopher M. Kellett

Abstract Lyapunov's second method is one of the most successful tools for analyzing stability properties of dynamical systems. If a control Lyapunov function is known, then asymptotic stabilizability of an equilibrium of the corresponding dynamical system can be concluded without the knowledge of an explicit solution of the dynamical system. Whereas necessary and sufficient conditions for the existence of nonsmooth control Lyapunov functions are known by now, constructive methods to generate control Lyapunov functions for given dynamical systems are not known up to the same extent. In this paper we build on previous work to compute (control) Lyapunov functions based on linear programming and mixed integer linear programming. In particular, we propose a mixed integer linear program based on a discretization of the state space where a continuous piecewise affine control Lyapunov function can be recovered from the solution of the optimization problem. Different to previous work, we incorporate a semiconcavity condition into the formulation of the optimization problem. Results of the proposed scheme are illustrated on the example of Artstein's circles and on a two-dimensional system with two inputs. The underlying optimization problems are solved in Gurobi (2016, http://www.gurobi.com).

R. Baier · L. Grüne
University of Bayeuth, Chair of Applied Mathematics, Bayreuth, Germany
e-mail: robert.baier@uni-bayreuth.de; lars.gruene@uni-bayreuth.de

P. Braun (✉)
University of Bayeuth, Chair of Applied Mathematics, Bayreuth, Germany

University of Newcastle, School of Electrical Engineering and Computing, Callaghan, NSW, Australia
e-mail: philipp.braun@uni-bayreuth.de; philipp.braun@newcastle.edu.au

C. M. Kellett
University of Newcastle, School of Electrical Engineering and Computing, Callaghan, NSW, Australia
e-mail: chris.kellett@newcastle.edu.au

© Springer Nature Switzerland AG 2018
P. Giselsson, A. Rantzer (eds.), *Large-Scale and Distributed Optimization*, Lecture Notes in Mathematics 2227,
https://doi.org/10.1007/978-3-319-97478-1_12

Keywords Control Lyapunov functions · Mixed integer programming · Dynamical systems

AMS Subject Classifications 93D30, 90C11, 93D05, 93D20

12.1 Introduction

Lyapunov's second method [20] is one of the most successful tools for analyzing stability properties of dynamical systems. This largely stems from the fact that Lyapunov's second method provides an approach to ascertaining stability that does not depend on examining solutions of the control system directly. Rather, it relies on finding an energy-like function, called a Lyapunov function, and examining how the derivative of the Lyapunov function evolves along system solutions. While the concept of Lyapunov functions was initially defined for dynamical systems without inputs, the concept was extended to control Lyapunov functions in the context of dynamical systems with inputs by Artstein in the early 1980s [2]. Similar to stability concepts of dynamical systems without inputs where the existence of a Lyapunov function is equivalent to asymptotic stability of an equilibrium or an equilibrium set, the existence of a control Lyapunov function is necessary and sufficient for asymptotic stabilizability of dynamical systems with inputs. However, whereas the existence of a Lyapunov function implies the existence of a smooth Lyapunov function, a similar property does not hold in the context of control Lyapunov functions. Illustrative examples of dynamical systems, known as Brockett's integrator [5] and Artstein's circles [2] in the literature, show that there are dynamical systems, which are asymptotically stabilizable to the origin but which do not admit a smooth control Lyapunov function. This gap was closed by using tools from nonsmooth analysis and considering control Lyapunov functions defined through nonsmooth generalizations of gradients, e.g., using the Dini derivative in Sontag's work [23] or the proximal subgradient [8] used by Clarke. Using definitions of nonsmooth control Lyapunov functions, existence results for asymptotically stabilizable systems were provided by Clarke et al. [8, 9], Rifford [22], and Kellett and Teel [18, 19].

Whereas the question on existence of Lyapunov functions and control Lyapunov functions is basically answered by now, constructive methods to find Lyapunov and control Lyapunov functions are limited. A comprehensive review of approaches for numerical computation of Lyapunov functions can be found in [10]. One such approach to construct Lyapunov functions for ordinary differential equations originating in [16] and [21] and further explored in [13] is based on linear programming. In these contributions, continuous piecewise affine Lyapunov functions for ordinary differential equations are constructed. Based on a discretization of the state space, a finite dimensional optimization problem representing the decrease condition of the Lyapunov function is obtained. If the corresponding linear problem is feasible, the coefficients of a piecewise affine Lyapunov function can be recovered from an optimal solution of the optimization problem.

This approach to construct continuous piecewise affine Lyapunov functions has been extended in several papers. In [4] the linear programming approach is extended to compute Lyapunov functions for differential inclusions, i.e., Lyapunov functions for dynamical systems with strongly asymptotically stable equilibria. In [14] the linear program is replaced by formulas used in the proofs of classical converse Lyapunov theorems by Massera and Yoshizawa [17] to compute the coefficients of the piecewise affine Lyapunov function, thereby reducing the numerical burden. In [3], the method is further extended to a mixed integer linear programming formulation with the ability to construct continuous piecewise affine control Lyapunov functions for dynamical systems which admit a smooth control Lyapunov function.

However, as argued earlier, system dynamics like Artstein's circles or Brockett's integrator are not covered by this approach. In this paper we further extend the approach to be able to construct local control Lyapunov functions based on the solution of finite dimensional optimization problems and drop the assumption of an existing smooth control Lyapunov function. In particular we propose a mixed integer linear program which returns a continuous piecewise affine control Lyapunov function when the program is feasible. A critical constraint introduced herein is the inclusion of semiconcavity conditions in the formulation of the optimization problem.

The approach described in this paper has two distinct applications. First, by explicitly computing a control Lyapunov function for a dynamical system the approach can be used to show that a dynamical system is asymptotically controllable to an equilibrium point or set. Second, if a system is known to be asymptotically controllable but a control Lyapunov function is not known, the mixed integer problem can be used to compute a control Lyapunov function, from which a (robust) stabilizing state feedback can be constructed. However, since a control Lyapunov function is only obtained in the case that the mixed integer problem is feasible, the approach only provides a sufficient condition to prove asymptotic controllability. If the optimization problem is infeasible, a finer discretization of the state space can be used, which might lead to feasibility and thus provide a control Lyapunov function. Yet, it remains an open question, and is beyond the scope of this paper, if asymptotic controllability implies feasibility of the proposed mixed integer program if the discretization of the state space is chosen fine enough.

A different approach for the construction of piecewise affine and piecewise quadratic Lyapunov functions for the class of piecewise affine systems can be found in work of Johansson [15]. There, linear matrix inequalities and convex programming are used to construct Lyapunov functions for the class of piecewise affine systems. Compared to our work the focus is on Lyapunov functions instead of control Lyapunov functions, which allows the use of convex optimization problems without the need of integer variables, but therefore cannot be applied to the dynamics of Artstein's circles or Brockett's integrator.

The paper is structured as follows. In Sect. 12.2, the notation for dynamical systems and a stability result based on nonsmooth control Lyapunov functions is presented. Section 12.3 discusses the triangulation of the state space and introduces continuous piecewise affine functions. Section 12.4 discusses the decrease condition of control Lyapunov functions in the context of continuous piecewise affine functions using the Dini derivative. Additionally, the role of semiconcavity is discussed here. The section concludes with a finite dimensional optimization problem providing a control Lyapunov function on a compact domain excluding a neighborhood around the equilibrium. The finite dimensional optimization problem is approximated by a mixed integer problem in Sect. 12.5. In Sect. 12.6 the mixed integer problem is solved in Gurobi for Artstein's circles and a two-dimensional control system with two inputs. The corresponding control Lyapunov functions are visualized before the paper concludes in Sect. 12.7.

Throughout the paper the following notation is used. $\mathcal{P}$ denotes the class of continuous positive functions

$$\mathcal{P} = \{\rho : \mathbb{R}_{\geq 0} \to \mathbb{R} \mid \rho \text{ continuous}, \rho(r) > 0 \ \forall r > 0 \text{ and } \rho(0) = 0\}.$$

The classes of functions $\mathcal{K}$, $\mathcal{K}_\infty$ and $\mathcal{L}$ are defined as

$$\mathcal{K} = \{\alpha : \mathbb{R}_{\geq 0} \to \mathbb{R} \mid \alpha \text{ continuous, strictly increasing and } \alpha(0) = 0\},$$

$$\mathcal{K}_\infty = \{\alpha \in \mathcal{K} \mid \lim_{r \to \infty} \alpha(r) = \infty\},$$

$$\mathcal{L} = \{\sigma : \mathbb{R}_{\geq 0} \to \mathbb{R} \mid \sigma \text{ continuous, strictly decreasing and } \lim_{t \to \infty} \sigma(t) = 0\}.$$

Finally, the class of $\mathcal{KL}$ functions is defined as

$$\mathcal{KL} = \{\beta : \mathbb{R}_{\geq 0}^2 \to \mathbb{R} \mid \beta \text{ continuous}, \beta(\cdot, t) \in \mathcal{K} \ \forall t \geq 0, \ \beta(r, \cdot) \in \mathcal{L} \ \forall r \geq 0\}.$$

For a set $\mathbb{A} \subset \mathbb{R}^n$ we denote its convex hull as $\text{conv}(\mathbb{A})$. The 2-norm and the ∞-norm of a vector $x \in \mathbb{R}^n$ are defined as $\|x\|_2 = \sqrt{\sum_{i=1}^n x_i^2}$ and $\|x\|_\infty = \max_{i=1,\dots,n} |x_i|$, respectively. For a matrix $A \in \mathbb{R}^{m \times n}$, $n, m \in \mathbb{N}$, we use the matrix norms $\|A\|_\infty = \max_{i=1,\dots,m} \sum_{j=1}^n |a_{ij}|$ and $\|A\|_2 = \sqrt{\lambda_{\max}}$ where $\lambda_{\max}$ denotes the largest eigenvalue of $A^T A$. For $x \in \mathbb{R}^n$ and $r > 0$, $B_r(x) = \{y \in \mathbb{R}^n \mid \|x - y\|_2 < r\}$ denotes a ball of radius r centered around x. The interior of a set is defined as $\text{int}(\mathbb{A}) = \{x \in \mathbb{A} \mid \exists \varepsilon > 0 \text{ such that } B_\varepsilon(x) \subset \mathbb{A}\}$. The relative interior of a convex set $\mathbb{A} \subset \mathbb{R}^n$ is denoted by $\text{relint}(\mathbb{A}) = \{x \in \mathbb{A} \mid \forall y \in \mathbb{A} \ \exists \lambda > 1 \text{ such that } \lambda x + (1 - \lambda)y \in \mathbb{A}\}$.

12.2 Mathematical Setting

As motivated in the introduction, we are interested in the construction of local control Lyapunov functions for nonlinear dynamical systems

$$\dot{x}(t) = f(x(t), u(t)) \qquad (\text{a.a. } t \in \mathbb{R}_{\geq 0}) \tag{12.1a}$$

$$u(t) \in \mathbb{U} \qquad (\text{a.a. } t \in \mathbb{R}_{\geq 0}) \tag{12.1b}$$

defined through a Lipschitz continuous function $f : \mathbb{X} \times \mathbb{U} \to \mathbb{R}^n$. The sets $\mathbb{X} \subset \mathbb{R}^n$ and $\mathbb{U} \subset \mathbb{R}^m$ denote convex and compact subsets of the state space and input space containing the origin. As already implicitly used in (12.1), we assume that solutions of the dynamical systems exist for all $t \in \mathbb{R}_{\geq 0}$. Moreover, we assume that the origin is an equilibrium of the dynamical system (12.1) in $\mathbb{X}$, i.e., we assume that $0 = f(0, 0)$ holds.

Alternatively, the dynamical system (12.1) can be represented as a differential inclusion

$$\dot{x} \in F(x) = \text{conv} \left(\bigcup_{u \in \mathbb{U}} \{f(x, u)\} \right) \tag{12.2}$$

using the set-valued map $F : \mathbb{X} \rightrightarrows \mathbb{R}^n$.

For an initial state $x \in \mathbb{X}$ we denote the set of solutions with $x(0) = x$ by $\mathcal{S}(x)$. A particular solution is denoted by $\phi(\cdot, x) \in \mathcal{S}(x)$, i.e., $\phi(\cdot, x)$ satisfies

$$\frac{d}{dt} \phi(t, x) \in F(\phi(t, x))$$

for almost all $t \in \mathbb{R}_{\geq 0}$. When it is necessary to explicitly include the input u, we write $\phi(\cdot, x, u)$ instead of $\phi(\cdot, x)$.

A control Lyapunov function characterizes the stability properties of an equilibrium or an equilibrium set. Here we consider stability of the following form, which is equivalent to asymptotic stability [1].

Definition 1 (Weak $\mathcal{KL}$-Stability) The origin of the system is said to be weakly $\mathcal{KL}$-stable for all $x \in \mathbb{R}^n$ if there exists $\beta \in \mathcal{KL}$ so that there exists $\phi \in \mathcal{S}(x)$ satisfying

$$\|\phi(t, x)\|_2 \leq \beta(\|x\|_2, t)$$

for all $t \geq 0$.

Definition 1, as well as the definition of asymptotic stability, is based on the explicit knowledge of the solutions $\phi(\cdot, x)$. Alternatively, stability can be characterized based on a so-called control Lyapunov function for a system with right-hand side $f(x, u)$ or a differential inclusion given by the set-valued map F.

For a smooth control Lyapunov function $V : \mathbb{R}^n \to \mathbb{R}$ stability is derived based on the sign of the directional derivate $\langle \nabla V(x), w \rangle$, $w \in F(x)$, representing the time derivate of the control Lyapunov function $\frac{d}{dt} V(\phi(t, x))$ with respect to a particular solution $\phi(\cdot, x) \in \mathcal{S}(x)$.

The following sections will be dedicated to the numerical construction of continuous piecewise affine control Lyapunov functions. Since continuous piecewise affine functions are not necessarily differentiable, the directional derivative cannot be used in our setting. In the literature, Clarke's subgradient, the Dini derivative or proximal subgradients are used as weak variants of the gradient in a given direction to handle nonsmooth control Lyapunov functions. The lower (right) Dini derivative of a Lipschitz continuous function $V : \mathbb{R}^n \to \mathbb{R}$ in direction $w \in \mathbb{R}^n$ is defined as

$$DV(x; w) = \liminf_{t \searrow 0} \frac{1}{t}(V(x + tw) - V(x)).$$

Observe that the definition of the Dini derivative extends the definition of the directional derivative, i.e., for a smooth function V, the directional derivative and the Dini derivative coincide

$$DV(x; w) = \langle \nabla V(x), w \rangle.$$

While we restrict our attention to control Lyapunov functions in the Dini sense here, we refer the reader to [7] for a discussion on control Lyapunov functions using proximal gradients, which in the context of more general classes of functions (i.e., those that are not necessarily continuous piecewise affine) lead to more general results.

With these definitions in mind we can state the main result connecting weak $\mathcal{KL}$-stability and the existence of a control Lyapunov function [7, 22].

Theorem 1 *The origin is weakly $\mathcal{KL}$-stable if and only if there exists a semiconcave (and thus Lipschitz continuous) control Lyapunov function $V : \mathbb{R}^n \to \mathbb{R}_{\geq 0}, \rho \in \mathcal{P},$ $\alpha_1, \alpha_2 \in \mathcal{K}_\infty$ so that*

$$\alpha_1(\|x\|_2) \leq V(x) \leq \alpha_2(\|x\|_2) \qquad and \tag{12.3}$$

$$\min_{w \in F(x)} DV(x; w) \leq -\rho(\|x\|_2) \tag{12.4}$$

for all $x \in \mathbb{R}^n$.

Semiconcavity is discussed in Sect. 12.4.2. As mentioned in the introduction, V cannot be assumed to be smooth in general. Thus, the consideration of nonsmooth control Lyapunov functions and tools such as nonsmooth generalizations of gradients are required for the numerical computation of control Lyapunov functions rather than unnecessary complications.

Remark 1 Observe that for a compact and convex set $\mathbb{X}$ excluding a neighborhood around the origin; i.e., $\mathbb{X} \backslash B_r(0),\ r > 0$, the decrease condition (12.4) can be rewritten as

$$\min_{w \in F(x)} DV(x; w) \leq -\delta$$

where $\delta > 0$ is defined as $\delta = \min_{x \in \mathbb{X} \backslash B_r(0)} \rho(\|x\|_2)$. Since ρ is continuous and $\mathbb{X} \backslash B_r(0)$ is compact, the minimum is attained and larger than zero.

12.3   Continuous Piecewise Affine Functions

As candidates for control Lyapunov functions we consider continuous piecewise affine functions defined on a discretization of the state space (as in [4, 13, 14, 16, 21] and [3] for strong or smooth control Lyapunov functions). The necessary definitions and notations are introduced in this section.

12.3.1   Discretization of the State Space

The construction of a Lipschitz continuous piecewise affine control Lyapunov function in this paper is based on a triangulation of the domain $\mathbb{X}$. To this end we assume that $\mathbb{X}$ is the union of the simplices of a simplicial triangulation

$$\mathcal{T} = \{T_\nu : \nu = 1, \dots, N + K\}, \qquad N, K \in \mathbb{N}, \qquad (12.5)$$

with N simplices not including the origin (i.e., $0 \notin \{T_\nu : \nu = 1, \dots, N\}$), K simplices defining a neighborhood around the origin (i.e., $B_\varepsilon(0) \subset \bigcup_{\nu=N+1}^{N+K} T_\nu$ for $\varepsilon > 0$), and

$$\bigcup_{\nu=1,\dots,N+K} T_\nu = \mathbb{X}. \qquad (12.6)$$

All vertices of all simplices (including the origin) are denoted by $p_k, k = 1, \dots, I + 1$, where we assume without loss of generality $p_{I+1} = 0$; i.e., p_{I+1} denotes the origin in $\mathbb{X}$. Each simplex T_ν is the convex hull of $n+1$ affinely independent vertices $T_\nu = \mathrm{conv}(\{p_{\nu_0}, \dots, p_{\nu_n}\})$. Furthermore, we assume that the following assumptions on the triangulation are satisfied as a base for the computation of a control Lyapunov function.

Assumption 1 *Let $\mathbb{X} \subset \mathbb{R}^n$ be convex and compact and assume that $0 \in \mathbb{X}$. For the computation of a control Lyapunov function we assume that the triangulation (12.5) satisfies (12.6) and the additional conditions:*

1. *The intersection of two simplices is either empty or a common face of both simplices.*
2. *It holds that $p_{I+1} = 0$ and $0 \notin T_\nu$ for all $\nu = 1, \dots, N$ and $0 \in T_\nu$ for all $\nu = N + 1, \dots, N + K$.*

The first point of the assumption ensures that two simplices are not overlapping on their interior or intersect only on parts of their faces at the boundary. The second point of the assumption ensures that a neighborhood around the origin can be excluded in the triangulation and only simplices T_ν with $\nu > N$ may contain the origin. This is a necessary property for the numerical computation of a control Lyapunov function which is made more precise in Sect. 12.4.1. A possible (regular)

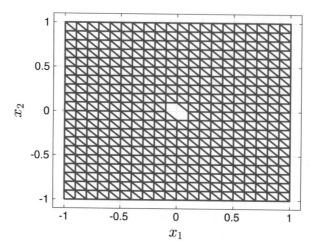

Fig. 12.1 Regular triangulation of the domain $\mathbb{X} = [-1, 1]^2$ excluding a neighborhood around the origin. Each simplex/triangle is uniquely determined by the convex hull of three vertices

triangulation of the domain $\mathbb{X} = [-1, 1]^2$ excluding a neighborhood around the origin is visualized in Fig. 12.1.

The convex hull of a subset of vertices $\{p_{v_0}, \ldots, p_{v_n}\}$ defining a simplex T_v is called face of T_v. The convex hull of exactly n (of $n + 1$) vertices defines a facet of T_v. The union of the $n + 1$ facets of a simplex T_v describe the boundary of T_v.

For a fixed vertex $p_k, k \in \{1, \ldots, I\}$, we denote by $p_{k_j}, j \in \{0, \ldots, I_{p_k}\}$, the set of vertices connected to p_k through an edge. Similarly, for a fixed simplex T_v, the simplices $T_{v_j}, j = 0, \ldots, n$ denote the set of simplices which have a common facet with T_v. The unique vertex $p_{v_j} \in \{p_{v_0}, \ldots, p_{v_n}\}$ satisfies $p_{v_j} \in T_v$ and $p_{v_j} \notin T_{v_j}$ by definition for $j = 0, \ldots, n$. For a facet $T_{v_j} \cap T_v, j \in \{0, \ldots, n\}$ defined through the vertices $\{p_{v_k} | k \in \{0, \ldots, n\}, k \neq j\}$, we define the barycenter of the facet as

$$q_{v_j} = \frac{1}{n} \sum_{k=0, k\neq j}^{n} p_{v_k}. \tag{12.7}$$

A fixed vertex p_k connected to its neighboring vertices $p_{k_j}, k \in \{0, \ldots, I_{p_k}\}$, is visualized in Fig. 12.2. Similarly, a fixed triangle T_v and the triangles $T_{v_j}, j = 0, \ldots, n$, sharing a facet with T_v are illustrated in Fig. 12.3.

12.3.2 Continuous Piecewise Affine Functions

For a given triangulation satisfying Assumption 1 we define a piecewise affine function $V : \mathbb{X} \to \mathbb{R}_{\geq 0}$,

$$V|_{T_v}(x) = V_v(x) = a_v^T x + b_v$$

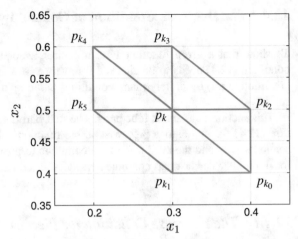

Fig. 12.2 An arbitrary vertex p_k, $k \in \{1, \ldots, I\}$, from the triangulation visualized in Fig. 12.1. The vertex p_k is connected to six other vertices p_{k_j}, $j \in \{0, \ldots, 5\}$ in this case

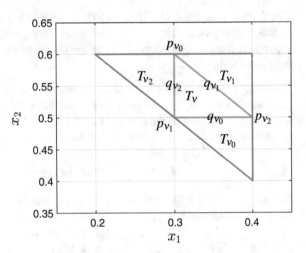

Fig. 12.3 An arbitrary simplex T_ν, $\nu \in \{1, \ldots, N\}$, of the triangulation visualized in Fig. 12.1 and all simplices T_{ν_j}, $j \in \{0, 1, 2\}$, which share a facet with T_ν. The point q_{ν_j}, $j \in \{0, 1, 2\}$, denotes the center of the facet $T_\nu \cap T_{\nu_j}$. The vertices p_{ν_j}, $j \in \{0, 1, 2\}$, which define the simplex T_ν, are ordered such that $p_{\nu_j} \notin T_{\nu_j}$

for all $x \in T_\nu$ and for all $\nu \in \{1, \ldots, N + K\}$. On a fixed simplex T_ν the function V is uniquely described through $a_\nu \in \mathbb{R}^n$ and $b_\nu \in \mathbb{R}$. To obtain continuity of V and to make sure that V is well-defined, additionally the condition

$$V|_{T_\nu}(p_k) = V|_{T_\mu}(p_k)$$

needs to be satisfied for all ν, $\mu \in \{1, \ldots, N+K\}$ and, $p_k \in T_\nu \cap T_\mu$, $k \in \{1, \ldots, I+1\}$. To compute the unknown coefficients a_ν and b_ν in such a way that V is a control Lyapunov function, an optimization problem is proposed in the following sections.

12.4 The Decrease Condition of Control Lyapunov Functions

To show that a given function V is a control Lyapunov function, the decrease condition (12.4) needs to be checked for every $x \in \mathbb{X}$. In this section we replace condition (12.4) by finitely many conditions based on a given triangulation of the set $\mathbb{X}$.

This section consists of four parts. The first part discusses the decrease condition (12.4) for continuous piecewise affine functions, the second part introduces semiconcavity and the third provides a condition to prevent the function V having local minima. The last part combines the ideas introduced in this section.

12.4.1 The Decrease Condition for Piecewise Affine Functions

The decrease condition (12.4) needs to be checked for every $x \in \mathbb{X}$. As a first result in this section we provide an estimate based on a given triangulation of the state space that reduces condition (12.4) to the center of the facets for states x in the interior of a simplex T_ν.

Theorem 2 *Let $V : \mathbb{X} \to \mathbb{R}$ be a continuous piecewise affine function defined on a triangulation $(T_\nu)_{\nu=1,\dots,N}$, satisfying Assumption 1 (and in particular, excluding a neighborhood around the origin). Let $f : \mathbb{X} \times \mathbb{U} \to \mathbb{R}^n$ be uniformly Lipschitz continuous in x, i.e., there exists an $L \in \mathbb{R}_{>0}$ such that*

$$\|f(x_1, u) - f(x_2, u)\|_2 \leq L \|x_1 - x_2\|_2$$

for all $x_1, x_2 \in \mathbb{X}$ uniformly in $u \in \mathbb{U}$. Let a constant $C \geq 0$ be given such that $\|a_\nu\|_2 \leq C$ for all $\nu \in \{1, \dots, N + K\}$ and let $h_\nu > 0$ be defined such that $T_\nu \subset \cup_{j=1}^{n+1} B_{h_\nu}(q_{\nu_j})$ holds (and q_{ν_j} is defined in Eq. (12.7)). Here, $a_\nu \in \mathbb{R}^n$ denotes the gradient $\nabla V_\nu(x) = a_\nu$ for all $x \in \text{int}(T_\nu)$, $\nu = 1, \dots, N + K$. If there exists $\delta > 0$ such that

$$\min_{u \in \mathbb{U}} \langle a_\nu, f(q_{\nu_j}, u) \rangle + CLh_\nu \leq -\delta \tag{12.8}$$

for all $j = 0, \dots, n + 0$, then the inequality

$$\min_{u \in \mathbb{U}} \langle a_\nu, f(x, u) \rangle \leq -\delta$$

holds for all $x \in T_\nu$.

Proof Let $\delta, h_\nu > 0$ be given and let $u_{\nu_j} \in \mathbb{U}$ be defined such that

$$\min_{u \in \mathbb{U}} \langle a_\nu, f(q_{\nu_j}, u_{\nu_j}) \rangle + CLh_\nu \leq -\delta$$

is satisfied for all $j = 0, \ldots, n$. Let $x \in T_\nu$ and let $j \in \{1, \ldots, n+1\}$ such that $x \in B_{h_\nu}(q_{\nu_j})$. Then the following estimate holds:

$$\min_{u \in \mathbb{U}} \langle a_\nu, f(x, u) \rangle$$

$$\leq \langle a_\nu, f(x, u_{\nu_j}) \rangle$$

$$= \left(\langle a_\nu, f(x, u_{\nu_j}) \rangle - \langle a_\nu, f(q_{\nu_j}, u_{\nu_j}) \rangle \right) + \langle a_\nu, f(q_{\nu_j}, u_{\nu_j}) \rangle$$

$$= \langle a_\nu, f(x, u_{\nu_j}) - f(q_{\nu_j}, u_{\nu_j}) \rangle + \langle a_\nu, f(q_{\nu_j}, u_{\nu_j}) \rangle$$

$$\leq \|a_\nu\|_2 \cdot \|f(x_{\nu_j}, u_{\nu_j}) - f(q_{\nu_j}, u_{\nu_j})\|_2 - CLh_\nu - \delta$$

$$\leq CL\|x - q_{\nu_j}\|_2 - CLh_\nu - \delta \leq -\delta$$

$\square$

Theorem 2 ensures a decrease of a solution $\phi(\cdot, x)$ in the interior of a simplex. Nevertheless, the theorem does not ensure that there exists a feasible decrease direction if x is on the boundary of the simplex. To ensure that the decrease condition is also satisfied on the boundary of a simplex, or in other words, when $\phi(\cdot, x)$ passes from one simplex to another, we discuss semiconcavity in the next subsection. In general, the speed of convergence towards the origin decreases to zero for $\|x\|_2 \to 0$ as one can see from the bound $\rho(\|x\|_2)$ in inequality (12.4). This implies that

$$\min_{\|x\|_2 = r} \min_{u \in \mathbb{U}} DV(x; f(x, u))$$

decreases to zero for $r \to 0$. Thus, inequality (12.8), including the positive error term CLh_ν can in general only hold on a domain excluding a neighborhood around the origin. However, this neighborhood around the origin can be made arbitrarily small if the triangulation is fine enough; i.e., h_ν is small enough.

Remark 2 Observe that even though we have assumed that f is uniformly Lipschitz continuous in Theorem 2, the Lipschitz constant L can be replaced by local constants L_ν on T_ν satisfying

$$\|f(x_1, u) - f(x_2, u)\|_2 \leq L_\nu \|x_1 - x_2\|_2$$

for all $x_1, x_2 \in T_\nu$, for all $u \in \mathbb{U}$ and for all $\nu = 1, \ldots, N$ to obtain smaller error terms $CL_\nu h_\nu$.

12.4.2   Semiconcavity Conditions

Theorem 2 provides a condition to check the decrease condition in the interior of a simplex. To ensure that the decrease condition (12.4) is also satisfied on the boundary of a simplex the concept of semiconcavity turns out to be useful. Here we follow the definitions and results provided in [6].

Definition 2 ([6, Def. 1.1.1]) Let $\mathbb{A} \subset \mathbb{R}^n$ be an open set. We say that a function $\phi : \mathbb{A} \to \mathbb{R}$ is semiconcave with linear modulus if it is continuous in $\mathbb{A}$ and there exists $\eta \geq 0$ such that

$$\phi(x) + \phi(y) - 2\phi\left(\frac{x+y}{2}\right) \leq \frac{\eta}{4} \|x - y\|^2 \tag{12.9}$$

for all $x, y \in \mathbb{A}$ with $\lambda x + (1 - \lambda)y \in \mathbb{A}$ for all $\lambda \in [0, 1]$. The constant η above is called a semiconcavity constant for u in $\mathbb{A}$.

Definition 2 is a weaker property than concavity. Observe that for a concave function inequality (12.9) holds for $\eta = 0$. Similar to concavity, there are equivalent conditions to identify semiconcave functions which form a subclass of DC (difference of convex) functions.

Proposition 1 ([6, Prop. 1.1.3]) *Given $\phi : \mathbb{A} \to \mathbb{R}$, with $\mathbb{A} \subset \mathbb{R}^n$ open and convex, and given $\eta \geq 0$, the following properties are equivalent:*

1. *ϕ is semiconcave with a linear modulus in $\mathbb{A}$ with a semiconcavity constant η.*
2. *ϕ satisfies*

$$\lambda\phi(x) + (1 - \lambda)\phi(y) - \phi(\lambda x + (1 - \lambda)y) \leq \eta \frac{\lambda(1 - \lambda)}{2} \|x - y\|^2$$

   for all x, y such that $\lambda x + (1 - \lambda)y \in \mathbb{A}$ and for all $\lambda \in [0, 1]$.
3. *There exist two functions $\phi_1, \phi_2 : \mathbb{A} \to \mathbb{R}$ such that $\phi = \phi_1 + \phi_2$, ϕ_1 is concave, ϕ_2 is twice continuously differentiable and satisfies $\|\nabla^2 \phi_2\|_\infty \leq \eta$.*
4. *ϕ can be represented as $\phi(x) = \inf_{i \in \mathcal{I}} \phi_i(x)$, where $(\phi_i)_{i \in \mathcal{I}}$ is a family of twice continuously differentiable functions such that $\|\nabla^2 \phi_i\|_\infty \leq C$ for all $i \in \mathcal{I}$.*

For a continuous piecewise affine function V defined on a triangulation, we give a condition to verify if V is semiconcave locally on two neighboring simplices sharing a common facet; i.e., we investigate semiconcavity on $V|_{T_\nu \cup T_\mu}$, $\nu, \mu \in \{1, \ldots, N\}$.

Lemma 1 Let $p_0, p_1, \ldots, p_n, p_{n+1} \in \mathbb{R}^n$ be a set of vertices such that the simplices T_ν and T_μ satisfy $T_\nu = \mathrm{conv}(\{p_0, p_1, \ldots, p_n\})$ and $T_\mu = \mathrm{conv}(\{p_1, \ldots, p_n, p_{n+1}\})$. We assume that

$$\mathrm{int}(T_\nu) \neq \emptyset, \qquad \mathrm{int}(T_\mu) \neq \emptyset \qquad and \qquad T_\nu \cap T_\mu = \mathrm{conv}(\{p_1, \ldots, p_n\})$$

holds. Additionally, we consider the functions $V_\nu : T_\nu \to \mathbb{R}$ and $V_\mu : T_\mu \to \mathbb{R}$ as

$$V_\nu(x) = a_\nu^T x + b_\nu, \qquad V_\mu(x) = a_\mu^T x + b_\mu$$

for $a_\nu, a_\mu \in \mathbb{R}^n$, $b_\nu, b_\mu \in \mathbb{R}$ such that $V_\nu(x) = V_\mu(x)$ for all $x \in T_\nu \cap T_\mu$. Then the piecewise affine function $V|_{T_\nu \cup T_\mu} : T_\nu \cup T_\mu \to \mathbb{R}$,

$$V|_{T_\nu \cup T_\mu}(x) = \begin{cases} V_\nu(x), & x \in T_\nu \\ V_\mu(x), & x \in T_\mu \end{cases} \tag{12.10}$$

is semiconcave if and only if the inequalities

$$(a_\nu - a_\mu)^T p_0 + (b_\nu - b_\mu) \leq 0 \quad and \quad (a_\mu - a_\nu)^T p_{n+1} + (b_\mu - b_\nu) \leq 0 \tag{12.11}$$

are satisfied.

Proof Let the inequalities (12.11) be satisfied. We define the functions $\tilde{V}_\nu, \tilde{V}_\mu : T_\nu \cup T_\mu \to \mathbb{R}$ by

$$\tilde{V}_\nu(x) = a_\nu^T x + b_\nu, \qquad \forall \, x \in T_\nu \cup T_\mu,$$

and

$$\tilde{V}_\mu(x) = \begin{cases} (a_\mu - a_\nu)^T x + (b_\mu - b_\nu), & x \in T_\mu \\ 0, & x \in T_\nu \end{cases}.$$

Then, by construction, it holds that $V|_{T_\nu \cup T_\mu} = \tilde{V}_\nu + \tilde{V}_\mu$, $\tilde{V}_\nu$ is twice continuously differentiable (with $\nabla^2 \tilde{V}_\nu = 0$) and $\tilde{V}_\mu$ is continuous (which in particular implies that $\tilde{V}_\mu(x) = 0$ for all $x \in T_\nu \cap T_\mu$). The second inequality in (12.11) implies that $\tilde{V}_\mu(x) \leq 0$ for all $x \in T_\mu$ and thus $\tilde{V}_\mu$ is concave, i.e., for all $x, y \in T_\nu \cup T_\mu$ such that $\lambda x + (1 - \lambda)y \in T_\nu \cup T_\mu$, $\lambda \in [0, 1]$, the function $\tilde{V}_\nu$ satisfies the condition $\tilde{V}_\nu(\lambda x + (1 - \lambda)y) \geq \lambda \tilde{V}_\nu(x) + (1 - \lambda)\tilde{V}_\nu(y)$. Hence, Proposition 1 implies semiconcavity of the function $V|_{T_\nu \cup T_\mu}$.

Conversely, let $V|_{T_\nu \cup T_\mu}$ be semiconcave. We extend the domain of the functions V_ν and V_μ by considering the functions $\tilde{V}_\nu, \tilde{V}_\mu : T_\nu \cup T_\mu \to \mathbb{R}$,

$$\tilde{V}_\nu(x) = a_\nu x + b_\nu \qquad and \qquad \tilde{V}_\mu(x) = a_\mu x + b_\mu.$$

According to point 4 of Proposition 1, $V|_{T_\nu \cup T_\mu}$ can be written as

$$V|_{T_\nu \cup T_\mu}(x) = \min\{\tilde{V}_\nu(x), \tilde{V}_\mu(x)\}.$$

Since $V|_{T_\nu \cup T_\mu}(p_0) = \tilde{V}_\nu(p_0)$ and $V|_{T_\nu \cup T_\mu}(p_{n+1}) = \tilde{V}_\mu(p_{n+1})$ the two inequalities $\tilde{V}_\nu(p_0) \leq \tilde{V}_\mu(p_0)$ and $\tilde{V}_\mu(p_{n+1}) \leq \tilde{V}_\nu(p_{n+1})$ hold. Since these inequalities are equivalent to those in (12.11) this completes the proof.

$\square$

Lemma 1 provides an easy condition to check semiconcavity of the function $V|_{T_\nu \cup T_\mu}$. Thus, according to Lemma 1, to verify if $V|_{T_\nu \cup T_\mu}$ is semiconcave, it is enough to check one (arbitrary) condition in (12.11). In Fig. 12.4 a non-semiconcave and in Fig. 12.5 a semiconcave continuous piecewise affine function using the notation of Lemma 12.4 is visualized. In addition to the function values, the decreasing directions on the intersection $T_\nu \cap T_\mu$ from the reference point $x = 0$ are visualized in the x_1-x_2-plane. Whereas in the case of a non-semiconcave function the possible decreasing directions are given by the intersection of the decreasing directions of the functions V_ν and V_μ (visualized by the green cone in Figs. 12.4 and 12.5), in the case of a semiconcave function V the decreasing directions are given by the union of the decreasing directions of V_ν and V_μ (i.e., the union of the blue, green and red cone in Fig. 12.5). The reason for this property is that in case of semiconcavity the decreasing directions of V_ν which are not in the intersection (i.e., which are in the blue cone and not in the green cone) point into the simplex T_ν. In the non-semiconcave case on the other hand, directions not in the intersection

Fig. 12.4 Visualization of a non-semiconcave continuous piecewise affine function V. In the x_1-x_2-plane the decreasing directions with respect to the origin $x = 0$ are visualized. The decreasing directions are the intersection of the decreasing directions of the two functions V_ν and V_μ

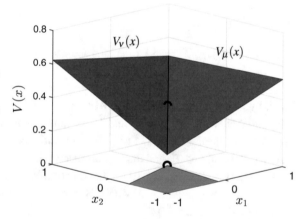

Fig. 12.5 Visualization of a semiconcave continuous piecewise affine function V. In the x_1-x_2-plane the decreasing directions with respect to the origin $x = 0$ are visualized. The decreasing directions are the union of the decreasing directions of the two functions V_ν and V_μ

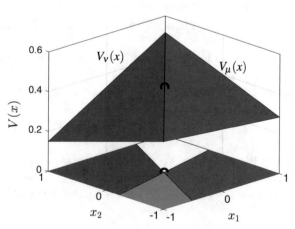

point out of the simplex T_ν and thus are not a valid direction of descent. The same observation holds from the point of view of the function V_μ and the simplex V_μ. These arguments are made more precise in the context of the construction of a control Lyapunov function in the following.

Consider again two simplices

$$T_\nu = \text{conv}(\{p_0, p_1, \ldots, p_n\}) \quad \text{and} \quad T_\mu = \text{conv}(\{p_1, \ldots, p_n, p_{n+1}\})$$

sharing a facet and let $x \in \text{relint}(T_\nu \cap T_\mu)$. By definition, we assume that the functions $V_\nu, V_\mu, V|_{T_\nu \cup T_\mu} : T_\nu \cup T_\mu \to \mathbb{R}$ are defined as

$$V_\nu(x) = a_\nu x + b_\nu,$$

$$V_\mu(x) = a_\mu x + b_\mu,$$

$$V|_{T_\nu \cup T_\mu}(x) = \begin{cases} V_\nu(x), & x \in T_\nu \\ V_\mu(x), & x \in T_\mu \end{cases}. \tag{12.12}$$

Additionally we assume that the assumptions of Theorem 2 are satisfied, i.e., there exists a $u_\nu \in \mathbb{U}$ such that $w_\nu = f(x, u_\nu)$ satisfies

$$\langle a_\nu, w_\nu \rangle \leq -\delta$$

for a given $\delta > 0$. Observe that for a function $V|_{T_\nu \cup T_\mu}$ defined in Eq. (12.12) the Dini derivative for $x \in \text{relint}(T_\nu \cap T_\mu)$ is given by

$$DV|_{T_\nu \cup T_\mu}(x; w) = \begin{cases} \langle a_\nu, w \rangle & \text{if } \exists \varepsilon > 0 \text{ such that } x + \varepsilon w \in T_\nu \\ \langle a_\mu, w \rangle & \text{if } \exists \varepsilon > 0 \text{ such that } x + \varepsilon w \in T_\mu \end{cases}.$$

To ensure that the decrease condition is satisfied on the boundary $\text{relint}(T_\nu \cap T_\mu)$ we can distinguish three cases:

1. If w_ν points to $\text{int}(T_\nu)$, then

$$DV|_{T_\nu \cup T_\mu}(x; w_\nu) = \langle a_\nu, w_\nu \rangle \leq -\delta$$

for all $x \in \text{relint}(T_\nu \cap T_\mu)$, which guarantees a decrease in a feasible direction.
2. If w_ν points to $\text{int}(T_\mu)$ we have to demand that additionally $\langle a_\mu, w_\nu \rangle \leq -\delta$ holds to ensure that

$$DV|_{T_\nu \cup T_\mu}(x; w_\nu) \leq -\delta$$

is satisfied for all $x \in \text{relint}(T_\nu \cap T_\mu)$.
3. If $V|_{T_\nu \cup T_\mu}$ is semiconcave and w_ν does not point to $\text{int}(T_\nu)$ (i.e., it points to $\text{int}(T_\mu)$) then $\langle a_\mu, w_\nu \rangle \leq -\delta$ is satisfied. This fact was already illustrated in Fig. 12.5 and is made more precise now. To this end, let $x \in \text{relint}(T_\nu \cap T_\mu)$ and let

w_ν point to $\text{int}(T_\mu)$. Let $\varepsilon > 0$ such that $x + \varepsilon w_\nu \in \text{int}(T_\mu)$. Due to the convexity of T_μ there exist $\lambda_i \geq 0$, for $i = 1, \ldots, n + 1$ such that $x + \varepsilon w_\nu = \sum_{i=1}^{n+1} \lambda_i p_i$ and $\sum_{i=1}^{n+1} \lambda_i = 1$. Then it holds that

$$V|_{T_\nu \cup T_\mu}(x + \varepsilon w_\nu) - V|_{T_\nu \cup T_\mu}(x) = V_\mu(x + \varepsilon w_\nu) - V_\mu(x)$$

$$= \sum_{i=1}^{n+1} \lambda_i V_\mu(p_i) - V_\mu(x)$$

$$= \sum_{i=1}^{n} \lambda_i V_\nu(p_i) + \lambda_{n+1} V_\mu(p_{n+1}) - V_\nu(x)$$

$$\leq \sum_{i=1}^{n} \lambda_i V_\nu(p_i) + \lambda_{n+1} V_\nu(p_{n+1}) - V_\nu(x)$$

$$= V_\nu(x + \varepsilon w_\nu) - V_\nu(x). \tag{12.13}$$

Since

$$V_\mu(x + \varepsilon w_\nu) - V_\mu(x) = \langle a_\mu, \varepsilon w_\nu \rangle, \quad \text{and}$$

$$V_\nu(x + \varepsilon w_\nu) - V_\nu(x) = \langle a_\nu, \varepsilon w_\nu \rangle,$$

(12.13) implies

$$V|_{T_\nu \cup T_\mu}(x + \varepsilon w_\nu) - V|_{T_\nu \cup T_\mu}(x) = \langle a_\mu, \varepsilon w_\nu \rangle \leq \langle a_\nu, \varepsilon w_\nu \rangle.$$

Thus, if w_ν satisfies $\langle a_\nu, w_\nu \rangle \leq -\delta$ and $V|_{T_\nu \cup T_\mu}$ is semiconcave then

$$\langle a_\mu, w_\nu \rangle \leq \langle a_\nu, w_\nu \rangle \leq -\delta$$

is satisfied as well.

Combining the three cases with the result obtained in Sect. 12.4.1 leads to the following condition ensuring that for a fixed simplex $T_\nu = \text{conv}(\{p_{\nu_0}, \ldots, p_{\nu_{n+1}}\})$ the decrease condition (12.4) is satisfied for all $x \in \text{int}(T_\nu)$ and for all x in the relative interior of the facets of T_ν. For every center q_{ν_j} of a facet $T_\nu \cap T_{\nu_j}$, $j = 0, \ldots, n$, there needs to be an input $u_{\nu_j} \in \mathbb{U}$ such that $w_{\nu_j} = f(q_{\nu_j}, u_{\nu_j})$ satisfies

$$\langle a_\nu, w_{\nu_j} \rangle + CLh_\nu \leq -\delta \tag{12.14a}$$

and

$$\left\{ \langle a_{\nu_j}, w_{\nu_j} \rangle + CLh_{\nu_j} \leq -\delta \quad \vee \quad \langle a_\nu - a_{\nu_j}, p_{\nu_j} \rangle + (b_\nu - b_{\nu_j}) \leq 0 \right\} \tag{12.14b}$$

simultaneously.

Remark 3 The additional semiconcavity condition in (12.14b) ensures the following: If the control Lyapunov functions is locally semiconcave on $T_\nu \cup T_\mu$, then we allow for different decrease directions in the two adjacent simplices T_ν and T_μ, but if the control Lyapunov function is not locally semiconcave, then we require the existence of a common decrease direction on $T_\nu \cup T_\mu$.

12.4.3 Local Minimum Condition

Condition (12.14) does not preclude the function V having a local minimum at some vertex p_k, $k = 1, \ldots, I$. In Fig. 12.6 such a situation is visualized. Even if condition (12.14) is satisfied locally on every facet, there is no feasible decrease direction at the vertex p_k in this case.

To ensure that the function V does not have local minima in the set of vertices $\{p_1, \ldots, p_I\}$ and does have a unique global minimum in the origin p_{I+1}, we enforce the constraints in the following lemma in addition to the constraints (12.14).

Lemma 2 *Let* $V : \mathbb{X} \to \mathbb{R}$, $\mathbb{X} \subset \mathbb{R}^n$, *be a continuous piecewise affine function, defined on a triangulation of the state space satisfying Assumption 1. Additionally, let $\delta > 0$ be fixed. If*

$$\forall \, p_k, k \in \{1, \ldots, I\} \, \exists \, p_{k_j}, \, j \in \{1, \ldots, I_{p_k}\} \text{ s.t. } V(p_k) - V(p_{k_j}) \geq \delta \qquad (12.15)$$

and

$$V(0) = V(p_{I+1}) = 0, \qquad (12.16)$$

then $p_{I+1} = 0$ *is the unique global minimum of* V *and* V *does not have local minima.*

Fig. 12.6 Local visualization of V centred around a vertex p_k, $k \in \{1, \ldots, I\}$. Even if the decrease condition (12.14) is satisfied for every center of an edge (or facet) connected to p_k, V can have a local minimum at p_k. In this case, there does not exist a feasible decrease direction of V at p_k; i.e., V is not a control Lyapunov function

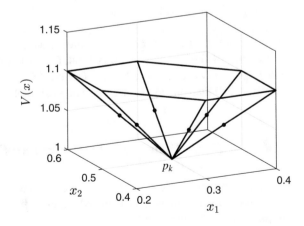

Fig. 12.7 Local visualization
of V centered around a vertex
p_k, $k \in \{1, \ldots, I\}$. If there
exists a p_{k_j}, $j \in \{1, \ldots, I_{p_k}\}$,
such that $V(p_k) < V(p_{k_j})$
and the decrease
condition (12.14) is satisfied
for every edge (or facet),
there exists a feasible
decrease direction for p_k.
Here, the green cone
visualizes the decrease
directions of V in p_k

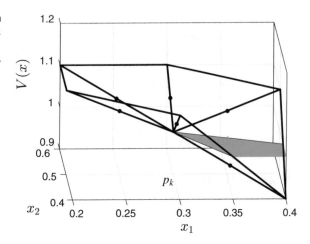

Remember that $I_{p_k} \in \mathbb{N}$ denotes the number of vertices connected to p_k through an edge, introduced in Sect. 12.3.1. Condition (12.15) includes vertices which are connected through an edge to the origin, but it does not include the origin.

Proof The statement about the local minima follows directly from (12.15). To illustrate that the statement about the global minimum is true, assume as a contrary, that there exists a p_k, $k \in \{1, \ldots, I\}$ such that $V(p_k) < 0 = V(0)$. Moreover, since the number of vertices is finite we can take the global minimum and assume without loss of generality that $V(p_k) \leq V(p_\ell)$ holds for all $\ell \in \{1, \ldots, I\}$, which is an immediate contradiction to (12.15). Thus, enforced through condition (12.15) and (12.16), V has its global minimum in $p_{I+1} = 0$ and V is positive definite.

□

Condition (12.15) not only prevents the existence of local minima, in combination with condition (12.14) it ensures that for all $x \in \mathbb{X} \backslash B_\varepsilon(0)$ there exists a feasible decreasing direction. Here $B_\varepsilon(0)$, $\varepsilon > 0$, denotes the neighborhood around the origin which is excluded from the control Lyapunov function computation. A vertex p_k, $k \in \{1, \ldots, I\}$ and its neighbors p_{k_j}, $j \in \{1, \ldots, I_{p_k}\}$ satisfying condition (12.15) for a given $\delta > 0$ are visualized in Fig. 12.7. If in addition condition (12.14) is satisfied on every facet, the decrease condition and the vertices with $V(p_{k_j}) < V(p_k)$ for $j \in \{1, \ldots, I_{p_k}\}$ ensure that it is impossible to get stuck in a vertex and there always exists a decreasing direction.

12.4.4 A Finite Dimensional Optimization Problem

Combining the results of this section, we can state the following nonlinear optimization problem from which a control Lyapunov function can be recovered if the optimal objective value is zero.

$$\min_{\substack{a_v \in \mathbb{R}^n \\ b_v \in \mathbb{R}}} \quad e_1 + e_2 \qquad (12.17\mathrm{a})$$
$$\scriptstyle v=1,\dots,N+K$$

subject to

$$\begin{cases} \langle a_v, w_{v_j} \rangle + CLh_v \leq -\delta_1 + e_1 \\ \left[\langle a_{v_j}, w_{v_j} \rangle + CLh_{v_j} + \delta_1 - e_1 \right] \cdot d^1_{v_j} \leq 0 \\ \left[\langle a_v - a_{v_j}, p_{v_j} \rangle + (b_v - b_{v_j}) \right] \cdot d^2_{v_j} \leq 0 \\ \|a_v\|_2 \leq C \\ d^1_{v_j} + d^2_{v_j} \geq 1 \\ w_{v_j} \in F(q_{v_j}),\ d^1_{v_j}, d^2_{v_j} \in \{0,1\} \end{cases} \quad \begin{cases} \forall v = 1, \dots, N \\ \forall j = 0, \dots, n \end{cases} \qquad (12.17\mathrm{b})$$

$$\begin{cases} -V(p_k) + \sum_{j=1}^{I_{p_k}} r_{k_j} V(p_{k_j}) \leq -\delta_2 + e_2 \\ r_{k_j} \in \{0,1\},\ \sum_{j=1}^{I_{p_k}} r_{k_j} \geq 1 \end{cases} \quad \begin{cases} \forall k = 1, \dots, I \\ \forall j = 1, \dots, I_{p_k} \end{cases} \qquad (12.17\mathrm{c})$$

$$\left\{ V_v(p_{v_\ell}) - V_{v_j}(p_{v_\ell}) = 0 \right. \qquad \begin{cases} \forall v = 1, \dots, N+K \\ \forall j = 0, \dots, n \\ \forall \ell = 0, \dots, n;\ j \neq \ell \end{cases}$$
$$(12.17\mathrm{d})$$

$$\left\{ q_{v_j} = \frac{1}{n} \sum_{k=0, k \neq j}^{n} p_{v_k} \right. \qquad \begin{cases} \forall v = 1, \dots, N+K \\ \forall j = 0, \dots, n \end{cases}$$
$$(12.17\mathrm{e})$$

$$\begin{cases} V(p_{I+1}) = V(0) = 0 \\ e_1 \geq 0,\ e_2 \geq 0 \end{cases} \qquad (12.17\mathrm{f})$$

Remark 4 In the objective function (12.17a) of the optimization problem, only the variables $a_v \in \mathbb{R}^n$, $b_v \in \mathbb{R}$, $v = 1, \dots, N + K$, are explicitly stated since a_v and b_v are the variables of interest to recover the control Lyapunov function. However, we note that additionally the continuous variables e_1, e_2, C and w_{v_j} as well as the binary variables d_{v_j} and r_{k_j} are optimization variables of the overall optimization problem (12.17).

The objective function (12.17a) is optional. Here, we minimize the slack variables $e_1 \geq 0$ and $e_2 \geq 0$ (see constraints (12.17f)) to ensure that the optimization problem is feasible. By choosing e_1 and e_2 large, a feasible solution of the optimization problem can be easily constructed. If $e_1 = e_2 = 0$ in the optimal solution, then the corresponding coefficients a_v, b_v, $v = 1, \dots, N + K$, define a continuous piecewise affine control Lyapunov function on $\mathbb{X}$ excluding a neighborhood around the origin.

The constraints (12.17b) describe the decrease condition discussed in Sect. 12.4.1 and the semiconcavity condition in Sect. 12.4.2. The a priori chosen parameter $\delta_1 > 0$ defines the minimal decrease. As already pointed out, the variable e_1 ensures that the constraints are feasible. Observe that the simplices containing the origin T_v, $v = N+1, \ldots, N+K$ are not included in the formulation of the constraints since a decrease cannot be guaranteed for these simplices. The constraints are nonlinear since the coefficients a, the directions w as well as the binary variables d are unknown. Moreover, C is not known in advance and depends on the coefficients a (see Theorem 2).

The constraints (12.17c) implement the local minimum condition of Sect. 12.4.3. The condition ensures that the origin $p_{I+1} = 0$ (which is excluded in the constraints) is the only minimum of the function V. Similar to δ_1, the parameter $\delta_2 > 0$ chosen in advance defines the minimal decrease of the function V in at least one direction. The variable e_2 ensures feasibility of the constraints.

The constraints (12.17d) ensure that the coefficients of V are chosen such that V is continuous. The constraints (12.17e) define the center points of the facets which are the reference points for the computation of a decrease condition in (12.17b). The constraints (12.17e) do not contain any unknowns, can be computed offline, and are only included for completeness here.

Finally the constraint (12.17f) sets $V(0) = 0$. Observe that this constraint in combination with (12.17d) ensures that $V(x) > 0$ for all $x \in \mathbb{X} \backslash \{0\}$. Additionally e_1 and e_2 are defined as positive optimization variables here. Alternatively the condition $e_1 \geq 0$ could be dropped to maximize the minimal decrease with the variable e_1.

In the next section we approximate the optimization problem (12.17) by a mixed integer linear program which can be implemented and solved by standard optimization software like Gurobi [12].

12.5 Reformulation as Mixed Integer Linear Programming Problem

To be able to compute a control Lyapunov function with Gurobi, the finite dimensional optimization problem (12.17) needs to be rewritten in form of a mixed integer linear program. Before we state the corresponding optimization problem we derive approximations for the parameters involved in the decrease condition (12.8) of Theorem 2.

12.5.1 Approximation of System Parameters and Reformulation of Nonlinear Constraints

The decrease condition (12.8) contains the constants L, h_v as well as the constant C depending on the norm of the gradient of V.

The constant h_v can be approximated by computing the maximal distance to the center of the simplex $T_v = \text{conv}(\{p_{v_0}, \ldots, p_{v_n}\})$; i.e., by computing

$$c_v = \frac{1}{n+1} \sum_{k=0}^{n} p_{v_k}$$

and defining

$$\tilde{h}_v = \max_{i=0,\ldots,n} \|c_v - p_{v_i}\|_2.$$

Then $B_{\tilde{h}_v}(q_{v_j})$ contains the facet under consideration and h_v can be replaced by $\tilde{h}_v$ in inequality (12.8) where $\tilde{h}_v$ is computed offline for all $v = 1, \ldots, N + K$.

If f is continuously differentiable, the Lipschitz constant on a simplex T_v can be computed by

$$L_v = \max_{\substack{u \in \mathbb{U} \\ x \in T_v}} \|Df(x, u)\|_2.$$

To simplify the computation we approximate the Lipschitz constant by computing the maximum over a rectangle $[\underline{r}, \overline{r}] \subset \mathbb{R}^n$ such that $T_v \subset [\underline{r}, \overline{r}]$ holds; i.e., we compute the approximated Lipschitz constants

$$\tilde{L}_v = \max_{\substack{u \in \mathbb{U} \\ x \in [\underline{r}, \overline{r}]}} \|Df(x, u)\|_2 \geq L_v$$

for all $v = 1, \ldots, N + K$ offline.

With these considerations, the additional condition on $\|a_v\|_2 \leq C$, and a given constant $\delta > 0$, the decrease condition (12.8) reads

$$\langle a_v, w_v \rangle + \tilde{h}_v \tilde{L}_v \|a_v\|_2 \leq -\delta \tag{12.18}$$

in the unknowns $a_v \in \mathbb{R}^n$ and $w_v \in F(q_v)$.

Since norms by definition satisfy the triangle inequality, constraints of the form $\|a_v\|_2 \leq C$ are convex and thus can be handled by Gurobi. However, to simplify the optimization problem we approximate the convex constraints by linear (convex) constraints. Thus, we assume that the entries of a_v are bounded; i.e., $\|a_v\|_\infty \leq a_{\max}$ for a $a_{\max} > 0$ given. Then it holds that $\|a_v\|_2 \leq \sqrt{n}\|a_v\|_\infty \leq \sqrt{n}a_{\max}$ and $\|a_v\|_2$ in inequality (12.18) can be replaced by $\tilde{C} = \sqrt{n}a_{\max}$.

To circumvent the nonlinearity in the term $\langle a_v, w_v \rangle$ we restrict the number of directions w_v to a finite number $\tilde{w}_{v_j}^1, \ldots, \tilde{w}_{v_j}^M \in F(q_{v_j})$, $M \in \mathbb{N}$ and replace (12.18) by $M + 1$ linear mixed integer constraints

$$\langle a_\nu, \tilde{w}^1_{\nu_j} \rangle + \tilde{C} \tilde{L}_\nu \tilde{h}_\nu \leq -\delta + (1 - d^1)\Gamma$$

$$\vdots$$

$$\langle a_\nu, \tilde{w}^M_{\nu_j} \rangle + \tilde{C} \tilde{L}_\nu \tilde{h}_\nu \leq -\delta + (1 - d^M)\Gamma$$

$$\sum_{\ell=1}^{M} d^\ell \geq 1, \qquad d^1, \dots, d^M \in \{0, 1\}.$$

Here, Γ denotes a large constant which ensures the inequalities are trivially satisfied for $d^\ell = 0$, $\ell = 1, \dots, M$. Inequality (12.18) is satisfied if at least one of the binary variables d^ℓ, $\ell \in \{1, \dots, M\}$ is equal to one.

12.5.2 The Mixed Integer Linear Programming Formulation

A mixed integer approximation of the optimization problem (12.17), using the ideas of Sect. 12.5.1, is given by:

$$\min_{\substack{a_\nu \in \mathbb{R}^n \\ b_\nu \in \mathbb{R} \\ \nu = 1, \dots, N+K}} e_1 + e_2 \tag{12.19a}$$

subject to

$$\begin{cases} \langle a_\nu, \tilde{w}^\ell_{\nu_j} \rangle + \tilde{C} \tilde{L}_\nu \tilde{h}_\nu \leq -\delta_1 + e_1 + (1 - d^\ell_{\nu_j})\Gamma \\ \langle a_{\nu_j}, \tilde{w}^\ell_{\nu_j} \rangle + \tilde{C} \tilde{L}_\nu \tilde{h}_\nu \leq -\delta_1 + e_1 + (1 - d^\ell_{\nu_j} + s_{\nu_j})\Gamma \\ \|a_\nu\|_\infty \leq a_{max} \\ \langle a_\nu - a_{\nu_j}, p_{\nu_j} \rangle + (b_\nu - b_{\nu_j}) \leq (1 - s_{\nu_j})\Gamma \\ \sum_{\ell=1}^{M} d^\ell_{\nu_j} \geq 1 \\ d^\ell_{\nu_j}, s_{\nu_j} \in \{0, 1\} \end{cases} \quad \begin{cases} \forall \nu = 1, \dots, N \\ \forall j = 0, \dots, n \\ \forall \ell = 1, \dots, M \end{cases}$$

$$\tag{12.19b}$$

$$\begin{cases} V(p_k) - V(p_{k_j}) \leq -\delta_2 \|p_k - p_{k_j}\|_2 + e_2 + (1 - r_{k_j})\Gamma \\ r_{k_j} \in \{0, 1\}, \ \sum_{j=1}^{I_{p_k}} r_{k_j} \geq 1 \end{cases} \quad \begin{cases} \forall k = 1, \dots, I \\ \forall j = 1, \dots, I_{p_k} \end{cases}$$

$$\tag{12.19c}$$

$$\begin{cases} V_\nu(p_{\nu_\ell}) - V_{\nu_j}(p_{\nu_\ell}) = 0 \end{cases} \quad \begin{cases} \forall \nu = 1, \dots, N + K \\ \forall j = 0, \dots, n \\ \forall \ell = 0, \dots, n, \ j \neq \ell \end{cases}$$

$$\left\{ q_{v_j} = \frac{1}{n} \sum_{k=0,k\neq j}^{n} p_{v_k} \right. \qquad \left\{ \begin{array}{l} \forall v = 1, \ldots, K \\ \forall j = 0, \ldots, n \end{array} \right.$$

$$\left\{ \begin{array}{l} V(p_{I+1}) = V(0) = 0 \\ e_1 \geq 0, \ e_2 \geq 0 \end{array} \right.$$

Remark 5 Similar to the optimization problem (12.17a) only the variables $a_v \in \mathbb{R}^n$, $b_v \in \mathbb{R}$, $v = 1, \ldots, N + K$, are explicitly stated in the objective function of the optimization problem (12.19). Here additionally the continuous variables e_1 and e_2, and the binary variables $d_{v_j}^{\ell}$, s_{v_j}, r_{k_j} are optimization variables of the overall mixed integer linear program (12.19).

In the bound for a_v the maximum norm is used which enables a linear reformulation in contrary to the used Euclidean norm in (12.17b). The constraints (12.19b) implement the ideas derived in Sect. 12.5.1. The semiconcavity condition is handled in a similar way to (12.18) by introducing the additional binary variables $s_{v_j} \in \{0, 1\}$ for $v = 1, \ldots, N$ and $j = 0, \ldots, n$. The same holds for the constraints (12.20) using the binary variables $r_{k_j} \in \{0, 1\}$ for all $k = 1, \ldots, I$ and $j = 1, \ldots, I_{p_k}$. By using $\delta_2 \| p_k - p_{k_j} \|_2$ instead of δ_2, different conditions in the case of nonuniform discretizations can be considered directly. The remaining constraints stay unchanged.

12.6 Numerical Examples

To illustrate the results we visualize the solution of the optimization problem (12.19) for two dynamical systems. For the simulations, the constant Γ is defined as $\Gamma = 1000$. The bound on the slope of the control Lyapunov function is defined as $a_{\max} = 4$ in the mixed integer linear program. Additionally, the lower and upper bound $|b_v| \leq 10$, are used for all $v = \{1, \ldots, N + K\}$ but it does not have an impact on the optimal solution of the optimization problem. The mixed integer linear programs are solved in Gurobi [12]. For prototyping, additionally the software package CVX was used [11].

12.6.1 Artstein's Circles

The dynamical system known as Artstein's circles is described by the dynamics

$$\dot{x}(t) = f(x(t), u(t)) = \begin{pmatrix} (-x_1(t)^2 + x_2(t)^2)u(t) \\ -2x_1(t)x_2(t)u(t) \end{pmatrix} \tag{12.20}$$

and $u \in \mathbb{U} = [-1, 1]$. Additionally we restrict our attention to the domain $\mathbb{X} = [-1, 1]^2$. The example is named after the mathematician Zvi Artstein. The term circles in the name stems from the shape of solution trajectories which are visualized in Fig. 12.8. All trajectories $\phi(\cdot, x)$ lie on circles where the radius of the circles is determined by the initial state x. The sign of the input determines the orientation of the solutions. By choosing the input $u = 1$ for $x \in \mathbb{R}_{\geq 0} \times \mathbb{R}$ and $x \in \mathbb{R}_{<0} \times \mathbb{R}$ one can show that all solutions $\phi(\cdot, x, u)$ satisfy $\phi(t, x, u) \to 0$ for $t \to \infty$ and thus the system is weakly $\mathcal{KL}$-stable according to Definition 1. However, there does not exist a continuous feedback asymptotically stabilizing the origin which additionally implies that Artstein's circles do not admit a smooth control Lyapunov function [2]. Nevertheless, a Lipschitz continuous control Lyapunov function according to Theorem 1 exists.

The results of the optimization problem (12.19) are visualized in Figs. 12.9, 12.10, 12.11 and 12.12. Figure 12.9 shows the triangulation of the state space $\mathbb{X} = [-1, 1]^2$. The maximal distance between two neighboring vertices ranges

Fig. 12.8 Visualization of solutions of the dynamical system known as Artstein's circles. All trajectories $\phi(\cdot, x)$ lie on circles with a radius determined by the initial condition x. With the input u only the orientation of the solution trajectory can be changed

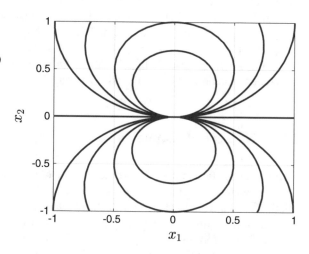

Fig. 12.9 Visualization of a nonuniform discretization of the state space $\mathbb{X} = [-1, 1]^2$. A neighborhood $B_{0.05}(0)$ is excluded from the discretization. Blue indicates $u = -1$ and red indicates $u = 1$ as possible inputs satisfying the decrease condition on the control Lyapunov function obtained from the solution of the corresponding optimization problem

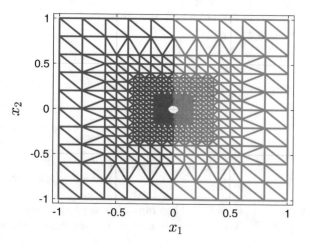

Fig. 12.10 Visualization of the decreasing directions $\tilde{w}$ leading to the control Lyapunov function obtained from the optimization problem (12.19). The directions are computed on the center of the facets. Blue corresponds to $u = -1$, red indicates $u = 1$

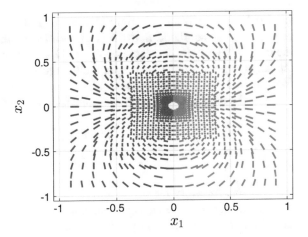

Fig. 12.11 Visualization of the continuous piecewise affine control Lyapunov function obtained from the solution of the optimization problem (12.19). Along the x_2-axis the control Lyapunov function is semiconcave

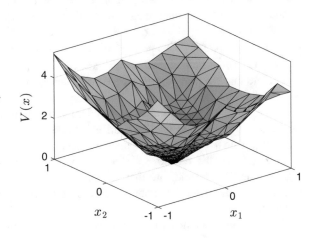

Fig. 12.12 Different view point for the control Lyapunov function in Fig. 12.11

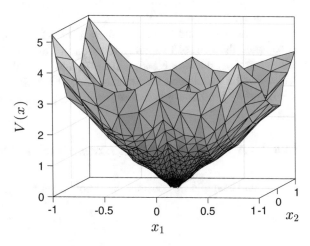

from $0.2 \cdot \sqrt{2}$ far from the origin to a minimal distance of 0.0125 close to the origin. The neighborhood $B_{0.05}(0)$ is excluded from the visualization to indicate the domain where the decrease condition (12.4) does not need to hold. Including the neighborhood around the origin, the triangulation consists of $N + K = 1448$ triangles and $I + 1 = 745$ vertices.

Since the input $u \in [-1, 1]$ is one-dimensional, two directions

$$\tilde{w}_{v_j}^1 = f(q_{v_j}, -1)$$

$$\tilde{w}_{v_j}^2 = f(q_{v_j}, 1)$$

are sufficient for every facet $j = 0, 1, 2$ of a fixed triangle T_v, $v \in \{1, \ldots, K\}$, to cover all possible directions. The blue and red color in Fig. 12.9 indicate which directions lead to a decreasing direction. Blue indicates that $\tilde{w}_{v_j}^1$ (i.e., $u = -1$ and $d_{v_j}^1 = 1$ holds for the binary variable) is a decreasing direction for the control Lyapunov function obtained from the solution of (12.19). Vice-versa, red indicates that $\tilde{w}_{v_j}^2$ (i.e., $u = -1$ and $d_{v_j}^2 = 1$) is a decreasing direction. On the x_2-axis the semiconcavity condition is satisfied (i.e., the corresponding variables s_{v_j} satisfy $s_{v_j} = 1$). Thus, both directions $\tilde{w}^1$ and $\tilde{w}^2$ are feasible decreasing directions here. Note that we have not enforced in the setting of the optimization problem in Sect. 12.5.2 that the semiconcavity condition is satisfied on the x_2-axis. A different implementation of the optimization problem, a different solver or a different order of the constraints could lead to a different control Lyapunov function where the switching from $u = -1$ to $u = 1$ does not need to be on the x_2-axis.

In Fig. 12.10 the decrease directions computed on the facets are visualized. Again, blue and red indicate that $\tilde{w}^1$ and $\tilde{w}^2$, respectively, provide a decreasing direction for the control Lyapunov function returned by the mixed integer linear program.

The control Lyapunov function V is visualized in Figs. 12.11 and 12.12 from different angles. The control Lyapunov function clearly shows the shape of a semiconcave function along the x_2-axis.

The constants δ_1 and δ_2 are set to $\delta_1 = \delta_2 = 0.1$ in the optimization problem used to compute the control Lyapunov function in Figs. 12.11 and 12.12. The optimal solution returns the variable $e_1 = 0.0972$ and $e_2 = 0$. Thus, a decrease of $\delta_1 = 0.1$ cannot be guaranteed for all $x \in [-1, 1]^2 \setminus B_{0.05}(0)$. However, a decrease of $-\delta_1 + e_1 = -0.0028$ is guaranteed for all $x \in [-1, 1]^2 \setminus B_{0.05}(0)$.

12.6.2 A Two-Dimensional Example with Two Inputs

As a second example we consider the simple dynamical system

$$\dot{x} = \begin{pmatrix} x_2 u_1 \\ u_2 \end{pmatrix}$$

with two inputs $u \in \mathbb{R}^2$, $|u_1| + |u_2| \leq 1$. The dynamical system is discussed in [7, Section 6]. By using the input $u_1 = -\text{sign}(x_2)\,\text{sign}(x_1)$ (and $u_2 = 0$) for $x_2 \neq 0$ the state x_1 can be steered to $x_1 = 0$ in finite time. If $x_1 = 0$ holds the input $u_2 = -\text{sign}(x_2)$ and $u_1 = 0$ steers x_2 to the origin in finite time while keeping x_1 constant. If $x_1 \neq 0$ and $x_2 = 0$ initially, the input $u_1 = 0$, $u_2 \neq 0$ for a fixed amount of time, steers the solution to a state already covered in the discussion. Thus the origin is stabilizable for all $x \in \mathbb{R}^2$.

For the optimization problem we concentrate again on the domain defined by $\mathbb{X} = [-1, 1]^2 \backslash B_{0.05}(0)$. Since $u \in \mathbb{U} \subset \mathbb{R}^2$ is two-dimensional, a finite number of inputs does not cover all possible directions and we have to pick a finite number of directions to be able to solve the mixed integer problem (12.19). In Figs. 12.13 and 12.14 a control Lyapunov function and the corresponding decreasing directions $\tilde{w}^i$, $i = 1, \ldots, 4$, using the inputs

$$u \in \left\{ \begin{pmatrix} -1 \\ 0 \end{pmatrix}, \begin{pmatrix} 1 \\ 0 \end{pmatrix}, \begin{pmatrix} 0 \\ -1 \end{pmatrix}, \begin{pmatrix} 0 \\ 1 \end{pmatrix} \right\}$$

are visualized. In Figs. 12.15 and 12.16 the results using the decreasing directions $\tilde{w}^i$, $i = 1, \ldots, 8$, with

$$u \in \left\{ \begin{pmatrix} -1 \\ 0 \end{pmatrix}, \begin{pmatrix} 1 \\ 0 \end{pmatrix}, \begin{pmatrix} 0 \\ -1 \end{pmatrix}, \begin{pmatrix} 0 \\ 1 \end{pmatrix}, \begin{pmatrix} -0.5 \\ -0.5 \end{pmatrix}, \begin{pmatrix} 0.5 \\ 0.5 \end{pmatrix}, \begin{pmatrix} -0.5 \\ 0.5 \end{pmatrix}, \begin{pmatrix} 0.5 \\ -0.5 \end{pmatrix} \right\}$$

are shown. For the numerical computations, the same bounds on the coefficients a_ν, b_ν and the same discretization of the state space as in Fig. 12.9 for Artstein's circles are used. The parameters are again defined as $\delta_1 = \delta_2 = 0.1$. Here, the optimal solutions of the optimization problem with $e_1 = e_2 = 0$ show that a control Lyapunov function has been found satisfying the decrease condition with a minimal decrease of -0.1 for all $x \in [-1, 1]^2 \backslash B_{0.05}(0)$ in the case with four possible inputs as well as in the case with eight different inputs.

Fig. 12.13 Visualization of the continuous piecewise affine control Lyapunov function obtained from the solution of the optimization problem (12.19) using four possible inputs

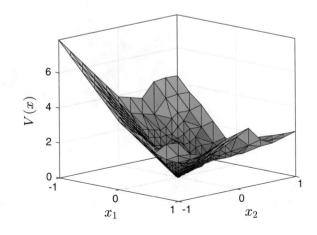

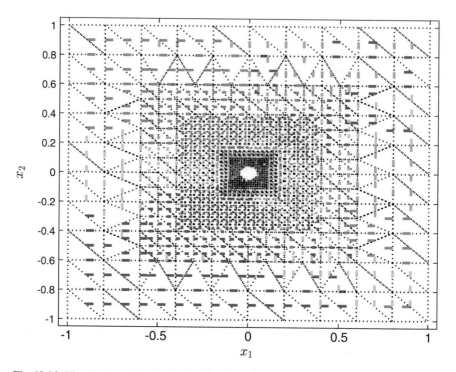

Fig. 12.14 Visualization of the decreasing directions $\tilde{w}$ leading to the control Lyapunov function obtained from the optimization problem (12.19). The directions are computed on the center of the facets. Blue corresponds to $u = [-1\ 0]^T$, red indicates $u = [1\ 0]^T$, green indicates $u = [0\ -1]^T$ and cyan indicates $u = [0\ 1]^T$

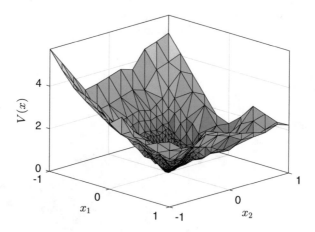

Fig. 12.15 Visualization of the continuous piecewise affine control Lyapunov function obtained from the solution of the optimization problem (12.19) using eight possible inputs

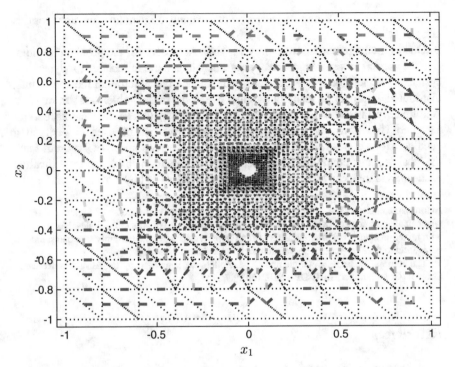

Fig. 12.16 Visualization of the decreasing directions $\tilde{w}$ leading to the control Lyapunov function obtained from the optimization problem (12.19). The different colors indicate the directions $\tilde{w}_1, \ldots, \tilde{w}_8$

In Figs. 12.14 and 12.16 on facets where two directions are shown, the control Lyapunov function is locally semiconcave and both directions lead to a decrease. (This can also be observed on the x_2-axis in Fig. 12.10 on the example of Artstein's circles.)

The control Lyapunov functions in Figs. 12.13 and 12.15 differ quite drastically. This example shows the degree of freedom in the design of control Lyapunov functions but also shows the difficulty in the numerical computation. Since the optimization problem contains a large number of binary variables which only have a local impact in the computation of the control Lyapunov functions, the optimization problem becomes intractable if the number of simplices and the number of inputs is large, at least in the current implementation.

12.7 Conclusions

In this paper we present a framework to compute continuous piecewise affine control Lyapunov functions for dynamical systems including systems which do not admit smooth control Lyapunov functions. To the best of our knowledge, this is the first paper to numerically compute piecewise affine control Lyapunov functions for this class of systems.

Due to the large number of binary variables necessary for the problem formulation the current mixed integer linear program is in general intractable for dynamical systems of dimension $n \geq 3$. However, due to the local structure of the constraints, distributed optimization techniques might overcome these problems in future work. Additionally, instead of using the optimization problem (12.19) to construct control Lyapunov functions, the constraints developed here can be used to verify if a candidate for a control Lyapunov function satisfies the conditions of Theorem 1 and the local decrease condition. Compared to the optimization problem, the verification is cheap, and thus can be used also for control systems of higher dimension.

Acknowledgements The authors acknowledge the numerical experiments of Sigurdur Hafstein who calculated a control Lyapunov function for the special case of Artstein's circles. His approach with mixed-integer programming differs from the presented work but certainly was one source of motivation to develop another approach based on local semiconcavity.

The authors, P. Braun, L. Grüne and C. M. Kellett, are supported by the Australian Research Council (Grant number: ARC-DP160102138).

References

1. V. Andriano, A. Bacciotti, G. Beccari, Global stability and external stability of dynamical systems. Nonlinear Anal. Theory Methods Appl. **28**(7), 1167–1185 (1997)
2. Z. Artstein, Stabilization with relaxed controls. Nonlinear Anal. Theory Methods Appl. **7**(11), 1163–1173 (1983)
3. R. Baier, S.F. Hafstein, Numerical computation of control Lyapunov functions in the sense of generalized gradients, in *MTNS 2014: Proceedings 21st International Symposium on Mathematical Theory of Networks and Systems*, Groningen, 7–11 July 2014 (University of Groningen, Groningen, 2014), pp. 1173–1180
4. R. Baier, L. Grüine, S. Hafstein, Linear programming based Lyapunov function computation for differential inclusions. Discrete Contin. Dyn. Syst. Ser. B **17**(1), 33–56 (2012)
5. R.W. Brockett, Asymptotic stability and feedback stabilization, in *Differential Geometric Control Theory* (Birkhauser, Boston, 1983), pp. 181–191
6. P. Cannarsa, C. Sinestrari, *Semiconcave Functions, Hamilton-Jacobi Equations, and Optimal Control*, vol. 58 (Springer Science and Business Media, Berlin, 2004)
7. F. Clarke, Lyapunov functions and discontinuous stabilizing feedback. Ann. Rev. Control **35**(1), 13–33 (2011)
8. F.H. Clarke, Y.S. Ledyaev, E.D. Sontag, A.I. Subbotin, Asymptotic controllability implies feedback stabilization. IEEE Trans. Autom. Control **42**(10), 1394–1407 (1997)
9. F.H. Clarke, Y.S. Ledyaev, R.J. Stern, Asymptotic stability and smooth Lyapunov functions. J. Differ. Equ. **149**, 69–114 (1998)
10. P. Giesl, S. Hafstein, Review on computational methods for Lyapunov functions. Discrete Contin. Dyn. Syst. Ser. B **20**(8), 2291–2331 (2015)
11. M. Grant, S. Boyd, CVX: Matlab software for disciplined convex programming, version 2.1. http://cvxr.com/cvx, Mar 2014
12. I. Gurobi Optimization, Gurobi optimizer reference manual, 2016. http://www.gurobi.com
13. S.F. Hafstein, An algorithm for constructing Lyapunov functions. Electron. J. Differ. Equ. **8**, 1–101 (2007)

14. S.F. Hafstein, C.M. Kellett, H. Li, Computing continuous and piecewise affine Lyapunov functions for nonlinear systems. J. Comput. Dyn. **2**(2), 227–246 (2016)
15. M. Johansson, *Piecewise Linear Control Systems: A Computational Approach* (Springer, Berlin, 2003)
16. P. Julian, J. Guivant, A. Desages, A parametrization of piecewise linear Lyapunov functions via linear programming. Int. J. Control **72**(7–8), 702–715 (1999)
17. C.M. Kellett, Classical converse theorems in Lyapunov's second method. Discrete Contin. Dynam. Syst. Ser. B **20**(8), 2333–2360 (2015)
18. C.M. Kellett, A.R. Teel, Uniform asymptotic controllability to a set implies locally Lipschitz control-Lyapunov function, in *Proceedings of the 39th IEEE Conference on Decision and Control*, Sydney (2000)
19. C.M. Kellett, A.R. Teel, Weak converse Lyapunov theorems and control-Lyapunov functions. SIAM J. Control Optim. **42**(6), 1934–1959 (2004)
20. A.M. Lyapunov, The general problem of the stability of motion. Int. J. Control **55**(3), 531–534 (1992) (Original in Russian, Math. Soc. of Kharkov 1892)
21. S.F. Marinósson, Lyapunov function construction for ordinary differential equations with linear programming. Dyn. Syst. **17**(2), 137–150 (2002)
22. L. Rifford, Existence of Lipschitz and semiconcave control-Lyapunov functions. SIAM J. Control Optim. **39**(4), 1043–1064 (2000)
23. E.D. Sontag, A Lyapunov-like characterization of asymptotic controllability. SIAM J. Control Optim. **21**(3), 462–471 (1983)

Chapter 13
Convergence of an Inexact Majorization-Minimization Method for Solving a Class of Composite Optimization Problems

Amir Beck and Dror Pan

Abstract We suggest a majorization-minimization method for solving nonconvex minimization problems. The method is based on minimizing at each iterate a properly constructed *consistent majorizer* of the objective function. We describe a variety of classes of functions for which such a construction is possible. We introduce an inexact variant of the method, in which only approximate minimization of the consistent majorizer is performed at each iteration. Both the exact and the inexact algorithms are shown to be descent methods whose accumulation points have a property which is stronger than standard stationarity. We give examples of cases in which the exact method can be applied. Finally, we show that the inexact method can be applied to a specific problem, called *sparse source localization*, by utilizing a fast optimization method on a smooth convex dual of its subproblems.

Keywords Majorization minimization · Nonconvex programming · Composite model

AMS Subject Classifications 90C26, 90C30

A. Beck (✉)
School of Mathematical Sciences, Tel Aviv University, Tel Aviv, Israel
e-mail: becka@tauex.tau.ac.il

D. Pan
Faculty of Industrial Engineering and Management, Technion - Israel Institute of Technology, Haifa, Israel
e-mail: dror.pan@campus.technion.ac.il

© Springer Nature Switzerland AG 2018
P. Giselsson, A. Rantzer (eds.), *Large-Scale and Distributed Optimization*, Lecture Notes in Mathematics 2227,
https://doi.org/10.1007/978-3-319-97478-1_13

13.1 Introduction

In this chapter we consider the general optimization problem

$$\min \left\{ F(\mathbf{x}) : \mathbf{x} \in \mathbb{R}^n \right\}, \tag{13.1}$$

where $F : \mathbb{R}^n \rightarrow (-\infty, \infty]$ is a proper, closed extended real-valued function satisfying that its domain $\mathrm{dom}(F) := \{\mathbf{x} \in \mathbb{R}^n : F(\mathbf{x}) < \infty\}$ is a convex subset of $\mathbb{R}^n$. In addition, we assume that F is *directionally differentiable*, that is, for any $\mathbf{x}, \mathbf{y} \in \mathrm{dom}(F)$, the directional derivative of F at $\mathbf{x}$ in the direction $\mathbf{y} - \mathbf{x}$,

$$F'(\mathbf{x}; \mathbf{y} - \mathbf{x}) := \lim_{t \to 0^+} \frac{F(\mathbf{x} + t(\mathbf{y} - \mathbf{x})) - F(\mathbf{x})}{t}$$

exists (finite or infinite). For the sake of simplicity of exposition, all the spaces are Euclidean $\mathbb{R}^n$ spaces with the endowed dot product, but all the results hold trivially for general Euclidean spaces. In this context, the gradient of a differentiable function $F : \mathbb{R}^n \rightarrow \mathbb{R}$, denoted by ∇F, is the vector of all partial derivatives $\nabla F(\mathbf{x}) := \left(\frac{\partial F}{\partial x_1}(\mathbf{x}), \ldots, \frac{\partial F}{\partial x_n}(\mathbf{x}) \right)^T$.

The optimization method that we suggest for solving (13.1) is based on the general *majorization-minimization (MM)* scheme. At each iteration, a *consistent majorizer* is computed around the current iterate, and the next iterate is an exact or an approximate minimizer of that majorizer. A consistent majorizer is an upper bound on F that coincides with it up to first-order terms around a given point in its domain. Consistent majorizers and methods based on the MM-scheme have been extensively studied in the literature, see for example the book [13] as well as the review paper [12] and references therein for a variety of constructions of consistent majorizers.

A special focus in the literature is on the case where F is given by a composition $F = \varphi \circ \mathbf{f} + g$, where $\mathbf{f}$ is a mapping comprising m real-valued differentiable functions with Lipschitzian gradients, φ is a support function of a nonnegative compact and convex subset of $\mathbb{R}^m$ and g is a proper closed and convex function. An applicable method for various composite models is the *proximal Gauss-Newton* method (PGNM), also known as *prox-linear* method. Its general step is

$$x^{k+1} = \operatorname*{argmin}_{\mathbf{y}} \left\{ g(\mathbf{y}) + \varphi\left(\mathbf{f}(\mathbf{x}^k) + \mathbf{J}_{\mathbf{f}}(\mathbf{x}^k)(\mathbf{y} - \mathbf{x}^k) \right) + \frac{1}{2t} \|\mathbf{y} - \mathbf{x}^k\|^2 \right\},$$

for some parameter $t > 0$, which depends on the smoothness parameters of ∇f_i and the global Lipschitz constant of φ, whose finiteness is guaranteed as φ is a support function of a bounded set. The matrix $\mathbf{J}_{\mathbf{f}}(\mathbf{x})$ is the Jacobian of $\mathbf{f}$ at $\mathbf{x}$. For a convergence analysis of the method see for example [15]. The prox-linear method was further investigated and extended in the more recent works [9–11, 14]. We note that a special instance of the prox-linear method is the *proximal gradient* method aimed at solving the additive model $F = f + g$ where f is differentiable and g is proper closed and convex; see the references [5, 6, 16] for convergence analysis as well as extensions.

The exact version of the MM scheme that we consider can be seen as a generalization of the prox-linear method to a broader class of models. Our main goal is to establish convergence results that will hold for both the exact and inexact MM algorithms.

The chapter is organized as follows. In Sect. 13.2 we define explicitly the concept of a consistent majorizer of a function. We describe a variety of classes of functions for which consistent majorizers can be constructed. In Sect. 13.3 we introduce the concept of *strongly stationary* points of (13.1) with respect to a given consistent majorizer of F, and show that in the case of a nonconvex consistent majorizer, it might lead to a stronger condition than the usual stationarity/"no descent directions" property. The potential advantage of the new optimality condition is demonstrated on an example of minimizing a concave quadratic function over a box. In Sect. 13.4 we describe the MM method and its inexact variant, in which only approximate minimizers of the consistent majorizers are computed, and analyze their convergence properties. We also provide an implementable example demonstrating some practical advantages of the MM method over the gradient projection method. Finally, in Sect. 13.5 we study a class of problems consisting of minimizing the composition of a nondegenerate support function with a mapping comprising functions for which strongly convex majorizers are constructable. For this class, the inexact MM method is shown to be fully implementable, and its application on a specific problem which we call *the sparse source localization* is provided.

Notations Vectors are written in lower case boldface letters, matrices in upper case boldface, scalars and sets in italic. We denote $\mathbf{e} = (1, 1, \ldots, 1)^T \in \mathbb{R}^n$, and for a given vector $\mathbf{d} \in \mathbb{R}^n$ the matrix diag($\mathbf{d}$) is the diagonal matrix whose ith diagonal entry is d_i for $i = 1, \ldots, n$. For two symmetric matrices $\mathbf{A}, \mathbf{B}$ we write $\mathbf{A} \succeq \mathbf{B}$ ($\mathbf{A} \succ \mathbf{B}$) if $\mathbf{A} - \mathbf{B}$ is positive semidefinite (positive definite). The notation $\lambda_{\max}(\mathbf{A})$ corresponds to the maximal eigenvalue of the matrix $\mathbf{A}$. The set Δ_n is the unit simplex in $\mathbb{R}^n$, namely, $\Delta_n := \{\mathbf{x} \in \mathbb{R}^n : \sum_{i=1}^n x_i = 1, \mathbf{x} \geq \mathbf{0}\}$. The norm notation $\| \cdot \|$ denotes the Euclidean norm in $\mathbb{R}^n$, i.e., $\|\mathbf{x}\| := \|\mathbf{x}\|_2 \equiv \sqrt{\langle \mathbf{x}, \mathbf{x} \rangle}$. For a given closed convex set $B \subseteq \mathbb{R}^n$ the orthogonal projection on B is defined by $P_B(\mathbf{x}) := \underset{\mathbf{y} \in B}{\operatorname{argmin}} \|\mathbf{y} - \mathbf{x}\|$.

13.2 Consistent Majorizers

13.2.1 Directionally Differentiable Functions

We consider the minimization problem

$$\min \{ F(\mathbf{x}) : \mathbf{x} \in \mathbb{R}^n \}, \tag{13.2}$$

where $F : \mathbb{R}^n \to (-\infty, \infty]$ is a proper, closed extended real-valued function which is directionally differentiable, a simple notion that is defined below.

Definition 1 (Directionally Differentiable Functions) A function $F : \mathbb{R}^n \to (-\infty, \infty]$ is called **directionally differentiable** if it satisfies the following two properties:

- $\mathrm{dom}(F)$ is a convex set.
- For any $\mathbf{x}, \mathbf{y} \in \mathrm{dom}(F)$, the directional derivative $F'(\mathbf{x}; \mathbf{y} - \mathbf{x})$ exists (finite or infinite).

Example 1 (Additive Composite Model) Suppose that $F = f + g$, where $f : \mathbb{R}^n \to \mathbb{R}$ is anywhere differentiable and $g : \mathbb{R}^n \to (-\infty, \infty]$ is convex. The function F is indeed directionally differentiable since $\mathrm{dom}(F) = \mathrm{dom}(g)$ is convex and for any $\mathbf{x}, \mathbf{y} \in \mathrm{dom}(F)$, by the convexity of g, $g'(\mathbf{x}; \mathbf{y} - \mathbf{x})$ exists (finite or infinite) and thus also $F'(\mathbf{x}; \mathbf{y} - \mathbf{x})$ exists and is given by

$$F'(\mathbf{x}; \mathbf{y} - \mathbf{x}) = \nabla f(\mathbf{x})^T (\mathbf{y} - \mathbf{x}) + g'(\mathbf{x}; \mathbf{y} - \mathbf{x}).$$

Example 2 (dc Functions) Let $F = f - g$ where $f : \mathbb{R}^n \to (-\infty, \infty]$ and $g : \mathbb{R}^n \to \mathbb{R}$ are convex. Then $\mathrm{dom}(F) = \mathrm{dom}(f)$ and by the convexity of f and g both possess directional derivatives at all feasible directions, and $g'(\mathbf{x}; \mathbf{y} - \mathbf{x})$ is finite for all $\mathbf{y}, \mathbf{x} \in \mathrm{dom}(F)$. In particular, for any $\mathbf{x}, \mathbf{y} \in \mathrm{dom}(F)$:

$$F'(\mathbf{x}; \mathbf{y} - \mathbf{x}) = f'(\mathbf{x}; \mathbf{y} - \mathbf{x}) - g'(\mathbf{x}; \mathbf{y} - \mathbf{x}).$$

13.2.2 Definition

A basic ingredient in the analysis in this paper is the concept of a *consistent majorizer*.

Definition 2 (Consistent Majorizer) Given a directionally differentiable function $F : \mathbb{R}^n \to (-\infty, \infty]$, a function $h : \mathbb{R}^n \times \mathbb{R}^n \to (-\infty, \infty]$ is called a **consistent majorizer function of F** if the following properties hold:

(A) $h(\mathbf{y}, \mathbf{x}) \geq F(\mathbf{y})$ for any $\mathbf{x}, \mathbf{y} \in \mathbb{R}^n$.
(B) $h(\mathbf{y}, \mathbf{y}) = F(\mathbf{y})$ for any $\mathbf{y} \in \mathrm{dom}(F)$.

(C)   For any $\mathbf{x} \in \text{dom}(F)$, the function $h_{\mathbf{x}}(\mathbf{y}) := h(\mathbf{y}, \mathbf{x})$ is directionally differentiable and satisfies that $\text{dom}(h_{\mathbf{x}}) = \text{dom}(F)$ and

$$h'_{\mathbf{x}}(\mathbf{x}; \mathbf{z} - \mathbf{x}) = F'(\mathbf{x}; \mathbf{z} - \mathbf{x}) \text{ for any } \mathbf{z} \in \text{dom}(F).$$

(D)   For any $\mathbf{y} \in \text{dom}(F)$ the function $\mathbf{x} \mapsto -h(\mathbf{y}, \mathbf{x})$ is closed.[1]

It is simple to show that the sum of two consistent majorizers is also a consistent majorizer.

Theorem 1 *Let F_1 and F_2 be two directionally differentiable functions where at least one of them, say F_i, satisfies $F'_i(\mathbf{x}; \mathbf{y} - \mathbf{x}) \in \mathbb{R}$ for all $\mathbf{x}, \mathbf{y} \in \text{dom}(F_i)$. Suppose that h_1, h_2 are consistent majorizers of F_1 and F_2, respectively. Then $h_1 + h_2$ is a consistent majorizer of $F_1 + F_2$.*

Proof Follows directly by the definition of consistent majorizers and the facts that (i) the sum of two closed functions is a closed function and (ii) the directional derivative is additive in the sense that $(h_1 + h_2)'(\mathbf{x}; \mathbf{d}) = h'_1(\mathbf{x}; \mathbf{d}) + h'_2(\mathbf{x}; \mathbf{d})$ for any $\mathbf{x}, \mathbf{d} \in \mathbb{R}^n$ for which the relevant expressions are well-defined. $\qquad \square$

13.2.3   Examples

Below are several examples of consistent majorizers in several important settings.

Example 3 (dd) If $F : \mathbb{R}^n \to (-\infty, \infty]$ is a directionally differentiable function, then obviously, the function $h(\mathbf{y}, \mathbf{x}) = F(\mathbf{y}) + \frac{\eta}{2}\|\mathbf{y} - \mathbf{x}\|^2$ is a consistent majorizer of F for any $\eta \geq 0$.

Example 4 (Concave Differentiable)   Consider a function $f : \mathbb{R}^n \to \mathbb{R}$ which is concave and continuously differentiable. By the concavity of f, it follows that $f(\mathbf{y}) \leq f(\mathbf{x}) + \langle \nabla f(\mathbf{x}), \mathbf{y} - \mathbf{x} \rangle$ for any $\mathbf{x}, \mathbf{y} \in \mathbb{R}^n$ and therefore the function

$$h(\mathbf{y}, \mathbf{x}) = f(\mathbf{x}) + \langle \nabla f(\mathbf{x}), \mathbf{y} - \mathbf{x} \rangle$$

is a majorizer of f, meaning that property (A) holds. Property (B) holds since for any $\mathbf{y} \in \mathbb{R}^n$, $h(\mathbf{y}, \mathbf{y}) = f(\mathbf{y})$. The function $h_{\mathbf{x}}(\mathbf{y}) \equiv h(\mathbf{y}, \mathbf{x})$, as an affine function, is directionally differentiable and satisfies for any $\mathbf{z} \in \mathbb{R}^n$,

$$h'_{\mathbf{x}}(\mathbf{x}; \mathbf{z} - \mathbf{x}) = \langle \nabla f(\mathbf{x}), \mathbf{z} - \mathbf{x} \rangle = f'(\mathbf{x}; \mathbf{z} - \mathbf{x}),$$

---

[1] Which is the same as saying that the function $\mathbf{x} \mapsto h(\mathbf{y}, \mathbf{x})$ is upper semicontinuous.

establishing the validity of property (C). Since $f, \nabla f$ are continuous functions, it also holds that for a fixed $\mathbf{y}$, the function $\mathbf{x} \mapsto h(\mathbf{y}, \mathbf{x})$ is continuous over $\mathbb{R}^n$, and is in particular closed and thus property (D) holds.

Example 5 (Differentiable Concave+dd) Consider the function

$$F(\mathbf{x}) = f(\mathbf{x}) + g(\mathbf{x}),$$

where $f : \mathbb{R}^n \rightarrow \mathbb{R}$ is concave and continuously differentiable and $g : \mathbb{R}^n \rightarrow (-\infty, \infty]$ is proper and directionally differentiable. By Examples 3 and 4, $h_1(\mathbf{x}, \mathbf{y}) = f(\mathbf{x}) + \langle \nabla f(\mathbf{x}), \mathbf{y} - \mathbf{x} \rangle$ and $h_2(\mathbf{y}, \mathbf{x}) = g(\mathbf{y}) + \frac{\eta}{2} \|\mathbf{y} - \mathbf{x}\|^2$ are consistent majorizers of f and g respectively, and hence, by Theorem 1,

$$h(\mathbf{y}, \mathbf{x}) = f(\mathbf{x}) + \langle \nabla f(\mathbf{x}), \mathbf{y} - \mathbf{x} \rangle + \frac{\eta}{2} \|\mathbf{y} - \mathbf{x}\|^2 + g(\mathbf{y})$$

is a consistent majorizer of F for any $\eta \geq 0$.

Example 6 ($C^{1,1}$) Suppose that f is L-smooth ($L > 0$) on $\mathbb{R}^n$, meaning that

$$\|\nabla f(\mathbf{x}) - \nabla f(\mathbf{y})\| \leq L \|\mathbf{x} - \mathbf{y}\| \text{ for any } \mathbf{x}, \mathbf{y} \in \mathbb{R}^n.$$

The set of functions satisfying the above is denoted by $C_L^{1,1}$. By the descent lemma [8, Proposition A.24],

$$f(\mathbf{y}) \leq f(\mathbf{x}) + \langle \nabla f(\mathbf{x}), \mathbf{y} - \mathbf{x} \rangle + \frac{L}{2} \|\mathbf{x} - \mathbf{y}\|^2.$$

Thus, the function

$$h(\mathbf{y}, \mathbf{x}) = f(\mathbf{x}) + \langle \nabla f(\mathbf{x}), \mathbf{y} - \mathbf{x} \rangle + \frac{L}{2} \|\mathbf{x} - \mathbf{y}\|^2$$

is a majorizer of f, meaning that it satisfies property (A) in the definition of consistent majorizers. It is very simple to show that properties (B), (C) and (D) also hold and hence h is a consistent majorizer of f.

Example 7 ($C^{1,1}$+dd) Consider the function

$$F(\mathbf{x}) = f(\mathbf{x}) + g(\mathbf{x}),$$

where $f : \mathbb{R}^n \rightarrow \mathbb{R}$ is L-smooth and g is a directionally differentiable function. By Examples 3 and 6 along with Theorem 1, it follows that

$$h(\mathbf{y}, \mathbf{x}) = f(\mathbf{x}) + \langle \nabla f(\mathbf{x}), \mathbf{y} - \mathbf{x} \rangle + g(\mathbf{y}) + \frac{L}{2} \|\mathbf{x} - \mathbf{y}\|^2$$

is a consistent majorizer of F.

Table 13.1 Consistent majorizers of typical additive models

Model	Assumptions	Consistent majorizer $h(\mathbf{y}, \mathbf{x})$
$f + g$	$f - C^1$, concave $g - $ dd	$f(\mathbf{x}) + \langle \nabla f(\mathbf{x}), \mathbf{y} - \mathbf{x} \rangle + g(\mathbf{y}) + \frac{\eta}{2} \|\mathbf{y} - \mathbf{x}\|^2 \ (\eta \geq 0)$
$f + g$	$f - C_L^{1,1}(L > 0)$ $g - $ dd	$f(\mathbf{x}) + \langle \nabla f(\mathbf{x}), \mathbf{y} - \mathbf{x} \rangle + g(\mathbf{y}) + \frac{L}{2} \|\mathbf{y} - \mathbf{x}\|^2$

Table 13.1 summarizes the above examples.

Example 8 (Polynomials) Consider a polynomial function

$$F(\mathbf{x}) = \sum_{i=1}^{m} f_i(\mathbf{x}),$$

where $f_i : \mathbb{R}^n \to \mathbb{R}$ are *monomials*, that is,

$$f_i(\mathbf{x}) = a_i x_1^{p_{i,1}} x_2^{p_{i,2}} \cdots x_n^{p_{i,n}}, \quad i = 1, \ldots, m,$$

where $a_1, a_2, \ldots, a_m$ are real numbers, and $p_{i,j} \in \mathbb{N} \cup \{0\}$ for all $i \in \{1, \ldots, m\}$, $j \in \{1, \ldots, n\}$. A consistent majorizer of F can be constructed as the sum of majorizers of the monomials $f_1, \ldots, f_m$, invoking Theorem 1 (as for all i f_i is a differential real-valued function). We now show how one can define a consistent majorizer of a monomial. Given $\mathbf{x}, \mathbf{y} \in \mathbb{R}^n$, by the Taylor formula we write the monomial as a polynomial in $\mathbf{y}$, developed around $\mathbf{x}$. Then we upper bound each non-pure[2] monomial of the obtained polynomial by a sum of pure monomials, through repeatedly applying the inequality

$$\alpha a b \leq \frac{1}{2} |\alpha| (a^2 + b^2), \tag{13.3}$$

which holds for any real numbers α, a, b.

We now demonstrate a construction of consistent majorizers on two numerical examples of third degree monomials.

$$\begin{aligned}
f_1(\mathbf{y}) &:= y_1^2 y_2 \\
&= x_1^2 x_2 + 2 x_1 x_2 (y_1 - x_1) + x_1^2 (y_2 - x_2) + x_2 (y_1 - x_1)^2 \\
&\quad + 2 x_1 (y_1 - x_1)(y_2 - x_2) + (y_1 - x_1)^2 (y_2 - x_2) \\
&\leq x_1^2 x_2 + 2 x_1 x_2 (y_1 - x_1) + x_1^2 (y_2 - x_2) + x_2 (y_1 - x_1)^2 \\
&\quad + 2 |x_1| \cdot \frac{1}{2} \left((y_1 - x_1)^2 + (y_2 - x_2)^2 \right) + \frac{1}{2} \left((y_1 - x_1)^4 + (y_2 - x_2)^2 \right) \\
&=: h_1(\mathbf{y}, \mathbf{x}).
\end{aligned}$$

---

[2] A monomial is called *pure* if $\exists j \ \forall k \neq j \ p_{i,k} = 0$.

$$f_2(\mathbf{y}) := y_1 y_2 y_3$$

$$= x_1 x_2 x_3 + x_2 x_3 (y_1 - x_1) + x_1 x_3 (y_2 - x_2) + x_1 x_2 (y_3 - x_3)$$
$$+ x_3 (y_1 - x_1)(y_2 - x_2) + x_2 (y_1 - x_1)(y_3 - x_3) + x_1 (y_2 - x_2)(y_3 - x_3)$$
$$+ (y_1 - x_1)(y_2 - x_2)(y_3 - x_3)$$
$$\leq x_1 x_2 x_3 + x_2 x_3 (y_1 - x_1) + x_1 x_3 (y_2 - x_2) + x_1 x_2 (y_3 - x_3)$$
$$+ |x_3| \cdot \frac{1}{2} \left((y_1 - x_1)^2 + (y_2 - x_2)^2 \right) + |x_2| \cdot \frac{1}{2} \left((y_1 - x_1)^2 + (y_3 - x_3)^2 \right)$$
$$+ |x_1| \cdot \frac{1}{2} \left((y_2 - x_2)^2 + (y_3 - x_3)^2 \right)$$
$$+ \frac{1}{2}(y_1 - x_1)^2 + \frac{1}{4}(y_2 - x_2)^4 + \frac{1}{4}(y_3 - x_3)^4$$
$$=: h_2(\mathbf{y}, \mathbf{x}),$$

where the upper bound on $(y_1 - x_1)(y_2 - x_2)(y_3 - x_3)$ is obtained by applying (13.3) twice

$$(y_1 - x_1)(y_2 - x_2)(y_3 - x_3) \leq \frac{1}{2}(y_1 - x_1)^2 + \frac{1}{2} \left((y_2 - x_2)^2 (y_3 - x_3)^2 \right)$$
$$\leq \frac{1}{2}(y_1 - x_1)^2 + \frac{1}{2} \cdot \left(\frac{1}{2}(y_2 - x_2)^4 + \frac{1}{2}(y_3 - x_3)^4 \right)$$
$$= \frac{1}{2}(y_1 - x_1)^2 + \frac{1}{4}(y_2 - x_2)^4 + \frac{1}{4}(y_3 - x_3)^4.$$

By the construction, property (A) of consistent majorizers is satisfied. Since $\mathbf{y} \mapsto f_i(\mathbf{y})$ and $\mathbf{y} \mapsto h_i(\mathbf{y}, \mathbf{x})$ are polynomials having the same constant and linear terms in their Taylor expansion around $\mathbf{x}$, properties (B) and (C) hold as well. Property (D) is also satisfied, as $\mathbf{x} \mapsto -h_i(\mathbf{y}, \mathbf{x})$ is continuous in $\mathbf{x}$, and thus closed. Finally, $\mathbf{y} \mapsto h_i(\mathbf{y}, \mathbf{x})$ are differentiable for all i, and in particular have finite directional derivatives at any point and in any direction. Thus, by repeatedly applying Theorem 1 we obtain that the function $h(\mathbf{y}, \mathbf{x}) := \sum_{i=1}^m h_i(\mathbf{y}, \mathbf{x})$ is a consistent majorizer of F.

An important property of the consistent majorizer of the form constructed above is its *separability*. It comprises n pure monomials, each depending on one variable. This property facilitates its minimization over a box in $\mathbb{R}^n$.

Example 9 (Quadratic Forms) Let

$$F(\mathbf{x}) = \mathbf{x}^T \mathbf{Q} \mathbf{x}$$

for some $\mathbf{Q} \in \mathbb{S}^n$. For any $\mathbf{x}, \mathbf{y} \in \mathbb{R}^n$ one has

$$F(\mathbf{y}) = \mathbf{y}^T \mathbf{Q} \mathbf{y} = \mathbf{x}^T \mathbf{Q} \mathbf{x} + 2(\mathbf{Q}\mathbf{x})^T (\mathbf{y} - \mathbf{x}) + (\mathbf{y} - \mathbf{x})^T \mathbf{Q}(\mathbf{y} - \mathbf{x}).$$

Let Λ be a **diagonal** matrix satisfying $\Lambda \succeq Q$. Denote

$$h(\mathbf{y}, \mathbf{x}) := \mathbf{x}^T Q \mathbf{x} + 2(Q\mathbf{x})^T (\mathbf{y} - \mathbf{x}) + (\mathbf{y} - \mathbf{x})^T \Lambda (\mathbf{y} - \mathbf{x}). \tag{13.4}$$

Then, for all $\mathbf{x}, \mathbf{y} \in \mathbb{R}^n$, the inequality $h(\mathbf{y}, \mathbf{x}) \geq F(\mathbf{y})$ holds, $h(\mathbf{y}, \mathbf{y}) = F(\mathbf{y})$, and $\nabla h_{\mathbf{x}}(\mathbf{x}) = \nabla F(\mathbf{x})$. The function $-h(\mathbf{y}, \mathbf{x})$ is also continuous in $\mathbf{x}$, and hence closed. Thus, h is a consistent majorizer of F. In addition, h is **separable** in the components of $\mathbf{y} = (y_1, y_2, \ldots, y_n)^T$. Denote $\mathbf{e} := (1, 1, \ldots, 1)^T$. We mention two possible options (out of many) for choosing the diagonal matrix $\Lambda := \mathrm{diag}(\bar{\lambda})$.

1. Defining $\bar{\lambda}$ as an optimal solution of the following SDP:

$$(\text{SDP}) \quad \min_{\lambda \in \mathbb{R}^n} \left\{ \mathbf{e}^T \lambda : \mathrm{diag}(\lambda) \succeq Q \right\}.$$

2. Setting $\bar{\lambda} := \lambda_{\max}(Q) \cdot \mathbf{e}$.

13.2.4   Consistent Majorizers of Composite Functions

Our objective in this section is to show how consistent majorizers of composite functions of the form

$$F(\mathbf{x}) = \varphi(f_1(\mathbf{x}), f_2(\mathbf{x}), \ldots, f_m(\mathbf{x})), \tag{13.5}$$

can be computed under certain assumptions in case where consistent majorizers of the functions $f_1, f_2, \ldots, f_m$ are available. We will use the notation $\mathbf{f}(\mathbf{x}) = (f_1(\mathbf{x}), f_2(\mathbf{x}), \ldots, f_m(\mathbf{x}))^T$, so that

$$F(\mathbf{x}) = \varphi(\mathbf{f}(\mathbf{x})).$$

The construction of consistent majorizers of F relies on Lemma 1 below that presents an expression for directional derivatives of functions of this form, but first, we explicitly write the required assumptions on φ and $\mathbf{f}$.

Assumption 1

(A) $\varphi(\mathbf{x}) = \sigma_C(\mathbf{x}) := \max_{\mathbf{y} \in C} \langle \mathbf{x}, \mathbf{y} \rangle$, where $C \subseteq \mathbb{R}_+^m$ is a nonnegative compact convex set.

(B)   The functions $f_1, f_2, \ldots, f_m : \mathbb{R}^n \to \mathbb{R}$ are closed and directionally differentiable with $f_i'(\mathbf{x}; \mathbf{d}) \in \mathbb{R}$ for all $i \in \{1, \ldots, m\}$ and $\mathbf{x}, \mathbf{d} \in \mathbb{R}^n$.

An interesting example of a function satisfying property (A) of Assumption 1 is $\varphi(\mathbf{x}) = \max\{x_1, x_2, \ldots, x_n\}$, which corresponds to the choice $C = \Delta_n$, (with $m = n$). An interesting example of a composition $F = \varphi \circ \mathbf{f}$ where (A) and (B) are satisfied is $F(\mathbf{x}) = \|\mathbf{x}\|_1$ which corresponds to $C = (\Delta_2)^n$ (with $m = 2n$) and $\mathbf{f}(\mathbf{x}) = (x_1, -x_1, x_2, -x_2, \ldots, x_n, -x_n)^T$.

Remark 1 (Properties of φ) Note that the fact that φ is a support function of a compact set implies that it is real-valued convex, subadditive and positively homogenous. The fact that the underlying set is nonnegative implies that the function is in addition nondecreasing in the sense that $\mathbf{x} \leq \mathbf{y}$ implies that $\varphi(\mathbf{x}) \leq \varphi(\mathbf{y})$.

In the following lemma we use the following notation: if the m functions $s_1, s_2, \ldots, s_m$ have a directional derivative at $\mathbf{x}$ in the direction $\mathbf{d}$, then the corresponding directional derivative of the vector-valued function $\mathbf{s} = (s_1, s_2, \ldots, s_m)^T$ is denoted by $\mathbf{s}'(\mathbf{x}; \mathbf{d})$ and is the m-dimensional column vector given by

$$\mathbf{s}'(\mathbf{x}; \mathbf{d}) = (s_i'(\mathbf{x}; \mathbf{d}))_{i=1}^m.$$

Lemma 1 *Let*

$$S(\mathbf{x}) = \varphi(\mathbf{s}(\mathbf{x})), \quad \mathbf{x} \in \mathbb{R}^n,$$

where

- $\varphi : \mathbb{R}^m \to \mathbb{R}$ *is a convex, subadditive and positively homogenous function.*
- $\mathbf{s} = (s_1, s_2, \ldots, s_m)^T$ *is a function from $\mathbb{R}^n$ to $\mathbb{R}^m$.*

Let $\mathbf{x}, \mathbf{d} \in \mathbb{R}^n$ and suppose that $\mathbf{s}$ is differentiable at $\mathbf{x}$ in the direction $\mathbf{d}$ with $s_i'(\mathbf{x}; \mathbf{d}) \in \mathbb{R}$ for all i. Then S has a directional derivative at $\mathbf{x}$ in the direction $\mathbf{d}$ which is given by

$$S'(\mathbf{x}; \mathbf{d}) = \varphi'(\mathbf{s}(\mathbf{x}); \mathbf{s}'(\mathbf{x}; \mathbf{d})). \tag{13.6}$$

Proof Note that by the fact that the components of $\mathbf{s}$ have a directional derivative at $\mathbf{x}$ in the direction $\mathbf{d}$, it follows that there exists a function $\mathbf{o} : \mathbb{R}^+ \to \mathbb{R}^m$ satisfying $\lim_{t \to 0^+} \frac{\mathbf{o}(t)}{t} = \mathbf{0}$ for which

$$\mathbf{s}(\mathbf{x} + t\mathbf{d}) = \mathbf{s}(\mathbf{x}) + t\mathbf{s}'(\mathbf{x}; \mathbf{d}) + \mathbf{o}(t).$$

By the subadditivity and positive homogeneity of φ, it follows that

$$\frac{\varphi(\mathbf{s}(\mathbf{x} + t\mathbf{d})) - \varphi(\mathbf{s}(\mathbf{x}))}{t} = \frac{\varphi(\mathbf{s}(\mathbf{x}) + t\mathbf{s}'(\mathbf{x}; \mathbf{d}) + \mathbf{o}(t)) - \varphi(\mathbf{s}(\mathbf{x}))}{t}$$

$$\leq \frac{\varphi(\mathbf{s}(\mathbf{x}) + t\mathbf{s}'(\mathbf{x}; \mathbf{d})) - \varphi(\mathbf{s}(\mathbf{x}))}{t} + \varphi\left(\frac{\mathbf{o}(t)}{t}\right). \tag{13.7}$$

Similarly,

$$\frac{\varphi(s(x + td)) - \varphi(s(x))}{t} \geq \frac{\varphi(s(x) + ts'(x; d)) - \varphi(s(x))}{t} - \varphi\left(-\frac{o(t)}{t}\right).$$
(13.8)

By the definition of the function o, $\lim_{t \to 0^+} \frac{o(t)}{t} = 0$, and thus, by the continuity of φ (as it is a real-valued convex function), it follows that $\lim_{t \to 0^+} \varphi\left(\frac{o(t)}{t}\right) = \lim_{t \to 0^+} \varphi\left(-\frac{o(t)}{t}\right) = \varphi(0) = 0$. It therefore follows by (13.7) and (13.8) that

$$S'(x; d) = \lim_{t \to 0^+} \frac{\varphi(s(x + td)) - \varphi(s(x))}{t} = \lim_{t \to 0^+} \frac{\varphi(s(x) + ts'(x; d)) - \varphi(s(x))}{t}$$

$$= \varphi'(s(x); s'(x; d)). \qquad \qquad \square$$

Equipped with Lemma 1, we can now show how to construct a consistent majorizer of the function F given in (1) out of consistent majorizers of $f_1, f_2, \ldots, f_m$.

Theorem 2 *Let*

$$F(x) = \varphi(f_1(x), f_2(x), \ldots, f_m(x)),$$

where φ and $\mathbf{f}$ satisfy the properties in Assumption 1. Assume that for any $i \in \{1, 2, \ldots, m\}$ the function h_i is a consistent majorizer of f_i. Then the function

$$H(y, x) = \varphi(h_1(y, x), h_2(y, x), \ldots, h_m(y, x))$$

is a consistent majorizer of F.

Proof We will show that the four properties in the definition of consistent majorizers hold:

(A) By the monotonicity of φ (see Remark 1) and the fact that h_i is a majorizer of f_i for any i, it follows that for any $x, y \in \mathbb{R}^n$,

$$H(y, x) = \varphi(h_1(y, x), h_2(y, x), \ldots, h_m(y, x))$$

$$\geq \varphi(f_1(y), f_2(y), \ldots, f_m(y)) = F(y),$$

establishing property (A) for the pair (F, H).

(B) Follows by the following simple computation:

$$H(y, y) = \varphi(h_1(y, y), h_2(y, y), \ldots, h_m(y, y))$$

$$= \varphi(f_1(y), f_2(y), \ldots, f_m(y)) = F(y).$$

(C) For a given $\mathbf{x} \in \text{dom}(F)$, define the functions $h_{i,\mathbf{x}}(\mathbf{y}) := h_i(\mathbf{y}, \mathbf{x}), i = 1, 2, \ldots, m$ and the function

$$H_{\mathbf{x}}(\mathbf{y}) = H(\mathbf{y}, \mathbf{x}) = \varphi(h_{1,\mathbf{x}}(\mathbf{y}), h_{2,\mathbf{x}}(\mathbf{y}), \ldots, h_{m,\mathbf{x}}(\mathbf{y})).$$

We need to prove that for any $\mathbf{x}, \mathbf{z} \in \text{dom}(F)$, $H'_{\mathbf{x}}(\mathbf{x}; \mathbf{z} - \mathbf{x}) = F'(\mathbf{x}; \mathbf{z} - \mathbf{x})$. Indeed, by Lemma 1 invoked with $\mathbf{s} = \mathbf{f}$, it follows that

$$F'(\mathbf{x}; \mathbf{z} - \mathbf{x}) = \varphi'(\mathbf{f}(\mathbf{x}); \mathbf{f}'(\mathbf{x}; \mathbf{z} - \mathbf{x})). \tag{13.9}$$

Finally, for any $\mathbf{x}, \mathbf{z} \in \text{dom}(F)$, invoking Lemma 1 once more with $s_i(\mathbf{y}) := h_{i,\mathbf{x}}(\mathbf{y})$, and taking into account that h_i is a consistent majorizer of f_i for any i, we obtain

$$\begin{aligned}
H'_{\mathbf{x}}(\mathbf{x}; \mathbf{z} - \mathbf{x}) &= \varphi'((h_{1,\mathbf{x}}(\mathbf{x}), \ldots, h_{m,\mathbf{x}}(\mathbf{x}))^T; (h'_{1,\mathbf{x}}(\mathbf{x}; \mathbf{z} - \mathbf{x}), \ldots, h'_{m,\mathbf{x}}(\mathbf{x}; \mathbf{z} - \mathbf{x}))^T) \\
&= \varphi'((f_1(\mathbf{x}), \ldots, f_m(\mathbf{x}))^T; (f'_1(\mathbf{x}; \mathbf{z} - \mathbf{x}), \ldots, f'_m(\mathbf{x}; \mathbf{z} - \mathbf{x}))^T) \\
&= \varphi'(\mathbf{f}(\mathbf{x}); \mathbf{f}'(\mathbf{x}; \mathbf{z} - \mathbf{x})) \\
&= F'(\mathbf{x}; \mathbf{z} - \mathbf{x}).
\end{aligned}$$

(D) For a fixed $\mathbf{y} \in \mathbb{R}^n$ we need to show that $\mathbf{x} \mapsto -H(\mathbf{y}, \mathbf{x})$ is closed. Specifically, let $\mathbf{x} \in \mathbb{R}^n$ and $\varepsilon > 0$ be fixed; we need to establish the existence of $\delta > 0$ such that

$$H(\mathbf{y}, \tilde{\mathbf{x}}) < H(\mathbf{y}, \mathbf{x}) + \varepsilon$$

for all $\tilde{\mathbf{x}}$ such that $\|\tilde{\mathbf{x}} - \mathbf{x}\| < \delta$. That would show the equivalent assertion that $\mathbf{x} \mapsto H(\mathbf{y}, \mathbf{x})$ is upper semicontinuous.

Indeed, by the continuity of φ, for any $\mathbf{z} \in \mathbb{R}^m$ there exists $\delta_{\mathbf{z}} > 0$ such that if $\|\tilde{\mathbf{z}} - \mathbf{z}\|_\infty < \delta_{\mathbf{z}}$, then $|\varphi(\tilde{\mathbf{z}}) - \varphi(\mathbf{z})| < \varepsilon$. In particular, this holds for

$$\mathbf{z} := (h_1(\mathbf{y}, \mathbf{x}), \ldots, h_m(\mathbf{y}, \mathbf{x}))^T.$$

Since for any $i \in \{1, \ldots, m\}$ the function $\mathbf{x} \mapsto -h_i(\mathbf{y}, \mathbf{x})$ is closed, there exists $\delta_i > 0$ such that if $\|\tilde{\mathbf{x}} - \mathbf{x}\| < \delta_i$, then

$$h_i(\mathbf{y}, \tilde{\mathbf{x}}) < h_i(\mathbf{y}, \mathbf{x}) + \delta_{\mathbf{z}}.$$

Define $\delta := \min\{\delta_1, \ldots, \delta_m\}$, and let $\tilde{\mathbf{x}}$ satisfy $\|\tilde{\mathbf{x}} - \mathbf{x}\| < \delta$. There exists two sets of indices

$$\begin{aligned}
I_{\tilde{\mathbf{x}}} &:= \{i \in \{1, \ldots, m\} : h_i(\mathbf{y}, \tilde{\mathbf{x}}) \le h_i(\mathbf{y}, \mathbf{x})\}, \\
J_{\tilde{\mathbf{x}}} &:= \{i \in \{1, \ldots, m\} : h_i(\mathbf{y}, \mathbf{x}) < h_i(\mathbf{y}, \tilde{\mathbf{x}}) < h_i(\mathbf{y}, \mathbf{x}) + \delta_{\mathbf{z}}\}
\end{aligned}$$

satisfying $I_{\tilde{\mathbf{x}}} \cup J_{\tilde{\mathbf{x}}} = \{1, 2, \ldots, m\}$ and $I_{\tilde{\mathbf{x}}} \cap J_{\tilde{\mathbf{x}}} = \emptyset$. Define a vector $\mathbf{u} \in \mathbb{R}^m$ as follows,

$$u_i = \begin{cases} h_i(\mathbf{y}, \mathbf{x}), & i \in I_{\tilde{\mathbf{x}}}, \\ h_i(\mathbf{y}, \tilde{\mathbf{x}}), & i \in J_{\tilde{\mathbf{x}}}. \end{cases}$$

By the monotonicity of φ it follows that

$$H(\mathbf{y}, \tilde{\mathbf{x}}) = \varphi(h_1(\mathbf{y}, \tilde{\mathbf{x}}), \ldots, h_m(\mathbf{y}, \tilde{\mathbf{x}})) \leq \varphi(\mathbf{u}). \tag{13.10}$$

In addition, by the construction, $\|\mathbf{u} - \mathbf{z}\|_\infty < \delta_{\mathbf{z}}$. Thus,

$$\varphi(\mathbf{u}) < \varphi(\mathbf{z}) + \varepsilon = H(\mathbf{y}, \mathbf{x}) + \varepsilon, \tag{13.11}$$

and the result follows by a summation of (13.10) and (13.11). □

Example 10 Suppose that

$$F(\mathbf{x}) = \max\{f_1(\mathbf{x}), f_2(\mathbf{x}), \ldots, f_m(\mathbf{x})\} + g(\mathbf{x}),$$

where $f_1, f_2, \ldots, f_m : \mathbb{R}^n \to \mathbb{R}$ are $C^{1,1}$ functions and $g : \mathbb{R}^n \to (-\infty, \infty]$ is proper closed and convex. We assume specifically that $f_i \in C_{L_i}^{1,1}$ $(L_i > 0)$ for any i. Then by Example 6,

$$h_i(\mathbf{y}, \mathbf{x}) = f_i(\mathbf{x}) + \langle \nabla f_i(\mathbf{x}), \mathbf{y} - \mathbf{x} \rangle + \frac{L_i}{2}\|\mathbf{x} - \mathbf{y}\|^2$$

is a consistent majorizer of f_i, and thus, by Theorem 2, which can be invoked since f_i are directionally differentiable and $\varphi = \sigma_{\Delta_n}$, it follows that the function $(\mathbf{y}, \mathbf{x}) \mapsto \max_{i=1,2,\ldots,m}\{h_i(\mathbf{y}, \mathbf{x})\}$ is a consistent majorizer of $\mathbf{x} \mapsto \max_{i=1,2,\ldots,m} f_i(\mathbf{x})$. Consequently, by Theorem 1, it follows that

$$H(\mathbf{y}, \mathbf{x}) := \max_{i=1,2,\ldots,m}\{h_i(\mathbf{y}, \mathbf{x})\} + g(\mathbf{y})$$

is a consistent majorizer of F.

Example 11 Let

$$F(\mathbf{x}) = \sum_{i=1}^m |f_i(\mathbf{x})|,$$

where $f_1, f_2, \ldots, f_m : \mathbb{R}^n \to \mathbb{R}$ are differentiable convex functions. Note that F can be rewritten as

$$F(\mathbf{x}) = \sum_{i=1}^m \max\{f_i(\mathbf{x}), -f_i(\mathbf{x})\},$$

meaning that $F = \varphi \circ \mathbf{t}$, where

$$\varphi(\mathbf{w}) = \sum_{i=1}^{m} \max\{w_{2i-1}, w_{2i}\},$$

$$t_{2i-1}(\mathbf{x}) = f_i(\mathbf{x}), \quad i = 1, 2, \ldots, m,$$

$$t_{2i}(\mathbf{x}) = -f_i(\mathbf{x}), \quad i = 1, 2, \ldots, m.$$

Since $-f_i$ is concave, it follows that $(\mathbf{y}, \mathbf{x}) \mapsto -f_i(\mathbf{x}) - \langle \nabla f_i(\mathbf{x}), \mathbf{y} - \mathbf{x} \rangle$ is a consistent majorizer of $-f_i$ (Example 4); in addition, $(\mathbf{y}, \mathbf{x}) \mapsto f_i(\mathbf{y})$ is a consistent majorizer of $\mathbf{x} \mapsto f_i(\mathbf{x})$ (Example 3 with $\eta = 0$). Thus, by Theorem 2, which can be invoked since the functions $t_1, t_2, \ldots, t_{2m}$ are directionally differentiable and $\varphi = \sigma_{(\Delta_2)^m}$, it follows that the function

$$H(\mathbf{y}, \mathbf{x}) = \sum_{i=1}^{m} \max\{f_i(\mathbf{y}), -f_i(\mathbf{x}) - \langle \nabla f_i(\mathbf{x}), \mathbf{y} - \mathbf{x} \rangle\}$$

is a consistent majorizer of F. It is interesting to note that this majorizer is a convex function w.r.t. $\mathbf{y}$.

13.3 Stationarity Measures and Conditions

Stationarity is a fundamental concept in optimization problems. For the optimization problem (13.2), perhaps the most natural stationarity condition of a given point is the following.

Definition 3 Let F be a proper, closed, directionally differentiable function. A point $\mathbf{x}^* \in \text{dom}(F)$ is called a **stationary** point of problem (13.2) if it satisfies

$$F'(\mathbf{x}^*; \mathbf{y} - \mathbf{x}^*) \geq 0 \quad \text{for all } \mathbf{y} \in \text{dom}(F). \tag{13.12}$$

Stationarity is a well-known necessary optimality condition for problem (13.2), and it becomes also sufficient in the convex case, as stated in the following simple lemma. For the convenience of the reader we provide its proof in the appendix.

Lemma 2 *Let F be a proper, closed, directionally differentiable function. If $\mathbf{x}^*$ is a local minimizer of (13.2), then it is a stationary point. If, in addition, F is convex, then any stationary point $\mathbf{x}^*$ of (13.2) is a global minimizer.*

Most of the known first-order methods are designed such that their limit points would satisfy (13.12). Their analysis in many cases is based on some stationarity measure, which is a nonnegative function that vanishes exactly at stationary points.

See e.g., [4, 5, 10, 17] and references therein for the wide usage of stationarity measures in analysis of first-order optimization algorithms.

In this section our main goal is to introduce stationarity measures that are based on consistent majorizers of F, the objective function of problem (13.2). At this point we introduce an additional property that will be assumed to be satisfied by consistent majorizers.

Assumption 2 *For any* $\mathbf{x} \in \mathbb{R}^n$ *the value* $\min_\mathbf{y} h(\mathbf{y}, \mathbf{x})$ *is finite.*

Assumption 2 does not require that $\min_\mathbf{y} h(\mathbf{y}, \mathbf{x})$ is attained; however, we always use the notation "min" rather than "inf". Now let $h : \mathbb{R}^n \times \mathbb{R}^n \to (-\infty, \infty]$ be a consistent majorizer of F such that Assumption 2 is satisfied. Define the function $S_{F,h} : \mathbb{R}^n \to (-\infty, \infty]$ by

$$S_{F,h}(\mathbf{x}) := F(\mathbf{x}) - \min_\mathbf{y} h(\mathbf{y}, \mathbf{x}).$$

By Assumption 2, the function $S_{F,h}$ is well defined, and its domain coincides with $\text{dom}(F)$. Though $S_{F,h}$ depends on F and on the consistent majorizer h, from now on we simply denote

$$S \equiv S_{F,h},$$

omitting the subscripts F and h whenever they are clear from the context. The following lemma establishes the main properties of S.

Lemma 3 *Let F be a proper, closed, directionally differentiable function, and h be a consistent majorizer of F. Suppose that Assumption 2 holds. Then the function S satisfies the following properties:*

1. *$S(\mathbf{x}) \geq 0$ for any $\mathbf{x} \in \text{dom}(F)$.*
2. *Any $\mathbf{x} \in \mathbb{R}^n$ and $\mathbf{p} \in \text{argmin}_\mathbf{y} h(\mathbf{y}, \mathbf{x})$ satisfy*

$$F(\mathbf{x}) - F(\mathbf{p}) \geq S(\mathbf{x}).$$

3. *S is lower semicontinuous, that is, if $\mathbf{x}^k \to \tilde{\mathbf{x}}$ as $k \to \infty$, then*

$$S(\tilde{\mathbf{x}}) \leq \liminf_{k \to \infty} S(\mathbf{x}^k).$$

4. *$S(\mathbf{x}) = 0$ if and only if $\mathbf{x} \in \text{argmin}_\mathbf{y} h(\mathbf{y}, \mathbf{x})$.*
5. *If $S(\mathbf{x}) = 0$, then the inequality $F'(\mathbf{x}; \mathbf{y} - \mathbf{x}) \geq 0$ holds for any $\mathbf{y} \in \text{dom}(F)$, that is, $\mathbf{x}$ is a stationary point of (13.2).*
   If, in addition, $\mathbf{y} \mapsto h(\mathbf{y}, \mathbf{x})$ is a convex function of $\mathbf{y}$ for any $\mathbf{x} \in \text{dom}(F)$, then the converse is also true.

Proof

1. Let $\mathbf{x} \in \mathbb{R}^n$. Then by property (B) of consistent majorizers we get

$$S(\mathbf{x}) = F(\mathbf{x}) - \min_{\mathbf{y}} h(\mathbf{y}, \mathbf{x})$$

$$\geq F(\mathbf{x}) - h(\mathbf{x}, \mathbf{x}) = F(\mathbf{x}) - F(\mathbf{x}) = 0.$$

2. Let $\mathbf{x}$ and $\mathbf{p}$ be as in the assumption. Then by property (A) of consistent majorizers

$$F(\mathbf{x}) - F(\mathbf{p}) \geq F(\mathbf{x}) - h(\mathbf{p}, \mathbf{x}) = F(\mathbf{x}) - \min_{\mathbf{y}} h(\mathbf{y}, \mathbf{x}) = S(\mathbf{x}).$$

3. By property (D) of consistent majorizers, the function $-h(\mathbf{y}, \cdot)$ is closed for any $\mathbf{y} \in \mathrm{dom}(F)$. Notice that

$$S(\mathbf{x}) = F(\mathbf{x}) - \min_{\mathbf{y}} h(\mathbf{y}, \mathbf{x})$$

$$= F(\mathbf{x}) - \min_{\mathbf{y} \in \mathrm{dom}(F)} h(\mathbf{y}, \mathbf{x}) = F(\mathbf{x}) + \max_{\mathbf{y} \in \mathrm{dom}(F)} \{-h(\mathbf{y}, \mathbf{x})\}.$$

   S is closed (equivalently, lower semicontinuous) as the sum of the closed function and a pointwise maximum of closed functions.

4. $S(\mathbf{x}) = 0$ if and only if $F(\mathbf{x}) = \min_{\mathbf{y}} h(\mathbf{y}, \mathbf{x})$ and by property (B) of consistent majorizers, the latter is valid if and only if $h(\mathbf{x}, \mathbf{x}) = \min_{\mathbf{y}} h(\mathbf{y}, \mathbf{x})$, which is equivalent to $\mathbf{x} \in \mathrm{argmin}_{\mathbf{y}} h(\mathbf{y}, \mathbf{x})$.

5. A necessary condition for $\mathbf{x}$ to be a global minimizer of $h_{\mathbf{x}}(\mathbf{y}) \equiv h(\mathbf{y}, \mathbf{x})$ with respect to $\mathbf{y}$ (i.e., for $h_{\mathbf{x}}(\mathbf{x})$ to be the minimal value of $h_{\mathbf{x}}$) is (see Lemma 2)

$$(h_{\mathbf{x}})'(\mathbf{x}; \mathbf{y} - \mathbf{x}) \geq 0 \quad \forall \mathbf{y} \in \mathrm{dom}(F). \tag{13.13}$$

By property (C) of consistent majorizers, the condition (13.13) is equivalent to

$$F'(\mathbf{x}; \mathbf{y} - \mathbf{x}) \geq 0 \quad \forall \mathbf{y} \in \mathrm{dom}(F),$$

and the result follows.

If, in addition, the function $h_{\mathbf{x}}$ is convex in $\mathbf{y}$ for all $\mathbf{x} \in \mathrm{dom}(F)$, then the necessary condition (13.13) becomes also sufficient (see Lemma 2), namely, it also implies $\mathbf{x} \in \mathrm{argmin}_{\mathbf{y}} h(\mathbf{y}, \mathbf{x})$. $\square$

As one can see by Lemma 3, if a point $\tilde{\mathbf{x}} \in \mathrm{dom}(F)$ satisfies $S(\tilde{\mathbf{x}}) = 0$, it is stationary, but in the nonconvex case there might exist some stationary points with $S(\tilde{\mathbf{x}}) > 0$. This observation leads us to formulate a necessary optimality condition, based on a property which is stronger than stationarity.

Definition 4 Let F be a proper, closed, directionally differentiable function. We say that $\mathbf{x} \in \text{dom}(F)$ is a **strongly stationary** point of problem (13.2) with respect to a consistent majorizer h if $S(\mathbf{x}) = 0$.

The following lemma establishes a necessary optimality condition for the optimization problem (13.2).

Lemma 4 *Let F be a proper, closed, directionally differentiable function and h be a consistent majorizer of F. Suppose that Assumption 2 holds. Let $\mathbf{x}^* \in \text{dom}(F)$ be a global optimal solution for problem (13.2). Then $\mathbf{x}^*$ is a strongly stationary point with respect to any consistent majorizer h.*

Proof Assume otherwise, that is, $S(\mathbf{x}^*) > 0$. Then there exists $\mathbf{y} \in \text{dom}(F)$ such that $h(\mathbf{y}, \mathbf{x}^*) < h(\mathbf{x}^*, \mathbf{x}^*)$. Since $\mathbf{x}^*$ is a global minimizer of F over $\text{dom}(F)$, for any $\mathbf{y} \in \text{dom}(F)$ we have (utilizing properties (A) and (B) of consistent majorizers)

$$F(\mathbf{x}^*) \leq F(\mathbf{y}) \leq h(\mathbf{y}, \mathbf{x}^*) < h(\mathbf{x}^*, \mathbf{x}^*) = F(\mathbf{x}^*),$$

which yields a contradiction. □

By Lemma 4, any global minimizer of (13.2) is a strongly stationary point with respect to any consistent majorizer, and by Lemma 3 any such point is also a stationary point. These two observations might help in solving specific problems of the setting (13.2) if, for example, a certain algorithm can be shown to converge to a strongly stationary point rather than just to a stationary point, it might have better chances of converging to a global solution. The choice of the majorizer can affect the number of strongly stationary points.

Example 12 (Minimizing a Concave Quadratic form Over a Box) Consider the optimization problem with the objective function defined in Example 9 for some $\mathbf{Q} \preceq 0$, and box constraints. That is, the minimization problem is given by

$$(\text{PQ}) \quad \min_{\mathbf{x} \in \mathbb{R}^n} \left\{ F(\mathbf{x}) := \mathbf{x}^T \mathbf{Q} \mathbf{x} : \quad -\mathbf{e} \leq \mathbf{x} \leq \mathbf{e} \right\}.$$

A concave function attains its minimal value over a compact convex set at least on one of its extreme points. Therefore, F attains its minimal value over $[-1, 1]^n$ at a vector in $\{-1, 1\}^n$. A well known combinatorial optimization problem that can be reformulated in the form of (PQ) is the MAXCUT problem; see e.g., [7, Section 3.4.1] and references therein.

For problem (PQ), the stationary points are the vectors $\mathbf{x}^* \in [-1, 1]^n$ satisfying

$$F'(\mathbf{x}^*; \mathbf{x} - \mathbf{x}^*) \geq 0 \quad \forall \mathbf{x} \in [-1, 1]^n,$$

or equivalently,

$$\langle \mathbf{Q}\mathbf{x}^*, \mathbf{x} - \mathbf{x}^* \rangle \geq 0 \quad \forall \mathbf{x} \in [-1, 1]^n.$$

We give two numerical examples with $n = 5, 7$ by setting $\mathbf{Q} := \mathbf{Q}_j$ for $j = 1, 2$, where

$$
\mathbf{Q}_1 \equiv \begin{pmatrix} -24 & 2 & -8 & 0 & -5 \\ 2 & -26 & 0 & -6 & 1 \\ -8 & 0 & -22 & -7 & 0 \\ 0 & -6 & -7 & -18 & 5 \\ -5 & 1 & 0 & 5 & -34 \end{pmatrix} \qquad \mathbf{Q}_2 \equiv \frac{1}{2} \begin{pmatrix} -24 & 2 & -8 & 0 & -5 & 0 & -6 \\ 2 & -26 & 0 & -6 & 1 & -1 & -3 \\ -8 & 0 & -22 & -7 & 0 & 4 & -1 \\ 0 & -6 & -7 & -18 & 5 & -1 & 1 \\ -5 & 1 & 0 & 5 & -34 & 0 & -3 \\ 0 & -1 & 4 & -1 & 0 & -28 & -7 \\ -6 & -3 & -1 & 1 & -3 & -7 & -32 \end{pmatrix}.
$$

Since $\mathbf{Q}_1, \mathbf{Q}_2 \prec \mathbf{0}$, at least one global minimizer must be a vertex of $[-1, 1]^n$. Therefore, we can reduce the discussion to the 2^n vertices $\{-1, 1\}^n$.

For each vertex $\mathbf{x} \in \{-1, 1\}^n$ we checked whether it is a stationary point. It is simple to show that stationarity in this case can be easily verified, utilizing the following explicit test. Denote by $\mathbf{q}_i$ the ith column of $\mathbf{Q}$. A vector $\mathbf{x}^* \in [-1, 1]^n$ is a stationary point of (PQ) if and only if for each $i = 1, \ldots, n$ one of the following holds:

- $\mathbf{q}_i^T \mathbf{x}^* \leq 0$ and $x_i^* = 1$,
- $\mathbf{q}_i^T \mathbf{x}^* \geq 0$ and $x_i^* = -1$,
- $\mathbf{q}_i^T \mathbf{x}^* = 0$.

We also checked for each vertex whether it is a strongly stationary point with respect to the majorizers described in Example 9 from Sect. 13.2.3. Note that utilizing (13.4), a consistent majorizer of F is given by

$$
h(\mathbf{y}, \mathbf{x}) := \sum_{i=1}^n h_i(y_i, \mathbf{x}) + \mathbf{x}^T \mathbf{Q} \mathbf{x},
$$

where

$$
h_i(y_i, \mathbf{x}) := 2(\mathbf{q}_i^T \mathbf{x})(y_i - x_i) + \bar{\lambda}_i (y_i - x_i)^2, \quad i = 1, \ldots, n,
$$

with $\bar{\lambda}_1, \bar{\lambda}_2, \ldots, \bar{\lambda}_n$ being the diagonal entries of a given diagonal matrix Λ satisfying $\Lambda \succeq \mathbf{Q}$. Since h is a separable sum of functions in the variables y_i, for a given $\mathbf{x} \in [-1, 1]^n$ the test whether $S(\mathbf{x}) = 0$ amounts to computing n numbers $y_1^*, y_2^*, \ldots, y_n^*$ satisfying the conditions

$$
y_i^* \in \underset{y_i \in [-1, 1]}{\arg\min}\ h_i(y_i, \mathbf{x}), \quad i = 1, \ldots, n,
$$

and testing whether $h(\mathbf{y}^*, \mathbf{x}) = h(\mathbf{x}, \mathbf{x})$, where $\mathbf{y}^* := (y_1^*, y_2^*, \ldots, y_n^*)^T$.

Table 13.2 contains the number of stationary (W) and strongly stationary (S) points out of the 2^n vertices. The column (G) indicates how many vertices are global optimal solutions of (PQ). The results show that in these examples the standard stationarity condition almost does not rule out any of the vertices. Strong stationarity

Table 13.2 Stationarity and optimality of vertices

Q	$\Lambda = \mathrm{diag}(\bar{\lambda})$	n	$m = 2^n$	W	S	G
Q_1	$\bar{\lambda}$ by (SDP)	5	32	32	12	4
Q_1	$\bar{\lambda} = \lambda_{\max}(Q_1)$	5	32	32	20	4
Q_2	$\bar{\lambda}$ by (SDP)	7	128	124	42	2
Q_2	$\bar{\lambda} = \lambda_{\max}(Q_2)$	7	128	124	86	2

is a more restrictive condition, and its strictness depends on the chosen consistent majorizer.

13.4 The Inexact Majorization-Minimization Method

13.4.1 The General Scheme

We introduce now the main algorithm proposed for solving problem (13.2). Let F be a directionally differentiable function, and let h be a given consistent majorizer of F. For the first variant of the algorithm we need to make the following assumption that is more restrictive than Assumption 2.

Assumption 3 For any $\mathbf{x} \in \mathbb{R}^n$ the function $h_{\mathbf{x}}(\mathbf{y}) \equiv h(\mathbf{y}, \mathbf{x})$ has at least one global minimizer.

Whenever Assumption 3 holds, and a minimizer of $h_{\mathbf{x}}$ can be computed **exactly** for any $\mathbf{x} \in \mathrm{dom}(F)$, the general scheme for the so-called *majorization-minimization (MM)* method described below is well-defined.

Algorithm 1 *Majorization-Minimization (MM) algorithm for solving (13.2)*

- *Pick an arbitrary $\mathbf{x}^0 \in \mathrm{dom}(F) \subseteq \mathbb{R}^n$.*
- *For $k = 0, 1, \ldots$ compute a vector*

$$\mathbf{x}^{k+1} \in \underset{\mathbf{x}}{\operatorname{argmin}}\, h(\mathbf{x}, \mathbf{x}^k).$$

The choice of the specific minimizer in iterations where more than one minimizer of $h_{\mathbf{x}^k}$ exist can be made arbitrarily, or, in some cases, according to some pre-specified policy. By part 2 of Lemma 3, the sequence generated by Algorithm 1 has a decrease guarantee of

$$F(\mathbf{x}^k) - F(\mathbf{x}^{k+1}) \geq S(\mathbf{x}^k) \quad \text{for all } k = 0, 1, \ldots.$$

In many cases, either Assumption 3 does not hold, or, it does, but an exact minimizer of $h_{\mathbf{x}^k}$ cannot be computed. In such cases, we formulate Algorithm 2, which is an *inexact* version of Algorithm 1. Let $\gamma \in (0, 1]$ be a given parameter. Algorithm 2 is based on the ability to compute vectors that achieve a decrease of at least γ times $S(\mathbf{x}^k)$, which is the decrease that exact minimization of $h_{\mathbf{x}^k}$ would have guaranteed. We still assume that Assumption 2 holds (but not necessarily Assumption 3) whenever we seek to apply Algorithm 2 with $\gamma < 1$. The choice $\gamma := 1$ corresponds to the exact version (Algorithm 1) as the only vectors that satisfy (13.14) for $\gamma = 1$ are exact minimizers of $h_{\mathbf{x}^k}$. Thus, $\gamma = 1$ requires the validity of Assumption 3.

Algorithm 2 *Inexact Majorization-Minimization (IMM) algorithm for solving (13.2)*

- *Input: $\gamma \in (0, 1]$.*
- *Pick an arbitrary $\mathbf{x}^0 \in \mathrm{dom}(F) \subseteq \mathbb{R}^n$.*
- *For $k = 0, 1, \ldots,$ set $\mathbf{x}^{k+1}$ to be any vector satisfying*

$$F(\mathbf{x}^k) - h(\mathbf{x}^{k+1}, \mathbf{x}^k) \geq \gamma \cdot S(\mathbf{x}^k). \tag{13.14}$$

In the context of the IMM method, for any $\mathbf{x} \in \mathbb{R}^n$, a vector $\mathbf{y}$ satisfying

$$F(\mathbf{x}) - h(\mathbf{y}, \mathbf{x}) \geq \gamma \cdot S(\mathbf{x})$$

is called *an approximate γ-vector* at $\mathbf{x}$. In this terminology, $\mathbf{x}^{k+1}$ is also chosen as an approximate γ-vector at $\mathbf{x}^k$. The inexact minimization criterion (13.14) indeed guarantees a decrease of

$$F(\mathbf{x}^k) - F(\mathbf{x}^{k+1}) \geq \gamma \cdot S(\mathbf{x}^k) \quad \text{for all } k = 0, 1, \ldots,$$

as follows by property (A) of consistent majorizers. The following is an example of a simple case where the **exact** method (Algorithm 1) can be implemented. In particular, the constructed consistent majorizer satisfies Assumption 3.

Example 13 Let $f : \mathbb{R}^3 \to \mathbb{R}$ given by

$$f(\mathbf{x}) := 2x_1^2 x_2 + 5x_2^3 + 5x_1 x_3^2 + 8x_3^3,$$

and $B := [-100, 1000] \times [-78, 802] \times [-123, 77] \subseteq \mathbb{R}^3$. Then, following Example 8, a consistent majorizer of f can be given by

$$h(\mathbf{y}, \mathbf{x}) := 5y_2^3 + 8y_3^3 + 2x_1^2 x_2 + 4x_1 x_2(y_1 - x_1) + 2x_1^2(y_2 - x_2)$$

$$+ 2x_2(y_1 - x_1)^2 + 2|x_1| \cdot \left((y_1 - x_1)^2 + (y_2 - x_2)^2 \right) + (y_1 - x_1)^4$$

$$+ (y_2 - x_2)^2$$

$$+5x_1x_3^2 + 10x_1x_3(y_3 - x_3) + 5x_3^2(y_1 - x_1)$$

$$+5x_1(y_3 - x_3)^2 + 5|x_3| \cdot \left((y_1 - x_1)^2 + (y_3 - x_3)^2\right) + 2.5(y_3 - x_3)^4$$

$$+2.5(y_1 - x_1)^2.$$

Consider now the optimization problem of minimizing f over the box-shaped domain B. In the setting of Example 1, we set $F := f + g$, where $g : \mathbb{R}^3 \to (-\infty, \infty]$ is the indicator function of B, that is,

$$g(\mathbf{x}) := \begin{cases} 0, & \mathbf{x} \in B, \\ \infty, & \mathbf{x} \notin B. \end{cases}$$

The constrained problem can therefore be recast as

$$\min_{\mathbf{x}} F(\mathbf{x}). \tag{13.15}$$

The function $H(\mathbf{y}, \mathbf{x}) := h(\mathbf{y}, \mathbf{x}) + g(\mathbf{y})$ is a consistent majorizer of F. Hence, the exact Algorithm 1 for solving (13.15) solves at each iteration k the minimization problem

$$\min_{\mathbf{y} \in B} h(\mathbf{y}, \mathbf{x}^k) \equiv \min_{\mathbf{y}} H(\mathbf{y}, \mathbf{x}^k),$$

which amounts to solving the three univariate minimization problems

$$\min_{y_1 \in [-100,1000]} 4x_1^k x_2^k (y_1 - x_1^k) + 2x_2^k (y_1 - x_1^k)^2 + 2|x_1^k|(y_1 - x_1^k)^2$$
$$+ (y_1 - x_1^k)^4 + 5(x_3^k)^2(y_1 - x_1^k) + 5|x_3^k|(y_1 - x_1^k)^2$$
$$+ 2.5(y_1 - x_1^k)^2,$$

$$\min_{y_2 \in [-78,802]} 5y_2^3 + 2(x_1^k)^2(y_2 - x_2^k) + 2|x_1^k|(y_2 - x_2^k)^2 + (y_2 - x_2^k)^2,$$

$$\min_{y_3 \in [-123,77]} 8y_3^3 + 10x_1^k x_3^k (y_3 - x_3^k) + 5x_1^k (y_3 - x_3^k)^2 + 5|x_3^k|(y_3 - x_3^k)^2$$
$$+ 2.5(y_3 - x_3^k)^4.$$

Each of the above problems can be solved by any solver that calculates roots of univariate polynomials, applied on each derivative. The obtained roots and the edge points of the intervals are the candidates among which the minimizers are those corresponding to the lowest function values.

13.4.2 Convergence Analysis of the IMM Method

We are now able to formulate the main convergence results of Algorithm IMM for a pre-determined fixed parameter $\gamma \in (0, 1]$.

Theorem 3 (Convergence of IMM (Algorithm 2)) *Let F be a proper, closed, directionally differentiable function. Consider the minimization problem (13.2) along with h being a given consistent majorizer of F. Let $\gamma \in (0, 1)$ be given. Suppose that Assumption 2 holds, and let $\{\mathbf{x}^k\}_{k \geq 0}$ be the sequence generated by the IMM method (Algorithm 2). Then the following properties hold.*

1. For any $k = 0, 1, \ldots,$

$$F(\mathbf{x}^k) - F(\mathbf{x}^{k+1}) \geq \gamma \cdot S(\mathbf{x}^k).$$

2. $F(\mathbf{x}^k) \geq F(\mathbf{x}^{k+1})$ for any $k = 0, 1, \ldots,$ and $F(\mathbf{x}^k) > F(\mathbf{x}^{k+1})$ if $S(\mathbf{x}^k) > 0$.
3. Any accumulation point $\mathbf{x}^*$ of the sequence $\{\mathbf{x}^k\}_{k \geq 0}$ is strongly stationary, that is, $S(\mathbf{x}^*) = 0$.
4. For any $K \in \mathbb{N}$ and an accumulation point $\mathbf{x}^*$ of the sequence $\{\mathbf{x}^k\}_{k \geq 0}$ one has

$$\text{(N)}\quad \min\{S(\mathbf{x}^0), S(\mathbf{x}^1), \ldots, S(\mathbf{x}^{K-1})\} \leq \frac{F(\mathbf{x}^0) - F(\mathbf{x}^*)}{\gamma \cdot K}.$$

5. If $\gamma = 1$, and Assumption 3 holds, then properties 1–4 remain valid.

Proof

1. By (13.14) and property (A) of consistent majorizers,

$$F(\mathbf{x}^k) - F(\mathbf{x}^{k+1}) \geq F(\mathbf{x}^k) - h(\mathbf{x}^{k+1}, \mathbf{x}^k) \geq \gamma \cdot S(\mathbf{x}^k).$$

2. By part 1 of Lemma 3 $S(\mathbf{x}^k) \geq 0$, so the monotonicity follows directly by the previous assertion. A strict decrease when $S(\mathbf{x}^k) > 0$ is guaranteed since $\gamma > 0$.
3. Since the sequence $\{F(\mathbf{x}^k)\}$ is non-increasing, it either has a limit $F^* > -\infty$, or it tends to $-\infty$ as $k \to \infty$.
   Case 1. $\{F(\mathbf{x}^k)\}_{k \geq 0}$ has a finite limit F^*. In this case we have $F(\mathbf{x}^k) - F(\mathbf{x}^{k+1}) \to F^* - F^* = 0$ as $k \to \infty$, and by the inequalities $F(\mathbf{x}^k) - F(\mathbf{x}^{k+1}) \geq \gamma \cdot S(\mathbf{x}^k) \geq 0$ it follows that

$$S(\mathbf{x}^k) \to 0.$$

Let $\{\mathbf{x}^{k_l}\}_{l \geq 1}$ a convergent subsequence of the generated sequence, and denote its limit by $\mathbf{x}^*$. Then, by parts 1 and 3 of Lemma 3 we have

$$0 \leq S(\mathbf{x}^*) \leq \liminf_{l \to \infty} S(\mathbf{x}^{k_l}) = \lim_{k \to \infty} S(\mathbf{x}^k) = 0,$$

and thus, $S(\mathbf{x}^*) = 0$.

Case 2. $F(\mathbf{x}^k) \rightarrow -\infty$ as $k \rightarrow \infty$. We will show by contradiction that the sequence $\{\mathbf{x}^k\}_{k\geq 0}$ has no accumulation points. Let $\left\{\mathbf{x}^{k_l}\right\}_{l\geq 1}$ a convergent subsequence, that is, $\mathbf{x}^{k_l} \rightarrow \mathbf{x}^*$ as $l \rightarrow \infty$. Then since F is closed

$$\liminf_{l\to\infty} F(\mathbf{x}^{k_l}) \geq F(\mathbf{x}^*) > -\infty,$$

contradicting the fact that $F(\mathbf{x}^k) \rightarrow -\infty$. Thus, no accumulation points exist in such a case, and the result holds trivially.

4. Again, by part 1 of the current theorem and Lemma 3, part 1,

$$F(\mathbf{x}^k) - F(\mathbf{x}^{k+1}) \geq \gamma \cdot S(\mathbf{x}^k) \geq 0 \quad \forall k \geq 0,$$

for any $K \in \mathbb{N}$ we get by summing over $k = 0, \ldots, K$ the inequalities

$$F(\mathbf{x}^0) - F(\mathbf{x}^*) \geq F(\mathbf{x}^0) - F(\mathbf{x}^K) = \sum_{k=0}^{K-1}(F(\mathbf{x}^k) - F(\mathbf{x}^{k+1})) \geq \gamma \cdot \sum_{k=0}^{K-1} S(\mathbf{x}^k)$$

$$\geq \gamma \cdot \min_{k\in\{0,\ldots,K-1\}} \{S(\mathbf{x}^k)\} \cdot K,$$

where the leftmost inequality $F(\mathbf{x}^K) \geq F(\mathbf{x}^*)$ holds by the monotonicity of $\{F(\mathbf{x}^k)\}_{k\geq 0}$. Thus,

$$\min_{k\in\{0,\ldots,K-1\}} \{S(\mathbf{x}^k)\} \cdot \gamma \cdot K \leq F(\mathbf{x}^0) - F(\mathbf{x}^*) \quad \forall K \in \mathbb{N},$$

from which (N) readily follows.

5. Under Assumption 3, the iterates where $\gamma = 1$ (Algorithm 1) are well-defined. The property $F(\mathbf{x}^k) - F(\mathbf{x}^{k+1}) \geq S(\mathbf{x}^k)$ is satisfied for any $k = 0, 1, \ldots$ by part 2 of Lemma 3. The other properties follow by the same arguments as in the case $0 < \gamma < 1$. □

At this point, it seems unclear how to verify condition (13.14) in cases where the inexact method is employed ($\gamma < 1$) since $S(\mathbf{x}^k)$ is not actually computed. In the next section we discuss some specific models on which Algorithm 2 is shown to be implementable. When $\gamma = 1$, assuming that exact minimizers of $h_{\mathbf{x}^k}$ are computable, the implementation of Algorithm 1 is clear, up to properly choosing a stopping criteria, and deciding on a rule for determining which minimizer of $h_{\mathbf{x}^k}$ should be taken when multiple minimizers exist.

Example 14 (Example 13 Revisited) We implemented the MM method (Algorithm 1) on problem (13.15) with 100 independent initial guesses $\mathbf{x}^0$ being randomly generated from a uniform distribution in B. For the sake of comparison, we also implemented the *gradient projection (GP)* method on the same 100 initial points. The GP is a first-order optimization method, whose accumulation points are guaranteed to be stationary; see e.g., [4, Section 9.4]. If a constant stepsize $t > 0$ is used, the general update step of the GP method is given by

$$\mathbf{x}^{k+1} = P_B(\mathbf{x}^k - t\nabla f(\mathbf{x}^k)),$$

where P_B is the orthogonal projection operator on the box B. We roughly pre-estimated the smoothness parameter L, which is a positive number satisfying the condition $\|\nabla f(\mathbf{y}) - \nabla f(\mathbf{x})\| \leq L\|\mathbf{x} - \mathbf{y}\|$ for all $\mathbf{x}, \mathbf{y} \in B$. We used the estimate $L \approx 7250$, and then set the constant stepsize $t := 1/L$. We stopped both algorithms (MM and GP) at the first iterate k for which the inequality

$$F(\mathbf{x}^k) - F(\mathbf{x}^{k+1}) < 10^{-7}$$

held true. In the MM method (Algorithm 1), whenever multiple minimizers were found for a univariate subproblem (in a variable y_i), we took the minimizer whose distance from x_i^k was maximal.

We compared the results of the two methods. To test the results in terms of the problem's objective, we also computed the global optimal value of (13.15), by applying the solver SCIP (see [1] and references therein). The solver found the global minimizer $\mathbf{x}^* = (1000, -78, 0)^T$ with an optimal value $F^* = -158372760$. Table 13.3 presents the following results regarding the 100 runs of each method (with the same 100 initial points).

- P-Glo: number of runs (out of 100) in which the method reached a global optimal solution, that is, with value F^*.
- IT-min, IT-max, IT-ave: minimal, maximal and average numbers of iterations (among 100 runs) till the method stopped.
- ITG-min, ITG-max, ITG-ave: minimal, maximal and average iteration numbers only among the runs in which a global solution was reached.

In addition, in 28 of 100 runs the MM method yielded a final output with a better (lower) objective value than the GP, while in **all** those 100 runs its final output was not worse than GP in objective value. Moreover, for each of the 100 final outputs of the GP method we also tested the performance of MM initialized at that output. In 27 cases we found that a run of MM initialized at those points yielded a better final output (a vector having a lower objective value).

Remark 2 It might be possible that the better chances of achieving a global optimum by MM are related to the phenomena demonstrated in Example 12 where

Table 13.3 Chances of reaching a global solution and iteration numbers of GP and MM

Method	P-Glo	IT-min	IT-max	IT-ave	ITG-min	ITG-max	ITG-ave
GP	56	827	69,079	5353.84	1360	30371	4524.3
MM	75	3	42	18.53	4	42	20.81

it was demonstrated that strongly stationary points can be much less common than standard stationary points. While accumulation points of Algorithm 1 are always strongly stationary by part 3 of Theorem 3, those obtained by first-order methods such as Algorithm GP are only guaranteed to be (standard) stationary points. Since by Lemma 4 global minimizers of (13.2) must be strongly stationary points, Algorithm 1 seems to be more likely to reach one of them.

13.5 Applying the IMM Method on Compositions of Strongly Convex Majorizers

The subproblem that is being approximately solved at each iteration k of the IMM method for solving (13.2) is

$$(H_k) \quad \min_{\mathbf{y}} H(\mathbf{y}, \mathbf{x}^k),$$

where H denotes a consistent majorizer of F. In this section we treat the case where F is given as a composition $F = \varphi \circ \mathbf{f}$, which is the setting described in Sect. 13.2.4. We introduce an algorithm that we relate to as the "inner" method. For each iterate k, it computes an approximate minimizer of problem (H_k) within the tolerance required to ensure the convergence properties established in Theorem 3. Let

$$F(\mathbf{x}) = \varphi(f_1(\mathbf{x}), f_2(\mathbf{x}), \ldots, f_m(\mathbf{x})), \tag{13.16}$$

$$H(\mathbf{y}, \mathbf{x}) = \varphi(h_1(\mathbf{y}, \mathbf{x}), h_2(\mathbf{y}, \mathbf{x}), \ldots, h_m(\mathbf{y}, \mathbf{x})), \tag{13.17}$$

where $f_1, f_2, \ldots, f_m : \mathbb{R}^n \rightarrow \mathbb{R}$ and φ satisfy Assumption 1. For any $i \in \{1, 2, \ldots, m\}$ let the function h_i be a consistent majorizer of f_i which satisfies Assumption 2. By Theorem 2, the function H is a consistent majorizer of F, and since all the functions h_i satisfy Assumption 2, it follows by the monotonicity of φ, that $H = \varphi \circ \mathbf{f}$ also satisfies Assumption 2. Recalling that by Assumption 1 we have $\varphi = \sigma_C$ for some convex compact set $C \subseteq \mathbb{R}^m_+$, we further use the following notation. Denote for any given $\lambda \in C$ and $\mathbf{x} \in \mathbb{R}^n$

$$q(\lambda, \mathbf{x}) := \min_{\mathbf{y}} \lambda^T \mathbf{h}(\mathbf{y}, \mathbf{x}),$$

where $\mathbf{h}(\mathbf{y}, \mathbf{x}) \equiv (h_1(\mathbf{y}, \mathbf{x}), h_2(\mathbf{y}, \mathbf{x}), \ldots, h_m(\mathbf{y}, \mathbf{x}))^T$. It should be noted that the minimum in the definition of $q(\lambda, \mathbf{x})$ is finite for any $\mathbf{x} \in \mathbb{R}^n$ as h_i satisfies Assumption 2 for all $i \in \{1, \ldots, m\}$, and $\lambda \in \mathbb{R}^m_+$.

For any given $\mathbf{x} \in \mathbb{R}^n$ we consider the two functions $H_\mathbf{x}(\mathbf{y}) \equiv H(\mathbf{y}, \mathbf{x})$ and $q_\mathbf{x}(\boldsymbol{\lambda}) \equiv q(\boldsymbol{\lambda}, \mathbf{x})$ as "primal" and "dual", respectively. In addition, we denote

$$Q_\mathbf{x} := \max_{\boldsymbol{\lambda} \in C} q_\mathbf{x}(\boldsymbol{\lambda}),$$

and recall that by (13.17) and Assumption 1(A) $H(\mathbf{y}, \mathbf{x}) = \max_{\boldsymbol{\lambda} \in C} \boldsymbol{\lambda}^T \mathbf{h}(\mathbf{y}, \mathbf{x})$. In the setting of this section, for any $\mathbf{x} \in \mathbb{R}^n$ one has $\text{dom}(H_\mathbf{x}) = \text{dom}(F) = \mathbb{R}^n$. The following theorem provides the theoretical basis of the proposed inner method.

Theorem 4 (Strong Duality) *Let $C \subseteq \mathbb{R}_+^m$ be nonempty, convex and compact, $f_i : \mathbb{R}^n \to \mathbb{R}$ be closed and directionally differentiable for all $i \in \{1, \dots, m\}$, and let F be defined by (13.16), with $\varphi \equiv \sigma_C$. Assume that h_i is a consistent majorizer of f_i which satisfies Assumption 2 for any $i \in \{1, \dots, m\}$. Assume further that for any $i \in \{1, \dots, m\}$ and $\mathbf{x} \in \mathbb{R}^n$ the function $\mathbf{y} \mapsto h_i(\mathbf{y}, \mathbf{x})$ is convex. Let H be defined by (13.17). Then for any $\mathbf{x} \in \mathbb{R}^n$ it holds that*

$$Q_\mathbf{x} = M_\mathbf{x} \quad \left[:= \min_\mathbf{y} H(\mathbf{y}, \mathbf{x}) \right].$$

Proof Let $\mathbf{x} \in \mathbb{R}^n$ be given. We utilize the classical min-max theorem of Sion [18]. The set C is convex and compact; $\mathbb{R}^n$ is convex and closed. For each $\boldsymbol{\lambda} \in C \subseteq \mathbb{R}_+^m$ the function $\boldsymbol{\lambda}^T \mathbf{h}(\mathbf{y}, \mathbf{x})$ is convex in $\mathbf{y}$ as a nonnegative linear combination of convex functions, and for each $\mathbf{y} \in \mathbb{R}^n$ the function $\boldsymbol{\lambda}^T \mathbf{h}(\mathbf{y}, \mathbf{x})$ is concave in $\boldsymbol{\lambda}$ as an affine function. Thus, by Sion's min-max theorem [18, Theorem 3.4], it follows that

$$M_\mathbf{x} = \min_\mathbf{y} H(\mathbf{y}, \mathbf{x}) = \min_\mathbf{y} \max_{\boldsymbol{\lambda} \in C} \boldsymbol{\lambda}^T \mathbf{h}(\mathbf{y}, \mathbf{x}) = \max_{\boldsymbol{\lambda} \in C} \min_\mathbf{y} \boldsymbol{\lambda}^T \mathbf{h}(\mathbf{y}, \mathbf{x}) = \max_{\boldsymbol{\lambda} \in C} q(\boldsymbol{\lambda}, \mathbf{x}) = Q_\mathbf{x}.$$

$\square$

The equality $Q_\mathbf{x} = M_\mathbf{x}$ enables to formulate a criterion ensuring that a tested vector $\tilde{\mathbf{x}} \in \mathbb{R}^n$ is an approximate γ-vector at $\mathbf{x}$.

Lemma 5 (Stopping Criteria) *Consider problem (13.2), where F is given by (13.16) for $f_1, \dots, f_m$ and $\varphi = \sigma_C$ that satisfy Assumption 1. Let H be defined by (13.17) for some consistent majorizers $h_1, \dots, h_m$ of $f_1, \dots, f_m$, respectively, which satisfy Assumption 2. Let $\mathbf{x} \in \mathbb{R}^n$, and $\gamma \in (0, 1]$. Assume that a vector $\tilde{\mathbf{x}} \in \mathbb{R}^n$ and a vector $\tilde{\boldsymbol{\lambda}} \in C$ satisfy the inequality*

$$H_\mathbf{x}(\tilde{\mathbf{x}}) - q_\mathbf{x}(\tilde{\boldsymbol{\lambda}}) \leq \frac{1 - \gamma}{\gamma} \left(F(\mathbf{x}) - H_\mathbf{x}(\tilde{\mathbf{x}}) \right). \tag{13.18}$$

Then $\tilde{\mathbf{x}}$ is a γ-approximate vector at $\mathbf{x}$:

$$F(\mathbf{x}) - F(\tilde{\mathbf{x}}) \geq F(\mathbf{x}) - H_\mathbf{x}(\tilde{\mathbf{x}}) \geq \gamma \cdot S(\mathbf{x}).$$

Proof By Theorem 4 and the definitions of $M_{\mathbf{x}}, Q_{\mathbf{x}}$, we obtain

$$H_{\mathbf{x}}(\tilde{\mathbf{x}}) \geq M_{\mathbf{x}} = Q_{\mathbf{x}} \geq q_{\mathbf{x}}(\tilde{\lambda}).$$

Thus, along with (13.18),

$$H_{\mathbf{x}}(\tilde{\mathbf{x}}) - M_{\mathbf{x}} \leq H_{\mathbf{x}}(\tilde{\mathbf{x}}) - q_{\mathbf{x}}(\tilde{\lambda}) \leq \frac{1-\gamma}{\gamma} \left(F(\mathbf{x}) - H_{\mathbf{x}}(\tilde{\mathbf{x}}) \right).$$

Rearrangement yields

$$-\gamma M_{\mathbf{x}} \leq (1-\gamma)F(\mathbf{x}) - H_{\mathbf{x}}(\tilde{\mathbf{x}}),$$

or, equivalently,

$$H_{\mathbf{x}}(\tilde{\mathbf{x}}) \leq M_{\mathbf{x}} + (1-\gamma)(F(\mathbf{x}) - M_{\mathbf{x}}) = M_{\mathbf{x}} + (1-\gamma) \cdot S(\mathbf{x}).$$

By property (A) of consistent majorizers along with the definition of S, it follows that

$$\begin{aligned} F(\mathbf{x}) - F(\tilde{\mathbf{x}}) &\geq F(\mathbf{x}) - H_{\mathbf{x}}(\tilde{\mathbf{x}}) \\ &\geq F(\mathbf{x}) - M_{\mathbf{x}} - (1-\gamma) \cdot S(\mathbf{x}) \\ &= S(\mathbf{x}) - (1-\gamma) \cdot S(\mathbf{x}) = \gamma \cdot S(\mathbf{x}). \end{aligned} \qquad \square$$

Lemma 5 covers the result of part 2 of Lemma 3 for $\gamma = 1$; in this case $\tilde{\mathbf{x}}$ is an exact minimizer of $H_{\mathbf{x}}$. Notice that the verification of (13.18) does not require to know the value $S(\mathbf{x})$. To complete the description of the implementation of Algorithm 2 with $\gamma < 1$ in this case we need to explain how we calculate vectors $\tilde{\lambda} \in C$ and $\tilde{\mathbf{x}} \in \mathbb{R}^n$ satisfying (13.18). We further assume two additional assumptions on C and on the consistent majorizers $h_1, \ldots, h_m$.

Assumption 4 (Strongly Convex Components) *There exists a number $\sigma > 0$, such that for any $i \in \{1, \ldots, m\}$ and $\mathbf{x} \in \mathbb{R}^n$ the function $\mathbf{y} \mapsto h_i(\mathbf{y}, \mathbf{x})$ is σ-strongly convex.*

Assumption 5 (Nondegeneracy of the Composition) $\mathbf{0} \notin C$.

Assumptions 4 and 5 guarantee that for each $\lambda \in C$ and $\mathbf{x} \in \mathbb{R}^n$ the function $\mathbf{y} \mapsto \lambda^T \mathbf{h}(\mathbf{y}, \mathbf{x})$ is strongly convex with convexity parameter uniformly bounded away from zero over $\lambda \in C$. Indeed, $\lambda^T \mathbf{h}(\mathbf{y}, \mathbf{x})$ is a nonnegative linear combination of σ-strongly convex functions, where at least one coefficient is positive. The convexity parameter is lower bounded by $\sigma \cdot \min_{\lambda \in C} \lambda^T \mathbf{e}$, where the latter quantity is positive since $C \subseteq \mathbb{R}_+^m$ is compact, and $\mathbf{0} \notin C$. In particular, they imply that H satisfies Assumption 2. We further note that Assumption 4 is not very restrictive, as its does not relate for the model itself, but rather only to the constructed majorizers $h_1, \ldots, h_m$.

The above two assumptions are needed for establishing smoothness properties on $q_{\mathbf{x}}$, which in turn, enable to apply some fast first-order optimization methods on the dual problem $\max_{\lambda \in C} q_{\mathbf{x}}(\lambda)$. Such a method can be utilized to compute an approximate minimizer of the primal problem $\min_{\mathbf{y}} H_{\mathbf{x}}(\mathbf{y})$. For $\mathbf{x} = \mathbf{x}^k$ the latter is the subproblem needed to be solved approximately at the kth iterate of the IMM method (Algorithm 2). The following shows how two vectors $\tilde{\mathbf{x}}$ and $\tilde{\lambda}$ satisfying (13.18) can be obtained given a vector $\tilde{\lambda}$ whose corresponding objective value is close in some sense to $Q_{\mathbf{x}}$.

Proposition 1 *Let $C \subseteq \mathbb{R}^m_+$ be compact and convex set, and let $f_1, \ldots, f_m$ and $\varphi \equiv \sigma_C$ satisfy Assumption 1. Let F be defined by (13.16) and H be defined by (13.17) for some consistent majorizers $h_1, \ldots, h_m$ of $f_1, \ldots, f_m$. Suppose that Assumptions 4 and 5 hold. Let $\mathbf{x} \in \mathbb{R}^n$ be a point satisfying $S(\mathbf{x}) > 0$. Denote $l_C := \min_{\lambda \in C} \lambda^T \mathbf{e}$. For any $\lambda \in C$ let $\mathbf{y}_\lambda = \underset{\mathbf{y}}{\operatorname{argmin}} \lambda^T \mathbf{h}(\mathbf{y}, \mathbf{x})$.*

1. For any $\lambda \in C$ the inequality

$$\frac{\sigma \cdot l_C}{2} \|\mathbf{y}^* - \mathbf{y}_\lambda\|^2 \leq Q_{\mathbf{x}} - q_{\mathbf{x}}(\lambda)$$

holds, where $\mathbf{y}^* \in \underset{\mathbf{y}}{\operatorname{argmin}} H_{\mathbf{x}}(\mathbf{y})$.

2. For any $\gamma \in (0, 1)$ there exists some $\varepsilon_\gamma > 0$ such that if $Q_{\mathbf{x}} - q_{\mathbf{x}}(\lambda) < \varepsilon_\gamma$, then

$$H_{\mathbf{x}}(\mathbf{y}_\lambda) - q_{\mathbf{x}}(\lambda) < \frac{1 - \gamma}{\gamma}(F(\mathbf{x}) - H_{\mathbf{x}}(\mathbf{y}_\lambda)).$$

Proof To show part 1, first we denote for all $\lambda \in C$ and $\mathbf{y} \in \mathbb{R}^n$

$$k_{\mathbf{x},\lambda}(\mathbf{y}) := \lambda^T \mathbf{h}(\mathbf{y}, \mathbf{x}).$$

By Assumption 5, $l_C > 0$. The function $k_{\mathbf{x},\lambda}$ is $(\sigma \cdot l_C)$-strongly convex by Assumption 4. Thus, $\mathbf{y}_\lambda$ exists and is unique, and for any $\mathbf{y} \in \mathbb{R}^n$ it holds that

$$k_{\mathbf{x},\lambda}(\mathbf{y}) \geq k_{\mathbf{x},\lambda}(\mathbf{y}_\lambda) + \frac{(\sigma \cdot l_C)}{2} \|\mathbf{y} - \mathbf{y}_\lambda\|^2. \tag{13.19}$$

In addition,

$$k_{\mathbf{x},\lambda}(\mathbf{y}^*) = \lambda^T \mathbf{h}(\mathbf{y}^*, \mathbf{x}) \leq \max_{\lambda \in C} \lambda^T \mathbf{h}(\mathbf{y}^*, \mathbf{x}) = H_{\mathbf{x}}(\mathbf{y}^*) = Q_{\mathbf{x}}, \tag{13.20}$$

where the last equality is the result of Theorem 4 (Assumption 2 holds for all h_i by Assumption 4). Therefore, combining (13.19), (13.20) and the fact that $q_{\mathbf{x}}(\lambda) = \min_{\mathbf{y}} k_{\mathbf{x},\lambda}(\mathbf{y}) = k_{\mathbf{x},\lambda}(\mathbf{y}_\lambda)$, we conclude that

$$\frac{(\sigma \cdot l_C)}{2} \|\mathbf{y}^* - \mathbf{y}_\lambda\|^2 \leq k_{\mathbf{x},\lambda}(\mathbf{y}^*) - k_{\mathbf{x},\lambda}(\mathbf{y}_\lambda) \leq Q_{\mathbf{x}} - k_{\mathbf{x},\lambda}(\mathbf{y}_\lambda) = Q_{\mathbf{x}} - q_{\mathbf{x}}(\lambda).$$

Let us now show part 2. Since $S(\mathbf{x}) > 0$, the vector $\mathbf{x}$ is not a minimizer of $H_{\mathbf{x}}$, and thus,

$$F(\mathbf{x}) = H_{\mathbf{x}}(\mathbf{x}) > H_{\mathbf{x}}(\mathbf{y}^*).$$

Denote

$$\varepsilon_1 := \frac{1}{2}(F(\mathbf{x}) - H_{\mathbf{x}}(\mathbf{y}^*)).$$

Since $H_{\mathbf{x}}$ is convex with $\text{dom}(H_{\mathbf{x}}) = \mathbb{R}^n$, it is continuous at $\mathbf{y}^*$, so there exists $\delta_H > 0$ such that if $\|\mathbf{y} - \mathbf{y}^*\| < \delta_H$, then

$$|H_{\mathbf{x}}(\mathbf{y}) - H_{\mathbf{x}}(\mathbf{y}^*)| < \varepsilon_1.$$

In particular, for all such $\mathbf{y}$ it holds that

$$H_{\mathbf{x}}(\mathbf{y}) < H_{\mathbf{x}}(\mathbf{y}^*) + \frac{1}{2}(F(\mathbf{x}) - H_{\mathbf{x}}(\mathbf{y}^*))$$
$$= \frac{1}{2}F(\mathbf{x}) + \frac{1}{2}H_{\mathbf{x}}(\mathbf{y}^*)$$
$$= \frac{1}{2}F(\mathbf{x}) + \frac{1}{2}F(\mathbf{x}) - \varepsilon_1 = F(\mathbf{x}) - \varepsilon_1,$$

or, equivalently,

$$F(\mathbf{x}) - H_{\mathbf{x}}(\mathbf{y}) > \varepsilon_1. \tag{13.21}$$

Denote $\varepsilon_{\gamma,1} := \frac{1-\gamma}{2\gamma}\varepsilon_1$, and let $\delta_{H,\gamma} > 0$ be such that if $\|\mathbf{y} - \mathbf{y}^*\| < \delta_{H,\gamma}$, then

$$|H_{\mathbf{x}}(\mathbf{y}) - H_{\mathbf{x}}(\mathbf{y}^*)| < \frac{\varepsilon_{\gamma,1}}{2}.$$

Now, if $\lambda \in C$ satisfies

$$Q_{\mathbf{x}} - q_{\mathbf{x}}(\lambda) < \varepsilon_\gamma := \min\left\{\frac{\sigma \cdot l_C}{2}\delta_{H,\gamma}^2, \frac{1-\gamma}{2\gamma}\varepsilon_1\right\}, \tag{13.22}$$

then by part 1 it follows that $\|\mathbf{y}^* - \mathbf{y}_\lambda\| < \delta_{H,\gamma}$, and thus,

$$|H_{\mathbf{x}}(\mathbf{y}_\lambda) - H_{\mathbf{x}}(\mathbf{y}^*)| < \varepsilon_{\gamma,1}.$$

In particular,

$$H_{\mathbf{x}}(\mathbf{y}_\lambda) - Q_{\mathbf{x}} = H_{\mathbf{x}}(\mathbf{y}_\lambda) - H_{\mathbf{x}}(\mathbf{y}^*) < \varepsilon_{\gamma,1} = \frac{1-\gamma}{2\gamma}\varepsilon_1,$$

and by (13.22),

$$Q_{\mathbf{x}} - q_{\mathbf{x}}(\lambda) < \frac{1-\gamma}{2\gamma}\varepsilon_1.$$

Summation of the above two inequalities yields

$$H_{\mathbf{x}}(\mathbf{y}_\lambda) - q_{\mathbf{x}}(\lambda) < \frac{1-\gamma}{\gamma}\varepsilon_1 \overset{(13.21)}{<} \frac{1-\gamma}{\gamma}(F(\mathbf{x}) - H_{\mathbf{x}}(\mathbf{y}_\lambda)). \qquad\qquad \square$$

By Proposition 1, if $\lambda \in C$ satisfies $Q_{\mathbf{x}} - q_{\mathbf{x}}(\lambda) < \varepsilon_\gamma$, one can choose $\tilde{\mathbf{x}} := \mathbf{y}_\lambda$ and $\tilde{\lambda} := \lambda$, and (13.18) holds. This means in particular that any iterative method for solving the dual problem

$$\max_{\lambda \in C} q_{\mathbf{x}}(\lambda) \qquad\qquad (13.23)$$

whose generated sequence $\{\lambda^k\}_{k \geq 0}$ satisfies

$$q_{\mathbf{x}}(\lambda^k) \to Q_{\mathbf{x}} \qquad\qquad (13.24)$$

will eventually produce two vectors $\tilde{\mathbf{x}} := \mathbf{y}_{\lambda^k}$ and $\tilde{\lambda} := \lambda^k$ that satisfy the stopping criteria (13.18).

One possible method that satisfies the convergence condition (13.24) is a variant of the *fast gradient projection (FGP)* algorithm, first formulated in [2, Sect. 8], and later recalled in [19]. To apply the FGP method we first need to establish some smoothness properties on the dual function $q_{\mathbf{x}}$ over C. Unlike other accelerated gradient projection schemes, e.g., FISTA [6], the proposed FGP algorithm does not evaluate the gradient $\nabla q_{\mathbf{x}}$ on vectors that are not included in C. Thus, for the FGP method to be well defined, and for its convergence properties to hold, we need to show that for any $\mathbf{x} \in \mathbb{R}^n$, there exists some $K_{\mathbf{x}} > 0$, such that the function $q_{\mathbf{x}}$ is $K_{\mathbf{x}}$-smooth over C. The smoothness of $q_{\mathbf{x}}$ is proven in the following proposition. Here we make an assumption that replaces Assumptions 4 and 5 and is somewhat stronger; we essentially assume strong convexity of the function $\mathbf{y} \mapsto \lambda^T \mathbf{h}(\mathbf{y}, \mathbf{x})$ for any λ that belongs to an open set containing C, and not just to C. Note that in cases where this assumption cannot be verified (but rather only Assumptions 4 and 5), some other method should be applied, as $q_{\mathbf{x}}$ might not be smooth at boundary points of C.

Proposition 2 *Let F be defined by (13.16) for some $f_1, \ldots, f_m$ and $\varphi = \sigma_C$ which satisfy Assumption 1. Let $\mathbf{x} \in \mathbb{R}^n$ be fixed. Let $h_1, \ldots, h_m$ be consistent majorizers of $f_1, \ldots, f_m$ respectively which satisfy Assumption 2, and which are convex in y. Assume that for a given $\sigma > 0$ the function $\mathbf{y} \mapsto \lambda^T \mathbf{h}(\mathbf{y}, \mathbf{x})$ is σ-strongly convex for any $\lambda \in \tilde{C}$, where $\tilde{C}$ is an open set satisfying $C \subseteq \tilde{C}$. Then the following properties hold.*

1. *[3, Thm 6.3.3]. For any $\lambda \in \tilde{C}$ the function $q_{\mathbf{x}}$ is differentiable at λ and*

$$\nabla q_{\mathbf{x}}(\lambda) = \mathbf{h}(\mathbf{y}_\lambda, \mathbf{x}).$$

2. *If, in addition, the mapping $\mathbf{h}$ is C^1, then there exists $L_{\mathbf{x}} > 0$ such that*

$$\|\nabla q_{\mathbf{x}}(\bar{\lambda}) - \nabla q_{\mathbf{x}}(\lambda)\| \le \frac{L_{\mathbf{x}}^2}{\sigma}\|\bar{\lambda} - \lambda\|$$

for all $\bar{\lambda}, \lambda \in C$.

Proof A full proof of part 1 can be found in [3, pp. 278–279] for the case where the domain for the minimization in $\mathbf{y}$ is compact. This assumption is used for the establishment of the existence of a minimizer $\mathbf{y}_\lambda$ for each λ. However, under the assumption of strong convexity, such a minimizer always exists even over $\mathbb{R}^n$. The uniqueness of such a minimizer also follows by the strong convexity of $\mathbf{y} \mapsto \lambda^T \mathbf{h}(\mathbf{y}, \mathbf{x})$. The same strong convexity property also enables to show that $\mathbf{y}_\lambda \mapsto \mathbf{y}_{\bar{\lambda}}$ when $\lambda \mapsto \bar{\lambda}$, which is crucial for the result when the domain is not compact.

Let us prove the second part. Since for any $\lambda \in C$ the function $\mathbf{y} \mapsto k_{\mathbf{x},\lambda}(\mathbf{y}) := \lambda^T \mathbf{h}(\mathbf{y}, \mathbf{x})$ is σ-strongly convex, it follows that for all $\lambda, \bar{\lambda} \in C$ one has

$$k_{\mathbf{x},\lambda}(\mathbf{y}_{\bar{\lambda}}) \ge k_{\mathbf{x},\lambda}(\mathbf{y}_\lambda) + \frac{\sigma}{2}\|\mathbf{y}_{\bar{\lambda}} - \mathbf{y}_\lambda\|^2$$

$$= \lambda^T \mathbf{h}(\mathbf{y}_\lambda, \mathbf{x}) + \frac{\sigma}{2}\|\mathbf{y}_{\bar{\lambda}} - \mathbf{y}_\lambda\|^2$$

$$= \lambda^T \nabla q_{\mathbf{x}}(\lambda) + \frac{\sigma}{2}\|\mathbf{y}_{\bar{\lambda}} - \mathbf{y}_\lambda\|^2,$$

where the last equality is valid by part 1, stating that $q_{\mathbf{x}}$ is differentiable in $\tilde{C} \supseteq C$ with $\nabla q_{\mathbf{x}}(\lambda) = \mathbf{h}(\mathbf{y}_\lambda, \mathbf{x})$. In addition, as

$$k_{\mathbf{x},\lambda}(\mathbf{y}_{\bar{\lambda}}) = \lambda^T \mathbf{h}(\mathbf{y}_{\bar{\lambda}}, \mathbf{x}) = \lambda^T \nabla q_{\mathbf{x}}(\bar{\lambda}),$$

we get

$$\lambda^T \nabla q_{\mathbf{x}}(\bar{\lambda}) \ge \lambda^T \nabla q_{\mathbf{x}}(\lambda) + \frac{\sigma}{2}\|\mathbf{y}_{\bar{\lambda}} - \mathbf{y}_\lambda\|^2.$$

Similarly, by changing roles of λ and $\bar{\lambda}$ we also get

$$\bar{\lambda}^T \nabla q_{\mathbf{x}}(\lambda) \ge \bar{\lambda}^T \nabla q_{\mathbf{x}}(\bar{\lambda}) + \frac{\sigma}{2}\|\mathbf{y}_{\bar{\lambda}} - \mathbf{y}_\lambda\|^2.$$

Summing the last two inequalities yields (after rearrangement)

$$(\lambda - \bar{\lambda})^T (\nabla q_{\mathbf{x}}(\bar{\lambda}) - \nabla q_{\mathbf{x}}(\lambda)) \geq \sigma \|\mathbf{y}_{\bar{\lambda}} - \mathbf{y}_{\lambda}\|^2. \tag{13.25}$$

In addition, since $\mathbf{h} \in C^1(\mathbb{R}^n, \mathbb{R}^m)$, its Jacobian matrix is continuous, and by Weierstrass theorem its norm is bounded on compact sets. We will now show that the set $U := \{\mathbf{y}_{\lambda} : \lambda \in C\} \subseteq \mathbb{R}^n$ is bounded. By Assumption 2, the monotonicity of φ and Theorem 4, $Q_{\mathbf{x}} = M_{\mathbf{x}}$ is finite. In addition, by part 1, $q_{\mathbf{x}}$ is differentiable over $\tilde{C}$ and thus, also continuous over $\tilde{C}$. Thus, the term $Q_{\mathbf{x}} - q_{\mathbf{x}}(\lambda)$ is bounded, and by Proposition 1, so is $\|\mathbf{y}^* - \mathbf{y}_{\lambda}\|$, and hence, U is bounded. In particular, $\bar{U} := \mathrm{cl}(U)$ is compact. Denote

$$L_{\mathbf{x}} := \max_{\mathbf{y} \in \bar{U}} \|\mathbf{J_h}(\mathbf{y})\| < \infty,$$

where $\mathbf{J_h}(\mathbf{y}, \mathbf{x})$ denotes the Jacobian matrix of the mapping $\mathbf{y} \mapsto \mathbf{h}(\mathbf{y}, \mathbf{x})$. Then, for any $\lambda, \bar{\lambda} \in C$ we have

$$\|\nabla q_{\mathbf{x}}(\bar{\lambda}) - \nabla q_{\mathbf{x}}(\lambda)\| = \|\mathbf{h}(\mathbf{y}_{\bar{\lambda}}) - \mathbf{h}(\mathbf{y}_{\lambda})\| \leq L_{\mathbf{x}} \|\mathbf{y}_{\bar{\lambda}} - \mathbf{y}_{\lambda}\|. \tag{13.26}$$

By (13.25), (13.26) and the Cauchy-Schwartz inequality, we obtain that

$$\|\bar{\lambda} - \lambda\| \cdot \|\nabla q_{\mathbf{x}}(\bar{\lambda}) - \nabla q_{\mathbf{x}}(\lambda)\| \geq (\lambda - \bar{\lambda})^T (\nabla q_{\mathbf{x}}(\bar{\lambda}) - \nabla q_{\mathbf{x}}(\lambda))$$
$$\geq \sigma \|\mathbf{y}_{\bar{\lambda}} - \mathbf{y}_{\lambda}\|^2$$
$$\geq \frac{\sigma}{L_{\mathbf{x}}^2} \|\nabla q_{\mathbf{x}}(\bar{\lambda}) - \nabla q_{\mathbf{x}}(\lambda)\|^2,$$

or, equivalently,

$$\|\nabla q_{\mathbf{x}}(\bar{\lambda}) - \nabla q_{\mathbf{x}}(\lambda)\| \leq \frac{L_{\mathbf{x}}^2}{\sigma} \|\bar{\lambda} - \lambda\|. \qquad \square$$

For any $\mathbf{x} \in \mathbb{R}^n$, the function $q_{\mathbf{x}}$ is $(L_{\mathbf{x}}^2/\sigma)$-smooth over C by Proposition 2, and concave as a minimum of linear functions. Our description of the FGP method is based on its formulation in [19, p. 12, eq. 37–39]. Algorithm 3 below explicitly describes the FGP method with a constant stepsize setting, applied on the dual problem $\max_{\lambda \in C} q_{\mathbf{x}}(\lambda)$ for a given $\mathbf{x}$, where for simplicity we denote $L_q := \frac{L_{\mathbf{x}}^2}{\sigma}$. Algorithm 3 is therefore guaranteed to generate a sequence $\{\lambda_l\}_{l \geq 0}$ such that $q_{\mathbf{x}}(\lambda_l) \to Q_{\mathbf{x}}$ as $l \to \infty$. See [19] for the complete details. As a result, for a large enough l, it reaches vectors $\tilde{\lambda} := \lambda_l \in C$ and $\tilde{\mathbf{x}} := \mathbf{y}_{\lambda_l}$, such that (13.18) is satisfied.

Algorithm 3 *FGP for finding an approximate solution $\tilde{\lambda}$ for the dual problem*

- *Pick arbitrary $\lambda^0, \eta^0 \in C$ and $\theta_0 = 1$.*
- *For $l = 0, 1, \ldots,$ until a stopping criterion (13.18) holds for $\tilde{\lambda} := \lambda^l$ and $\tilde{x} := y_{\tilde{\lambda}}$, compute*

$$\mu^l = (1 - \theta_l)\lambda^l + \theta_l\eta^l,$$

$$\eta^{l+1} = P_C\left(\eta^l + \frac{1}{\theta_l L_q}\nabla q_x(\mu^l)\right),$$

$$\lambda^{l+1} = (1 - \theta_l)\lambda^l + \theta_l\eta^{l+1},$$

$$\theta_{l+1} = \frac{1}{2}\left(\sqrt{\theta_l^4 + 4\theta_l^2} - \theta_l^2\right).$$

Example 15 (Sparse Source Localization) Consider a scenario where one seeks to find the best approximate solution for the system

$$\|x - a_i\| \approx d_i, \quad i = 1, \ldots, m,$$

where $x \in \mathbb{R}^n$ is the unknown location of a radiating source, and $a_1, a_2, \ldots, a_m \in \mathbb{R}^n$ are m different known anchor points in $\mathbb{R}^n$. At each anchor point there exists a sensor that measures the distance from the source, but returns just a noisy measurement d_i, where in most applications $n = 2$ or $n = 3$. The system can be reformulated as an optimization problem called *sparse source localization (SSL)*, where the approximation is in terms of the minimum sum of absolute values of the errors in the **squared** measurements:

$$\min_{x \in \mathbb{R}^n} \left\{ F(x) := \sum_{i=1}^m \left| \|x - a_i\|^2 - d_i^2 \right| \right\}.$$

As in Example 11, the objective F can be rewritten by $F = \varphi \circ f$, where $\varphi \equiv \sigma_C$ with $C := (\Delta_2)^m \subseteq \mathbb{R}_+^{2m}$, and for all $i \in \{1, \ldots, m\}$

$$f_{2i-1}(x) := \|x - a_i\|^2 - d_i^2, \qquad f_{2i}(x) := -\|x - a_i\|^2 + d_i^2.$$

Note that f_{2i-1} is strongly convex, and f_{2i} is concave. Thus, a consistent majorizer of F can be defined by $H(y, x) := \varphi \circ h(y, x)$, where for all $i \in \{1, \ldots, m\}$ we define

$$h_{2i-1}(y, x) := f_{2i-1}(y),$$

and

$$h_{2i}(\mathbf{y}, \mathbf{x}) := f_{2i}(\mathbf{x}) + \nabla f_{2i}(\mathbf{x})^T (\mathbf{y} - \mathbf{x}) + \eta \|\mathbf{y} - \mathbf{x}\|^2$$

for some $\eta > 0$.

For all i, the functions h_{2i-1} and h_{2i} are both strongly convex in $\mathbf{y}$ for any $\mathbf{x} \in \mathbb{R}^n$ with parameters 2 and 2η, respectively, and thus, Assumption 4 holds with the choice $\sigma := \min\{2, 2\eta\}$. In addition, Assumption 5 also holds, as $\mathbf{0} \notin C$. In this case, even the assumption required for the smoothness of the dual function in Proposition 2 is satisfied. Indeed, one can take

$$\tilde{C} := \bigcup_{\lambda \in C} \{\tilde{\lambda} \in \mathbb{R}^{2m} : \|\tilde{\lambda} - \lambda\|_\infty < \varepsilon\}$$

for some number ε satisfying $0 < \varepsilon < \min\left\{\frac{1}{2+\eta}, \frac{\eta}{1+2\eta}\right\}$. Some algebraic manipulations can show that indeed, for any $\tilde{\lambda} \in \tilde{C}$ the function $\mathbf{y} \mapsto \tilde{\lambda}^T \mathbf{h}(\mathbf{y}, \mathbf{x})$ is strongly convex with parameter bounded below by the positive number $2m\tilde{\sigma}$, where

$$\tilde{\sigma} := \min\{1 - \varepsilon(2 + \eta), \eta - \varepsilon(1 + 2\eta)\}.$$

We now show how at each iteration of Algorithm 2 one can apply the FGP method (Algorithm 3) on the dual problem. Denote by $P_C : \mathbb{R}^{2m} \to C$ the orthogonal projection on C. As C is a cartesian product of m copies of the two-dimensional unit simplex, $P_C(\lambda)$ can be calculated as m independent projections $P_{\Delta_2} : \mathbb{R}^2 \to \Delta_2$. Each of those projections is given by

$$P_{\Delta_2}(\lambda_1, \lambda_2) = \begin{cases} (1, 0)^T, & \lambda_2 < \lambda_1 - 1, \\ (0, 1)^T, & \lambda_2 > \lambda_1 + 1, \\ 0.5(1 - \lambda_2 + \lambda_1, 1 + \lambda_2 - \lambda_1)^T, & |\lambda_2 - \lambda_1| \le 1, \end{cases}$$

and thus, $P_C(\lambda) = \left(P_{\Delta_2}(\lambda_1, \lambda_2)^T, \ldots, P_{\Delta_2}(\lambda_{2m-1}, \lambda_{2m})^T\right)^T$. We denote by $L_q := \frac{L_\mathbf{x}^2}{2m\tilde{\sigma}}$ the smoothness parameter of $q_\mathbf{x}$ guaranteed by Proposition 2. Since L_q is not known in general, it can be estimated through a backtracking procedure.

Acknowledgements The research of Amir Beck was partially supported by the Israel Science Foundation Grant 1821/16.

Appendix: A Proof of Lemma 2

We provide a proof of Lemma 2. The necessity is proven very similarly to the proof given in [4, Theorem 9.2] for the case where F is continuously differentiable.

Proof Let $\mathbf{x}^*$ be a local minimizer of problem (13.2). Assume to the contrary that $\mathbf{x}^*$ is not a stationary point of (13.2). Then, recalling that F is directionally differentiable (dd), there exists $\mathbf{y} \in \text{dom}(F)$ such that $F'(\mathbf{x}^*; \mathbf{y} - \mathbf{x}^*) < 0$. By the definition of a directional derivative, it follows that there exists a number $0 < \delta < 1$ such that $F(\mathbf{x}^* + t(\mathbf{y} - \mathbf{x}^*)) < F(\mathbf{x}^*)$ for all $0 < t < \delta$. Since $\text{dom}(F)$ is convex (as F is dd), we have $\mathbf{x}^* + t(\mathbf{y} - \mathbf{x}^*) = (1 - t)\mathbf{x}^* + t\mathbf{y} \in \text{dom}(F)$ for all $0 < t < \delta$, contradicting the local minimality of $\mathbf{x}^*$.

As for the sufficiency part when F is convex, let $\mathbf{x}^*$ be a stationary point of (13.2), and assume to the contrary that $\mathbf{x}^*$ is not a global minimizer of (13.2). Then there exists $\mathbf{y} \in \text{dom}(F)$ such that $F(\mathbf{y}) < F(\mathbf{x}^*)$. By the stationarity of $\mathbf{x}^*$ and the convexity of F, we obtain

$$0 \leq F'(\mathbf{x}^*; \mathbf{y} - \mathbf{x}^*) = \lim_{t \to 0^+} \frac{F(\mathbf{x}^* + t(\mathbf{y} - \mathbf{x}^*)) - F(\mathbf{x}^*)}{t}$$

$$= \lim_{t \to 0^+} \frac{F(t\mathbf{y} + (1 - t)\mathbf{x}^*) - F(\mathbf{x}^*)}{t} \leq \lim_{t \to 0^+} \frac{tF(\mathbf{y}) + (1 - t)F(\mathbf{x}^*) - F(\mathbf{x}^*)}{t}$$

$$= \lim_{t \to 0^+} \frac{t(F(\mathbf{y}) - F(\mathbf{x}^*))}{t} = F(\mathbf{y}) - F(\mathbf{x}^*) < 0,$$

which is a contradiction. $\qquad\square$

References

1. T. Achterberg, T. Berthold, T. Koch, K. Wolter, Constraint integer programming: a new approach to integrate CP and MIP. ZIB-Report 08-01 (2008)
2. A. Auslender, M. Teboulle, Interior gradient and proximal methods for convex and conic optimization. SIAM J. Optim. **16**(3), 697–725 (2006)
3. M.S. Bazaraa, H.D. Sherali, C.M. Shetty, *Nonlinear Programming: Theory and Algorithms*, 3rd edn. (Wiley-Interscience [Wiley], Hoboken, 2006), p. xvi+853. ISBN: 978-0-471- 48600-8; 0-471-48600-0
4. A. Beck, *First-Order Methods in Optimization*. MOS-SIAM Series on Optimization (Society for Industrial and Applied Mathematics, Philadelphia, 2017)
5. A. Beck, *Introduction to Nonlinear Optimization Theory Algorithms, and Applications with MATLAB*. MOS-SIAM Series on Optimization, vol. 19 (Society for Industrial and Applied Mathematics, Philadelphia, 2014), p. xii+282. ISBN: 978-1-611973-64-8
6. A. Beck, M. Teboulle, A fast iterative shrinkage-thresholding algorithm for linear inverse problems. SIAM J. Imag. Sci. **2**(1), 183–202 (2009)
7. A. Ben-Tal, A. Nemirovski, *Lectures on Modern Convex Optimization* (Society for Industrial and Applied Mathematics, Philadelphia, 2001)

8. D.P. Bertsekas, *Nonlinear Programming*. Athena Scientific Optimization and Computation Series, 2nd edn. (Athena Scientific, Belmont, 1999), p. xiv+777. ISBN: 1-886529-00-0
9. J. Bolte, Z. Chen, E. Pauwels. The multiproximal linearization method for convex composite problems (2017, Preprint)
10. C. Cartis, N.I.M. Gould, P.L. Toint, On the evaluation complexity of composite function minimization with applications to nonconvex nonlinear programming. SIAM J. Optim. **21**(4), 1721–1739 (2011). https://epubs.siam.org/doi/abs/10.1137/11082381X
11. D. Drusvyatskiy, C. Paquette, Efficiency of minimizing compositions of convex functions and smooth maps (2017, Preprint)
12. D.R. Hunter, K. Lange, A tutorial on MM algorithms. Am Stat. **58**(1), 30–37 (2004). ISSN: 0003-1305
13. K. Lange, *MM Optimization Algorithms* (Society for Industrial and Applied Mathematics, Philadelphia, 2016), p. ix+223. ISBN: 978-1-611974-39-3
14. A.S. Lewis, S.J. Wright, A proximal method for composite minimization. Math. Program. **158**(1–2, Ser A), 501–546 (2016). ISSN: 0025-5610
15. Y. Nesterov, *Introductory Lectures on Convex Optimization, a Basic Course*. Applied Optimization, vol. 87 (Kluwer Academic Publishers, Boston, 2004)
16. Y. Nesterov, Modified gauss-newton scheme with worst case guarantees for global performances. Optim. Methods Softw. **22**(3), 469–483 (2007)
17. Y. Nesterov, Gradient methods for minimizing composite functions. Math. Program. **140**(1, Ser B), 125–161 (2013)
18. M. Sion, On general minmax theorems. Pac. J. Math. **8**(1), 171–176 (1958)
19. P. Tseng, Approximation accuracy gradient methods, and error bound for structured convex optimization. Math. Program. **125**(2, Ser B), 263–295 (2010). ISSN: 0025-5610

LECTURE NOTES IN MATHEMATICS

 Springer

Editors in Chief: J.-M. Morel, B. Teissier;

Editorial Policy

1. Lecture Notes aim to report new developments in all areas of mathematics and their applications – quickly, informally and at a high level. Mathematical texts analysing new developments in modelling and numerical simulation are welcome.

   Manuscripts should be reasonably self-contained and rounded off. Thus they may, and often will, present not only results of the author but also related work by other people. They may be based on specialised lecture courses. Furthermore, the manuscripts should provide sufficient motivation, examples and applications. This clearly distinguishes Lecture Notes from journal articles or technical reports which normally are very concise. Articles intended for a journal but too long to be accepted by most journals, usually do not have this "lecture notes" character. For similar reasons it is unusual for doctoral theses to be accepted for the Lecture Notes series, though habilitation theses may be appropriate.

2. Besides monographs, multi-author manuscripts resulting from SUMMER SCHOOLS or similar INTENSIVE COURSES are welcome, provided their objective was held to present an active mathematical topic to an audience at the beginning or intermediate graduate level (a list of participants should be provided).

   The resulting manuscript should not be just a collection of course notes, but should require advance planning and coordination among the main lecturers. The subject matter should dictate the structure of the book. This structure should be motivated and explained in a scientific introduction, and the notation, references, index and formulation of results should be, if possible, unified by the editors. Each contribution should have an abstract and an introduction referring to the other contributions. In other words, more preparatory work must go into a multi-authored volume than simply assembling a disparate collection of papers, communicated at the event.

3. Manuscripts should be submitted either online at www.editorialmanager.com/lnm to Springer's mathematics editorial in Heidelberg, or electronically to one of the series editors. Authors should be aware that incomplete or insufficiently close-to-final manuscripts almost always result in longer refereeing times and nevertheless unclear referees' recommendations, making further refereeing of a final draft necessary. The strict minimum amount of material that will be considered should include a detailed outline describing the planned contents of each chapter, a bibliography and several sample chapters. Parallel submission of a manuscript to another publisher while under consideration for LNM is not acceptable and can lead to rejection.

4. In general, **monographs** will be sent out to at least 2 external referees for evaluation.

   A final decision to publish can be made only on the basis of the complete manuscript, however a refereeing process leading to a preliminary decision can be based on a pre-final or incomplete manuscript.

   Volume Editors of **multi-author works** are expected to arrange for the refereeing, to the usual scientific standards, of the individual contributions. If the resulting reports can be forwarded to the LNM Editorial Board, this is very helpful. If no reports are forwarded or if other questions remain unclear in respect of homogeneity etc, the series editors may wish to consult external referees for an overall evaluation of the volume.

5. Manuscripts should in general be submitted in English. Final manuscripts should contain at least 100 pages of mathematical text and should always include

   – a table of contents;
   – an informative introduction, with adequate motivation and perhaps some historical remarks: it should be accessible to a reader not intimately familiar with the topic treated;
   – a subject index: as a rule this is genuinely helpful for the reader.
   – For evaluation purposes, manuscripts should be submitted as pdf files.

6. Careful preparation of the manuscripts will help keep production time short besides ensuring satisfactory appearance of the finished book in print and online. After acceptance of the manuscript authors will be asked to prepare the final LaTeX source files (see LaTeX templates online: https://www.springer.com/gb/authors-editors/book-authors-editors/manuscriptpreparation/5636) plus the corresponding pdf- or zipped ps-file. The LaTeX source files are essential for producing the full-text online version of the book, see http://link.springer.com/bookseries/304 for the existing online volumes of LNM). The technical production of a Lecture Notes volume takes approximately 12 weeks. Additional instructions, if necessary, are available on request from lnm@springer.com.

7. Authors receive a total of 30 free copies of their volume and free access to their book on SpringerLink, but no royalties. They are entitled to a discount of 33.3 % on the price of Springer books purchased for their personal use, if ordering directly from Springer.

8. Commitment to publish is made by a *Publishing Agreement*; contributing authors of multiauthor books are requested to sign a *Consent to Publish form*. Springer-Verlag registers the copyright for each volume. Authors are free to reuse material contained in their LNM volumes in later publications: a brief written (or e-mail) request for formal permission is sufficient.

Addresses:
Professor Jean-Michel Morel, CMLA, École Normale Supérieure de Cachan, France
E-mail: moreljeanmichel@gmail.com

Professor Bernard Teissier, Equipe Géométrie et Dynamique,
Institut de Mathématiques de Jussieu – Paris Rive Gauche, Paris, France
E-mail: bernard.teissier@imj-prg.fr

Springer: Ute McCrory, Mathematics, Heidelberg, Germany,
E-mail: lnm@springer.com